Differential Equations of Linear Elasticity of Homogeneous Media

Stress, Strain, Deformation, Deflection, Rotation,
Twisting, Vibration, Torsion, Bending
Shear, Tension, Compression

By Mohamed F. El-Hewie

2013

TABLE OF CONTENTS

TABLE OF FIGURES

CHAPTER 1: STATICS OF STRESS

Figure 1-1. Force acting on solid body where inter-molecular forces are assumed constant and uniform acting to keep the molecules of the solid body at steady spatial configuration.

Figure 1-1. Surfaces of a hypothetical infinitesimal element of volume with sides taken parallel to the three Cartesian axes.

Figure 1-3. Trapezoid cross section in a dam of height H, upper width A, lower width B. The dam resists water in a river with depth D on the vertical side.

Figure 1-4. Cartesian directions of areas in solid body where stresses are described. Notice the similarity between pressure acting on fluid (liquid or gas) with stresses acting on solid bodies.

Figure 1-5. Managing stresses in solids similar to pressure in fluids

Figure 1-6. Element of volume in cylindrical polar coordinates showing radial and azimuthal variation of normal and shear stresses of the coordinate infinitesimal elements dr and dθ.

Figure 1-7. Spherical polar coordinates of element volume for determining Navier's equilibrium differential equations.

Figure 1-8. Depiction of stresses over element volume along the radial axis.

Figure 1-9. Depiction of stresses over element volume along the polar direction.

Figure 1-10. Depiction of stresses over element volume along the azimuth direction.

Figure 1-21. Reciprocity of stress as mutually perpendicular planes.

Figure 1-12. Paralleloid rigid bar exposed to pre-determined stresses.

Figure 1-13. Geometry of an arbitrarily oriented plane with normal vector (l,m,n). On this plane, we will transfer the effects of the nine stresses on the three Cartesian planes interesting at the Origin O, such that we get three net stresses acting in this surface

Figure 1-14. Geometry of an arbitrarily oriented plane with normal vector (l_1,m_1,n_1), where, (l_1,m_1,n_1) in Figure 1-14 corresponds to (l,m,n) in Figure 1-13.

CHAPTER 2: GEOMETRY OF STRAIN

Figure 2-3. Deformation in cylindrical polar coordinates.

Figure 2-4. Contributions of elongation, translation, and rotation in the deformation of length segments

Figure 2-5. Analytic geometry of plane equations. Determining the equation of a plane, normal vector, and perpendicular plane, line of intersection, before and after homogeneous deformation.

Figure 2-6. Geometry of bending a rod the xOy plane.

Figure 2-7. Analytic geometry of plane equations. Determining the equation of a plane, normal vector, and perpendicular plane, line of intersection, before and after homogeneous deformation.

CHAPTER 3: THE MATERIAL VOLUMETRIC HOOKE'S LAW

Figure 3-1. Tensile stress-strain curve for aluminum. Test specimen diameter, 12.7 mm (0.5 in.). Gage length: 203.2 mm (8 in.). For this aluminum alloy (WA.002 1060-O) rod, the nominal tensile strength, 67.2 MPa (9.75 ksi). True tensile strength, 86.2 MPa (12.5 k Figure 3-1. Tensile stress-strain curve for aluminum. Test specimen diameter, 12.7 mm (0.5 in.). Gage length: 203.2 mm (8 in.). For this aluminum alloy (WA.002 1060-O) rod, the nominal tensile strength, 67.2 MPa (9.75 ksi). True tensile strength, 86.2 MPa (12.5 ksi). Nominal yield strength (0.2% offset), 21 MPa (3.0 ksi). Elongation (in 50.8 mm, or 2 in.), 42.7%. Reduction of area, 91%.

Figure 3-2. Relative compressive stress plotted against surface displacements of concrete columns of various height-to-diameter ratios. (Effect of Length on Compressive Strain Softening of Concrete, Journal of Engineering Mechanics, Vol. 123, No. 1, January 1997, pp. 25-35, Figure 7.)

Figure 3-3. Compositional variation of Poisson ratio v of iron-nickel alloys. (source FIGURE 8. H. M. Ledbetter and R. P. Reed, "Elastic Properties of Metals and Alloys, 1. Iron, Nickel, and Iron. Nickel Alloys" J. Phys. Chem. Ref. Data, Vol. 2, No 3 19 Figure 3-8. Compositional variation of Poisson ratio v of iron-nickel alloys. (Source FIGURE 8. H. M. Ledbetter and R. P. Reed, "Elastic Properties of Metals and Alloys, 1. Iron, Nickel, and Iron. Nickel Alloys" J. Phys. Chem. Ref. Data, Vol. 2, No 3 1973)

Figure 3-4. Compositional variation of Young's modulus E o f Iron-Nickel alloys (source FIGURE 5. H. M. Ledbetter and R. P. Reed, "Elastic Properties of Metals and Alloys, 1. Iron, Nickel, and Iron. Nickel Alloys" J. Phys. Chem. Ref. Data, Vol. 2, No 3 1973)

Figure 3-5. Compositional variation of shear modulus G of iron-nickel alloys. (Source FIGURE 6. H. M. Ledbetter and R. P. Reed, "Elastic Properties of Metals and Alloys, 1. Iron, Nickel, and Iron. Nickel Alloys" J. Phys. Chem. Ref. Data, Vol. 2, No 3 1973).

Figure 3-6. Young's elongation and Poisson's contraction

CHAPTER 8: BI-HARMONIC EQUATION

No figures

CHAPTER 9: TORSION OF PRISMATICAL BARS

CHAPTER 10: GENERAL SOLUTION OF ELASTICITY PROBLEMS

CHAPTER 11: THIN SLAB. SOLUTION BY PLANE APPROXIMATION

Figure 11-1. Prismatical plate with center of coordinates taken at its half thickness, z-axis perpendicular downward and xOy plane on the widest plane of the plate.

Figure 11-2. Cross section in the prismatical plate showing the assumption made on the middle plane and the vertical stresses and strains.

Figure 11-3. Explanation of the terms in as Sophie Germain's equation of approximate bending of a plate.

Figure 11-4. Four configurations of loading and bending of circular slab

Figure 11-5. Deflection of uniformly loaded rectangular slab.

Figure 11-6. Maximum shear forces of uniformly loaded rectangular slab.

Figure 11-7. Maximum torsion forces of uniformly loaded rectangular slab.

Figure 11-8. Graphing rotational moment on the cross section of the rectangular plate

Figure 11-9. Resultant shear forces on thin slab.

Figure 11-10. Graphing a single Fourier's sine product on the cross-section of the rectangular plate.

Figure 11-11. Boundary conditions in Levy's problem and solution.

Figure 11-12. Graphing a single Fourier's product of trigonometric and hyperbolic functions on the cross-section of the rectangular plate.

CHAPTER 12: VARIATIONAL METHOD OF SOLUTION IN PLANAR ELASTICITY

Figure 12-1. Torsion of prismatical rod solved by Castigliano's variation method.

Figure 12-2. First and second theorems of minimum, one determines peaks and bottoms, the other determines reflection.

Figure 12-3. Rectangular prism loaded on two opposite faces.

Figure 12-4. Fourier's series expansions for rectangular prism showing the vanishing of the functions on the six faces of the rectangular prism.

Figure 12-5. Graphing the system of equations (12-31.1) shows the closeness of solution as the index "m" increases.

INTRODUCTION

The transmission of forces from without to within solid medium comprises a mathematical challenge of utmost complexity. The sources of difficulties in modeling the behavior of solid medium under forcible stress vary in the scope of challenge as follows:

1. Surface indeterminate conditions

Difficulties arising from completeness of solution such that the **boundary conditions** are either insufficient to allow the determination of the unknowns or vice versa. This stems from solving the **equations of static equilibrium** on the surface of the body, where forces could only decompose to **three projections** on the surface of the body, while the immediate interior of the medium to the surface generates **nine components of stresses** from the three external projections of forces. This difficulty is reduced slightly by a reciprocity principle that renders the six shear stresses to three canonical relationships of the form $\tau_{xy} = \tau_{yx}$. That leaves us with six unknown internal stresses that must be determined from only three given external stresses. Hence, the problem of equilibrium in solid medium is deemed **indeterminate**.

2. Medium indeterminate relationships

The next logical problem arises from probing the **properties of solid matter** that could help determine the three extra-unknown stresses. Two main physical constants, namely **Young's modulus** and **Poisson's ratio**, offer great help in describing the flow of matter continuum in two perpendicular directions inside the medium. Young's modulus links longitudinal strain to normal stress on a thin isolated strip of matter. The thinning of the longitudinal strip by dilatational strain is linked to transverse shear plane by Poisson's ratio. However, such ideal mathematical vision of converting longitudinal dilatation to transverse thinning is complicated with the three problems: (i) **linearity** of Young's and Poisson's constants with the value of applied stress; (ii) **homogeneity** of the medium could allow use of specific practical formula for the two material constants; and (iii) **isotropy** of the medium which could permit modeling the preferred directions of distributions of the external three stresses into internal six stresses.

3- Spatial indeterminate continuity

Having defined the scope within which we could approximate of solutions, regarding the nature of the solid medium and its interaction with surface stresses, we face the problem of **continuity of medium,** which is expressed in terms of geometrical variables of displacements, strains, and twisting. We should not lose sight that our numerical representation of **geometrical variables** must be kept in check such that ruptures, voids, or discontinuities are excluded from our mathematical expressions. Continuity of material is ensured by equating the tangents of displacement functions on both sides on each point in the medium, when arriving final solution.

4. Fixing and loading indeterminate conditions

The fourth mathematical challenge in modeling elasticity of solids lies in the **geometry of the shape** of the medium and the **conditions of loading and support**. Both geometrical problems determine the manner of transmission of external stresses onto the medium and following behavior of the medium. There are indefinite configurations of loading, support, and body shapes, which render boundary conditions particular to each problem. For example, a freely supported beam allows sliding and bending over the support, while built-in support does not permit either sliding or bending. Also, the load distributions on the external surfaces of the body pose difficulties as to the immediate violation of linearity of elastic propertied where the load is concentrated and integrated effects of continuous load distribution versus lumped forces.

5. Inertial rotational indeterminate equilibrium

The fifth challenge plays into the interaction of aggregation of particles of masses distributed arbitrarily around the axes or points where resultant moments of couples tend to rotate or twist the body. Here, gross shape of body determines the viable **moment of inertia**, which depends mostly of the distribution of masses around arbitrarily determined axes of rotation. In other words, the **same geometrical shape** rotates with different moments of inertia depending on location and orientation of axes of the equilibrium of net forces and couples.

The above five challenges behooved all investigators who tackled problems of continuum of matter to make various **approximations** in order to attain some crude sense of the behavior of solid matter under the effect of various loading and support conditions.

Even though such hard labor of attempting to model **structural deformation** of solids under the stress of forces lasted over two hundred years, it continues to encounters great difficulties of complex material properties and designs. Yet, the mathematical equations presented in this book comprise very powerful instrument in outlining the state of motion of solids with complex **force functions**.

First, the readers will learn the particular nature of projection of stresses within a solid medium, which differs from projection of forces in that stresses apply to surfaces while forces are limited to direction and magnitude of force. Thus, within a **solid medium**, each direction of the three Cartesian coordinates has three surfaces, making total of **nine stresses** corresponding to **three projections** of a force vector. The transition from solid state to liquid state reduces the governing equations of elastic deformation into fluid flow, where nine surfaces of solid medium lose any spatial fixation. Particularly, the six shearing stresses in solid medium coalesce to single **viscous force**, the three normal stresses to one **pressure force**.

Second, the learner will grow aware of the essence of **Newton's three laws** in the complex motion of three-dimensional aggregations of particles. Each of the three laws is demonstrated exclusively in vibration, static equilibrium, and deformation, yet intertwined within complex differential operations. When integrated, the three laws offer a governing expression for elastic potential that enables us to minimize force function for **maximal stability**.

Third, the learner is introduced to unavoidable limitation of **plane section approximation**, which allows us to obtain simple algebraic expressions with desired level of correctness and completeness for distributions of deformation, stresses, strains, bending, twisting, shear, and torsion. Yet, such simplification compromises the unpredictable rupture and plastic deformation in spots where plane section is violated, due to either extreme stress or extreme strain.

Fourth, the learner will be left with a very powerful mathematical tool of an integrated **conservation of elastic energy, external work-done, and internal vibrational and virtual energies**, which enables the reader to weigh the effects of various geometrical, physical, and equilibrium on the final state of energy of the system. The reader will be able to make simplifications that render the equations of energy conservations of elastic deformation applicable to liquids, gases, or any infinitesimal material particles. For example, the **Bernoulli equation** for flowing fluids is obtained by direct substitution in equation of conservation of elastic energy in solids by making proper assumption.

Mohamed F. El-Hewie

New Jersey, USA
July 29, 2013

.

CHAPTER 1
STATICS OF STRESS

1.1. Static balance of forces in elastic medium

1. Solids possess unique intermolecular (or interatomic) forces that maintain **the shape** of the molecules of the material in space such that applied forces **(external** or **internal)** also maintain the direction of their **vectorial summation** with intermolecular forces.

Example 1

In Figures 1-1 and 1-2, a body subjected to force along the z-axis. Identify the resulting stresses in the body in terms of the force vector.

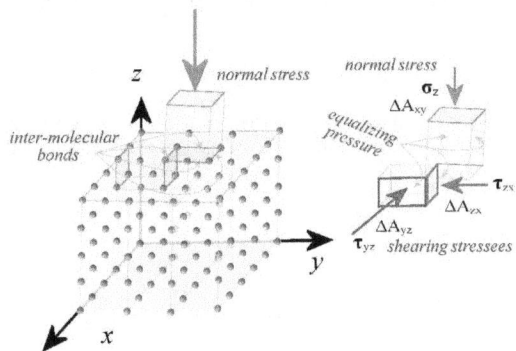

Figure 1-1. Force acting on solid body where inter-molecular forces are assumed constant and uniform acting to keep the molecules of the solid body at steady spatial configuration.

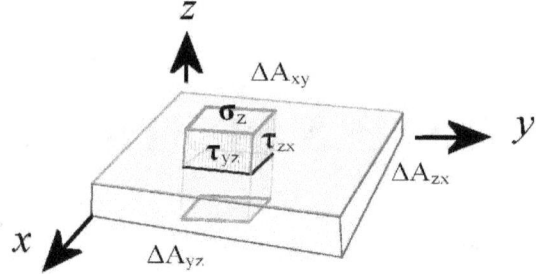

Figure 1-2. Surfaces of a hypothetical infinitesimal element of volume with sides taken parallel to the three Cartesian axes.

Solution

A force directed on x-axis applied on a surface of a solid body, which normal directed in y-axis is described as forces $\mathbf{F}_x$ applied on xz-plane of normal $\hat{\mathbf{y}}$.

(i) Due to the unique spatial constancy of intermolecular solid configuration, such force imposes **normal stress** on said surface determined by

$$\sigma_x = \lim_{\Delta A \to 0} \frac{\Delta F_x}{\Delta A_{yz}} \qquad (1\text{-}1.1)$$

(ii) In addition to normal stress σ_x, such force generates perpendicular stresses, called **shearing stresses,** inside the body (under the surface within the body) transmitted through the perpendicular intermolecular bonds such that

$$\tau_{xy} = \lim_{\Delta A \to 0} \frac{\Delta F_x}{\Delta A_{xy}} \qquad (1\text{-}1.2)$$

$$\tau_{xz} = \lim_{\Delta A \to 0} \frac{\Delta F_x}{\Delta A_{xz}} \qquad (1\text{-}1.3)$$

(iii) If we were dealing with **gas** or **liquid**, in lieu of solid body, the three stresses would have been equal to the **total pressure** of the gas, which is uniform in all direction, such that

$$P_{gas} = \sigma_x = \tau_{xz} = \tau_{zx} \qquad (1\text{-}1.4)$$

(iv) Note that in equations (1-1.2) and (1-1.3), we divided ΔF_x by A_{xy} and $A_{xz,}$ which lie in the two perpendicular planes on x-axes, as if we concluded that ΔF_x caused **pressure** with equal effects on all directions on the infinitesimal volume of matter.

(v) Despite the fact that σ_x is purely normal stress on the surface yz, yet σ_x creates shearing stresses τ_{xy} and τ_{xz} deeper in the body.

2. **Force.** In the chapter, forces are represented by vector (arrow indicating the direction and scalar value indicating the magnitude of the force). The method of application of the force is not discussed. In the case of **normal stress**, the scenario is clear since most of daily applications of forces entail perpendicular actions on surfaces such as standing of a floor, pushing an object. However, **shearing stresses** are <u>daunting to imagine</u> when applied to surfaces. One must account for friction or some sort of mechanism by which the force acts tangential to the surface.

In fact, **no shearing stresses can be induced without normal stresses acting on the orthogonal planes where shearing exists**.

When fluids act on solids, such as airlift on an airplane or water pressure on a dam, the fluid forces impart normal stresses on the external surfaces solid body, which in turn, generate internal shearing stresses in the rigid body. In other words, shearing stresses are unique to rigidity, viscosity, or resistance of molecules of a body to force.

Example 2

The cross section of a dam shown, in Figure 1-3, is shaped as a trapezoid of height H, upper width A, lower width B, and mass M, stands perpendicular to the water flow of a river of depth D. Assume that the water speed is zero, i.e., stagnant flow. Determine the stresses on two perpendicular cross sections of the dam.

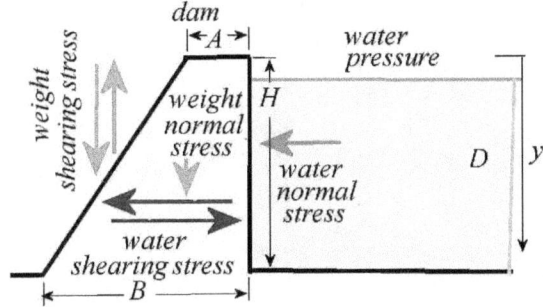

Figure 1-3. Trapezoid cross section in a dam of height H, upper width A, lower width B. The dam resists water in a river with depth D on the vertical side.

Solution

The dam is subjected to the following stresses:

(1) On the side of the dam exposed to the water stream, **normal stress** is directed along the direction of flow of water, perpendicular to the surface of the dam and increasing in magnitude from the top of water to the bottom. We could add more complexity to the problem by assuming that the upper layers of water possess greater speeds, and hence kinetic energy than the lower layers. Yet, that only changes the magnitude of normal stress, not its direction.

$$\sigma_{horizontal} = -\rho g(y - H + D) \qquad (1-1.5)$$

Where, ρ is the water density, y is the depth of water measured from the top surface of the dam, g is the gravitational acceleration. The negative sign accounts for the depth below the surface. y changes from 0 to $-D$.

(2) Inside the body of the dam, the horizontal planes are exposed to **shearing stresses** exerted by the normal stresses on the orthogonal surface exposed to water. The shearing stresses try to slice the dam along its horizontal planes. Note that we already equalized the normal stress by the shearing stress only at the microscopic of infinitesimal volumes but maintained the three distinctive orthogonal directions of the projections of the stress vector.

$$\tau_{horizontalplanes} = \rho g(y - H + D) \qquad (1-1.6)$$

(3) Two similar but orthogonal stresses are added by the weight of the dam. The **normal stress** is vertical and increases from top to bottom due to the weight of the dam. The shearing stresses attempt to slice the dam vertically against the resistance of the bottom rock bed.

The density of the dam is calculated as follows: $= \dfrac{M}{volume} = \dfrac{M}{LH\dfrac{A+B}{2}}$

$$\sigma_{vertical} = -g\dfrac{M}{LH\dfrac{A+B}{2}}y \qquad (1-1.7)$$

Where, L is an assumed length of the dam.

1.2. Navier's Partial differential equations of stress

(i) Cartesian coordinates

Let ρ = **density** of an elastic at any given point P(x,y,z) within the medium, having the dimension of kg/cm^3. Let **F** = **resultant force** at P(x,y,z) generated by the static balances of all external forces on the body of the medium. Let **A** = **cross section** perpendicular to **F** at P. Figure 1-4.

The **stress** τ(x,y,z) at P is defined as

22

$$\tau(x, y, z) = \lim_{\Delta A \to 0} \frac{\Delta F}{\Delta A}, \qquad\qquad \frac{kg.cm}{sec^2 .cm^2} \qquad\qquad (1\text{-}2.1)$$

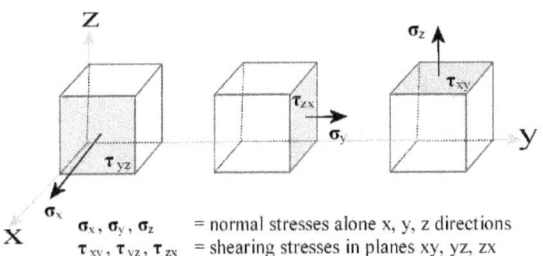

$\sigma_x, \sigma_y, \sigma_z$ = normal stresses alone x, y, z directions
$\tau_{xy}, \tau_{yz}, \tau_{zx}$ = shearing stresses in planes xy, yz, zx

Figure 1-4. Cartesian directions of areas in solid body where stresses are described. Notice the similarity between pressure acting on fluid (liquid or gas) with stresses acting on solid bodies.

The **net resultant stress** $\tau(x,y,z)$ imposes nine effects on the three Cartesian planes, which normals directed along the directions OX, OY, OZ.

The three stresses σ_x, σ_y σ_z along those axes are the **normal stresses** perpendicular to the planes with normals $\hat{x}, \hat{y}$, and $\hat{z}$.

The six stresses $2\tau_{xy}$, $2\tau_{yz}$, $2\tau_{zx}$ parallel to those directions are the **shearing stresses** parallel to the planes with normals $\hat{z}$, $\hat{x}$, and $\hat{y}$.

The internal balance of stresses in solid media is governed by **Newton's third law** of action and reaction. Thus, each projection of the stress vector is equal in magnitude and opposite in direction to the anti-parallel projection, along the same axis. Thus,

$$\sigma_x(x, y, z) = -\sigma_{-x}(x, y, z) \perp \text{to the infinitesimal area } \delta y \delta z \qquad (1\text{-}2.2)$$

$$\sigma_y(x, y, z) = -\sigma_{-y}(x, y, z) \perp \text{to the infinitesimal area } \delta x \delta z \qquad (1\text{-}2.3)$$

$$\sigma_z(x, y, z) = -\sigma_{-z}(x, y, z) \perp \text{to the infinitesimal area } \delta x \delta y \qquad (1\text{-}2.4)$$

The differential change of stresses between points $P(x,y,z)$ and $P(x+\delta x,\ y+\delta y,\ z+\delta z)$, along the three Cartesian coordinates, is given by

$$\sigma_x(x, y, z) \quad \text{and} \quad \sigma_x(x, y, z) + \frac{\partial \sigma_x(x, y, z)}{\partial x} \delta x \qquad (1\text{-}3.1)$$

$$\tau_{xy}(x, y, z) \quad \text{and} \quad \tau_{xy}(x, y, z) + \frac{\partial \tau_{xy}(x, y, z)}{\partial y} \delta y \qquad (1\text{-}3.2)$$

$$\tau_{xz}(x, y, z) \quad \text{and} \quad \tau_{xz}(x, y, z) + \frac{\partial \tau_{xz}(x, y, z)}{\partial z} \delta z \qquad (1\text{-}3.3)$$

23

Thus, the **net external force** applied to the volume $\delta v = \delta x \delta y \delta z$ along the **x-axis** is given by multiplying the projections of stress vector by the perpendicular areas $\delta y \delta z$ of the infinitesimal element δv.

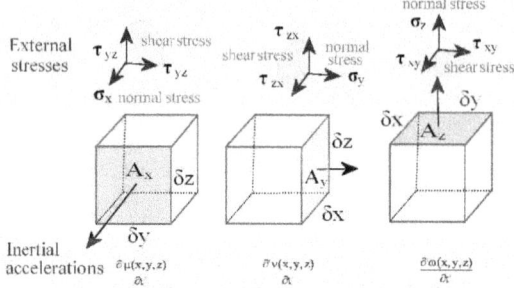

Figure 1-5. Managing stresses in solids similar to pressure in fluids

We assume the existence of **net internal body force** $F_i(x,y,z)$ **per unit mass** represented

$$F_i(x,y,z) = iX_i(x,y,z) + jY_i(x,y,z) + kZ_i(x,y,z) \qquad (1\text{-}3.4)$$

arising from the properties of the elastic volume $\delta v = \delta x \delta y \delta z$.

Thus, the net internal force along the **x-axis** is given by multiplying the x-projection of $F_i(x,y,z)$ by the mass of the infinitesimal element δv.

$$X_i \rho \, \delta x \, \delta y \delta z \qquad (1\text{-}3.5)$$

Therefore, the equation of motion of the element δv, shown in Figure 1-5, is governed by **Newton's second law**, where **net inertial force** $\overline{\overline{u}}(\overline{\overline{\mu}}, \overline{\overline{v}}, \overline{\overline{\omega}})$ per unit mass, Figure 1-5, balances the net resultants of external and internal forces

$$
\begin{aligned}
&\left(\sigma_x(x,y,z) + \frac{\partial \sigma_x(x,y,z)}{\partial x} dx - \sigma_x(x,y,z) \right) dy dz \\
&+ \left(\tau_{xy}(x,y,z) + \frac{\partial \tau_{xy}(x,y,z)}{\partial y} dy - \tau_{xy}(x,y,z) \right) dx dz \\
&+ \left(\tau_{xz}(x,y,z) + \frac{\partial \tau_{xz}(x,y,z)}{\partial z} dz - \tau_{xz}(x,y,z) \right) dx dy \\
&+ X_i \rho dx dy dz \quad = \quad \rho dx dy dz \frac{\partial^2 u(x,y,z)}{\partial t^2}
\end{aligned}
\qquad (1\text{-}3.6)
$$

Similarly, we have two equations for the net internal force along the **y** and **z –axes**.

Therefore, equation (1-3.6) yield the three **Navier's partial differential equations of equilibrium of stresses** of the three forces acting on the solid body as follows

$$\frac{\partial \sigma_x(x,y,z)}{\partial x} + \frac{\partial \tau_{xy}(x,y,z)}{\partial y} + \frac{\partial \tau_{xz}(x,y,z)}{\partial z} + \rho X_i = \rho \frac{\partial^2 u(x,y,z)}{\partial t^2} \qquad (1\text{-}4.1)$$

$$\frac{\partial \tau_{yx}(x,y,z)}{\partial x} + \frac{\partial \sigma_y(x,y,z)}{\partial y} + \frac{\partial \tau_{yz}(x,y,z)}{\partial z} + \rho Y_i = \rho \frac{\partial^2 v(x,y,z)}{\partial t^2} \qquad (1\text{-}4.2)$$

$$\frac{\partial \tau_{zx}(x,y,z)}{\partial x} + \frac{\partial \tau_{zy}(x,y,z)}{\partial y} + \frac{\partial \sigma_z(x,y,z)}{\partial z} + \rho Z_i = \rho \frac{\partial^2 w(x,y,z)}{\partial t^2} \qquad (1\text{-}4.3)$$

Where, **Newton's second law of motion** is applied in the three Cartesian directions to entail the following forces affecting solid body equilibrium.

(1)   Nine effects of **external forces** represented by the normal σ's and shearing τ's stresses along the three perpendicular axes.
(2)   Three **internal effects** represented by F_i with different described effects such as body weight, elastic forces, sources or sinks of energy that could generate forces.
(3)   **Inertial effects** $\upsilon(\tilde{u}, \tilde{v}, \tilde{w})$ summing the net balance of forces on the body according to Newton's law of motion

Therefore,

Equations (1-4) contain the **nine** unknowns σ_i and τ_{ij} where j and i = x, y, and z, where i $\neq$ j. Solving for the nine unknowns requires the knowledge of the properties of matter in order to reduce the nine variables into three that can be solved by the three Navier's partial differential equations (1-4).

In this chapter will only use equation (1-4) to examine the validity of known distributions of stresses. The real use of the differential equations (1-4) will follow the chapters on **Hooke's law**

(ii) Cylindrical polar coordinates

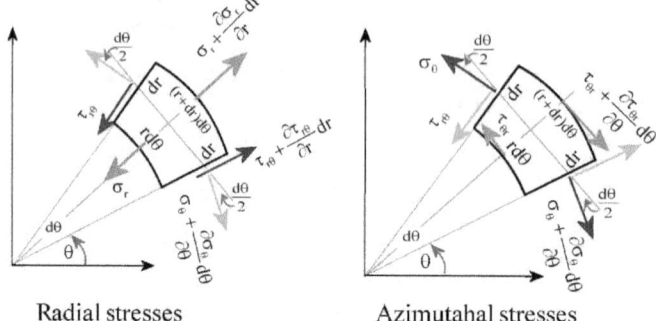

| Radial stresses | Azimutahal stresses |

Figure 1-6. Element of volume in cylindrical polar coordinates showing radial and azimuthal variation of normal and shear stresses of the coordinate infinitesimal elements dr and dθ.

In cylindrical polar coordinates, in Figure 1-6, the **net resultant stress** $\tau(r,\theta,z)$ imposes nine effects on the three Cartesian planes, which normals directed along the directions r, θ, and z.

The three stresses along those axes σ_r, σ_θ, σ_z are the **normal stresses** perpendicular to the planes with normals $\hat{r}, \hat{\theta}$, and $\hat{z}$.

The six stresses $2\tau_{r\theta}$, $2\tau_{rz}$, $2\tau_{\theta x}$ parallel to those directions are the **shearing stresses** parallel to the planes with normals $\hat{r}, \hat{\theta}$, and $\hat{z}$.

The internal balance of stresses in solid media is governed by **Newton's third law** of action and reaction. Thus, each projection of the stress vector is equal in magnitude and opposite in direction to the anti-parallel projection, along the same axis. Thus,

$$\sigma_r(r,\theta) = -\sigma_{-r}(r,\theta) \perp \text{ to the infinitesimal area } r\delta\theta\delta z \qquad (1\text{-}4.4)$$

$$\sigma_\theta(r,\theta) = -\sigma_{-\theta}(r,\theta) \perp \text{ to the infinitesimal area } \delta r\delta z \qquad (1\text{-}4.5)$$

$$\sigma_z(r,\theta) = -\sigma_{-z}(r,\theta) \perp \text{ to the infinitesimal area } r\delta\theta\delta r \qquad (1\text{-}4.6)$$

The differential change of stresses between points $P(r,\theta)$ and $P(r+\delta r,\ \theta+\delta\theta)$, along the two polar coordinates, is given by

$$\sigma_r(r,\theta) \text{ and } \sigma_r(r,\theta) + \frac{\partial\sigma_r(r,\theta)}{\partial r}\delta r \qquad (1\text{-}4.7)$$

$$\tau_{r\theta}(r,\theta) \text{ and } \tau_{r\theta}(r,\theta) + \frac{\partial\tau_{r\theta}(r,\theta)}{\partial\theta}\delta\theta \qquad (1\text{-}4.8)$$

$$\tau_{rz}(r,\theta) \text{ and } \tau_{rz}(r,\theta) + \frac{\partial\tau_{rz}(r,\theta)}{\partial z}\delta z \qquad (1\text{-}4.9)$$

Thus, the **net external force** applied to the volume $\delta v = r\delta r\delta\theta\delta z$ along the **r-axis** is given by multiplying the projections of stress vector by the perpendicular areas $r\delta\theta\delta z$ of the infinitesimal element δv.

Due to the **curved coordinates** in polar coordinates, the following considerations are taken into account:

1. As the **polar length** of the arbitrary volume element changes from r to r + dr, the **radial normal stress** σ_r changes in magnitude along the direction of r, without changing inclination from the radial direction.

2. As the **azimuthal length** of the arbitrary volume element changes from θ to $\theta + d\theta$, the **radial shear stress** $\tau_{r\theta}$ also changes in magnitude along the direction of r, without changing inclination from the radial direction.

3. As the **azimuthal length** of the arbitrary volume element changes from θ to $\theta + d\theta$, the **azimuthal normal stress** σ_θ changes in [both] magnitude and direction (from $\dfrac{d\theta}{2}$ to $-\dfrac{d\theta}{2}$) along the azimuthal direction of θ.

Thus, the equation of motion of the element δv, shown in Figure 1-6, is governed by **Newton's second law** of equilibrium of inertial forces along the r-axis as follows

$$\left(\left[\sigma_r(r,\theta,z)+\frac{\partial\sigma_r(r,\theta,z)}{\partial r}dr\right](r+dr)d\theta-\sigma_r(r,\theta,z)rd\theta\right)dz$$

$$+\left(\left[\tau_{r\theta}(r,\theta,z)+\frac{\partial\tau_{r\theta}(r,\theta,z)}{\partial\theta}d\theta\right]dr-\tau_{r\theta}(r,\theta,z)dr\right)dz$$

$$+\left(\left[\sigma_\theta(r,\theta,z)+\frac{\partial\sigma_\theta(r,\theta,z)}{\partial\theta}d\theta\right]dr\left(-\frac{d\theta}{2}\right)-\sigma_\theta(r,\theta,z)dr\frac{d\theta}{2}\right)dz \qquad (1\text{-}4.10)$$

$$+\left(\left[\tau_{rz}(r,\theta,z)+\frac{\partial\tau_{rz}(r,\theta,z)}{\partial z}dz\right]rdrd\theta-\tau_{rz}(r,\theta,z)rdrd\theta\right)$$

$$+\Pi_i\rho rdrd\theta dz \;=\; \rho rdrd\theta dz\frac{\partial^2 u_r(r,\theta,z)}{\partial t^2}$$

Similarly, the curved coordinates in polar coordinates affect the azimuthal equilibrium of forces as follows.

4. As the **polar length** of the arbitrary volume element changes from r to r + dr, the **azimuthal normal stress** σ_θ changes in magnitude along the direction of r, without changing inclination from the radial direction.

5. As the **azimuthal length** of the arbitrary volume element changes from θ to $\theta + d\theta$, the **azimuthal shear stress** $\tau_{r\theta}$ changes, not only magnitude but also in direction (from $\dfrac{d\theta}{2}$ to $-\dfrac{d\theta}{2}$).

6. As the **polar length** of the arbitrary volume element changes from r to r + d r, the **azimuthal shear stress** $\tau_{r\theta}$ changes only in magnitude along the direction of r, without changing inclination from the radial direction.

Therefore, the equation of motion of the element δv along the θ-axis as follows

$$
\begin{aligned}
&\left[\boldsymbol{\sigma}_\theta(r,\theta,z)+\frac{\partial\boldsymbol{\sigma}_\theta(r,\theta,z)}{\partial\theta}d\theta-\boldsymbol{\sigma}_\theta(r,\theta,z)\right]drdz \\
&+\left[\left(\boldsymbol{\tau}_{r\theta}(r,\theta,z)+\frac{\partial\boldsymbol{\tau}_{r\theta}(r,\theta,z)}{\partial\theta}d\theta\right)\frac{d\theta}{2}-\boldsymbol{\tau}_{r\theta}(r,\theta,z)\left(-\frac{d\theta}{2}\right)\right]drdz \\
&+\left(\left[\boldsymbol{\tau}_{r\theta}(r,\theta,z)+\frac{\partial\boldsymbol{\tau}_{r\theta}(r,\theta,z)}{\partial r}dr\right](r+dr)d\theta-\boldsymbol{\tau}_{r\theta}(r,\theta,z)rd\theta\right)dz && (1\text{-}4.11) \\
&+\left(\left[\boldsymbol{\tau}_{\theta z}(r,\theta,z)+\frac{\partial\boldsymbol{\tau}_{\theta z}(r,\theta,z)}{\partial z}dz\right]rdrd\theta-\boldsymbol{\tau}_{\theta z}(r,\theta,z)rdrd\theta\right) \\
&+\Omega_i\rho drd\theta dz \quad = \quad \rho drd\theta dz\frac{\partial^2 u_\theta(r,\theta,z)}{\partial t^2}
\end{aligned}
$$

Therefore, the equation of motion of the element δv along the θ-axis as follows

$$
\begin{aligned}
&\left[\boldsymbol{\sigma}_z(r,\theta,z)+\frac{\partial\boldsymbol{\sigma}_z(r,\theta,z)}{\partial z}dz-\boldsymbol{\sigma}_z(r,\theta,z)\right]rdrd\theta \\
&+\left[\left(\boldsymbol{\tau}_{z\theta}(r,\theta,z)+\frac{\partial\boldsymbol{\tau}_{z\theta}(r,\theta,z)}{\partial\theta}d\theta\right)-\boldsymbol{\tau}_{r\theta}(r,\theta,z)\right]drdz \\
&+\left(\left[\boldsymbol{\tau}_{zr}(r,\theta,z)+\frac{\partial\boldsymbol{\tau}_{zr}(r,\theta,z)}{\partial r}dr\right](r+dr)d\theta-\boldsymbol{\tau}_{zr}(r,\theta,z)rd\theta\right)dz && (1\text{-}4.12) \\
&+\Gamma_i\rho drd\theta dz \quad = \quad \rho drd\theta dz\frac{\partial^2 u_z(r,\theta,z)}{\partial t^2}
\end{aligned}
$$

Divide equations by rdrdθdz

$$
\frac{\partial\boldsymbol{\sigma}_r(r,\theta,z)}{\partial r}+\frac{\partial\boldsymbol{\tau}_{r\theta}(r,\theta,z)}{r\partial\theta}+\frac{\boldsymbol{\sigma}_r(r,\theta,z)-\boldsymbol{\sigma}_\theta(r,\theta,z)}{r}+\frac{\partial\boldsymbol{\tau}_{rz}(r,\theta,z)}{\partial z}+\Pi_i\rho=\rho\frac{\partial^2 u_r(r,\theta,z)}{\partial t^2} \quad (1\text{-}4.13)
$$

$$\frac{\partial \sigma_\theta(r,\theta,z)}{r\partial\theta} + \frac{\partial \tau_{r\theta}(r,\theta,z)}{\partial r} + \frac{\partial \tau_{\theta z}(r,\theta,z)}{\partial z} + \frac{2\tau_{r\theta}(r,\theta,z)}{r} + \Omega_i\rho = \rho\frac{\partial^2 u_\theta(r,\theta,z)}{\partial t^2} \qquad (1\text{-}4.14)$$

$$\frac{\partial \sigma_z(r,\theta,z)}{\partial z} + \frac{\partial \tau_{z\theta}(r,\theta,z)}{r\partial\theta} + \frac{\partial \tau_{zr}(r,\theta,z)}{\partial r} + \frac{\tau_{zr}(r,\theta,z)}{r} + \Gamma_i\rho = \rho\frac{\partial^2 u_z(r,\theta,z)}{\partial t^2} \qquad (1\text{-}4.15)$$

(iii) Spherical polar coordinates

Navier's equations of equilibrium are written in spherical polar coordinates, as follows. Figure 1-4,

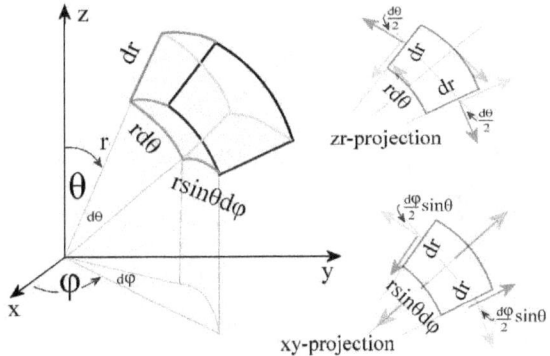

Figure 1-7. Spherical polar coordinates of element volume for determining Navier's equilibrium differential equations.

1. Static equilibrium of stresses along the r-axis

Figure 1-8 shows the various components of stresses in the radial direction, which are balanced as follows.

$$\left(\left[\sigma_r(r,\theta,\varphi)+\frac{\partial\sigma_r(r,\theta,\varphi)}{\partial r}dr\right](r+dr)(r+dr)\sin\theta d\theta d\varphi-\sigma_r(r,\theta,\varphi)r^2\sin\theta d\theta d\varphi\right)$$

$$+\left(\left[\tau_{r\theta}(r,\theta,\varphi)+\frac{\partial\tau_{r\theta}(r,\theta,\varphi)}{\partial\theta}d\theta\right]r\sin(\theta+d\theta)dr-\tau_{r\theta}(r,\theta,\varphi)r\sin\theta dr\right)d\varphi$$

$$+\left(\left[\tau_{r\varphi}(r,\theta,\varphi)+\frac{\partial\tau_{r\varphi}(r,\theta,\varphi)}{\partial\varphi}d\varphi\right]rdrd\theta-\tau_{r\varphi}(r,\theta,\varphi)rdrd\theta\right)$$

$$+\left(\left[\sigma_\theta(r,\theta,\varphi)+\frac{\partial\sigma_\theta(r,\theta,\varphi)}{\partial\theta}d\theta\right]dr\left(-\frac{d\theta}{2}\right)-\sigma_\theta(r,\theta,\varphi)dr\frac{d\theta}{2}\right)r\sin\theta d\varphi$$

$$+\left(\left[\sigma_\varphi(r,\theta,\varphi)+\frac{\partial\sigma_\varphi(r,\theta,\varphi)}{\partial\varphi}d\varphi\right]dr\left(-\frac{\sin\theta d\varphi}{2}\right)-\sigma_\varphi(r,\theta,\varphi)dr\left(\frac{\sin\theta d\varphi}{2}\right)\right)rd\theta$$

$$+\Pi_i\rho r^2\sin\theta drd\theta d\varphi=\rho r^2\sin\theta drd\theta d\varphi\frac{\partial^2 u(r,\theta,z)}{\partial t^2}$$

$$(1\text{-}4.16)$$

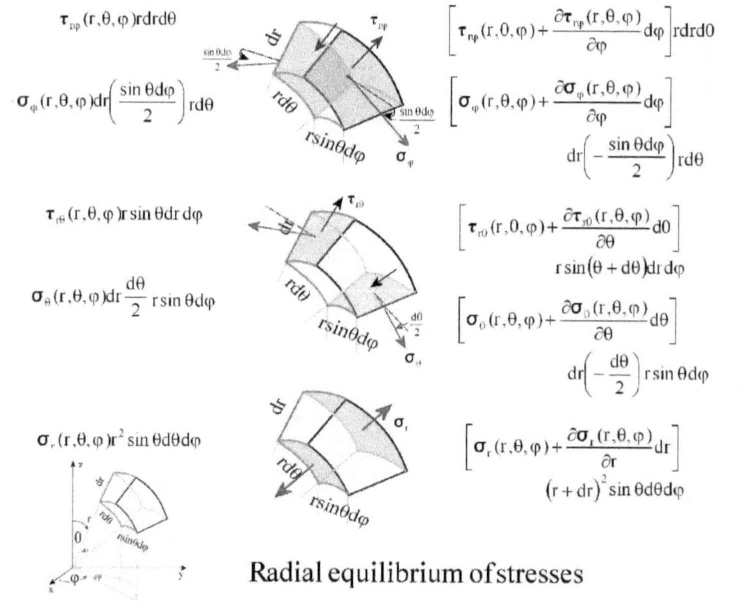

Radial equilibrium of stresses

Figure 1-8. Depiction of stresses over element volume along the radial axis.

2. Static equilibrium of stresses along the θ-curve, Figure 1-9

$$\left(\left[\tau_{r\theta}(r,\theta,\varphi)+\frac{\partial\tau_{r\theta}(r,\theta,\varphi)}{\partial r}dr\right](r+dr)^2 d\theta\sin\theta d\varphi-\tau_{r\theta}(r,\theta,\varphi)r^2\sin\theta d\varphi d\theta\right)$$

$$\left(\left[\tau_{r\theta}(r,\theta,\varphi)+\frac{\partial\tau_{r\theta}(r,\theta,\varphi)}{\partial\varphi}d\varphi\right]rdr d\theta\left(\sin\theta\frac{d\varphi}{2}\right)-\tau_{r\theta}(r,\theta,\varphi)rdr d\theta\left(-\sin\theta\frac{d\varphi}{2}\right)\right)$$

$$+\left(\left[\sigma_{\theta}(r,\theta,\varphi)+\frac{\partial\sigma_{\theta}(r,\theta,\varphi)}{\partial\theta}d\theta\right]r(\sin\theta+d\theta\cos\theta)d\varphi dr-\sigma_{\theta}(r,\theta,\varphi)r\sin\theta d\varphi dr\right)$$

$$+\left(\left[\sigma_{\varphi}(r,\theta,\varphi)+\frac{\partial\sigma_{\varphi}(r,\theta,\varphi)}{\partial\varphi}d\varphi\right]rdr d\theta\left(-\cos\theta\frac{d\varphi}{2}\right)-\sigma_{\varphi}(r,\theta,\varphi)rdr d\theta\left(\cos\theta\frac{d\varphi}{2}\right)\right)$$

$$+\left(\left[\tau_{\theta\varphi}(r,\theta,\varphi)+\frac{\partial\tau_{\theta\varphi}(r,\theta,\varphi)}{\partial\varphi}d\varphi\right]rdr d\theta-\tau_{\theta\varphi}(r,\theta,\varphi)rdr d\theta\right)$$

$$+\Omega_i\rho r^2\sin\theta dr d\theta d\varphi=\rho r^2\sin\theta dr d\theta d\varphi\frac{\partial^2 u_{\theta}(r,\theta,z)}{\partial t^2}$$

(1-4.17)

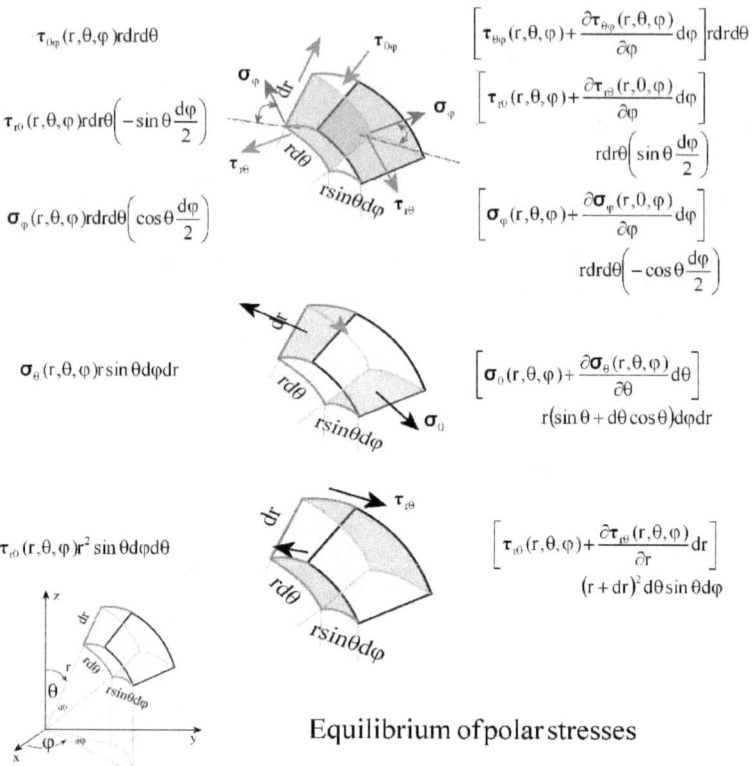

$$\tau_{\theta\varphi}(r,\theta,\varphi)rdrd\theta$$

$$\left[\tau_{\theta\varphi}(r,\theta,\varphi)+\frac{\partial\tau_{\theta\varphi}(r,\theta,\varphi)}{\partial\varphi}d\varphi\right]rdrd\theta$$

$$\tau_{r\theta}(r,\theta,\varphi)rdrd\theta\left(-\sin\theta\frac{d\varphi}{2}\right)$$

$$\left[\tau_{r\theta}(r,\theta,\varphi)+\frac{\partial\tau_{r\theta}(r,\theta,\varphi)}{\partial\varphi}d\varphi\right]$$
$$rdrd\theta\left(\sin\theta\frac{d\varphi}{2}\right)$$

$$\sigma_{\varphi}(r,\theta,\varphi)rdrd\theta\left(\cos\theta\frac{d\varphi}{2}\right)$$

$$\left[\sigma_{\varphi}(r,\theta,\varphi)+\frac{\partial\sigma_{\varphi}(r,\theta,\varphi)}{\partial\varphi}d\varphi\right]$$
$$rdrd\theta\left(-\cos\theta\frac{d\varphi}{2}\right)$$

$$\sigma_{\theta}(r,\theta,\varphi)r\sin\theta d\varphi dr$$

$$\left[\sigma_{\theta}(r,\theta,\varphi)+\frac{\partial\sigma_{\theta}(r,\theta,\varphi)}{\partial\theta}d\theta\right]$$
$$r(\sin\theta+d\theta\cos\theta)d\varphi dr$$

$$\tau_{r\theta}(r,\theta,\varphi)r^2\sin\theta d\varphi d\theta$$

$$\left[\tau_{r\theta}(r,\theta,\varphi)+\frac{\partial\tau_{r\theta}(r,\theta,\varphi)}{\partial r}dr\right]$$
$$(r+dr)^2 d\theta\sin\theta d\varphi$$

Equilibrium of polar stresses

Figure 1-9. Depiction of stresses over element volume along the polar direction.

3. Static equilibrium of stresses along the azimuth φ-curve, Figure 1-10

$$\left(\left[\boldsymbol{\tau}_{r\varphi}(r,\theta,\varphi)+\frac{\partial\boldsymbol{\tau}_{r\varphi}(r,\theta,\varphi)}{\partial r}dr\right](r+dr)^2\,d\theta\sin\theta\,d\varphi-\boldsymbol{\tau}_{r\varphi}(r,\theta,\varphi)r^2\sin\theta\,d\varphi\,d\theta\right)$$

$$\left(\left[\boldsymbol{\tau}_{r\varphi}(r,\theta,\varphi)+\frac{\partial\boldsymbol{\tau}_{r\varphi}(r,\theta,\varphi)}{\partial\varphi}d\varphi\right]rdr\theta\left(\sin\theta\frac{d\varphi}{2}\right)-\boldsymbol{\tau}_{r\varphi}(r,\theta,\varphi)rdr\theta\left(-\sin\theta\frac{d\varphi}{2}\right)\right)$$

$$+\left(\left[\boldsymbol{\sigma}_{\varphi}(r,\theta,\varphi)+\frac{\partial\boldsymbol{\sigma}_{\varphi}(r,\theta,\varphi)}{\partial\varphi}d\varphi\right]rdrd\theta-\boldsymbol{\sigma}_{\varphi}(r,\theta,\varphi)rdrd\theta\right) \qquad (1\text{-}4.18)$$

$$+\left(\left[\boldsymbol{\tau}_{\theta\varphi}(r,\theta,\varphi)+\frac{\partial\boldsymbol{\tau}_{\theta\varphi}(r,\theta,\varphi)}{\partial\theta}d\theta\right]rdr\sin(\theta+d\theta)d\varphi-\boldsymbol{\tau}_{\theta\varphi}(r,\theta,\varphi)r\sin\theta\,d\varphi\,dr\right)$$

$$+\Omega_i\rho r^2\sin\theta\,drd\theta d\varphi=\rho r^2\sin\theta\,drd\theta d\varphi\frac{\partial^2 u_\theta(r,\theta,z)}{\partial t^2}$$

Equilibrium of azimuthal stresses

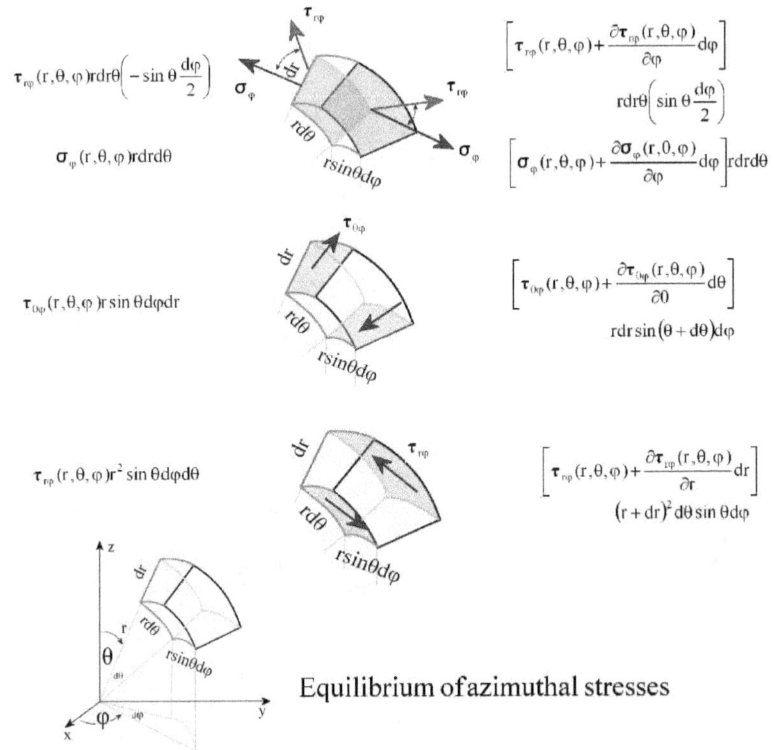

Figure 1-10. Depiction of stresses over element volume along the azimuth direction.

Substituting by

$$\sin(\theta + d\theta) = \sin\theta\cos d\theta + \sin d\theta\cos\theta$$
$$= \sin\theta + d\theta\cos\theta \qquad (1\text{-}4.19)$$

Arranging, we get

$$\frac{\partial\sigma_r(r,\theta,\varphi)}{\partial r} + \frac{\partial\tau_{r\theta}(r,\theta,\varphi)}{r\partial\theta} + \frac{1}{r}\begin{pmatrix} 2\sigma_r(r,\theta,\varphi) \\ -\sigma_\theta(r,\theta,\varphi) \\ -\sigma_\theta(r,\theta,\varphi) \\ +\tau_{r\theta}(r,\theta,\varphi)\cot\theta \end{pmatrix} + \frac{\partial\tau_{r\varphi}(r,\theta,\varphi)}{r\sin\theta\partial\varphi} + \Pi_i\rho = \rho\frac{\partial^2 u_r(r,\theta,\varphi)}{\partial t^2} \qquad (1\text{-}4.20)$$

$$\frac{\partial\tau_{r\theta}(r,\theta,\varphi)}{\partial r} + \frac{\partial\sigma_\theta(r,\theta,\varphi)}{r\partial\theta} + \frac{\partial\tau_{\theta\varphi}(r,\theta,\varphi)}{r\sin\theta\partial\varphi} + \frac{1}{r}\begin{pmatrix} 3\tau_{r\theta}(r,\theta,\varphi) \\ +\begin{pmatrix}\sigma_\theta(r,\theta,\varphi) \\ -\sigma_\varphi(r,\theta,\varphi)\end{pmatrix}\cot\theta \end{pmatrix} + \Omega_i\rho = \rho\frac{\partial^2 u_\theta(r,\theta,\varphi)}{\partial t^2}$$

$$(1\text{-}4.21)$$

$$\frac{\partial\tau_{r\varphi}(r,\theta,\varphi)}{\partial r} + \frac{\partial\sigma_\varphi(r,\theta,\varphi)}{r\sin\theta\partial\varphi} + \frac{\partial\tau_{\theta\varphi}(r,\theta,\varphi)}{r\partial\theta} + \frac{1}{r}\begin{pmatrix} 3\tau_{r\varphi}(r,\theta,\varphi) \\ +2\tau_{\theta\varphi}(r,\theta,\varphi)r\cot\theta \end{pmatrix} + \Gamma_i\rho = \rho\frac{\partial^2 u_\varphi(r,\theta,\varphi)}{\partial t^2}$$

$$(1\text{-}4.22)$$

1.3. Reciprocity of mutually perpendicular shearing stresses

The difference between **stress vectors** and **force vectors** is clearly elucidated by the **differential equations of equilibrium** of stresses, equations (1-4) which shows that stresses are **infinitesimal flow of forces rather than lumped, isolated forces acting ion the absence of matter**. Stresses require matter to be considered as normal and shearing, which imposes the differentials in equations (1-4).

For example, when we deal with shearing stresses, we do not ask about the perpendicular area upon which the force acts. Since stress differs from force, shearing stresses apply on area parallel to the direction normal stress. Figure 1-11 illustrates that idea.

34

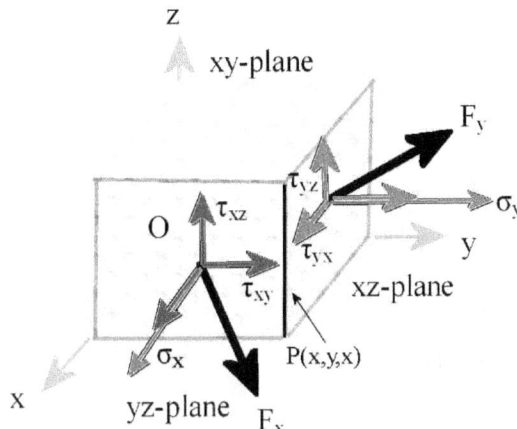

Figure 1-11. Reciprocity of stress as mutually perpendicular planes.

Let F_x and F_y the total forces acting on the yz-plane and xz-plane. Each force generates three stresses on each surface given by F_x (σ_x, τ_{xy}, τ_{xz}) and F_y (σ_y, τ_{yx}, τ_{yz}). Point $P(x,y,z)$ located on the intersection of yz-plane and xz-plane is exposed to the equilibrium of the two forces

$$\tau_{xy} = \tau_{yx}$$

Which is the called the **law of conjugation of shearing stresses.** Clearly, the component of shearing stresses that shares common edge between **two orthogonal planes** imposes reciprocally equal stress.

Example 3

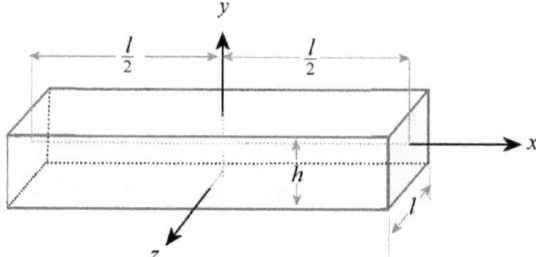

Figure 1-12. Parelleloid rigid bar exposed to pre-determined stresses.

The parelleloid shown in Figure 1-12 is exposed to the following stresses:

$$\sigma_x = \frac{12}{h^2}\left(M_o + \frac{ql^2}{8} - \frac{qh^2}{20}\right)y - \frac{6q}{h^3}x^2y + \frac{4q}{h^3}y^3 \qquad (1\text{-}5.1)$$

$$\sigma_y = -\frac{6q}{h^3}\left(\frac{y^3}{3} - \frac{h^2}{4}y + \frac{h^3}{12}\right) \qquad (1\text{-}5.2)$$

$$\tau_{xy} = -\frac{6q}{h^3}\left(\frac{h^2}{4} - y^2\right)x \qquad (1\text{-}5.3)$$

$$\tau_{yz} = \tau_{zx} = \sigma_z = 0 \qquad (1\text{-}5.4)$$

(i) Show that the above stresses satisfy the partial equations of stress, equations (1-4).
(ii) Find the forces acting on the six faces of the parelleloid.

Solution

(i) The Navier's partial equations of stress (1-4.1 through 3) contain derivatives of the stresses given in equations (1-5.1 through 4), which are obtained as follows.

$$\frac{\partial \sigma_x(x,y,z)}{\partial x} = \frac{\partial}{\partial x}\left[\frac{12}{h^2}\left(M_o + \frac{ql^2}{8} - \frac{qh^2}{20}\right)y - \frac{6q}{h^3}x^2y + \frac{4q}{h^3}y^3\right] = -\frac{12q}{h^3}xy \qquad (1\text{-}5.6)$$

$$\frac{\partial \tau_{xy}(x,y,z)}{\partial y} = \frac{\partial}{\partial y}\left[-\frac{6q}{h^3}\left(\frac{h^2}{4} - y^2\right)x\right] = \frac{12q}{h^3}yx \qquad (1\text{-}5.7)$$

$$\frac{\partial \tau_{xy}(x,y,z)}{\partial x} = \frac{\partial}{\partial y}\left[-\frac{6q}{h^3}\left(\frac{h^2}{4} - y^2\right)x\right] = -\frac{6q}{h^3}\left(\frac{h^2}{4} - y^2\right) \qquad (1\text{-}5.8)$$

$$\frac{\partial \sigma_y(x,y,z)}{\partial y} = \frac{\partial}{\partial y}\left[-\frac{6q}{h^3}\left(\frac{y^3}{3} - \frac{h^2}{4}y + \frac{h^3}{12}\right)\right] = -\frac{6q}{h^3}\left(y^2 - \frac{h^2}{4}\right) \qquad (1\text{-}5.9)$$

Substituting the derivatives from (1-5.6 through 9) into equations (1-4.1 through 3), we get

$$-\frac{12q}{h^3}xy + \frac{12q}{h^3}yx + 0 + 0 = 0 \tag{1-5.10}$$

$$-\frac{6q}{h^3}\left(\frac{h^2}{4} - y^2\right) + -\frac{6q}{h^3}\left(y^2 - \frac{h^2}{4}\right) + 0 + 0 = 0 \tag{1-5.11}$$

$$0 + 0 + 0 + 0 = 0 \tag{1-5.12}$$

Therefore, the given stresses satisfy the partial differential equations of stresses when internal forces are ignored and external accelerations are zeros.

(ii) Determining the forces on the faces of the paralleloid

(1) Stress and force at x = 0 in the x-direction on surface in the yz-plane

Substituting by x = 0 in equation (1-5.1), we get the normal stress in the x-direction acting at x = 0 on the surface yz, as follows.

$$\sigma_x = \frac{12}{h^2}\left(M_o + \frac{ql^2}{8} - \frac{qh^2}{20}\right)y + \frac{4q}{h^3}y^3 \tag{1-5.13}$$

Integrating the normal stress on surface yz, at x = 0, we get the **net force** on the surface yz, at x = 0, as follows.

$$\begin{aligned}
F_x &= \int_0^h \int_0^l \sigma_x dy dz = \int_0^h \int_0^l \left[\frac{12}{h^2}\left(M_o + \frac{ql^2}{8} - \frac{qh^2}{20}\right)y + \frac{4q}{h^3}y^3\right]dy dz \\
&= \left[\left[\frac{6}{h^2}\left(M_o + \frac{ql^2}{8} - \frac{qh^2}{20}\right)y^2 z + \frac{q}{h^3}y^4 z\right]_{y=0}^{y=l}\right]_{z=0}^{z=h} \\
&= \frac{6}{h^2}\left(M_o + \frac{ql^2}{8} - \frac{qh^2}{20}\right)l^2 h + \frac{q}{h^3}l^4 h \tag{1-5.14}
\end{aligned}$$

(2) Stress and force at x = l in the x-direction on surface in the yz-plane

Substituting by x = l in equation (1-5.1), we get the normal stress in the x-direction acting at x = l on the surface yz, as follows.

$$\sigma_x = \frac{12}{h^2}\left(M_o + \frac{ql^2}{8} - \frac{qh^2}{20}\right)y - \frac{6q}{h^3}l^2 y + \frac{4q}{h^3}y^3 \tag{1-5.15}$$

Integrating the normal stress on surface yz, at x = l, we get the **net force** on the surface yz, at x = l, as follows.

$$F_x \big|_{y=l} = \int_0^h \int_0^l \sigma_x \, dy \, dz = \int_0^h \int_0^l \left[\frac{12}{h^2}\left(M_o + \frac{ql^2}{8} - \frac{qh^2}{20} \right)y - \frac{6q}{h^3}l^2 y + \frac{4q}{h^3}y^3 \right] dy \, dz$$

$$= \left[\left[\left[\frac{6}{h^2}\left(M_o + \frac{ql^2}{8} - \frac{qh^2}{20} \right)y^2 - \frac{3q}{h^3}l^2 y^2 + \frac{q}{h^3}y^4 \right]_{y=0}^{y=l} z \right]_{z=0}^{z=h} \right]$$

$$= \frac{6}{h^2}\left(M_o + \frac{ql^2}{8} - \frac{qh^2}{20} \right)l^2 h - \frac{3q}{h^3}l^4 h + \frac{q}{h^3}l^4 h$$

$$= \frac{6}{h}\left(M_o + \frac{ql^2}{8} - \frac{qh^2}{20} \right)l^2 - \frac{2q}{h^2}l^4 \qquad (1\text{-}5.16)$$

(3) Stress and force at y = 0, in the y-direction on surface in the xz-plane

Substituting by y = 0, in equation (1-5.2), we get the **normal stress** in the y-direction acting at y = 0 on the surface xz, as follows.

$$\sigma_y = -\frac{6q}{h^3}\left(\frac{0}{3} - \frac{h^2}{4}0 + \frac{h^3}{12} \right) = -\frac{q}{2} \qquad (1\text{-}5.17)$$

Substituting by y = 0 in equation (1-5.3), we get the **shearing stress** in the x-direction acting at y = 0 on the surface xz, as follows.

$$\tau_{xy} = -\frac{6q}{h^3}\left(\frac{h^2}{4} - 0 \right)x = -\frac{3q}{2h}x \qquad (1\text{-}5.18)$$

Similarly, integrating the normal stress on surface xz, at y = 0, we get the **net force** on the surface xz, at y = 0, as follows.

Forces normal on the xz-plane at y = 0

$$F_{y\perp y} = \int_0^h \int_0^l \sigma_y \, dx \, dz = \int_0^h \int_0^l -\frac{q}{2} \, dx \, dz = \left[\left[-\frac{q}{2}xz \right]_{x=0}^{x=l} \right]_{z=0}^{z=h} = -\frac{q}{2}hl \qquad (1\text{-}5.19)$$

Forces parallel on the xz-plane at y = 0

$$F_{y//y} = \int_0^h \int_0^l \tau_{xy} \, dx \, dz = \int_0^h \int_0^l -\frac{3q}{2h}x \, dx \, dz = \left[\left[-\frac{3q}{4h}x^2 z \right]_{x=0}^{x=l} \right]_{z=0}^{z=h} = -\frac{3q}{4}l^2 \qquad (1\text{-}5.20)$$

(1-4) Stress and force at y = l in the y-direction on surface in the xz-plane

Substituting by y = l in equation (1-5.2), we get the **normal stress** in the y-direction acting at y = l on the surface xz, as follows.

$$\sigma_y = -\frac{6q}{h^3}\left(\frac{l^3}{3} - \frac{h^2}{4}l + \frac{h^3}{12}\right)$$

(1-5.21)

Substituting by y = l in equation (1-5.3), we get the **shearing stress** in the x-direction acting at y = l on the surface xz, as follows.

$$\tau_{xy} = -\frac{6q}{h^3}\left(\frac{h^2}{4} - l^2\right)x$$

(1-5.22)

Similarly, integrating the normal stress on surface xz, at y = l, we get the **net force** on the surface xz, at y = l, as follows.

Forces normal on the xz-plane at y = l

$$F_{y\perp y} = \int_0^h \int_0^l \sigma_y \, dxdz = \int_0^h \int_0^l -\frac{6q}{h^3}\left(\frac{l^3}{3} - \frac{h^2}{4}l + \frac{h^3}{12}\right)dxdz$$

$$= \left[\left[-\frac{6q}{h^3}\left(\frac{l^3}{3} - \frac{h^2}{4}l + \frac{h^3}{12}\right)xz\right]_{x=0}^{x=l}\right]_{z=0}^{z=h} = -\frac{6q}{h^2}\left(\frac{l^4}{3} - \frac{h^2}{4}l^2 + \frac{lh^3}{12}\right)$$

(1-5.23)

Forces parallel on the xz-plane at y = l

$$F_{y//y} = \int_0^h \int_0^l \tau_{xy} \, dxdz = \int_0^h \int_0^l -\frac{6q}{h^3}\left(\frac{h^2}{4} - l^2\right)xdxdz$$

$$= \left[\left[-\frac{3q}{h^3}\left(\frac{h^2}{4} - l^2\right)x^2z\right]_{x=0}^{x=l}\right]_{z=0}^{z=h} = -\frac{3q}{h^2}\left(\frac{l^2h^2}{4} - l^4\right)$$

(1-5.24)

1.4. Surface conditions for projection of stress

On an **arbitrary plane** oriented in the $\hat{n}\,(l,m,n)$ direction, where l,m,n are the direction cosines $\hat{n}$, the nine stresses imparted on the body surfaces can be lumped onto three stresses in three Cartesian projections on the n-directed plane as follows.

An area ΔA_n on the n-directed plane is projected onto the three areas ΔA_{xy}, ΔA_{yz}, and ΔA_{zx} by the relationships:

$$\Delta A_{yz} = l\, \Delta A_n$$

(1-6.1)

39

$$\Delta A_{zx} = m \ \Delta A_n \qquad (1\text{-}6.2)$$
$$\Delta A_{xy} = n \ \Delta A_n \qquad (1\text{-}6.3)$$

Lumping the three projections of stresses on each Cartesian direction, we get

$$X_n \ \Delta A_n = \sigma_x \ \Delta A_{yz} + \tau_{xy} \ \Delta A_{zx} + \tau_{xz} \ \Delta A_{xy} \qquad (1\text{-}7.1)$$
$$Y_n \ \Delta A_n = \tau_{yx} \ \Delta A_{yz} + \sigma_y \ \Delta A_{zx} + \tau_{yz} \ \Delta A_{xy} \qquad (1\text{-}7.2)$$
$$Z_n \ \Delta A_n = \tau_{zx} \ \Delta A_{yz} + \tau_{zy} \ \Delta A_{zx} + \sigma_z \ \Delta A_{xy} \qquad (1\text{-}7.3)$$

Substituting from equations (1-6) in (1-7), we get

$$X_n = \sigma_x \ l + \tau_{xy} \ m + \tau_{xz} \ n \qquad (1\text{-}8.1)$$
$$Y_n = \tau_{yx} \ l + \sigma_y \ m + \tau_{yz} \ n \qquad (1\text{-}8.2)$$
$$Z_n = \tau_{zx} \ l + \tau_{zy} \ m + \sigma_z \ n \qquad (1\text{-}8.3)$$

Equations (1-8) are fundamental **surface conditions** for projection of stresses on arbitrarily oriented planes. The nine projections of stresses from the three surface planes xOy, yOz, and xOz into their three resultants on the plane oriented in the $\hat{n}\,(l,m,n)$ directions show the difference between projecting **linear forces** versus **arial stresses**.

Equations (1-8) allow us to determine three important stresses:

(i) **Total stress** P_n on the **arbitrary plane** oriented in the $\hat{n}\,(l,m,n)$ direction, by the use of Pythagorean Theorem as follows

$$P_n^2 = X_n^2 + Y_n^2 + Z_n^2 \qquad (1\text{-}9)$$

(ii) **Normal stress** N_n on the **arbitrary plane** oriented in the $\hat{n}\,(l,m,n)$ direction, by the use of the directional cosines of the normal, as follows.

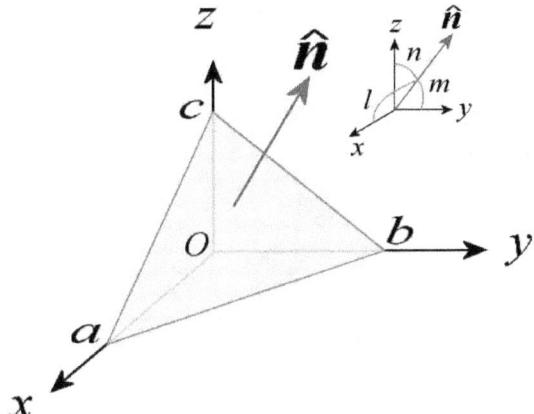

Figure 1-14. Geometry of an arbitrarily oriented plane with normal vector (l, m, n). On this plane, we will transfer the effects of the nine stresses on the three Cartesian planes interesting at the Origin O, such that we get three net stresses acting in this surface

From Figure 1-13, we can write

$$N_n = X_n \, l + Y_n \, m + Z_n \, n_1 \tag{1-10}$$

Substituting from equations (1-8) into equation (1-9) by the values of the projections X_n, Y_n, and Z_n and remembering the conjugation of $\tau_{xy} = \tau_{yx}$, $\tau_{zy} = \tau_{yz}$, and $\tau_{xz} = \tau_{zx}$, we get

$$N_n = \sigma_x l^2 + \sigma_y m^2 + \sigma_z n^2 + 2(\tau_{xy} lm + \tau_{yz} mn + \tau_{zx} nl) \tag{1-11}$$

Equation (1-11) represents **stress ellipsoidal surface**, which will be used to determine the **principal surface** of constant principal normal stresses throughout a rigid body.

(iii) **Total shearing stress** T_n on the **arbitrary plane** oriented in the $\hat{n}(l, m, n)$ direction, by the use of the Pythagorean Theorem, as follows.

From equations (1-9) and (1-11), we get

$$T_n^2 = P_n^2 - N_n^2 \tag{1-12}$$

From equations (1-8.1 through 3), (1-11) and (1-12), we get

41

$$T_n^2 = \left(\sigma_x l + \tau_{xy} m + \tau_{xz} n\right)^2 + \left(\tau_{yx} l + \sigma_y m + \tau_{yz} n\right)^2 + \left(\tau_{zx} l + \tau_{zy} m + \sigma_z n\right)^2$$
$$- \left[\sigma_x l^2 + \sigma_y m^2 + \sigma_z l^2 + 2\left(\tau_{xy} ml + \tau_{yz} nm + \tau_{zz} nl\right)\right]^2 \tag{1-13}$$

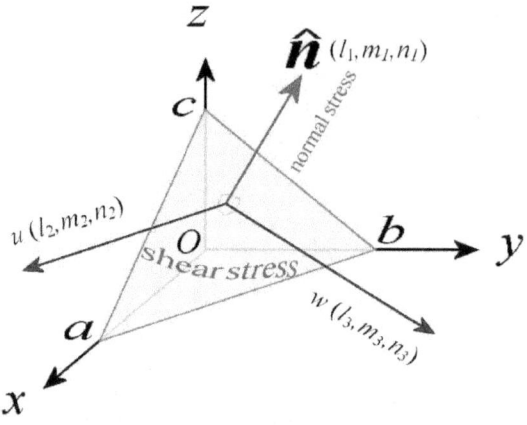

Figure 1-14. Geometry of an arbitrarily oriented plane with normal vector (l_1, m_1, n_1), where, (l_1, m_1, n_1) in Figure 1-14 corresponds to (l, m, n) in Figure 1-13.

(iv) Similarly, the **shear stresses** U_n and W_n are the sum of the projections of (1-8.1 through 3) on the two axes u (l_2, m_2, n_2) and w (l_3, m_3, n_3), shown in Figure 1-14.

$$U_n = X_n l_2 + Y_n m_2 + Z_n n_2 \tag{1-13.1}$$

Thus, substituting from (1-8.1 through 3) by the values of the projections X_n, Y_n, and Z_n and again remembering the conjugation of $\tau_{xy} = \tau_{yx}$, $\tau_{zy} = \tau_{yz}$, and $\tau_{xz} = \tau_{zx}$, we get

$$U_n = \sigma_x l_1 l_2 + \sigma_y m_1 m_2 + \sigma_z n_1 n_2 + \tau_{xy}\left(l_1 m_2 + l_2 m_1\right)$$
$$+ \tau_{yz}\left(n_1 m_2 + n_2 m_1\right) + \tau_{zx}\left(l_1 n_2 + l_2 n_1\right) \tag{1-13.2}$$

Where, (l_1, m_1, n_1) in Figure 1-14 corresponds to (l, m, n) in Figure 1-13

$$W_n = X_n l_3 + Y_n m_3 + Z_n n_3 \tag{1-13.3}$$

Also,

$$W_n = \sigma_x l_1 l_3 + \sigma_y m_1 m_3 + \sigma_z n_1 n_3 + \tau_{xy}\left(l_1 m_3 + l_3 m_1\right)$$
$$+ \tau_{yz}\left(n_1 m_3 + n_3 m_1\right) + \tau_{zx}\left(l_1 n_3 + l_3 n_1\right) \tag{1-13.4}$$

Equations (1-10), (1-13.2) and (1-13.4) provide the three projections N_n, U_n, and W_n on the arbitrarily oriented plane in terms of the directional cosines l_i m_i n_i (i = 1, 2, 3) and the projections of the stresses X_n, Y_n, and Z_n.

Example 4

If all forces are located on the Oxy plane alone, find the normal stress and shearing stress at the arbitrarily oriented plane *ab* shown in Figure 1-15.

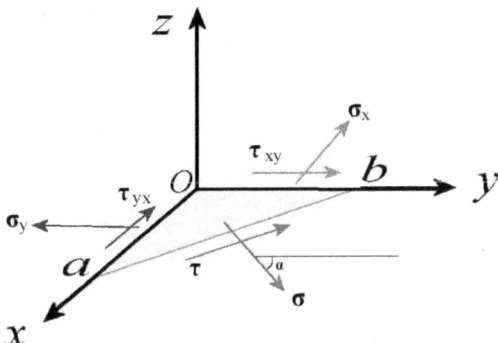

Figure 1-15. Geometry of an arbitrarily oriented plane ab with normal vector $(\sin\alpha, \cos\alpha, 0)$. Note that ab is the projection of the vertical plane on the xy-plane.

Solution

Since all projections of stresses on the z-plane disappear, then $\sigma_z = \tau_{xz} = \tau_{yz} = 0$. Thus equations (1-11) and (1-13.2) give

$$N_n = \sigma_x l_1^2 + \sigma_y m_1^2 + 2\tau_{xy} l_1 m_1 \tag{1-14.1}$$

$$U_n = \sigma_x l_1 l_2 + \sigma_y m_1 m_2 + \tau_{xy}\left(l_1 m_2 + l_2 m_1\right) \tag{1-14.2}$$

Where, the directional cosines of the **normal** on the plane *ab* are

$$l_1 = \sin\alpha \tag{1-14.3.1}$$
$$m_1 = \cos\alpha \tag{1-14.3.2}$$
$$n_1 = \cos\pi/2 = 0 \tag{1-14.3.3}$$

And, the directional cosines of the line *ab* are

$$l_2 = -\sin(\pi/2-\alpha) = -\cos\alpha \qquad (1\text{-}14.3.4)$$
$$m_2 = \cos(\pi/2-\alpha) = \sin\alpha \qquad (1\text{-}14.3.5)$$
$$n_2 = \cos\pi/2 = 0 \qquad (1\text{-}14.3.6)$$

Therefore, equations (1-14.1 and 14.2) become

$$\sigma = N_n = \sigma_x\sin^2\alpha + \sigma_y\cos^2\alpha + 2\tau_{xy}\cos\alpha\sin\alpha \qquad (1\text{-}14.5)$$

$$\tau = U_n = -\sigma_x\sin\alpha\cos\alpha + \sigma_y\sin\alpha\cos\alpha - \tau_{xy}\sin^2\alpha + \tau_{xy}\cos^2\alpha$$

$$= (\sigma_y - \sigma_x)\ \sin\alpha\cos\alpha + (\cos^2\alpha - \sin^2\alpha)\tau_{xy} \qquad (1\text{-}14.6)$$

Equations (1-14.5 and 6) show that the projections of stresses σ_x and τ_x on the inclined plane are **arial projections**, unlike the projections of vectors, which are **linear projections**.

The common mistake is to decompose stresses as decomposing forces, when stresses are pressure forces affecting areas, while vectorial forces affect linear points.

(1) The condition for **vanishing shear stress** on the plane ab is obtained by equating τ by zero, in equation (1-14.6). Thus

$$(\sigma_y - \sigma_x)\ \sin\alpha\cos\alpha + (\cos^2\alpha - \sin^2\alpha)\tau_{xy} = 0 \qquad (1\text{-}14.7.1)$$

Or

$$\frac{(\sigma_y - \sigma_x)}{\tau_{xy}} = -(\cot\alpha - \tan\alpha) = -\frac{(\cos\alpha\cos\alpha - \sin\alpha\sin\alpha)}{\sin\alpha\cos\alpha} = -\frac{2\cos2\alpha}{\sin2\alpha} = -2\cot2\alpha$$

Or,

$$\frac{\tau_{xy}}{(\sigma_y - \sigma_x)} = -\frac{1}{2}\tan2\alpha \qquad (1\text{-}14.7.2)$$

Substituting from (1-14.7.2) by τ_{xy} into (1-14.5), we get

$$\sigma = \sigma_x\sin^2\alpha + \sigma_y\cos^2\alpha + (\sigma_x - \sigma_y)\tan2\alpha\cos\alpha\sin\alpha \qquad (1\text{-}14.8)$$

(2) The condition for **vanishing normal stress** on the plane ab is obtained by equating σ by zero, in equation (1-14.5). Thus

$$\tau_{xy} = -\frac{1}{2}(\sigma_x\tan\alpha + \sigma_y\cot\alpha) \qquad (1\text{-}14.9)$$

(3) Inversely, the condition for vanishing τ_{xy} is obtained from equations (1-14.5) and (1-14.6) as follows

$$\sigma = \sigma_x\sin^2\alpha + \sigma_y\cos^2\alpha + 0 \qquad (1\text{-}14.10)$$

$$\tau = \frac{(\sigma_y - \sigma_x)}{2}) \; \sin 2\alpha \qquad\qquad (1\text{-}14.11)$$

By using the trigonometric relations

$$\sin^2\alpha + \cos^2\alpha = 1 \qquad \text{and} \qquad \cos^2\alpha = \frac{1}{2}\cos 2\alpha$$

Equation (1-14.10) can be rewritten as follows

$$\sigma = \sigma_x + (\sigma_y - \sigma_x)\cos^2\alpha = \sigma_x + \frac{1}{2}(\sigma_y - \sigma_x)\cos 2\alpha \qquad\qquad (1\text{-}14.12.1)$$

Or

$$\sigma = \sigma_x + \sqrt{\left(\frac{1}{2}(\sigma_y - \sigma_x)\right)^2 - \tau^2} \qquad\qquad (1\text{-}14.12.2)$$

Squaring and arranging (1-14.12.2) we get

$$(\sigma - \sigma_x)^2 + \tau^2 = \frac{1}{4}(\sigma_y - \sigma_x)^2 \qquad\qquad (1\text{-}14.13)$$

Equation (1-14.13) is a relationship between τ and σ when the x and y axes are the **principal axes** of vanishing shearing stresses. Equation (1-14.13) represents a circle of radius $\frac{1}{2}(\sigma_y - \sigma_x)$ center on the σ-axis at σ_x and $\tau = 0$. Figure 1-16.

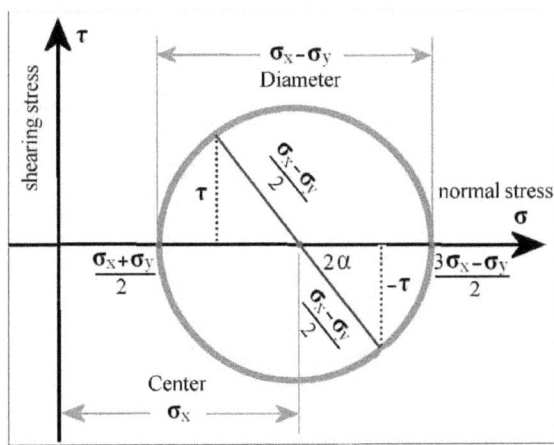

Figure 1-16. Mohr's Circle Principle

45

Graphing equation (1-14.13) shows the relationship between the normal and shear stresses on the plane ab in terms of the normal stresses on sides Ox and Oy, taken as principal planes, where no shearing stresses act on Ox or Oy.

Equation (1-14.13) is used to calculate τ and σ, when given σ_x and at σ_y, which are substituted in equation (1-14.11) to determined the angle α.

Mohr's Circle Principle, Figure 1-16

Mohr's circle principle is the two dimensional graphical representation of the state of stress at a point. The abscissa and ordinate of each point on the circle are the normal stress σ and shear stress τ components, and acting on a particular cut plane with a unit vector with components. The circumference of the circle is the locus of points that represent the state of stress on individual planes at all their orientations. Mohr's circle is used to determine the normal and shear stress for any oriented, Figure 1-15, for any values of maximum and minimum compressive stresses.

Example 5

Use the (1-14.10) and (1-14.11) to find the total stress at an arbitrarily oriented section in the xy-plane of the paralleloid bar

Solution

From equation (1-5.1), put $x = 0$ to obtain $\sigma_x(0,y)$ which is used in equations (1-14.10) and (1-14.11) to get

$$\sigma_x(0,y) = \frac{12}{h^2}\left(M_o + \frac{ql^2}{8} - \frac{qh^2}{20}\right)y + \frac{4q}{h^3}y^3 \qquad (1\text{-}15.1)$$

Also, in from equation (1-5.2), put $y = 0$ to obtain $\sigma_y(x,0)$ which is also used in equations (1-14.10) and (1-14.11) to get

$$\sigma_y(x,0) = -\frac{q}{2} \qquad (1\text{-}15.2)$$

Total stress on the plane ab is obtained from (1-14.0) and (1-14.11) by the Pythagorean Theorem

$$\sqrt{\sigma^2 + \tau^2} = \sqrt{\left(\sigma_x \sin^2\alpha + \sigma_y \cos^2\alpha\right)^2 + \left(\frac{(\sigma_y - \sigma_x)}{2}\right)\sin2\alpha\right)^2}$$

$$= \sqrt{\left(\sigma_x + \frac{1}{2}(\sigma_y - \sigma_x)\cos2\alpha\right)^2 + \left(\frac{(\sigma_y - \sigma_x)}{2}\right)\sin2\alpha\right)^2}$$

$$= \sqrt{\sigma_x^2 + \sigma_x\left(\sigma_y - \sigma_x\right)\cos 2\alpha + \frac{1}{4}\left(\sigma_y - \sigma_x\right)^2} \qquad (1\text{-}15.3)$$

Equations (1-15.1), (1-15.2), and (1-15.3) describe the distribution of total stress at lines arbitrarily at angle with the y-axis, in the xy-plane

Example 6

A dam cross section yOS is exposed to water pressure on the side yOz (z is perpendicular to the plane of the paper). Figure 1-17.

i. Find the normal stress and shearing stress on the opposite side. SOz in terms of the angle θ between the two faces yOz and SOz and the profile of water pressure.

ii. Find the stresses at the sections *aa* and *bb*

iii. Find the condition when the face SOz is free of load

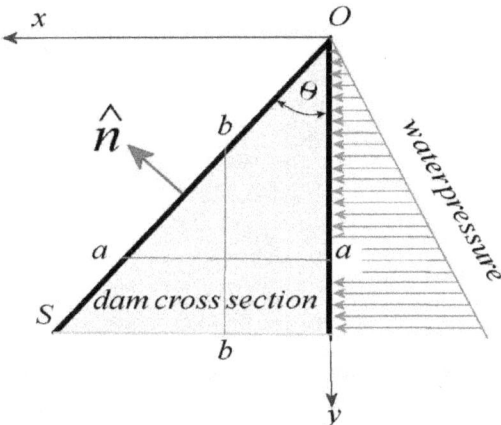

Figure 1-17. Triangular cross-section in a dam resisting pressure water in a river oriented in the x-direction, with depth along the y-axis on the vertical side.

Solution

(1) Equations of stresses

47

Since all forces, external and internal are linear functions of coordinates we will assume the following forms of the three internal strains

$$\sigma_x = A\ x + B\ y \qquad (1\text{-}16.1a)$$
$$\sigma_y = C\ x + D\ y \qquad (1\text{-}16.1b)$$
$$\tau_{xy} = E\ x + F\ y \qquad (1\text{-}16.1c)$$

(2) Determining constants from boundary conditions

a. Surface yOz at $x = 0$, the pressure on the surface $yOz = P_o y$

Therefore, boundary conditions on the yOz surface are:

$$\sigma_{-x} = -\sigma_x = P_o y \qquad (1\text{-}16.2a)$$
$$\sigma_{-y} = -\sigma_y = \rho g y \qquad (1\text{-}16.2b)$$
$$\tau_{-xy} = -\tau_{xy} = 0 \qquad (1\text{-}16.2c)$$

Thus, equations (1-16.1) become

$$\sigma_x = A\ 0 + B\ y = -P_o y \qquad (1\text{-}16.2d)$$
$$\sigma_y = C0 + D\ y = -\rho g y \qquad (1\text{-}16.2e)$$
$$\tau_{xy} = E\ 0 + F\ y = 0 \qquad (1\text{-}16.2f)$$

Thus,

$$B = -P_o \qquad (1\text{-}16.2g)$$
$$D = -\rho g \qquad (1\text{-}16.2h)$$
$$F = 0 \qquad (1\text{-}16.2i)$$

$$\sigma_x = A\ x - P_o\ y \qquad (1\text{-}16.1a)$$
$$\sigma_y = C\ x - \rho g\ y \qquad (1\text{-}16.1b)$$
$$\tau_{xy} = E\ x \qquad (1\text{-}16.1c)$$

b. Surface SOz is devoid of external forces, therefore the boundary conditions on the SOz side of the dam are

$$X_n = 0 \qquad (1\text{-}16.3a)$$
$$Y_n = 0 \qquad (1\text{-}16.3b)$$

The directional cosines of the normal to OSz are

$$l_1 - \cos\theta,$$
$$m_1 = -\sin\theta \quad \text{(directed in the negative direction of y-axis)}$$
$$n_1 = \cos \pi/2 \qquad (1\text{-}16.3c)$$

The directional cosines of the of the plane OSz, Figure 1-18, are

$$l_2 = \sin\theta, \qquad\qquad m_2 = \cos\theta, \qquad\qquad n_2 = \cos\pi/2 \qquad\qquad (1\text{-}16.3d)$$

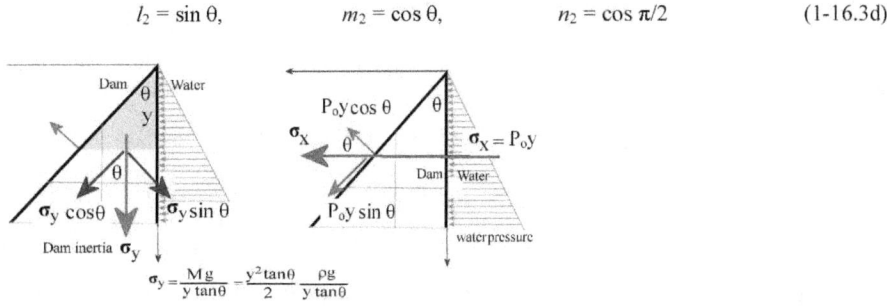

$$\sigma_y = \frac{Mg}{y\,\tan\theta} - \frac{y^2\tan\theta}{2}\,\frac{\rho g}{y\,\tan\theta}$$

(a) Dam inertia   (b) water pressure

Figure 1-18. Balances of forces in the rigid body of dam.

Surface equations (1-8.1 through 3) determine the stresses on the surfaces, which yield the remaining constants: A, E, and C, in equations (1-16)

$$x - y\tan\theta \qquad\qquad\qquad (1\text{-}16.3e)$$

$$
\begin{aligned}
X_n &= \sigma_x\,l_1 + \tau_{xy}\,m_1 + \tau_{xz}\,n_1 \\
&= \sigma_x\cos\theta - \tau_{xy}\sin\theta = 0 \\
&= (A\,x - P_o\,y)\cos\theta - (E\,x)\sin\theta = 0 \\
&= (A\,y\tan\theta - P_o\,y)\cos\theta - (E\,y\tan\theta)\sin\theta = 0
\end{aligned}
$$

$$A = E\tan\theta - P_o\cot\theta \qquad\qquad\qquad (1\text{-}16.3f)$$

$$
\begin{aligned}
Y_n &= \tau_{yx}\,l + \sigma_y\,m + \tau_{yz}\,n \\
&= (E\,y\tan\theta)\cos\theta - (C\,y\tan\theta - \rho g\,y)\sin\theta - 0
\end{aligned}
$$

$$E = C\tan\theta - \rho g \qquad\qquad\qquad (1\text{-}16.3g)$$

From equations (1-16.3f) and (1-16.3g), C is the last unknown constant in our problem.

$$
\begin{aligned}
\sigma_x &= ((C\tan\theta - \rho g)\tan\theta - P_o\cot\theta)\,x - P_o\,y && (1\text{-}16.1a) \\
\sigma_y &= C\,x - \rho g\ y && (1\text{-}16.1b) \\
\tau_{xy} &= (C\tan\theta - \rho g)\,x && (1\text{-}16.1c)
\end{aligned}
$$

(i) The terms $P_o y \cos \theta$ and $P_o y \sin \theta$ are contributions from the normal stress σ_x, equations (1-16.1) due to water pressure.

(ii) The terms $(\rho g y / 2)\cos \theta$ and $(\rho g y / 2)\sin \theta$ are contributions from the normal stress on the *aa* plane, equation (1-16.2), due to dam inertia.

In deriving equations (1-16.8) and (1-16.9), we calculated the average density of the upper section of the dam over plane *aa* by dividing mass by the volume in order to estimate σ_y.

Condition for <u>vanishing normal stress</u> on the surface SOz, from equation (1-16.8), we get

$$\sigma_n = P_o y \cos \theta - (\rho g y / 2) \sin \theta - 0$$

$$\rho_{water} / \rho_{dam} = (1-1/2) \tan \theta \tag{1-16.10}$$

Equation (1-16.10) determines the angle θ in terms of the ratio of water density to dam density.

$$\theta = 2 \tan^{-1} \left(\frac{\rho_{water}}{\rho_{dam}} \right) \tag{1-16.11}$$

Therefore, the physical properties of water and construction material of the dam determine the angular orientation of the two sides of the dam.

(2) Surface *aa* at depth y = Y.

The height of the dam at distance x on the *aa* plane is $h = (y \tan \theta - x) \cot \theta$. Therefore,

$$\sigma_y (x,Y) = \rho g(Y-x \cot\theta) \tag{1-16.2a}$$
$$\tau_{xy} (x,Y) = P_o Y \tag{1-16.2b}$$

(3) Surface *bb* at width x = X

$$\sigma_x(X,y) = P_o \, y \tag{1-16.3a}$$
$$\tau_{xy} (X,y) = \rho g(y-X \cot\theta) \tag{1-16.3b}$$

Example 7

In the preceding Example, best fitting profiles for stresses are as follows.

$$\sigma_x = -\gamma y \tag{1-17.1}$$

$$\sigma_y = \left(\frac{p}{\tan \beta} - \frac{2\gamma}{\tan^3 \beta} \right) x + \left(\frac{\gamma}{\tan^2 \beta} - p \right) y \tag{1-17.2}$$

$$\tau_{xy} = \tau_{yx} = -\frac{\gamma}{\tan^2\beta}x \qquad\qquad (1\text{-}17.3)$$

$$\tau_{yz} = \tau_{zy} = \tau_{zx} = \tau_{xz} = \sigma_z = 0 \qquad\qquad (1\text{-}17.4)$$

Show that those distributions of stresses on the two planes oriented in the x and y directions conform to the differential equation of equilibrium of stresses in a rigid body.

Solution

Substituting from equations (1-17.1 through 4) in equations (1-4.1 through 3), we get

$$\frac{\partial}{\partial x}(-\gamma y) + \frac{\partial}{\partial y}\left(-\frac{\lambda}{\tan^2\beta}x\right) = 0 + 0 = 0 \qquad\qquad (1\text{-}17.5)$$

$$\frac{\partial}{\partial x}\left(-\frac{\gamma}{\tan^2\beta}x\right) + \frac{\partial}{\partial y}\left[\left(\frac{p}{\tan\beta} - \frac{2\gamma}{\tan^3\beta}\right)x + \left(\frac{\gamma}{\tan^2\beta} - p\right)y\right] + p$$

$$= -\frac{\gamma}{\tan^2\beta} + \frac{\gamma}{\tan^2\beta} - p + p = 0 \qquad\qquad (1\text{-}17.6)$$

We assumed that $X_i = Z_i = 0$ and $Y_i = g\rho = p$, since gravity works in the vertical direction. Therefore, the stresses given by equations (1-17.1 through 4) constitute balanced forces acting over a rigid body.

1.5. Transfer of stresses from surface of known stresses to a point in the body

We have derived equation (1-11) that describes the normal stress N_n on an arbitrarily oriented plane, in terms of the Cartesian projections of stresses. In order to obtain the equation of principal stresses, we need to rotate the axes of equation (1-11), such that the shearing stresses vanish, which is the condition for **principal surface of stresses**.

In Figure 1-19, start by choosing arbitrary origin M in a body. Let a vector of length $\mathbf{p}(l,m,n)$ originates from M passes through unit area with outward normal $\hat{n}$.

The coordinates of the point at the tip of $\mathbf{p}$, Figure 1-5, are

$$\xi - pl,\ \eta - pm,\ \zeta - pn \qquad\qquad (1\text{-}18)$$

Substitute by the directional cosines from (1-18) in (1-11), the normal stress at point $P(\xi\ \eta,\ \zeta)$ becomes

$$N_n = \sigma_x\left(\frac{\xi}{p}\right)^2 + \sigma_y\left(\frac{\eta}{p}\right)^2 + \sigma_z\left(\frac{\zeta}{p}\right)^2 + 2\tau_{xy}\left(\frac{\xi\eta}{p^2}\right) + 2\tau_{yz}\left(\frac{\eta\zeta}{p^2}\right) + 2\tau_{xz}\left(\frac{\xi\zeta}{p^2}\right)$$

$$p^2 N_n = \sigma_x \xi^2 + \sigma_y \eta^2 + \sigma_z \zeta^2 + 2\left(\tau_{xy}\xi\eta + \tau_{yz}\eta\zeta + \tau_{xz}\xi\zeta\right) \tag{1-19}$$

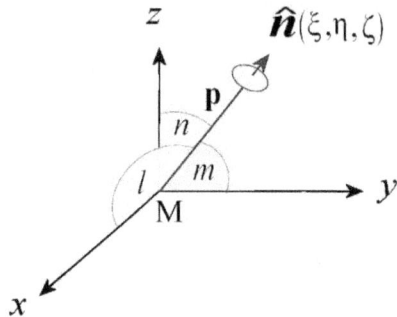

Figure 1-19. Configuration of the vector p of an arbitrary point P(ξ = pl, η = pm, ζ = pn)

Equation (1-19) is the **ellipsoidal equation of stress** as it represents the normal stress at the area surrounding the point at tip of the vector **p**(l,m,n). We will deal with the matrix representation and rotation of axes of equation (1-19). The ellipsoidal feature of stresses distribution arises from two main facts:

(1) Rigid bodies posses three distinctive surfaces (whereas fluids distribute forces equally between surfaces, negating the distinctive nature of rigid shapes). The three surfaces transmit **three directions of forces**.

(2) Forces must act on areas, which confers the **quadratic feature of the dimensions** of space where the stress applies.

(3) Within such **ellipsoidal distribution** of forces or stresses, there exists spherical stress distribution with respect to the principal surfaces of principal stresses. The spherical stress distribution represents the uniform distribution of stress, depending only of the **volume of matter,** whereas deviation form uniform stress distribution depends on **change in shape** of the rigid body. We reached that conclusion through our discussion of the octahedral distribution of stress with respect to the principal surface.

We have used a simplified version of equation (1-19) for planar forces acting in the xOy plane, equations (1-14). For that purpose, we obtained equation (1-14.13) that determine the stresses on a plane oriented arbitrarily on the coordinate axes Oy and Ox. We also graphed equation (1-14.13), in Figure 1-20, to show the conditions for vanishing shearing stresses on the arbitrary plane.

Here, equation (1-19) can be referred to rotated axes such that all shearing stress on the principal axes, of principal stresses, vanish. Thus, equation (1-19), reduces to

$$p^2 N_n = N_1 \xi^2 + N_2 \eta^2 + N_3 \zeta^2 \qquad (1-20)$$

The rationale for stating equation (1-20) from (1-19), without using the tedious technique of rotation of the three Cartesian axes, stems from our assumption that point M was chosen as an arbitrary point and N_n was derived for an arbitrarily oriented plane with respect to the Cartesian axes. Thus, we assumed that the principal surface, of principal stresses, is determined by putting

$$\tau_{xy} = \tau_{xz} = \tau_{yz} = 0.$$

Thus, we eliminated the mixed terms in equation (1-19).

1.6. Cauchy's quadratic or surface of normal stresses

The **surface of constant normal stress** or the **Cauchy's quadratic** is obtained from equation (1-19) given by assigning constants to surfaces, as follows.

$$\Phi (\xi, \eta, \zeta) = p^2 N_n = \sigma_x \xi^2 + \sigma_y \eta^2 + \sigma_z \zeta^2 + 2\tau_{xy} \xi\eta + 2\tau_{yz} \eta\zeta + 2\tau_{xz} \xi\zeta = \pm \ c^2 \qquad (1-21)$$

Where the negative sign denotes **compressive stress**, positive sign denotes **tensile stress**.
Equation (1-20) also yields an equivalent **Cauchy Quadratic** equation of principal surface of the form

$$N_1 \xi^2 + N_2 \eta^2 + N_3 \zeta^2 = \pm \ c^2 \qquad (1-22)$$

Where, N_1, N_2, and N_3, are the **principal stresses** acting on three mutually perpendicular principal areas at the given point P, where shear stresses vanish.

Signs of principal stresses

Case 1. All N's > 0 or all N's < 0, Figure 1-20

Equation (1-22) can be written as

$$\frac{\xi^2}{\left(\dfrac{c^2}{N_1}\right)} + \frac{\eta^2}{\left(\dfrac{c^2}{N_2}\right)} + \frac{\zeta^2}{\left(\dfrac{c^2}{N_3}\right)} = 1 \qquad (1-22.1)$$

Which is an equation of an ellipsoid with semi-axes

$$a_i = \frac{c}{\sqrt{N_i}}, \qquad\qquad i = 1, 2, 3$$

Thus, if all N's are positive, $N_n = c^2/p^2$, which comprise **tensile stresses** on ellipsoidal surfaces. Also, if all N's are negative, $N_n = -c^2/p^2$, which comprise **compressive stresses** on ellipsoidal surfaces.

Case 2. Some N's > 0 and some N's < 0, Figure 1-20

Assume that $N_3 < 0$, then Equation (1-22) can be written in three ways

One-sheet Hyperbola:

$$\frac{\xi^2}{\left(\frac{c^2}{N_1}\right)} + \frac{\eta^2}{\left(\frac{c^2}{N_2}\right)} - \frac{\zeta^2}{\left(\frac{c^2}{|N_3|}\right)} = 1 \tag{1-22.2}$$

Two-sheet Hyperbola:

$$-\frac{\xi^2}{\left(\frac{c^2}{N_1}\right)} - \frac{\eta^2}{\left(\frac{c^2}{N_2}\right)} + \frac{\zeta^2}{\left(\frac{c^2}{|N_3|}\right)} = 1 \tag{1-22.3}$$

Asymptomatic Cone:

$$N_1\xi^2 + N_2\eta^2 - |N_3|\zeta^2 = 1 \tag{1-22.4}$$

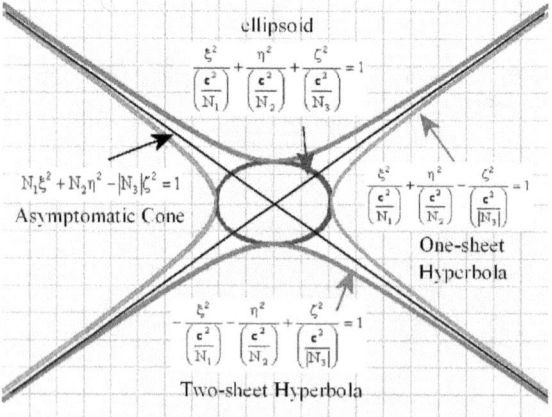

Figure 1-20. Degeneration of the stress ellipsoidal surface to four possible distributions.

1.7. Principal stress and principal plane

In equations (1-8), if the total stress acting on the arbitrary plane with $\hat{n}(l,m,n)$ was principal stress denoted by **N**, then its Cartesian projections are given by

$$X_n = \mathbf{N}\, l \qquad\qquad (1\text{-}23.1)$$
$$Y_n = \mathbf{N}\, m \qquad\qquad (1\text{-}23.2)$$
$$Z_n = \mathbf{N}\, n \qquad\qquad (1\text{-}23.3)$$

Substituting from (1-23) into (1-8), we get a relationship between the principal stress N and net resultant stress

$$\mathbf{N}\, l = \sigma_x\, l + \tau_{xy}\, m + \tau_{xz}\, n \qquad\qquad (1\text{-}24.1)$$
$$\mathbf{N}\, m = \tau_{yx}\, l + \sigma_y\, m + \tau_{yz}\, n \qquad\qquad (1\text{-}24.2)$$
$$\mathbf{N}\, n = \tau_{zx}\, l + \tau_{zy}\, m + \sigma_z\, n \qquad\qquad (1\text{-}24.3)$$

Therefore, we have the four equations that determine the **principal stress N** and the three directional cosines **l**, **m**, and **n** of the normal to the area where the principal stresses affect.

$$(\sigma_x - \mathbf{N})\, l + \tau_{xy}\, m + \tau_{xz}\, n = 0 \qquad\qquad (1\text{-}25.1)$$
$$\tau_{yx}\, l + (\sigma_y - \mathbf{N})\, m + \tau_{yz}\, n = 0 \qquad\qquad (1\text{-}25.2)$$
$$\tau_{zx}\, l + \tau_{zy}\, m + (\sigma_z - \mathbf{N})\, n = 0 \qquad\qquad (1\text{-}25.3)$$

and

$$l^2 + m^2 + n^2 = 1 \qquad\qquad (1\text{-}25.4)$$

The three linear simultaneous equations (1-25.1 through 3), can be written in matrix form as follows

$$\begin{bmatrix} \sigma_x - N & \tau_{xy} & \tau_{xz} \\ \tau_{yx} & \sigma_y - N & \tau_{yz} \\ \tau_{zx} & \tau_{zy} & \sigma_z - N \end{bmatrix} \begin{bmatrix} l \\ m \\ n \end{bmatrix} = 0 \qquad\qquad (1\text{-}25.5)$$

Where the determinant of the stress matrix must vanish since the directional cosines cannot vanish simultaneously. Thus,

$$\begin{vmatrix} \sigma_x - N & \tau_{xy} & \tau_{xz} \\ \tau_{yx} & \sigma_y - N & \tau_{yz} \\ \tau_{zx} & \tau_{zy} & \sigma_z - N \end{vmatrix} = 0 \qquad\qquad (1\text{-}25.6)$$

Or, $(\sigma_x - N)\begin{vmatrix} \sigma_y - N & \tau_{yz} \\ \tau_{zy} & \sigma_z - N \end{vmatrix} - \tau_{xy}\begin{vmatrix} \tau_{yx} & \tau_{yz} \\ \tau_{zx} & \sigma_z - N \end{vmatrix} + \tau_{xz}\begin{vmatrix} \tau_{yx} & \sigma_y - N \\ \tau_{zx} & \tau_{zy} \end{vmatrix} = 0$

Converting determinants into algebraic sums, we get

$(\sigma_x - N)[(\sigma_y - N)(\sigma_z - N) - \tau_{yz}\tau_{zy}] - \tau_{xy}[\tau_{yx}(\sigma_z - N) - \tau_{yz}\tau_{zx}] + \tau_{xz}[\tau_{yx}\tau_{zy} - (\sigma_y - N)\tau_{zx}] = 0$

Arranging the terms of cubic polynomial we get

$$-N^3 + N^2(\sigma_x + \sigma_y + \sigma_z) - N(\sigma_x\sigma_y + \sigma_x\sigma_z + \sigma_y\sigma_z - \tau_{yz}\tau_{zy} - \tau_{xy}\tau_{yx} - \tau_{xz}\tau_{zx})$$
$$+ \sigma_x\sigma_y\sigma_z + \tau_{xy}\tau_{yz}\tau_{zx} + \tau_{xz}\tau_{yx}\tau_{zy} - \tau_{xz}\tau_{zx}\sigma_y - \sigma_x\tau_{yz}\tau_{zy} - \tau_{xy}\tau_{yx}\sigma_z = 0 \qquad (1\text{-}25.7)$$

Equation (1-25.7) offers the following benefits:

(i) The cubic equation (1-25.7) has three real roots for the principal stress N on the principal surface defined by equations (1-23.1 through 3). After obtaining the three roots of (1-25.7) we substitute by the N_1, N_2, N_3, in (1-25.1 through 4) to **determine the directional cosines of the principal surface**. We have already done so in equation (1-14.7.2) by determining **tan 2α** in terms of τ_{xy}, σ_x, and σ_y.

(ii) In case of planar stresses, equation (1-25.7) is reduced to the quadratic equation.

$-N^3 + N^2(\sigma_x + \sigma_y + 0) - N(\sigma_x\sigma_y + 0 + 0 + 0 - \tau_{xy}\tau_{yx} + 0) + 0 + 0 + 0 - 0 - 0 - 0 = 0$ (1-25.8)

Or, $-N^3 + N^2(\sigma_x + \sigma_y) - N(\sigma_x\sigma_y - \tau_{xy}\tau_{yx}) = 0$

i.e., $-N^2 + N(\sigma_x + \sigma_y) - (\sigma_x\sigma_y - \tau_{xy}\tau_{yx}) = 0$ (1-25.9)

Which can also be written as

$$\begin{bmatrix} \sigma_x - N & \tau_{xy} & 0 \\ \tau_{yx} & \sigma_y - N & 0 \\ 0 & 0 & 0 \end{bmatrix}\begin{bmatrix} l \\ m \\ 0 \end{bmatrix} = 0 \qquad (1\text{-}25.9.1)$$

(iii) The roots of equation (1-25.7) are **invariant**, independent of the system of coordinates.

Example 8

A cylindrical tube exposed to torsion at its upper surface *ab*. Figure 1-21.

(i) Find the stresses at point P on the upper surface. State the condition for vanishing normal stress.

(ii) Find the stresses at point M on the lateral wall. State the condition of vanishing load.

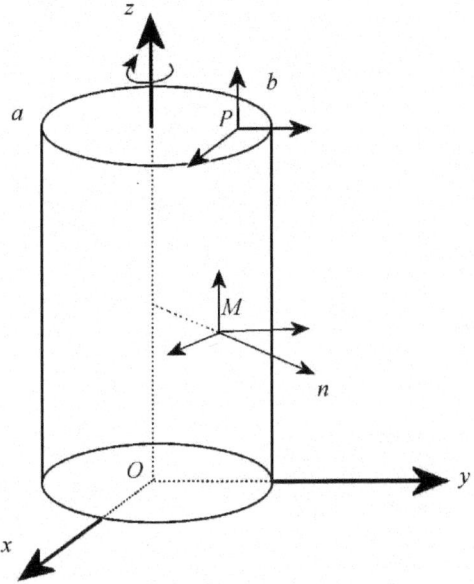

Figure 1-21. Solid circular cylindrical bar twisted at the top surface.

Solution

External forces

Figure 1-22 shows the geometry of stresses, which are defined as follows.

$$\tau_{xz} = \tau_{yz} = r\,Q \qquad\qquad (1\text{-}26.1)$$
$$\sigma_z = 0$$

(i) Point P(x, y, h) on the upper surface.

$$\tau_{xz} = \tau_{yz} = r\,Q$$
$$\sigma_z = 0$$

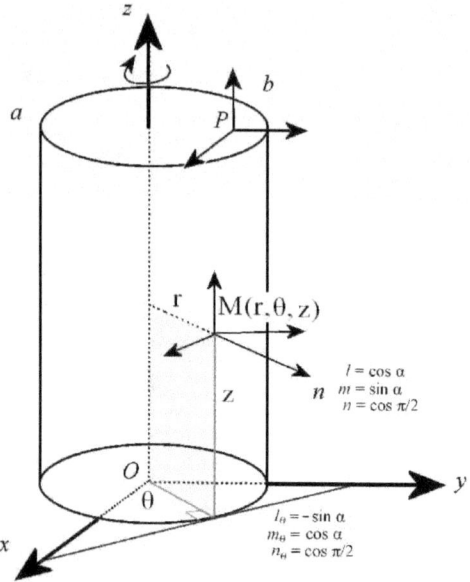

Figure 1-22. Radial and azimuthal planes at the lateral point M.

(ii) Point M(r, θ, z) on the lateral surface

Directional cosines of Normal to surface at M

$$l = \cos \alpha, \qquad m = \sin \alpha, \qquad n = \cos \pi/2. \qquad (1\text{-}26.2)$$

Directional cosines of Tangent to surface at M

$$l_\theta = -\sin \alpha, \qquad m_\theta = \cos \alpha, \qquad n_\theta = \cos \pi/2. \qquad (1\text{-}26.3)$$

Projections of stresses in cylindrical polar coordinates

$$\sigma_r = \tau_{xz} \cos \alpha + \tau_{yz} \sin \alpha \qquad (1\text{-}26.4)$$
$$\tau_{\theta z} = -\tau_{xz} \sin \alpha + \tau_{yz} \cos \alpha \qquad (1\text{-}26.5)$$

Normal stress vanished at M implies that

$$\sigma_r = \tau_{xz} \cos \alpha + \tau_{yz} \sin \alpha = 0$$
$$\tau_{xz} / \tau_{yz} = - \tan \alpha \qquad\qquad (1\text{-}26.6)$$

Shearing stress vanished at M implies that

$$\tau_{\theta z} = -\tau_{xz} \sin \alpha + \tau_{yz} \cos \alpha = 0 \qquad\qquad (1\text{-}26.7)$$
$$\tau_{xz} / \tau_{yz} = \cot \alpha \qquad\qquad (1\text{-}26.8)$$

Example 9

A cylindrical tube is bent by force Q and twisted by surface force T. Figure 1-23.

(i) Find the normal and shearing stresses at P on the lateral surface of the tube.
(ii) Find the conditions of vanishing stresses.

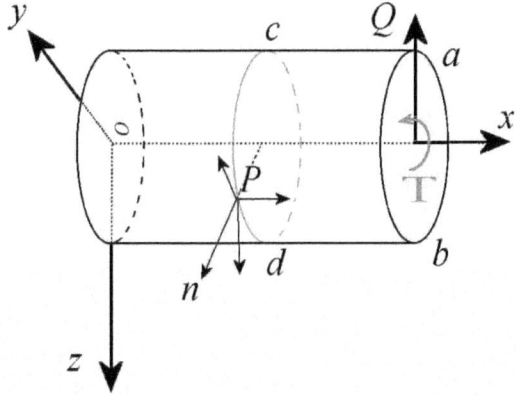

Figure 1-23 Circular cylindrical bar affected by stress that tends to twist and bend the bar at one end.

Solution

(i) External stresses

$$\tau_{zx} = -Q \qquad\qquad\qquad \tau_{yx} = T \qquad\qquad (1\text{-}27.1)$$

(ii) Point P(r, θ, x) on the lateral surface

Directional cosines of normal to surface at P

$$l = \cos \pi/2. \qquad\qquad m = \cos \alpha, \qquad n = \sin \alpha \qquad\qquad (1\text{-}27.2)$$

Directional cosines of tangent to surface at P. Figure 1-24.

$$l_\theta = \cos \pi/2. \qquad\qquad m_\theta = -\sin \alpha, \qquad n_\theta = \cos \alpha. \qquad\qquad (1\text{-}27.3)$$

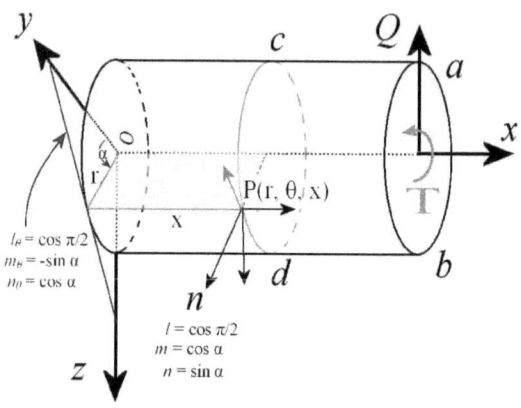

Figure 1-24. Radial and azimuthal planes at the lateral point P.

$$\sigma_r = n\,\tau_{zx} + m\,\tau_{yz} = -Q \sin \alpha + T \cos \alpha \qquad\qquad (1\text{-}27.4)$$
$$\tau_{\theta z} = n_\theta \tau_{zx} + m_\theta\,\tau_{yx} = Q \cos \alpha - T \sin \alpha \qquad\qquad (1\text{-}27.5)$$

Normal stress vanished at P implies that
$$\sigma_r = -Q \sin \alpha + T \cos \alpha = 0$$
$$T/Q = \tan \alpha \qquad\qquad (1\text{-}27.6)$$

Shearing stress vanished at P implies that
$$\tau_{\theta z} = Q \cos \alpha - T \sin \alpha = 0 \qquad\qquad (1\text{-}27.7)$$
$$T/Q = \cot \alpha \qquad\qquad (1\text{-}27.8)$$

1.8. Distribution of shearing stress on arbitrary plane in rigid body

From (1-12) gives the **total shearing T_n stress** on the **arbitrary plane** oriented in the $\hat{n}\,(l,m,n)$ direction as

$$T_n^2 = P_n^2 - N_n^2 \qquad\qquad (1\text{-}28.3)$$

60

From equations (1-9), (1-23), and (1-24), we get

$$T_n^2 = \left(N_1^2 l^2 + N_2^2 m^2 + N_3^2 n^2\right) - \left(N_1 l^2 + N_2 m^2 + N_3 n^2\right)^2 \qquad (1-28.4)$$

Since

$$n^2 = 1 - l^2 - m^2 \qquad (1-28.5)$$

Equation (1-28.4) can be reduced to a function in m and l, in order to determine the derivatives of T_n on w.r.t. l and m alone, as follows

$$T_n^2 = \left[N_1^2 l^2 + N_2^2 m^2 + N_3^2\left(1 - m^2 - l^2\right)\right] - \left[N_1 l^2 + N_2 m^2 + N_3\left(1 - m^2 - l^2\right)\right]^2$$

i.e.,

$$T_n^2 = \left[\left(N_1^2 - N_3^2\right)l^2 + \left(N_2^2 - N_3^2\right)m^2 + N_3^2\right] - \left[\left(N_1 - N_3\right)l^2 + \left(N_2 - N_3\right)m^2 + N_3\right]^2 \qquad (1-28.6)$$

In order to determine the maxima and minima of shearing stresses we need to determine the derivatives of T_n on w.r.t. l and m, which specify the changes of shearing from increasing, constant, or decreasing. Thus, finding the derivatives of T_n and equating them by zero, determines the directional cosines of the surfaces of **maximal shearing stresses**, as follows.

Differentiating T_n with respect to l, we get

$$\frac{\partial T_n^2}{\partial l} = 2T_n \frac{\partial T_n}{\partial l} = 2T_n\left[2\left(N_1^2 - N_3^2\right)l\right] - 8T_n\left[\left(N_1 - N_3\right)l^2 + \left(N_2 - N_3\right)m^2 + N_3\right]\left(N_1 - N_3\right)l$$

$$= 4T_n\left(N_1 - N_3\right)l\left[\left(N_1 - N_3\right)\left(1 - 2l^2\right) - 2\left(N_2 - N_3\right)m^2\right] = 0 \qquad (1-28.6.1)$$

And, with respect to m, we get

$$\frac{\partial T_n^2}{\partial m} = 2T_n \frac{\partial T_n}{\partial m} = 4T_n\left(N_2^2 - N_3^2\right)m - 8T_n\left[\left(N_1 - N_3\right)l^2 + \left(N_2 - N_3\right)m^2 + N_3\right]\left(N_2 - N_3\right)m$$

$$= 4T_n\left(N_2 - N_3\right)m\left[\left(N_2 - N_3\right)\left(1 - 2m^2\right) - 2\left(N_1 - N_3\right)l^2\right] = 0 \qquad (1-28.6.2)$$

Equations (1-28.6.1) and (1-28.6.2) give the following options:

(i) T_n vanishes, at principal surfaces of principal stresses.
(ii) $N_1 = N_3$ or $N_2 = N_3$.
(iii) $l = 0$ or $m = 0$
(iv)   The following relations between the three principal normal stresses N's and the two directional cosines m and l.

Therefore,

$$\left(N_1 - N_3\right)\!\left(1 - 2l^2\right) - 2\left(N_2 - N_3\right)m^2 = 0$$
$$\left(N_2 - N_3\right)\!\left(1 - 2m^2\right) - 2\left(N_1 - N_3\right)l^2 = 0$$

i.e.,

$$N_1 - N_3 - \left(N_1 - N_3\right)\!2l^2 - 2\left(N_2 - N_3\right)m^2 = 0 \qquad (1\text{-}28.6.3)$$

$$N_2 - N_3 - \left(N_2 - N_3\right)\!2m^2 - 2\left(N_1 - N_3\right)l^2 = 0 \qquad (1\text{-}28.6.4)$$

Subtracting (1-28.6.4) from (1-28.6.3), we get

$$N_1 - N_2 = 0 \qquad (1\text{-}28.6.5)$$

Substitute from equation (1-5), by $N_1 = N_2$, in equation (1-28.6.3), we get

$$N_1 - N_3 - \left(N_1 - N_3\right)\!2l^2 - 2\left(N_1 - N_3\right)m^2 = 0$$
$$\left(N_1 - N_3\right)\!\left(1 - 2l^2 - 2m^2\right) = 0$$

Equations (1-28.6.2) and (1-28.6.4), provide the following solutions

(i) $m = 0$

$$m = 0, \quad l^2 = \frac{1}{2} \quad l = \pm\sqrt{\frac{1}{2}} \quad n = \pm\sqrt{\frac{1}{2}} \qquad (1\text{-}28.6.6)$$

(ii) $l = 0$

$$m^2 = \frac{1}{2} \quad m = \pm\sqrt{\frac{1}{2}} \quad n = \pm\sqrt{\frac{1}{2}} \qquad (1\text{-}28.6.7)$$

(iii) $l \neq 0$ and $m \neq 0$

$$m^2 = \frac{1}{2} \quad m = \pm\sqrt{\frac{1}{2}} \quad l = \pm\sqrt{\frac{1}{2}} \quad n = 0 \qquad (1\text{-}28.6.8)$$

Therefore, the maximal total shearing stress in (1-28.6) becomes

$$T_1^2 = \left[\left(N_1^2 - N_3^2\right)\frac{l}{2} + N_3^2\right] - \left[\left(N_1 - N_3\right)\frac{l}{2} + N_3\right]^2$$
$$= \left[\left(N_1^2 + N_3^2\right)\frac{l}{2}\right] - \left[\left(N_1 + N_3\right)\frac{l}{2}\right]^2$$
$$= \left[\left(N_1^2 + N_3^2\right)\frac{l}{2}\right] - \frac{l}{4}\left(N_1^2 + 2N_1 N_3 + N_3^2\right)$$

$$= \left[\left(N_1^2 + N_3^2 \right) \frac{l}{4} \right] - \frac{l}{4} 2N_1N_3 = \left(\frac{N_1 - N_3}{2} \right)^2$$

$$T_1 = \pm \frac{N_1 - N_3}{2} \tag{1-28.6.9}$$

$$T_2 = \pm \frac{N_1 - N_3}{2} \tag{1-28.6.10}$$

Example 10

Find the total normal and shearing stresses on the eight-planes of **octahedron** equally inclined to the principal area of principal stress. Figure 1-25.

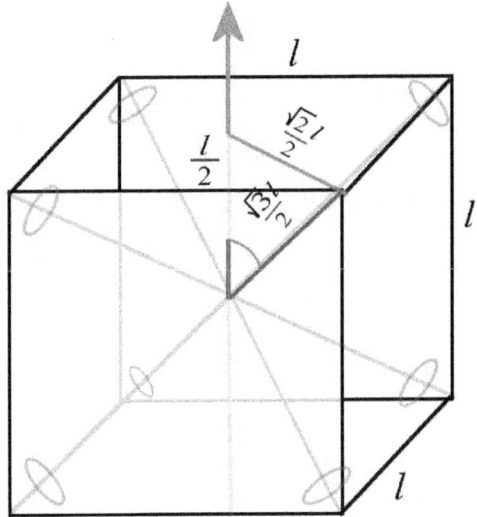

Figure 1-25. Octahederal surfaces around the principal plane of principal stresses showing the method of calculation of the directional cosines of the normals to the eight corners in terms of the equal lengths of the sides of the cube.

Solution

The eight corners of a cube are connected by four diagonals, each has two halves directed oppositely from the center of the cube. Each diagonal has three directional cosines $(\pm l, \pm m, \pm n)$. Therefore, we have $2^3 = 8$ cosines (two halves, each has three directional cosines)

63

The length of diagonal of the six surfaces of the cube measures $\sqrt{2}l$. The length of the four diagonals connecting the each opposite corner from the center measures $\sqrt{3}l$.

Therefore, the eight directional cosines are

$$l, m, n = \pm \frac{1}{\sqrt{3}} \tag{1-29.1}$$

The octahedral normal stress is determined by substituting the directional cosines of octahedron in the **ellipsoidal stress equation** (1-20), as follows

$$N_n = N_1 \frac{\xi^2}{p^2} + N_2 \frac{\eta^2}{p^2} + N_3 \frac{\zeta^2}{p^2} = N_1 \left(\frac{1}{\sqrt{3}}\right)^2 + N_2 \left(\frac{1}{\sqrt{3}}\right)^2 + N_3 \left(\frac{1}{\sqrt{3}}\right)^2$$

$$= \frac{1}{3}(N_1 + N_2 + N_3) = = \frac{1}{3}(\sigma_x + \sigma_y + \sigma_z) \tag{1-29.2}$$

Equation (1-29.2) represents the **averaging** of the three principal normal stresses, which is identical to the distribution of pressure in fluids. Hence, the octahedral stresses of the principal stresses are called the **hydrostatic pressure**, which possesses **spherical symmetry**.

The **shear stresses** at the octahedral corners are obtained similarly from equation (1-28.4), as follows

$$T_n^2 = \left(N_1^2 l^2 + N_2^2 m^2 + N_3^2 n^2\right) - \left(N_1 l^2 + N_2 m^2 + N_3 n^2\right)^2$$

$$= \left(N_1^2 \left(\frac{1}{\sqrt{3}}\right)^2 + N_2^2 \left(\frac{1}{\sqrt{3}}\right)^2 + N_3^2 \left(\frac{1}{\sqrt{3}}\right)^2\right)$$

$$- \left(N_1 \left(\frac{1}{\sqrt{3}}\right)^2 + N_2 \left(\frac{1}{\sqrt{3}}\right)^2 + N_3 \left(\frac{1}{\sqrt{3}}\right)^2\right)^2$$

$$= \frac{1}{3}\left(N_1^2 + N_2^2 + N_3^2\right) - \left(\frac{1}{3}(N_1 + N_2 + N_3)\right)^2$$

$$= \frac{1}{3}\left(N_1^2 + N_2^2 + N_3^2\right) - \frac{1}{9}\left(N_1^2 + N_2^2 + N_3^2 + 2N_1 N_2 + 2N_1 N_3 + 2N_3 N_2\right)$$

$$= \frac{1}{9}\left[2N_1^2 + 2N_2^2 + 2N_3^2 - \left(2N_1 N_2 + 2N_1 N_3 + 2N_3 N_2\right)\right] \tag{1-29.3}$$

$$= \frac{1}{9}\left[\left(N_1^2 - 2N_1 N_2 + N_2^2\right) + \left(N_3^2 - 2N_1 N_3 + N_1^2\right) + \left(N_2^2 - 2N_3 N_2 + N_3^2\right)\right]$$

$$= \frac{1}{9}\left[\left(N_1 - N_2\right)^2 + \left(N_2 - N_3\right)^2 + \left(N_1^2 - N_3\right)^2\right] \tag{1-29.3}$$

$$T_n = \frac{1}{3}\sqrt{\left(N_1 - N_2\right)^2 + \left(N_2 - N_3\right)^2 + \left(N_1^2 - N_3\right)}$$
(1-29.4)

1.9. Matrix and tensor representation of stress ellipsoidal surface[1]

The stress ellipsoidal surface in equations (1-11) and (1-19), graphed in Figure 1-26, can be written in matrix form as follows.

$$\mathbf{XSX^T = 0}$$
(1-30.1)

Where the matrices X, R, and X^T, the transpose of X, are written as follows

$$\mathbf{S} = \frac{1}{2}\begin{bmatrix} 2\sigma_x & 2\tau_{xy} & 2\tau_{xz} \\ 2\tau_{xy} & 2\sigma_y & 2\tau_{yz} \\ 2\tau_{xz} & 2\tau_{yz} & 2\sigma_z \end{bmatrix}$$
(1-30.2)

However, we will add another row and column, in order to allow for the translation vector of the origin of the Cartesian coordinates to the point (X_0, Y_0, Z_0) as follows.

$$\mathbf{S} = \frac{1}{2}\begin{bmatrix} 2\sigma_x & 2\tau_{xy} & 2\tau_{xz} & 0 \\ 2\tau_{xy} & 2\sigma_y & 2\tau_{yz} & 0 \\ 2\tau_{xz} & 2\tau_{yz} & 2\sigma_z & 0 \\ 0 & 0 & 0 & 0 \end{bmatrix}$$
(1-30.3)

$$\mathbf{X} = \begin{bmatrix} \xi & \eta & \zeta & 1 \end{bmatrix}$$
(1-30.4)

$$\mathbf{X^T} = \begin{bmatrix} \xi \\ \eta \\ \zeta \\ 1 \end{bmatrix}$$
(1-30.5)

Thus, final outlook of equation (1-19) written in the matrix form is as follows

---

[1] Salvatore Alfano and Meredith L. Greer, "Determining If Two Solid Ellipsoids Intersect", JOURNAL OF GUIDANCE, CONTROL, AND DYNAMICS, Vol. 26, No. 1, p. 106-110, January–February 2003

$$\frac{1}{2}\begin{bmatrix} \xi & \eta & \zeta & 1 \end{bmatrix} x \begin{bmatrix} 2\sigma_x & 2\tau_{xy} & 2\tau_{xz} & 0 \\ 2\tau_{xy} & 2\sigma_y & 2\tau_{yz} & 0 \\ 2\tau_{xz} & 2\tau_{yz} & 2\sigma_z & 0 \\ 0 & 0 & 0 & 0 \end{bmatrix} x \begin{bmatrix} \xi \\ \eta \\ \zeta \\ 1 \end{bmatrix} = 0 \tag{1-30.6}$$

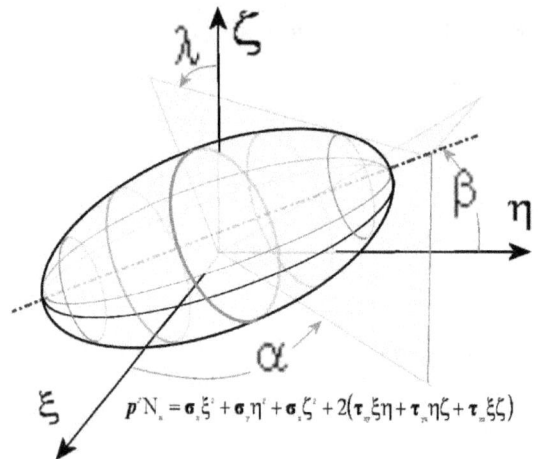

$$p'N_s = \sigma_s \xi^2 + \sigma_\eta \eta^2 + \sigma_\zeta \zeta^2 + 2(\tau_{\omega} \xi\eta + \tau_{\eta} \eta\zeta + \tau_{\kappa} \xi\zeta)$$

Figure 1-26. Stress ellipsoid oriented at angle α, β, and λ between three projections of the axes of the ellipsoidal and the Cartesian axes.

The **rotation of axes of the ellipsoidal surface,** which intends to eliminate the mixed terms of off-diagonal stress in equation (1-30.2), is carried out by three matrices, each rotates the surface around a Cartesian axis, as follows.

$$\mathbf{R}_\zeta = \begin{bmatrix} 1 & 0 & 0 & 0 \\ 0 & \cos\alpha & \sin\alpha & 0 \\ 0 & -\sin\alpha & \cos\alpha & 0 \\ 0 & 0 & 0 & 1 \end{bmatrix} \tag{1-30.7}$$

$$\mathbf{R}_\xi = \begin{bmatrix} \cos\beta & 0 & -\sin\beta & 0 \\ 0 & 1 & 0 & 0 \\ \sin\beta & 0 & \cos\beta & 0 \\ 0 & 0 & 0 & 1 \end{bmatrix} \tag{1-30.8}$$

$$\mathbf{R}_\eta = \begin{bmatrix} \cos\lambda & \sin\lambda & 0 & 0 \\ -\sin\lambda & \cos\lambda & 0 & 0 \\ 0 & 0 & 1 & 0 \\ 0 & 0 & 0 & 1 \end{bmatrix} \tag{1-30.9}$$

The three axial rotations can be combined into one rotation by the product

$$\mathbf{R} = \mathbf{R}_\xi \mathbf{R}_\zeta \mathbf{R}_\eta = \begin{bmatrix} \cos\beta\cos\lambda & \cos\beta\sin\lambda & -\sin\beta & 0 \\ \sin\alpha\sin\beta\cos\lambda - \cos\alpha\sin\lambda & \sin\alpha\sin\beta\sin\lambda + \cos\alpha\cos\lambda & \sin\alpha\cos\beta & 0 \\ \cos\alpha\sin\beta\cos\lambda + \sin\alpha\sin\lambda & \cos\alpha\sin\beta\sin\lambda - \sin\alpha\cos\lambda & \cos\alpha\cos\beta & 0 \\ 0 & 0 & 0 & 1 \end{bmatrix}$$

$$\tag{1-30.10}$$

The transpose of R is

$$\mathbf{R}^T = \begin{bmatrix} \cos\beta\cos\lambda & \sin\alpha\sin\beta\cos\lambda - \cos\alpha\sin\lambda & \cos\alpha\sin\beta\cos\lambda + \sin\alpha\sin\lambda & 0 \\ \cos\beta\sin\lambda & \sin\alpha\sin\beta\sin\lambda + \cos\alpha\cos\lambda & \cos\alpha\sin\beta\sin\lambda - \sin\alpha\cos\lambda & 0 \\ -\sin\beta & \sin\alpha\cos\beta & \cos\alpha\cos\beta & 0 \\ 0 & 0 & 0 & 1 \end{bmatrix}$$

$$\tag{1-30.11}$$

Thus, the rotated ellipsoidal surface of equation (1-30.1) takes the form

$$\mathbf{XRSR}^T\mathbf{X}^T = 0 \tag{1-30.12}$$

Similarly, **translation of the center** of the coordinates of equation (1-19) into the point (X_0, Y_0, Z_0) is accomplished by the tensor

$$\mathbf{T} = \begin{bmatrix} 1 & & & 0 \\ & 1 & & 0 \\ & & 1 & 0 \\ -X_0 & -Y_0 & -Z_0 & 1 \end{bmatrix} \tag{1-30.13}$$

Therefore,

$$\mathbf{XTST}^T\mathbf{X}^T = 0 \tag{1-30.14}$$

Example 11

67

Translate the origin of coordinates of equation (1-19) into the point (x_0, y_0, z_0). Show the effect of translating the origin of coordinates on the mixed terms in equation (1-19).

Solution

From equations (1-30.3 through 5), (1-30.13), and after executing the matrix first product **XT**, equation (1-30.14) becomes

$$\begin{bmatrix} \xi - X_0 & \eta - Y_0 & \zeta - Z_0 & 1 \end{bmatrix} x \begin{bmatrix} 2\sigma_x & 2\tau_{xy} & 2\tau_{xz} & 0 \\ 2\tau_{xy} & 2\sigma_y & 2\tau_{yz} & 0 \\ 2\tau_{xz} & 2\tau_{yz} & 2\sigma_z & 0 \\ 0 & 0 & 0 & 0 \end{bmatrix} x \begin{bmatrix} 1 & & & -X_0 \\ & 1 & & -Y_0 \\ & & 1 & -Z_0 \\ 0 & 0 & 0 & 1 \end{bmatrix} x \begin{bmatrix} \xi \\ \eta \\ \zeta \\ 1 \end{bmatrix} = 0 \qquad (1\text{-}30.15)$$

After executing the matrix product $\mathbf{XTST^T}$, we get

$$\left[\begin{pmatrix} (\xi - X_0)2\sigma_x \\ +(\eta - Y_0)2\tau_{xy} \\ +(\zeta - Z_0)2\tau_{xz} \end{pmatrix} \begin{pmatrix} (\xi - X_0)2\tau_{xy} \\ +(\eta - Y_0)2\sigma_y \\ +(\zeta - Z_0)2\tau_{yz} \end{pmatrix} \begin{pmatrix} (\xi - X_0)2\tau_{xz} \\ +(\eta - Y_0)2\tau_{yz} \\ +(\zeta - Z_0)2\sigma_z \end{pmatrix} \begin{array}{l} -X_0\begin{pmatrix} (\xi - X_0)2\sigma_x \\ +(\eta - Y_0)2\tau_{xy} \\ +(\zeta - Z_0)2\tau_{xz} \end{pmatrix} \\ -Y_0\begin{pmatrix} (\xi - X_0)2\tau_{xy} \\ +(\eta - Y_0)2\sigma_y \\ +(\zeta - Z_0)2\tau_{yz} \end{pmatrix} \\ -Z_0\begin{pmatrix} (\xi - X_0)2\tau_{xz} \\ +(\eta - Y_0)2\tau_{yz} \\ +(\zeta - Z_0)2\sigma_z \end{pmatrix} \end{array} \right] x \begin{bmatrix} \xi \\ \eta \\ \zeta \\ 1 \end{bmatrix} = 0$$

$$(1\text{-}30.15.1)$$

The final product $\mathbf{XTST^T X^T}$ is now,

$$\begin{aligned} &\xi\left((\xi - X_0)2\sigma_x + (\eta - Y_0)2\tau_{xy} + (\zeta - Z_0)2\tau_{xz}\right) \\ &+ \eta\left((\xi - X_0)2\tau_{xy} + (\eta - Y_0)2\sigma_y + (\zeta - Z_0)2\tau_{yz}\right) \\ &+ \zeta\left((\xi - X_0)2\tau_{xz} + (\eta - Y_0)2\tau_{yz} + (\zeta - Z_0)2\sigma_z\right) \\ &- X_0\left((\xi - X_0)2\sigma_x + (\eta - Y_0)2\tau_{xy} + (\zeta - Z_0)2\tau_{xz}\right) \\ &- Y_0\left((\xi - X_0)2\tau_{xy} + (\eta - Y_0)2\sigma_y + (\zeta - Z_0)2\tau_{yz}\right) \\ &- Z_0\left((\xi - X_0)2\tau_{xz} + (\eta - Y_0)2\tau_{yz} + (\zeta - Z_0)2\sigma_z\right) = 0 \end{aligned} \qquad (1\text{-}30.15.2)$$

In order to show that **the translation of the origin of the ellipsoid does not remove the mixed terms** $\xi\eta$, $\zeta\eta$, $\xi\zeta$, we will arrange the terms in (1-30.15.2) as follows

$$2\sigma_x\xi^2 + 2\sigma_y\eta^2 + 2\sigma_z\zeta^2 +$$
$$+\xi\left(-2\sigma_xX_0 - Y_02\tau_{xy} - Z_02\tau_{xz} - X_02\sigma_x - Y_02\tau_{xy} - Z_02\tau_{xz}\right)$$
$$+\eta\left(-2\tau_{xy}X_0 - Y_02\sigma_y - Z_02\tau_{yz} - X_02\tau_{xy} - Y_02\sigma_y - Z_02\tau_{yz}\right)$$
$$+\zeta\left(-2\tau_{xz}X_0 - Y_02\tau_{yz} - Z_02\sigma_z - X_02\tau_{xz} - Y_02\tau_{yz} - Z_022\sigma_z\right) \qquad (1\text{-}30.15.3)$$
$$+\xi\eta\left(2\tau_{xy} + 2\tau_{xy}\right) + \eta\zeta\left(2\tau_{yz} + 2\tau_{yz}\right) + \xi\zeta\left(2\tau_{xz} + 2\tau_{xz}\right)$$
$$+2\sigma_xX_0^{\,2} + 2\sigma_yY_0^{\,2} + 2\sigma_zZ_0^{\,2}$$
$$+4\left(X_0Y_0\tau_{xy} + X_0Z_02\tau_{xz} + Y_0Z_0\tau_{yz}\right) = 0$$

Form equation (1-30.15.3), we conclude that the translation of origin of coordinates of the ellipsoid only affects the **isolated coordinates** ξ, η, and ζ and **the constants**, where, X_0, Y_0, and Z_0 appear.

Example 12

Find the condition for removing the mixed terms $\xi\eta$, $\zeta\eta$, $\xi\zeta$, in equation (1-19) by the rotation of the coordinate axes of equation (1-19) by angles α, β, and λ.

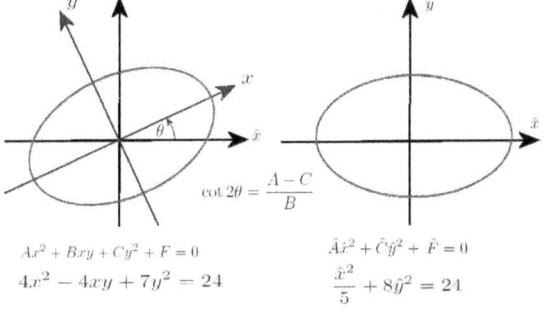

Figure 1-27. Rotation of axes of two-dimensional ellipse.

Solution

Equation (1-19) provides the constants needed to determine the angles α, β, and λ.

$$p^2N_n = \sigma_x\xi^2 + \sigma_y\eta^2 + \sigma_z\zeta^2 + 2\left(\tau_{xy}\xi\eta + \tau_{yz}\eta\zeta + \tau_{xz}\xi\zeta\right)$$

(1) <u>Rotation around ζ-axis, in the $\xi\eta$-plane</u>

As Figure 1-27 shows, and from equation (1-19), we get

$$\cot 2\alpha = \frac{\sigma_x - \sigma_y}{2\tau_{xy}} \tag{1-30.16.1}$$

$$\cos 2\alpha = \frac{(\sigma_x - \sigma_y)}{\sqrt{4\tau_{xy}^2 + (\sigma_x - \sigma_y)^2}} \tag{1-30.16.2}$$

$$\sin 2\alpha = \frac{2\tau_{xy}}{\sqrt{4\tau_{xy}^2 + (\sigma_x - \sigma_y)^2}} \tag{1-30.16.3}$$

From equation (1-30.7), the old coordinates are replaced by the new coordinates according to the transforms

$$\xi = \hat{\xi}\cos\alpha - \hat{\eta}\sin\alpha ,$$
$$\eta = \hat{\xi}\sin\alpha + \hat{\eta}\cos\alpha$$
$$\zeta = \hat{\zeta} \tag{1-30.16.4}$$

Thus, equation (1-19) becomes

$$p^2 N_n = \sigma_x \left[\hat{\xi}\cos\alpha - \hat{\eta}\sin\alpha\right]^2 + \sigma_y \left[\hat{\xi}\sin\alpha + \hat{\eta}\cos\alpha\right]^2 + \sigma_z \hat{\zeta}^2$$
$$+ 2\tau_{xy}\left[\hat{\xi}\cos\alpha - \hat{\eta}\sin\alpha\right]\left[\hat{\xi}\sin\alpha + \hat{\eta}\cos\alpha\right] \tag{1-30.16.5}$$
$$+ 2\zeta\tau_{yz}\left[\hat{\xi}\sin\alpha + \hat{\eta}\cos\alpha\right] + 2\zeta\tau_{xz}\left[\hat{\xi}\cos\alpha - \hat{\eta}\sin\alpha\right]$$

Arranging the terms, we get

$$p^2 N_n = \hat{\xi}^2\left(\sigma_x \cos^2\alpha + \sigma_y \sin^2\alpha + \tau_{xy}\cos\alpha\sin\alpha\right)$$
$$+ \hat{\eta}^2\left(\sigma_y \cos^2\alpha + \sigma_x \sin^2\alpha - \tau_{xy}\cos\alpha\sin\alpha\right) + \sigma_z \hat{\zeta}^2$$
$$+ 2\hat{\xi}\hat{\eta}\cos\alpha\sin\alpha\left(\sigma_y - \sigma_x\right) + 2\tau_{xy}\hat{\xi}\hat{\eta}\left(\cos^2\alpha - \sin^2\alpha\right)$$
$$+ 2\hat{\zeta}\hat{\xi}\left(\tau_{yz}\sin\alpha + \tau_{xz}\cos\alpha\right) + 2\hat{\zeta}\hat{\eta}\left(\tau_{yz}\cos\alpha - \tau_{xz}\sin\alpha\right)$$

Substituting by the trigonometric relations

(2 **cos** α **sin** α = **sin2α**) and
(**cos²α-sin²α = cos 2α**),

we get

$$p^2 N_n = \hat{\xi}^2\left(\sigma_x \cos^2\alpha + \sigma_y \sin^2\alpha + \tau_{xy}\cos\alpha\sin\alpha\right)$$
$$+ \hat{\eta}^2\left(\sigma_y \cos^2\alpha + \sigma_x \sin^2\alpha - \tau_{xy}\cos\alpha\sin\alpha\right) + \sigma_z\hat{\zeta}^2$$
$$+ 2\hat{\xi}\hat{\eta}\frac{\sin 2\alpha}{2}\left(\sigma_y - \sigma_x\right) + 2\tau_{xy}\hat{\xi}\hat{\eta}(\cos 2\alpha)$$
$$+ 2\hat{\zeta}\hat{\xi}\left(\tau_{yz}\sin\alpha + \tau_{xz}\cos\alpha\right) + 2\hat{\zeta}\hat{\eta}\left(\tau_{yz}\cos\alpha - \tau_{xz}\sin\alpha\right)$$

Using equations (1-30.16.2 and 3) to replace (**sin2α**) and (**cos 2α**) we get

$$p^2 N_n = \hat{\xi}^2\left(\sigma_x \cos^2\alpha + \sigma_y \sin^2\alpha + \tau_{xy}\cos\alpha\sin\alpha\right)$$
$$+ \hat{\eta}^2\left(\sigma_y \cos^2\alpha + \sigma_x \sin^2\alpha - \tau_{xy}\cos\alpha\sin\alpha\right) + \sigma_z\hat{\zeta}^2$$
$$+ \hat{\xi}\hat{\eta}\left(\frac{2\tau_{xy}}{\sqrt{4\tau_{xy}^2 + \left(\sigma_x - \sigma_y\right)^2}}\left(\sigma_y - \sigma_x\right) + 2\tau_{xy}\frac{\left(\sigma_x - \sigma_y\right)}{\sqrt{4\tau_{xy}^2 + \left(\sigma_x - \sigma_y\right)^2}}\right) \qquad (1\text{-}30.16.6)$$
$$+ 2\hat{\zeta}\hat{\xi}\left(\tau_{yz}\sin\alpha + \tau_{xz}\cos\alpha\right) + 2\hat{\zeta}\hat{\eta}\left(\tau_{yz}\cos\alpha - \tau_{xz}\sin\alpha\right)$$

In equation (1-30.16.6), the **mixed term in** $\hat{\xi}\hat{\eta}$ vanish since the signs of $(\sigma_x - \sigma_y)$ are opposite. Therefore, we get

$$p^2 N_n = \hat{\xi}^2\left(\sigma_x \cos^2\alpha + \sigma_y \sin^2\alpha + \tau_{xy}\cos\alpha\sin\alpha\right)$$
$$+ \hat{\eta}^2\left(\sigma_y \cos^2\alpha + \sigma_x \sin^2\alpha - \tau_{xy}\cos\alpha\sin\alpha\right) + \sigma_z\hat{\zeta}^2 \qquad (1\text{-}30.16.7)$$
$$+ 2\hat{\zeta}\hat{\xi}\left(\tau_{yz}\sin\alpha + \tau_{xz}\cos\alpha\right) + 2\hat{\zeta}\hat{\eta}\left(\tau_{yz}\cos\alpha - \tau_{xz}\sin\alpha\right)$$

This can be written in a more concise form as follows

$$p^2 N_n = \hat{\xi}^2 A + \hat{\eta}^2 B + \sigma_z\hat{\zeta}^2 + 2\hat{\zeta}\hat{\xi}\left(\tau_{yz}\sin\alpha + \tau_{xz}\cos\alpha\right) + 2\hat{\zeta}\hat{\eta}\left(\tau_{yz}\cos\alpha - \tau_{xz}\sin\alpha\right) \qquad (1\text{-}30.16.8)$$

Where,
$$A = \sigma_x \cos^2\alpha + \sigma_y \sin^2\alpha + \tau_{xy}\cos\alpha\sin\alpha \qquad (1\text{-}30.16.8a)$$
$$B = \sigma_y \cos^2\alpha + \sigma_x \sin^2\alpha - \tau_{xy}\cos\alpha\sin\alpha \qquad (1\text{-}30.16.8b)$$

Even though, we have succeeded in removing the missed term $\hat{\xi}\hat{\eta}$, **we still have two mixed terms in three variables.**

(2) Rotation around ξ-axis, in the $\zeta\eta$-plane

As Figure 1-27 shows, and from equation (1-30.16.8), we get

71

$$\cot 2\beta = \frac{B - \sigma_z}{2(\tau_{yz}\cos\alpha - \tau_{xz}\sin\alpha)} \tag{1-30.16.9}$$

$$\cos 2\beta = \frac{(B - \sigma_z)}{\sqrt{4(\tau_{yz}\cos\alpha - \tau_{xz}\sin\alpha)^2 + (B - \sigma_z)^2}} \tag{1-30.16.10}$$

$$\sin 2\beta = \frac{2(\tau_{yz}\cos\alpha - \tau_{xz}\sin\alpha)}{\sqrt{4(\tau_{yz}\cos\alpha - \tau_{xz}\sin\alpha)^2 + (B - \sigma_z)^2}} \tag{1-30.16.11}$$

Again, from equation (1-30.7), the old coordinates are replaced by the new coordinates according to the transforms

$$\hat{\xi} = \overline{\xi}$$
$$\hat{\eta} = \overline{\eta}\cos\beta - \overline{\zeta}\sin\beta$$
$$\hat{\zeta} = \overline{\eta}\sin\beta + \overline{\zeta}\cos\beta \tag{1-30.16.12}$$

Thus, equation (1-30.16.8) becomes

$$p^2 N_n = \overline{\xi}^2 A + (\overline{\eta}\cos\beta - \overline{\zeta}\sin\beta)^2 B + \sigma_z(\overline{\eta}\sin\beta + \overline{\zeta}\cos\beta)^2$$
$$+ 2\overline{\xi}(\overline{\eta}\sin\beta + \overline{\zeta}\cos\beta)(\tau_{yz}\sin\alpha + \tau_{xz}\cos\alpha) \tag{1-30.16.13}$$
$$+ 2(\overline{\eta}\sin\beta + \overline{\zeta}\cos\beta)(\overline{\eta}\cos\beta - \overline{\zeta}\sin\beta)(\tau_{yz}\cos\alpha - \tau_{xz}\sin\alpha)$$

Arranging and substituting by the trigonometric relations ($1\text{-}2\cos\alpha\sin\alpha = \sin 2\alpha$) and ($\cos^2\alpha - \sin^2\alpha = \cos 2\alpha$), we get

$$p^2 N_n = \overline{\xi}^2(\sigma_x\cos^2\alpha + \sigma_y\sin^2\alpha + \tau_{xy}\cos\alpha\sin\alpha)$$
$$+ \overline{\eta}^2[B\cos^2\beta + \sigma_z\sin^2\beta + 2\sin\beta\cos\beta(\tau_{yz}\cos\alpha - \tau_{xz}\sin\alpha)]$$
$$+ \overline{\zeta}^2[B\sin^2\beta + \sigma_z\cos^2\beta - 2\sin\beta\cos\beta(\tau_{yz}\cos\alpha - \tau_{xz}\sin\alpha)]$$
$$+ 2\overline{\eta}\overline{\zeta}\left\{\frac{\sin 2\beta}{2}[\sigma_z - B] + \cos 2\beta(\tau_{yz}\cos\alpha - \tau_{xz}\sin\alpha)\right\} \tag{1-30.16.14}$$
$$+ 2\overline{\xi}(\overline{\eta}\sin\beta + \overline{\zeta}\cos\beta)(\tau_{yz}\sin\alpha + \tau_{xz}\cos\alpha)$$

Using equations (1-30.16.9 and 10) to replace ($\sin 2\alpha$) and ($\cos 2\alpha$) we get

72

$$p^2 N_n = \hat{\xi}^2 \left(\sigma_x \cos^2 \alpha + \sigma_y \sin^2 \alpha + \tau_{xy} \cos\alpha \sin\alpha \right)$$
$$+ \overline{\eta}^2 \left[B\cos^2 \beta + \sigma_z \sin^2 \beta + 2\sin\beta\cos\beta \left(\tau_{yz} \cos\alpha - \tau_{xz} \sin\alpha \right) \right]$$
$$+ \overline{\zeta}^2 \left[B\sin^2 \beta + \sigma_z \cos^2 \beta - 2\sin\beta\cos\beta \left(\tau_{yz} \cos\alpha - \tau_{xz} \sin\alpha \right) \right] \qquad (1\text{-}30.16.15)$$
$$+ 2\hat{\xi} \left(\overline{\eta} \sin\beta + \overline{\zeta} \cos\beta \right) \left(\tau_{yz} \sin\alpha + \tau_{xz} \cos\alpha \right)$$

Again, this can be written in a more concise form as follows

$$p^2 N_n = \hat{\xi}^2 A + \overline{\eta}^2 C + \overline{\zeta}^2 D + 2\hat{\xi}\overline{\eta} \sin\beta \left(\tau_{yz} \sin\alpha + \tau_{xz} \cos\alpha \right)$$
$$+ 2\hat{\xi}\overline{\zeta} \cos\beta \left(\tau_{yz} \sin\alpha + \tau_{xz} \cos\alpha \right) \qquad (1\text{-}30.16.16)$$

Where,

$$D = B\sin^2 \beta + \sigma_z \cos^2 \beta - 2\sin\beta\cos\beta \left(\tau_{yz} \cos\alpha - \tau_{xz} \sin\alpha \right) \qquad (1\text{-}30.16.16a)$$

From equation (1-30.16.16), we define the **condition for eliminating the mixed terms** $\hat{\xi}\overline{\eta}$ and $\hat{\xi}\overline{\zeta}$ is

$$\tau_{yz} \sin\alpha + \tau_{xz} \cos\alpha = 0$$

Or

$$\tan\alpha = -\frac{\tau_{xz}}{\tau_{yz}} \qquad (1\text{-}30.16.16)$$

From (1-30.16.2), we get our initial definition of α as follows:

$$\tan\alpha = \sqrt{\frac{1-\cos 2\alpha}{1+\cos 2\alpha}}$$

$$\tan\alpha = \sqrt{\frac{1 - \dfrac{\left(\sigma_x - \sigma_y\right)}{\sqrt{4\tau_{xy}^2 + \left(\sigma_x - \sigma_y\right)^2}}}{1 + \dfrac{\left(\sigma_x - \sigma_y\right)}{\sqrt{4\tau_{xy}^2 + \left(\sigma_x - \sigma_y\right)^2}}}} = \sqrt{\frac{\sqrt{4\tau_{xy}^2 + \left(\sigma_x - \sigma_y\right)^2} - \left(\sigma_x - \sigma_y\right)}{\sqrt{4\tau_{xy}^2 + \left(\sigma_x - \sigma_y\right)^2} + \left(\sigma_x - \sigma_y\right)}} \qquad (1\text{-}30.16.17)$$

From equations (1-30.16.16) and (1-30.16.17), the condition for vanishing of the mixed terms is

$$-\frac{\tau_{xz}}{\tau_{yz}} = \sqrt{\frac{\sqrt{4\tau_{xy}^2 + (\sigma_x - \sigma_y)^2} - (\sigma_x - \sigma_y)}{\sqrt{4\tau_{xy}^2 + (\sigma_x - \sigma_y)^2} + (\sigma_x - \sigma_y)}}$$ (1-30.16.18)

Or, $$4\tau_{xy}^2 + (\sigma_x - \sigma_y)^2 = \frac{(\tau_{xz}^2 + \tau_{yz}^2)^2(\sigma_x - \sigma_y)^2}{(\tau_{xz}^2 - \tau_{yz}^2)^2}$$

i.e., $$\frac{2\tau_{xy}}{(\sigma_x - \sigma_y)} = \sqrt{\frac{(\tau_{xz}^2 + \tau_{yz}^2)^2}{(\tau_{xz}^2 - \tau_{yz}^2)^2} - 1}$$ (1-30.16.19)

Therefore,

$$p^2 N_n = \hat{\xi}^2 A + \overline{\eta}^2 C + \overline{\zeta}^2 D = \hat{\xi}^2 N_1 + \overline{\eta}^2 N_2 + \overline{\zeta}^2 N_3$$ (1-31)

Therefore, from equations (1-30.16.8a), (1-30.16.8b), (1-30.16.16a), we get

$$N_1 = \sigma_x \cos^2 \alpha + \sigma_y \sin^2 \alpha + \tau_{xy} \cos\alpha \sin\alpha$$ (1-31.1)

$$N_2 = \sigma_y \cos^2 \alpha + \sigma_x \sin^2 \alpha - \tau_{xy} \cos\alpha \sin\alpha$$ (1-31.2)

$$N_3 = (\sigma_y \cos^2 \alpha + \sigma_x \sin^2 \alpha - \tau_{xy} \cos\alpha \sin\alpha)\sin^2 \beta$$
$$+ \sigma_z \cos^2 \beta - 2\sin\beta \cos\beta(\tau_{yz} \cos\alpha - \tau_{xz} \sin\alpha)$$ (1-31.31)

Equations (1-31) are subjected to the following definitions:

(i)   The condition imposed on shearing and normal stresses, defined by equations (1-30.16.19)

(ii)   The definition of β, defined by (1-30.16.9).

(iii)   The definition of B, defined by (1-30.16.8b).

1.10. Composition of the general stress tensor

1.10.1. Spherical stress tensor

The effects of external forces on a rigid body create a **balance** between **shearing** and **normal** stresses. We have seen that, the octahederal directions, which are equally spaced from the point on the principal surface, are exposed to the average of the three orthogonal principal stresses. Therefore, we will propose the **spherical tensor** as follows.

74

Consider equation (1-30.2), where the normal stresses lie on the diagonal of the matrix as shown in Figure 1-28.

$$S = \frac{1}{2}\begin{bmatrix} 2\sigma_x & 2\tau_{xy} & 2\tau_{xz} \\ 2\tau_{xy} & 2\sigma_y & 2\tau_{yz} \\ 2\tau_{xz} & 2\tau_{yz} & 2\sigma_z \end{bmatrix}$$

normal stress   shearing stress   shearing stress

Figure 1-28 Stress ellipsoid oriented at angle α, β, and λ between three projections of the axes of the ellipsoidal and the Cartesian axes.

Therefore, we can describe the stress distribution in a rigid body in terms of two contributions:[2]

(1) **Principal normals** which represent fluid pressure (compression or tensile) and which represent the **diagonal matrix** of principal normal stresses, called **spherical tensor** and

(2) Deviation resulting from shearing stresses that depend on the shape of the material, called **stress deviator**.

$$S = \begin{pmatrix} \sigma_x - N_o & \tau_{xy} & \tau_{xz} \\ \tau_{xy} & \sigma_y - N_o & \tau_{yz} \\ \tau_{xz} & \tau_{yz} & \sigma_z - N_o \end{pmatrix} + \begin{pmatrix} N_o & 0 & 0 \\ 0 & N_o & 0 \\ 0 & 0 & N_o \end{pmatrix} \qquad (1\text{-}31)$$

General Stress   =   Stress Devaitor   +   Spherical Stress

The **general stress tensor**, in equation (1-31), is obtained from equation (1-25.6), which represents the relationships between principal stresses and applied stresses.

---

[2] Anderson, Orson L, "Conditions for the Derivation of the Stress Deviator Tensor" American Journal of Physics -- April 1952 -- Volume 20, Issue 4, pp. 236

1.10.2. Stress deviator tensor

In equation (1-31), we isolate the **spherical tensor** from the general tensor of stress, which becomes the **stress deviator**. Thus, we could investigate each effect separately, as will be shown later. The decomposition of the stress tensor into the **spherical stress tensor** and the **stress deviator** depends upon the postulate that the stress tensor is the tensor sum of two physically **independent stress constituents**. The decomposition is derived by two different methods, the first method based on energy arguments stemming from balance of forces, and the second method based on cause and effect or distribution of shearing and normal stresses.

Let us sort out the various feature of the stress deviator from the equation of general stress in equation (1-25.7). The general stress tensor, in equation (1-25.7), has the following three invariants (i.e., applied external or internal stresses):

$$\Theta = \sigma_x + \sigma_y + \sigma_z \tag{1-32.1}$$

$$\Phi = \sigma_x \sigma_y + \sigma_x \sigma_z + \sigma_y \sigma_z - \tau_{yz}\tau_{zy} - \tau_{xy}\tau_{yx} - \tau_{xz}\tau_{zx} \tag{1-32.2}$$

$$\Sigma = \sigma_x \sigma_y \sigma_z + \tau_{xy}\tau_{yz}\tau_{zx} + \tau_{xz}\tau_{yx}\tau_{zy} - \tau_{xz}\tau_{zx}\sigma_y - \sigma_x \tau_{yz}\tau_{zy} - \tau_{xy}\tau_{yx}\sigma_z \tag{1-32.3}$$

To generate stress deviator, we subtract the **average principal normal** (also called **hydrostatic stress**)

$$N_0 = \frac{1}{3}(N_1 + N_2 + N_3) \tag{1-32.4}$$

From the three normal stresses $(\sigma_x, \sigma_y, \sigma_x)$ of equations (1-32) are as follows

1.10.2.1. Vanishing deviator of the first invariant of the general stress tensor

(i) Deviator D_Θ of first invariant Θ :

$$D_\Theta = (\sigma_x - N_0) + (\sigma_y - N_0) + (\sigma_z - N_0) \tag{1-33.1}$$

Substitute by $N_0 = \frac{1}{3}(\sigma_x + \sigma_y + \sigma_x)$ from equation (1-29.2), into equation (1-33.1), we get

$$D_\Theta = \sigma_x + \sigma_y + \sigma_z - 3N_0 = \sigma_x + \sigma_y + \sigma_z - 3\left(\frac{\sigma_x + \sigma_y + \sigma_z}{3}\right) = 0 \tag{1-33.2}$$

$$D_\Theta = N_1 + N_2 + N_3 - 3N_0 = \left(\begin{array}{l}\text{Vanishing stresses, i.e., no deviation} \\ \text{from the hydrostatic stress } N_0.\end{array}\right)$$

Therefore, we conclude that **the deviator of the first invariant of general stress tensor vanishes** since the main (average) of the principal stresses balances out the applied stresses.

1.10.2.2. Squared deviations of the second invariant of the general stress tensor

(ii) Deviator D_Φ of second invariant Φ :

$$D_\Phi = (\sigma_x - N_0)(\sigma_y - N_0) + (\sigma_x - N_0)(\sigma_z - -N_0) + (\sigma_y - N_0)(\sigma_z - N_0)$$
$$- \tau_{yz}\tau_{zy} - \tau_{xy}\tau_{yx} - \tau_{xz}\tau_{zx}$$

$$D_\Phi = \sigma_x\sigma_y + \sigma_x\sigma_z + \sigma_y\sigma_z - 2N_0(\sigma_x + \sigma_y + \sigma_z) + 3N_0^2$$
$$- \tau_{yz}\tau_{zy} - \tau_{xy}\tau_{yx} - \tau_{xz}\tau_{zx}$$

(1-34.1)

Substitute by $N_0 = \dfrac{1}{3}(\sigma_x + \sigma_y + \sigma_z)$ from equation (1-29.2), into equation (1-34.1), we get

$$D_\Phi = \sigma_x\sigma_y + \sigma_x\sigma_z + \sigma_y\sigma_z - \frac{2}{3}(\sigma_x + \sigma_y + \sigma_z)^2 + 3\left(\frac{\sigma_x + \sigma_y + \sigma_z}{3}\right)^2$$
$$- \tau_{yz}\tau_{zy} - \tau_{xy}\tau_{yx} - \tau_{xz}\tau_{zx}$$

i.e.,
$$D_\Phi = \sigma_x\sigma_y + \sigma_x\sigma_z + \sigma_y\sigma_z - \frac{1}{3}(\sigma_x + \sigma_y + \sigma_z)^2 - \tau_{yz}\tau_{zy} - \tau_{xy}\tau_{yx} - \tau_{xz}\tau_{zx}$$

Arranging the terms in few successive steps, we get

$$D_\Phi = \frac{1}{3}\Big[3\sigma_x\sigma_y + 3\sigma_x\sigma_z + 3\sigma_y\sigma_z - \big(\sigma_x^2 + \sigma_y^2 + \sigma_z^2 + 2\sigma_y\sigma_z + 2\sigma_x\sigma_z + 2\sigma_x\sigma_y\big)\Big]$$
$$- \tau_{yz}\tau_{zy} - \tau_{xy}\tau_{yx} - \tau_{xz}\tau_{zx}$$

$$D_\Phi = \frac{1}{3}\Big[\sigma_x\sigma_y + \sigma_x\sigma_z + \sigma_y\sigma_z - \sigma_x^2 - \sigma_y^2 - \sigma_z^2\Big] - \tau_{yz}\tau_{zy} - \tau_{xy}\tau_{yx} - \tau_{xz}\tau_{zx}$$

$$D_\Phi = -\frac{1}{6}\Big[2\sigma_x\sigma_y - 2\sigma_x\sigma_z - 2\sigma_y\sigma_z + 2\sigma_x^2 + 2\sigma_y^2 + 2\sigma_z^2\Big]$$
$$- \tau_{yz}\tau_{zy} - \tau_{xy}\tau_{yx} - \tau_{xz}\tau_{zx}$$

$$D_\Phi = -\frac{1}{6}\Big[2\sigma_x\sigma_y - 2\sigma_x\sigma_z - 2\sigma_y\sigma_z + \big(\sigma_x^2 + \sigma_y^2\big) + \big(\sigma_x^2 + \sigma_z^2\big) + \big(\sigma_y^2 + \sigma_z^2\big)\Big]$$
$$- \tau_{yz}\tau_{zy} - \tau_{xy}\tau_{yx} - \tau_{xz}\tau_{zx}$$

$$D_\Phi = -\frac{1}{6}\Big[\big(\sigma_x^2 - 2\sigma_x\sigma_y + \sigma_y^2\big) + \big(\sigma_x^2 - 2\sigma_x\sigma_z + \sigma_z^2\big) + \big(\sigma_y^2 - 2\sigma_y\sigma_z + \sigma_z^2\big)\Big]$$
$$- \tau_{yz}\tau_{zy} - \tau_{xy}\tau_{yx} - \tau_{xz}\tau_{zx}$$

Finally, we arrive at the deviator of the second invariant as follows

$$D_{\Phi} = -\frac{1}{6}\left[\left(\sigma_x - \sigma_y\right)^2 + \left(\sigma_x - \sigma_z\right)^2 + \left(\sigma_y - \sigma_z\right)^2\right]$$
$$-\tau_{yz}\tau_{zy} - \tau_{xy}\tau_{yx} - \tau_{xz}\tau_{zx} \tag{1-34.2}$$

Equation (1-34.2) bears the features of equation (1-29.3), of **total shearing stress** T_n^2, when the concerned plane in (1-34.2) is the principal plane, such that

$$N_1 = \sigma_x, \quad N_2 = \sigma_y \quad N_3 = \sigma_z, \quad \tau_{yz} = \tau_{zy} = \tau_{xy} = \tau_{yx} = \tau_{xz} = \tau_{zx} = 0$$

$$D_{\Phi} = -\frac{1}{6}\left[\left(N_1 - N_2\right)^2 + \left(N_1 - N_3\right)^2 + \left(N_2 - N_3\right)^2\right] \tag{1-34.3}$$

Therefore, we conclude that **the deviator of second invariant of the general stress tensor is shearing stress**.

********** Auxiliary Algebraic Proof ********

In order to describe the deviator of the second invariant of the general stress tensor, equations (1-19) and (1-30.2) with respect to the hydrostatic stress N_0, we first rearrange equation (1-34.3) to create two terms by **adding and subtracting** N_0 to each stress as follows

$$D_{\Phi} = -\frac{1}{6}\left[\begin{array}{l}\left(\left(N_1 - N_0\right) - \left(N_2 - N_0\right)\right)^2 + \left(\left(N_1 - N_0\right) - \left(N_3 - N_0\right)\right)^2 \\ + \left(\left(N_3 - N_0\right) - \left(N_2 - N_0\right)\right)^2\end{array}\right] \tag{1-34.3.1}$$

Executing the squaring of the three bracketed terms, we get

$$D_{\Phi} = -\frac{1}{6}\left[\begin{array}{l}\left(N_1 - N_0\right)^2 + \left(N_2 - N_0\right)^2 - 2\left(N_1 - N_0\right)\left(N_2 - N_0\right) \\ + \left(N_1 - N_0\right)^2 + \left(N_3 - N_0\right)^2 - 2\left(N_1 - N_0\right)\left(N_3 - N_0\right) \\ + \left(N_3 - N_0\right)^2 + \left(N_2 - N_0\right)^2 - 2\left(N_3 - N_0\right)\left(N_2 - N_0\right)\end{array}\right]$$

Which, upon associating together similar algebraic terms, becomes

$$D_{\Phi} = -\frac{1}{6}\left[\begin{array}{l}2\left(N_1 - N_0\right)^2 + 2\left(N_2 - N_0\right)^2 + 2\left(N_3 - N_0\right)^2 \\ -\left(N_1 - N_0\right)\left(N_2 - N_0\right) - \left(N_1 - N_0\right)\left(N_2 - N_0\right) \\ -\left(N_1 - N_0\right)\left(N_3 - N_0\right) - \left(N_1 - N_0\right)\left(N_3 - N_0\right) \\ -\left(N_3 - N_0\right)\left(N_2 - N_0\right) - \left(N_3 - N_0\right)\left(N_2 - N_0\right)\end{array}\right] \tag{1-34.3.2}$$

which, upon taking common factors, becomes

78

$$D_\Phi = -\frac{1}{6}\begin{bmatrix} 2(N_1 - N_0)^2 + 2(N_2 - N_0)^2 + 2(N_3 - N_0)^2 \\ -(N_1 - N_0)(N_2 + N_3 - 2N_0) \\ -(N_3 - N_0)(N_1 + N_2 - 2N_0) \\ -(N_2 - N_0)(N_3 + N_1 - 2N_0) \end{bmatrix} \qquad (1\text{-}34.3.3)$$

Substitute by $N_0 = \frac{1}{3}(N_1 + N_2 + N_3)$ from equation (1-29.2) into equation (1-34.3.3), such that $N_1 = 3N_0 - N_2 - N_3$ or $N_1 - N_0 = 2N_0 - N_2 - N_3$, and so on for N_2-N_0 and N_3-N_0, we get

$$D_\Phi = -\frac{1}{6}\begin{bmatrix} 2(N_1 - N_0)^2 + 2(N_2 - N_0)^2 + 2(N_3 - N_0)^2 \\ -(N_1 - N_0)(N_0 - N_1) \\ -(N_3 - N_0)(N_0 - N_3) \\ -(N_2 - N_0)(N_0 - N_2) \end{bmatrix} \qquad (1\text{-}34.3.4)$$

Or,

$$D_\Phi = -\frac{1}{6}\left[3(N_1 - N_0)^2 + 3(N_2 - N_0)^2 + 3(N_3 - N_0)^2 \right] \qquad (1\text{-}34.3.5)$$

i.e.,

$$D_\Phi = -\frac{1}{2}\left[(N_1 - N_0)^2 + (N_2 - N_0)^2 + (N_3 - N_0)^2 \right] \qquad (1\text{-}34.3.5)$$

$$D_\Phi = -\frac{3}{2}\left[\frac{(N_1 - N_0)^2 + (N_2 - N_0)^2 + (N_3 - N_0)^2}{3} \right] = -\frac{3}{2}\begin{pmatrix} \text{Average of sqaures of} \\ \text{deviations of } N_1, N_2, \text{and } N_3, \\ \text{from hydrostate stess, } N_0 \end{pmatrix}$$

********** End of Auxiliary Algebraic Proof ********

From equation (1-34.3.5), we conclude that **the deviator of the second invariant of the general stress tensor** of equation (1-19) and (1-30.2) represents (1-3/2) **the average (division by 3) of the square deviations of the three projections of the main principal stress, N_1, N_2, N_3, from the hydrostatic stress N_0.**

1.10.2.3. Cubic deviations of the third invariant of the general stress tensor

(iii) Deviator D_Ξ of second invariant Ξ :

Again, we will assume that the concerned plane in equation (1-33.3), of the third invariant of the general stress tensor, to be the principal plane, such that

79

$$N_1 = \sigma_x, \quad N_2 = \sigma_y \quad N_3 = \sigma_z, \quad \tau_{yz} = \tau_{zy} = \tau_{xy} = \tau_{yx} = \tau_{xz} = \tau_{zx} = 0$$

Therefore, the deviator of the third invariant (obtained by subtracting the main normal stress from the three applied normal stresses of equation (1-32.3) and removing the shearing stresses by the assumption of principal plane made above, is as follows

$$D_\Sigma = \left(\sigma_x - N_0\right)\left(\sigma_y - N_0\right)\left(\sigma_z - N_0\right) = \left(N_1 - N_0\right)\left(N_2 - N_0\right)\left(N_3 - N_0\right) \qquad (1\text{-}35.1)$$

We will take the trouble of arranging the algebraic terms of (1-35.1) in order to be able to describe the behavior of the deviator of the **third invariant of the general stress tensor**, as follows.

******* Auxiliary Algebraic Manipulation ***************

Substitute by $N_0 = \dfrac{1}{3}\left(N_1 + N_2 + N_3\right)$ from equation (1-29.2) into equation (1-34.1), such that $N_1 = 3N_0 - N_2 - N_3$ or $N_1 - N_0 = 2N_0 - N_2 - N_3$, we get

$$D_\Sigma = \left(2N_0 - N_2 - N_3\right)\left(N_2 - N_0\right)\left(N_3 - N_0\right) \qquad (1\text{-}35.1.1)$$

With algebraic association of the term $2N_0$ with each of N_2 and N_3, the first term of the right hand side takes different form as follows

$$D_\Sigma = -\left[\left(N_2 - N_0\right) + \left(N_3 - N_0\right)\right]\left(N_2 - N_0\right)\left(N_3 - N_0\right) \qquad (1\text{-}35.1.2)$$

We will repeatedly resort to this manipulation until we arrive at the required form by recognizing the following substitutions, obtained from the relation $N_0 = \dfrac{1}{3}\left(N_1 + N_2 + N_3\right)$ from equation (1-29.2),

$$\begin{aligned}
N_1 - N_0 &= -\left(N_2 - N_0\right) - \left(N_3 - N_0\right) \\
N_2 - N_0 &= -\left(N_3 - N_0\right) - \left(N_1 - N_0\right) \\
N_3 - N_0 &= -\left(N_1 - N_0\right) - \left(N_2 - N_0\right)
\end{aligned} \qquad (1\text{-}35.1.3)$$

Performing algebraic distribution on equation (1-35.1.2), we get

$$D_\Sigma = -\left(N_2 - N_0\right)^2\left(N_3 - N_0\right) - \left(N_2 - N_0\right)\left(N_3 - N_0\right)^2 \qquad (1\text{-}35.1.4)$$

We need to alter the unsquared terms in N_2 and N_3 onto equivalent terms that produce cubic terms, when multiplied by the squared brackets. To do so, we substitute from (1-35.1.3) into (1-35.1.4) to get

$$D_\Sigma = -\left[\left(N_2 - N_0\right)^2\left(-\left(N_2 - N_0\right) - \left(N_1 - N_0\right)\right) + \left(N_3 - N_0\right)^2\left(-\left(N_1 - N_0\right) - \left(N_3 - N_0\right)\right)\right]$$

80

Or, $$D_\Sigma = \left[(N_2 - N_0)^3 + (N_3 - N_0)^3 + (N_1 - N_0)(N_2 - N_0)^2 + (N_1 - N_0)(N_3 - N_0)^2\right]$$

Hence, we get our first two cubic terms isolated as follows

$$D_\Sigma = (N_2 - N_0)^3 + (N_3 - N_0)^3 + (N_1 - N_0)\left[(N_2 - N_0)^2 + (N_3 - N_0)^2\right] \quad (1\text{-}35.1.5)$$

Again, we need to alter the squared terms in N_2 and N_3 to equivalent terms in N_1 alone. To do so, we substitute from (1-35.1.3) into (1-35.1.5) to get

$$D_\Sigma = \begin{bmatrix} (N_3 - N_0)^3 + (N_3 - N_0)^3 + \\ (N_1 - N_0)\left[(-(N_1 - N_0)-(N_3 - N_0))^2 + (N_3 - N_0)^2\right] \end{bmatrix} \quad (1\text{-}35.1.6)$$

which, upon performing the squaring of the mixed term, becomes

$$D_\Sigma = \begin{bmatrix} (N_2 - N_0)^3 + (N_3 - N_0)^3 + \\ (N_1 - N_0)\left[2(N_1 - N_0)(N_3 - N_0)+(N_1 - N_0)^2 + 2(N_3 - N_0)^2\right] \end{bmatrix}$$

which, upon taking common factor from the mixed term, becomes

$$D_\Sigma = \begin{bmatrix} (N_1 - N_0)^3 + (N_2 - N_0)^3 + (N_3 - N_0)^3 \\ + 2(N_1 - N_0)(N_3 - N_0)\left[(N_1 - N_0)+(N_3 - N_0)\right] \end{bmatrix} \quad (1\text{-}35.1.7)$$

Again, we need to alter the term in N_3 in the mixed last term to equivalent terms in N_1 such that Our final substitution from (1-35.1.3) into (1-35.1.7) reduce the last term onto $-(N_0-N_3)$ to give

$$D_\Sigma = \begin{bmatrix} (N_1 - N_0)^3 + (N_2 - N_0)^3 + (N_3 - N_0)^3 \\ + 2(N_1 - N_0)(N_3 - N_0)\left[-(N_2 - N_0)\right] \end{bmatrix}$$

i.e.,
$$D_\Sigma = (N_1 - N_0)^3 + (N_2 - N_0)^3 + (N_3 - N_0)^3 - 2(N_1 - N_0)(N_3 - N_0)(N_2 - N_0) \quad (1\text{-}35.1.8)$$

What seems like simple algebraic manipulation turned tedious due to the **circular pattern** of equation (1-35.1.3). Here, equation (1-35.1.8) ends such circular manipulation since the last term is replaced by $-2D_\Sigma$ from equation (1-35.1.1) (note the three different signs between equation (1-35.1) and (1-35.1.8)) to give

$$D_\Sigma = (N_1 - N_0)^3 + (N_2 - N_0)^3 + (N_3 - N_0)^3 - 2D_\Sigma \quad (1\text{-}35.1.8)$$

Or, $$3D_\Sigma = (N_1 - N_0)^3 + (N_2 - N_0)^3 + (N_3 - N_0)^3 \quad (1\text{-}35.1.9)$$

Finally, we get our desired form of the **deviator of the third invariant of the general stress tensor** as

$$D_\Sigma = \frac{1}{3}\left[\left(N_1 - N_0\right)^3 + \left(N_2 - N_0\right)^3 + \left(N_3 - N_0\right)^3\right] \qquad (1\text{-}35.2)$$

$$D_\Sigma = \frac{1}{3}\left[\left(N_1 - N_0\right)^3 + \left(N_2 - N_0\right)^3 + \left(N_3 - N_0\right)^3 = \begin{pmatrix}\text{Average of cubes of} \\ \text{deviations of } N_1, N_2, \text{and } N_3, \\ \text{from hydrostatic stess, } N_0\end{pmatrix}\right]$$

******* End of Auxiliary Algebraic Manipulation ***************

From equation (1-35.2), we conclude that **the deviator of the third invariant of the general stress tensor** of equation (1-19) and (1-30.2) represents **the average (division by 3) of the cubic deviations of the three projections of the main principal stress, N_1, N_2, N_3, from the hydrostatic stress N_0.**

1.11. Exercises on Geometry of Stress

Example 13

Find the equation of the planes passing through the points $P(1-1,-1,3)$, $Q(1-4,1,-2)$ and $R(-1,-1,1)$ and the planes normal to each line and the common plane PRQ.

Solution

The **vector representation of the lines** extending between points P and Q and Q and R are given by

$$\overline{PQ} = \langle 4 - 1,1 + 1,-2 - 3 \rangle = \langle 3,2,-5 \rangle \qquad (1\text{-}36.1)$$

$$\overline{QR} = \langle -1 - 4,-1 - 1,1 + 2 \rangle = \langle -5,-2,3 \rangle \qquad (1\text{-}36.2)$$

The **normal** on the lines $\overline{PQ}$ and $\overline{QR}$ is given by

$$\overline{n}_1 = \overline{PQ} \times \overline{QR} = \begin{vmatrix} \hat{i} & \hat{j} & \hat{k} \\ 3 & 2 & -5 \\ -5 & -2 & 3 \end{vmatrix} = \langle -4,16,4 \rangle = \langle a,b,c \rangle \qquad (1\text{-}36.3)$$

i.e., $\qquad \overline{n}_1 = \langle -4,16,4 \rangle$

The equation of the plane is constructed by any of the three points P, Q, or R and the normal to the plane

$$ax + by + cz + d = 0 \qquad (1\text{-}36.4)$$

Substituting from (1-36.3) into (1-36.4), we get

$$-4(1) + 16(-1) + 4(3) + d = -8 + d = 0$$

Or, $\qquad -4x + 16y + 4z + 8 = 0 \qquad (1\text{-}36.5)$

The plane passing by PQ and normal to (1-36.5) is given by

$$\bar{n}_2 = \bar{n} \times \overline{PQ} = \begin{vmatrix} \hat{i} & \hat{j} & \hat{k} \\ -4 & 16 & 4 \\ 3 & 2 & -5 \end{vmatrix} = \langle -80-8, -(20-12), -8-48 \rangle = \langle -11, -8, -56 \rangle = -8\langle 11,1,7 \rangle$$

$$\bar{n}_2 = -8\langle 11,1,7 \rangle \qquad (1\text{-}36.6)$$

The plane passing by PQ and **normal** to (1-36.6) is given by

$$a(x - x_P) + b(y - y_P) + c(z - z_P) = 0$$

$$11(x-1) + (y+1) + 7(z-3) = 0$$

$$11x + y + 7z - 11 + 1 - 21 = 0$$

$$11x + y + 7z - 31 = 0 \qquad (1\text{-}36.7)$$

****
Proof

$$\overline{n_\uparrow} \bullet \bar{n} = 11(-4) + 1(16) + 4(7) = -44 + 16 + 28 = 0$$

Subtract plane 5 from plane 4, we get

$$11x + y + 7z - 31 = 0$$

$$-4x + 16y + 4z + 8 = 0$$

$$(16 \times 11x + 16y + 112z - 16 \times 31 = 0$$

$$-4x + 16y + 4z + 8 = 0$$

$$180x + 108z - 16 \times 31 - 8 = -180x + 108z - 8(62 + 1) = 0$$

$$5x + 3z - 14 = 0$$

Which is the line: $\overline{PQ} = \langle 4-1,1+1,-2-3 \rangle = \langle 3,2,-5 \rangle$

Equation of the line: $\overline{PQ} = \langle 4-1,1+1,-2-3 \rangle = \langle 3,2,-5 \rangle$

$$\frac{x-1}{3} = \frac{y+1}{2} = \frac{z-3}{-5}$$

$$-5(x-1) = 3(z-3)$$

$$5x + 3z - 14 = 0$$

CHAPTER 2

GEOMETRY OF STRAIN

2.1. Cauchy's equations for displacement, elongation, shear, and rotational strains

The following basics underline the study of **material strain**:

1. A fixed elastic body incurs **deformation** of its interior material points when exposed to **stress**.

2. **Linear deformation** comprises **elongation** of distances between the material points of the elastic body <u>along</u> the three Cartesian directions and is called **extension strain** or **principal strain**. This is denoted by ε_{xx}, ε_{yy}, ε_{zz}

3. **Angular deformation** comprises of elongation of distances between the material points of the elastic body <u>across</u> the three Cartesian directions and is called **shearing strain**. This is denoted by α_{xy}, α_{yx}, $\alpha_{yz,}$ α_{zy}, α_{zx}, $\alpha_{xz.}$

4. Therefore, the deformation of an elastic body is described by the changes in the **vector of displacement of material points** (u, v, w), of an elastic body, in reference to the **space vector of coordinate system** (x, y, z).

5. An **elastic body** is still described as **rigid** in the relative sense, since the material points of an elastic body are displaced by stress, which is infinitesimal to the local point. **Fluid and gases** homogenize internal stresses into uniform pressure of molecules in constant motion.

6. Like the considerations taken in the previous chapter to distinguish between stress of material and applied force, we emphasize the fact that **strain** is a material phenomena comprising for **principal** and **shearing** components, in the case of rigid bodies.

Denote the three Cartesian projections of the **displacement vector of deformation** of an arbitrary point in an elastic rigid body by

$$U\,(x,y,z),\ v(x,y,z),\ \text{and}\ w(x,y,\ z). \tag{2-1}$$

Where u, v, and w are the projections of the material point vector along the Cartesian axes x, y, and z, respectively.

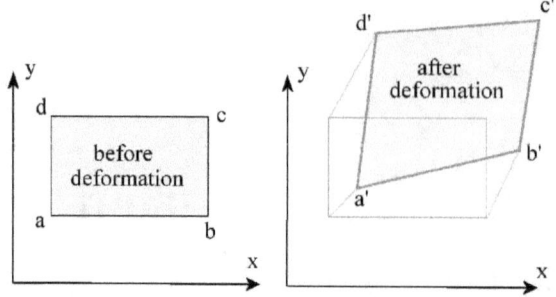

Figure 2-1. Projection of the deformation of rectangle abcd to a'b'c'd' on the xy-plane.

In Figure 2-1, assume that the arbitrary points **abcd** in the rigid body are **stressed** by some force. The generated deformation or **displacement vectors** possess the following projections over an infinitesimal volume **δx, δy**, and **δz**, shown in Figure 2-2, as follows.

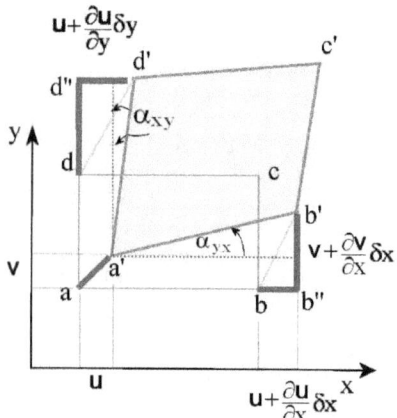

Figure 2-2. Change of strains with respect to coordinates. The displacements of the three points a' b'c'.

(i) **Extension strains**: strains along the three Cartesian directions:

$$\mathbf{u}(x,y,z) \quad \text{and} \quad \mathbf{u}(x,y,z) + \frac{\partial \mathbf{u}(x,y,z)}{\partial x}\delta x \qquad (2\text{-}2.1)$$

$$\mathbf{v}(x,y,z) \quad \text{and} \quad \mathbf{v}(x,y,z) + \frac{\partial \mathbf{v}(x,y,z)}{\partial y}\delta y \qquad (2\text{-}2.2)$$

$$\mathbf{w}(x,y,z) \quad \text{and} \quad \mathbf{w}(x,y,z) + \frac{\partial \mathbf{w}(x,y,z)}{\partial z} \delta z \qquad (2\text{-}2.3)$$

The derivatives of displacements in equations (2-2) are called **unit elongation** or **extension strains**, which are **dimensionless values**, denoted as follows

$$\varepsilon_{xx} = \frac{\partial \mathbf{u}(x,y,z)}{\partial x} \qquad (2\text{-}2.4)$$

$$\varepsilon_{yy} = \frac{\partial \mathbf{v}(x,y,z)}{\partial y} \qquad (2\text{-}2.5)$$

$$\varepsilon_{zz} = \frac{\partial \mathbf{w}(x,y,z)}{\partial z} \qquad (2\text{-}2.6)$$

In cylindrical polar coordinates, we could easily prove that elongation strains are given by

$$\varepsilon_{rr} = \frac{\partial \mathbf{u}(r,\theta,z)}{\partial r} \qquad (2\text{-}2.7)$$

$$\varepsilon_{\theta\theta} = \frac{1}{r} \frac{\partial \mathbf{v}(r,\theta,z)}{\partial \theta} + \frac{\mathbf{u}}{r} \qquad (2\text{-}2.8)$$

$$\varepsilon_{zz} = \frac{\partial \mathbf{w}(r,\theta,z)}{\partial z} \qquad (2\text{-}2.9)$$

and the shear strains by

$$\alpha_{r\theta} = \frac{\partial \mathbf{v}(r,\theta,z)}{\partial r} - \frac{\mathbf{v}(r,\theta,z)}{r} + \frac{1}{r} \frac{\partial \mathbf{u}(r,\theta,z)}{\partial \theta} \qquad (2\text{-}2.10)$$

We could prove equation (2-2.10) by analytical geometry or graphically as shown in Figure 2-3.

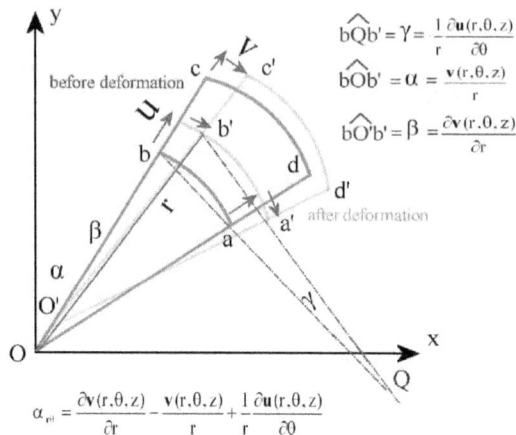

$$\widehat{bQb'} = \gamma = \frac{1}{r}\frac{\partial u(r,\theta,z)}{\partial \theta}$$

$$\widehat{bOb'} = \alpha = \frac{v(r,\theta,z)}{r}$$

$$\widehat{bO'b'} = \beta = \frac{\partial v(r,\theta,z)}{\partial r}$$

$$\alpha_{r\theta} = \frac{\partial v(r,\theta,z)}{\partial r} - \frac{v(r,\theta,z)}{r} + \frac{1}{r}\frac{\partial u(r,\theta,z)}{\partial \theta}$$

Figure 2-3. Deformation in cylindrical polar coordinates.

As with the case of stress equations (1-3) (also know as **Cauchy's equations**), strains refer to ratios of effects over (or per) distances.

The three unit elongations, of the three projections of displacements, along the directions of those displacements are called **linear elongation** and correspond to the **normal stresses**.

(ii) **Shearing strains**: strains perpendicular to each of the three Cartesian directions:

Shearing stresses induced angular strains by changing the projections of displacements, u (x,y,z), v(x,y,z), and w(x,y, z), on the orthogonal planes, such that

$$2\alpha_{xy} = \frac{\partial u(x,y,z)}{\partial y} + \frac{\partial v(x,y,z)}{\partial x} \tag{2-3.1}$$

$$2\alpha_{yz} = \frac{\partial v(x,y,z)}{\partial z} + \frac{\partial w(x,y,z)}{\partial y} \tag{2-3.2}$$

$$2\alpha_{zx} = \frac{\partial w(x,y,z)}{\partial x} + \frac{\partial u(x,y,z)}{\partial z} \tag{2-3.3}$$

The appearance of the two terms in equations (2-3) is illustrated in Figure 2-2, by the two angles α_{xy} and α_{yx} created due to the rotation of **ab** to **a'b'** and **ad** to **a'd'**.

It is well known in calculus, that the following approximation holds true form small angles, which is the case in infinitesimal analysis.

88

$$\alpha_{yx} = \frac{\dfrac{\partial v(x,y,z)}{\partial x}\delta x}{\delta x + \dfrac{\partial u(x,y,z)}{\partial x}\delta x} = \frac{\dfrac{\partial v(x,y,z)}{\partial x}}{1 + \dfrac{\partial u(x,y,z)}{\partial x}} \approx \frac{\partial v(x,y,z)}{\partial x} \qquad (2\text{-}4.1)$$

Similarly,

$$\alpha_{xy} = \frac{\dfrac{\partial u(x,y,z)}{\partial y}\delta y}{\delta y + \dfrac{\partial v(x,y,z)}{\partial y}\delta y} = \frac{\dfrac{\partial u(x,y,z)}{\partial y}}{1 + \dfrac{\partial v(x,y,z)}{\partial y}} \approx \frac{\partial u(x,y,z)}{\partial y} \qquad (2\text{-}4.2)$$

(iii) Rotational strains

In Figure 2-2, the rectangle **abcd** is deformed by the stress to **a'b'c'd'**, where the right angle dâb transforms to d'â'b' by the subtraction of the two angles α_{xy} and α_{yx}. Therefore, the **net rotation** of the deformed body a'b'c'd' around the z axis is actually $(\alpha_{yx} - \alpha_{xy})\ /\ 2$ (the average of angular rotation of the **abcd**). Therefore, from equations (2-3), we get

$$2\omega_x = \frac{\partial w(x,y,z)}{\partial y} - \frac{\partial v(x,y,z)}{\partial z} \qquad (2\text{-}5.1)$$

$$2\omega_y = \frac{\partial u(x,y,z)}{\partial z} - \frac{\partial w(x,y,z)}{\partial x} \qquad (2\text{-}5.2)$$

$$2\omega_z = \frac{\partial v(x,y,z)}{\partial x} - \frac{\partial u(x,y,z)}{\partial y} \qquad (2\text{-}5.3)$$

Thus, the **strain tensor** takes similar form of the stress tensor equation (2-30.2), as follows

$$\begin{pmatrix} \varepsilon_{xx} & \alpha_{xy} & \alpha_{xz} \\ \alpha_{yx} & \varepsilon_{yy} & \alpha_{yz} \\ \alpha_{zx} & \alpha_{zy} & \varepsilon_{zz} \end{pmatrix} \qquad (2\text{-}6)$$

Where diagonal strains ε_{xx}, ε_{yy}, and ε_{zz} are linear elongation- or extension- strains, off-diagonal strains α_{xy}, α_{yz}, α_{zx} are shearing strains.

2.2. General strain tensor

The nine components of the **general strain tensor**, equations (2-2), (2-3), and (2-5), can be rewritten in the usual tensor form which lends greater analytical utility. First, adding each of the three equations (2-3) to corresponding equations (2-5), we get

From (2-3.1), subtracting equation (2-5.3) gives

$$\frac{\partial \mathbf{u}(x,y,z)}{\partial y} = \alpha_{xy} - \omega_z \qquad (2\text{-}7.1)$$

From (2-3.2), subtracting (2-5.1) gives

$$\frac{\partial \mathbf{v}(x,y,z)}{\partial z} = \alpha_{yz} - \omega_x \qquad (2\text{-}7.2)$$

From (2-3.3), subtracting (2-5.2) gives

$$\frac{\partial \mathbf{w}(x,y,z)}{\partial x} = \alpha_{zx} - \omega_y \qquad (2\text{-}7.3)$$

From (2-3.1), adding (2-5.3) gives

$$\frac{\partial \mathbf{v}(x,y,z)}{\partial x} = \alpha_{xy} + \omega_z \qquad (2\text{-}7.4)$$

From (2-3.2), adding (2-5.1) gives

$$\frac{\partial \mathbf{w}(x,y,z)}{\partial y} = \alpha_{yz} + \omega_x \qquad (2\text{-}7.5)$$

From (2-3.3), adding (2-5.2) gives

$$\frac{\partial \mathbf{u}(x,y,z)}{\partial z} = \alpha_{zx} + \omega_y \qquad (2\text{-}7.6)$$

Therefore, equations (2-2) and (2-9) give the general strain tensor

$$\begin{pmatrix} \dfrac{\partial \mathbf{u}}{\partial x} & \dfrac{\partial \mathbf{u}}{\partial y} & \dfrac{\partial \mathbf{u}}{\partial z} \\[2mm] \dfrac{\partial \mathbf{v}}{\partial x} & \dfrac{\partial \mathbf{v}}{\partial y} & \dfrac{\partial \mathbf{v}}{\partial z} \\[2mm] \dfrac{\partial \mathbf{w}}{\partial x} & \dfrac{\partial \mathbf{w}}{\partial y} & \dfrac{\partial \mathbf{w}}{\partial z} \end{pmatrix} = \begin{pmatrix} \varepsilon_{xx} & \alpha_{xy} - \omega_z & \alpha_{zx} + \omega_y \\[2mm] \alpha_{xy} + \omega_z & \varepsilon_{yy} & \alpha_{yz} - \omega_x \\[2mm] \alpha_{zx} - \omega_y & \alpha_{yz} + \omega_x & \varepsilon_{zz} \end{pmatrix} \qquad (2\text{-}8.1)$$

$$= \begin{pmatrix} 0 & \alpha_{xy} & \alpha_{zx} \\ \alpha_{xy} & 0 & \alpha_{yz} \\ \alpha_{zx} & \alpha_{yz} & 0 \end{pmatrix} + \begin{pmatrix} \varepsilon_{xx} & 0 & 0 \\ 0 & \varepsilon_{yy} & 0 \\ 0 & 0 & \varepsilon_{zz} \end{pmatrix} + \begin{pmatrix} 0 & -\omega_z & \omega_y \\ \omega_z & 0 & -\omega_x \\ -\omega_y & \omega_x & 0 \end{pmatrix} \qquad (2\text{-}8.2)$$

Equation (2-8) sorts out the components of the general strain tensor into **shearing strain tensor, extension strain tensor,** and **rotational strain tensor.**

2.3. Homogeneous deformation

Assume that an elastic body responds to stress by equal deformation in all directions. We can describe this scenario as writing equation (2-1) as follows

Example 13

$$u(x,y,z) = u_0 + c_{11}\,x + c_{12}\,y + c_{13}\,z \qquad (2\text{-}9.1)$$
$$v(x,y,z) = v_0 + c_{21}\,x + c_{22}\,y + c_{23}\,z \qquad (2\text{-}9.2)$$
$$w(x,y,z) = w_0 + c_{31}\,x + c_{32}\,y + c_{33}\,z \qquad (2\text{-}9.3)$$

The components of homogenous strains of equations (2-9) are obtained from equations (2-2), (2-4) and (2-5) as follows.

$$
\begin{array}{lll}
\varepsilon_{xx} = c_{11} & 2\alpha_{xy} = c_{12} + c_{21} & 2\omega_z = c_{21} - c_{12} \\
\varepsilon_{yy} = c_{22} & 2\alpha_{yz} = c_{23} + c_{32} & 2\omega_x = c_{32} - c_{23} \\
\varepsilon_{zz} = c_{33} & 2\alpha_{zx} = c_{13} + c_{31} & 2\omega_y = c_{32} - c_{13}
\end{array}
\qquad (2\text{-}9.4)
$$

(i) Homogeneous strains with non-vanishing c's in equation (2-9.4).

Since strains obtained in equation (2-9.4) are constants, the deformation depends on the difference of constants in different directions. Thus, lines and planes retain their form (being lines and planes) but change their directional cosines (or slopes). Also, right angles between planes or lines change.

(i.a) Consider to the two equations of lines

$$
\begin{array}{llll}
x_1 = x_{01} + a_1 t, & y_1 = y_{01} + a_1 t, & z_1 = z_{01} + a_1 t & (2\text{-}9.4.1) \\
x_2 = x_{02} + a_2 t, & y_2 = y_{02} + a_2 t, & z_2 = z_{02} + a_2 t & (2\text{-}9.4.2)
\end{array}
$$

The net deformation in the in the x-direction = $(c_{11}+c_{12}+c_{13})\,\delta x = p_x\,\delta x$ (2-9.4.3)
The net deformation in the in the y-direction = $(c_{22}+c_{12}+c_{23})\,\delta y = p_y\,\delta y$ (2-9.4.4)
The net deformation in the in the z-direction = $(c_{33}+c_{13}+c_{23})\,\delta z = p_z\,\delta z$ (2-9.4.5)

Thus, the equations of the two deformed lines become

$$x'_1 = x_{01} + a_1 t + p_x\,\delta x, \qquad y'_1 = y_{01} + a_1 t + p_y\,\delta y, \qquad z'_1 = z_{01} + a_1 t + p_z\,\delta z \qquad (2\text{-}9.4.6)$$
$$x'_2 = x_{02} + a_2 t + p_x\,\delta x, \qquad y'_2 = y_{02} + a_2 t + p_y\,\delta y, \qquad z'_2 = z_{02} + at + p_z\,\delta z \qquad (2\text{-}9.4.7)$$

Therefore, even though the two deformed lines retain their slopes a_1 and a_2, they differ in the intersection of their parametric equations with the Cartesian axes.

(i.b) Consider to the two equations of surfaces

$$a_1\,x + b_1 y + c_1 z + d_1 = 0 \qquad (2\text{-}9.4.8)$$
$$a_2\,x + b_2 y + c_2 z + d_2 = 0 \qquad (2\text{-}9.4.9)$$

The two surfaces possess the **surface normals** $\hat{n}_1\langle a_1, b_1, c_1\rangle$ and $\hat{n}_2\langle a_2, b_2, c_2\rangle$, which will remain unchanged when homogeneous deformation is performed as in the case of equations (2-9.4.6 and 7).

(ii) Vanishing extension and shearing strains, in equation (2-9.4), or not deforming body:

From (2-9.4), we get

$$c_{12} = -c_{21}, \; c_{13} = -c_{31}, \; c_{32} = -c_{23} \qquad (2\text{-}9.5)$$

Thus, equations (2-9.1 through 3) become

$$u\,(x,y,z) = u_0 + c_{12}\,y - c_{31}\,z \qquad (2\text{-}9.6)$$
$$v(x,y,z) = v_0 - c_{12}\,x + c_{23}\,z \qquad (2\text{-}9.7)$$
$$w(x,y,z) = w_0 + c_{31}\,x - c_{23}\,y \qquad (2\text{-}9.8)$$

Equations (2-9.6 through 8) show that each of the three displacements is constant along its own coordinate (u is independent of x, v is independent of y, w is independent of z) and that only translations take place (u_0, v_0, w_0).

The occurrences of the perpendicular coordinates in equations (2-9.6 through 8) signify angular rotations. Since, by means of equations (2-5.1 through 3), we get

$$2\omega_z = c_{21} - c_{12} = 2c_{21}$$
$$2\omega_x = c_{32} - c_{23} = 2c_{32} \qquad (2\text{-}9.9)$$
$$2\omega_y = c_{32} - c_{13} = 2c_{32}$$

Therefore, we conclude that with vanishing normal and shearing stresses, the rigid body can **translate or rotate, but not deform**.

(iii) Vanishing rotational strains but unequal extension and shearing strains, in equation (2-9.4):

From equations (2-5), the vanishing rotational strains implies that

$$\frac{\partial \mathbf{w}(x,y,z)}{\partial y} = \frac{\partial \mathbf{v}(x,y,z)}{\partial z} \qquad (2\text{-}9.10)$$

$$\frac{\partial \mathbf{u}(x,y,z)}{\partial z} = \frac{\partial \mathbf{w}(x,y,z)}{\partial x} \qquad (2\text{-}9.11)$$

$$\frac{\partial \mathbf{v}(x,y,z)}{\partial x} = \frac{\partial \mathbf{u}(x,y,z)}{\partial y} \qquad (2\text{-}9.12)$$

First, equations (2-9.10 through 12) imply the existence of a **total differential**

$$u\mathrm{d}x + v\mathrm{d}y + w\mathrm{d}z \qquad (2\text{-}9.13)$$

where, u, v, and w are related by a certain function $\Phi(x,y,z)$ such that

$$\mathbf{u}(x,y,z) = \frac{\partial \Phi(x,y,z)}{\partial x} \quad \mathbf{v}(x,y,z) = \frac{\partial \Phi(x,y,z)}{\partial y} \quad \mathbf{w}(x,y,z) = \frac{\partial \Phi(x,y,z)}{\partial z} \qquad (2\text{-}9.14)$$

$\Phi(x,y,z)$ describes a condition of **pure deformation**, without rotation, governed by equations (2-9.10 through 12)

2.4. Equations of continuity of strain components (Saint –Venant's equation)

The six equations of strains, (2-2) and (2-3), must coalesce into three compatible Cartesian relationships of the components of strains such that all geometrical rules are satisfied in all directions of normal and shearing strains. It is evident from the those six equations that upon unifying the common factors of derivatives, we could obtain the **conditions of compatibility** of six components of stress, in the same manner we dealt with the equilibrium equations of stress, equations (2-4).

(i) First, to find the common derivatives between the **extension strains**, equations (2-2.4) and (2-2.5), we will

> differentiate ε_{xx} twice with respect to y, and
> differentiate ε_{yy} twice with respect to x.

Thus, we create the common derivates with respect to x and y, as follows.

$$\frac{\partial^2 \varepsilon_{xx}}{\partial y^2} = \frac{\partial^2}{\partial y^2}\left(\frac{\partial \mathbf{u}(x,y,z)}{\partial x}\right) = \frac{\partial^2}{\partial y \partial x}\left(\frac{\partial \mathbf{u}(x,y,z)}{\partial y}\right) \qquad (2\text{-}10.1)$$

$$\frac{\partial^2 \varepsilon_{yy}}{\partial x^2} = \frac{\partial^2}{\partial x^2}\left(\frac{\partial \mathbf{v}(x,y,z)}{\partial y}\right) = \frac{\partial^2}{\partial y \partial x}\left(\frac{\partial \mathbf{v}(x,y,z)}{\partial x}\right) \qquad (2\text{-}10.2)$$

Adding the two equations (2-10.1) and (2-10.2) and substituting by 2 α_{xy}, from (2-3.1), we get

$$\frac{\partial^2 \varepsilon_{xx}}{\partial y^2} + \frac{\partial^2 \varepsilon_{yy}}{\partial x^2} = \frac{\partial^2}{\partial y \partial x}\left[2\alpha_{xy}\right] \qquad (2\text{-}10.3)$$

From equations (2-10 and (2-4), we could easily discern the pattern of appearance of shearing stresses α's in differential equations of the extension strains ε's. We can perform the similar differentiations of the remaining equations of (2-2.4), to obtain the continuity equations as follows:

$$\frac{\partial^2 \varepsilon_{yy}}{\partial x^2} + \frac{\partial^2 \varepsilon_{xx}}{\partial y^2} = 2\frac{\partial^2 \alpha_{xy}}{\partial x \partial y} \qquad (2\text{-}11.1)$$

$$\frac{\partial^2 \varepsilon_{zz}}{\partial y^2} + \frac{\partial^2 \varepsilon_{yy}}{\partial z^2} = 2\frac{\partial^2 \alpha_{yz}}{\partial y \partial z} \qquad (2\text{-}11.2)$$

$$\frac{\partial^2 \varepsilon_{xx}}{\partial z^2} + \frac{\partial^2 \varepsilon_{zz}}{\partial x^2} = 2\frac{\partial^2 \alpha_{zx}}{\partial z \partial x} \qquad (2\text{-}11.3)$$

(ii) Second, to find the common derivatives between **shearing strains,** equations (2-3), we will

differentiate α_{xy} with respect to z,
differentiate α_{xz} with respect to y, and
differentiate α_{zy} with respect to x.

Thus, we get

$$2\frac{\partial \alpha_{xy}}{\partial z} = \frac{\partial^2 \mathbf{u}(x,y,z)}{\partial y \partial z} + \frac{\partial^2 \mathbf{v}(x,y,z)}{\partial z \partial x} \qquad (2\text{-}12.1)$$

$$2\frac{\partial \alpha_{yz}}{\partial x} = \frac{\partial^2 \mathbf{v}(x,y,z)}{\partial z \partial x} + \frac{\partial^2 \mathbf{w}(x,y,z)}{\partial x \partial y} \qquad (2\text{-}12.2)$$

$$2\frac{\partial \alpha_{zx}}{\partial y} = \frac{\partial^2 \mathbf{w}(x,y,z)}{\partial x \partial y} + \frac{\partial^2 \mathbf{u}(x,y,z)}{\partial y \partial z} \qquad (2\text{-}12.3)$$

Adding equations (2-12.1) and (2-12.2), then from the sum, subtracting (2-12.3), we get

$$\frac{\partial \alpha_{xy}}{\partial z} + \frac{\partial \alpha_{yz}}{\partial x} - \frac{\partial \alpha_{zx}}{\partial y} = \frac{\partial^2 \mathbf{v}(x,y,z)}{\partial z \partial x} \qquad (2\text{-}12.4)$$

We could link the derivatives of the shearing strains in (2-12.4) with the derivatives of extension strains in equation (2-2), by

differentiating (2-12.4) with respect to y and
differentiating (2-2.5) with respect to x and z

94

Thus, equation (2-12.4) becomes

$$\frac{\partial^2 \alpha_{xy}}{\partial y \partial z} + \frac{\partial^2 \alpha_{yz}}{\partial x \partial y} - \frac{\partial^2 \alpha_{zx}}{\partial y^2} = \frac{\partial^3 \mathbf{v}(x,y,z)}{\partial z \partial x \partial y} \qquad (2\text{-}12.5)$$

And equation (2-2.5) becomes

$$\frac{\partial^2 \varepsilon_{yy}}{\partial x \partial z} = \frac{\partial^3 \mathbf{v}(x,y,z)}{\partial x \partial y \partial z}. \qquad (2\text{-}12.6)$$

From (2-12.5) and (2-12.6), we establish link between the derivatives of extension strains and shearing strains as follows

$$\frac{\partial^2 \alpha_{xy}}{\partial y \partial z} + \frac{\partial^2 \alpha_{yz}}{\partial x \partial y} - \frac{\partial^2 \alpha_{zx}}{\partial y^2} = \frac{\partial^2 \varepsilon_{yy}}{\partial x \partial z} \qquad (2\text{-}12.7)$$

Similarly, we can derive the other two equations, from equations (2-12.1 through 3) in the same manner we obtained (2-12.7), such that

$$\frac{\partial}{\partial x}\left(\frac{\partial \alpha_{xy}}{\partial z} + \frac{\partial \alpha_{xz}}{\partial y} - \frac{\partial \alpha_{yz}}{\partial x} \right) = \frac{\partial^2 \varepsilon_{xx}}{\partial y \partial z} \qquad (2\text{-}13.1)$$

$$\frac{\partial}{\partial y}\left(\frac{\partial \alpha_{xy}}{\partial z} + \frac{\partial \alpha_{yz}}{\partial x} - \frac{\partial \alpha_{zx}}{\partial y} \right) = \frac{\partial^2 \varepsilon_{yy}}{\partial x \partial z} \qquad (2\text{-}13.2)$$

$$\frac{\partial}{\partial z}\left(\frac{\partial \alpha_{yz}}{\partial x} + \frac{\partial \alpha_{xz}}{\partial y} - \frac{\partial \alpha_{xy}}{\partial z} \right) = \frac{\partial^2 \varepsilon_{zz}}{\partial x \partial y} \qquad (2\text{-}13.3)$$

2.5. Distribution of strain in a rigid body

The propagation of strain from initial points of deformation to neighboring points was discussed in terms of three tensors comprising the differential derivatives of infinitesimal changes of material dimensions, equations (2-8.2). We now need to describe the integration of those derivatives such that we describe **strain at arbitrary points in the rigid body**. Such integration is implied not only by the differential formulation of strains, but also by the multiplicity of directional strains.

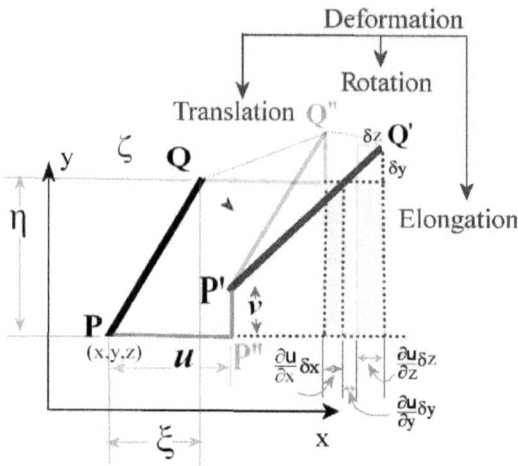

Figure 2-4. Contributions of elongation, translation, and rotation in the deformation of length segments

Before being displaced by deformation, let a line segment $\overline{PQ}$ passes between the material points

$$P(x,y,z) \text{ and } Q(x + \xi, y + \eta, z + \zeta), \qquad (2\text{-}16.1)$$

Deformation displaces initial point **P** to new point **P'** such that

$$P(x,y,z) \qquad \rightarrow \qquad P' \, (x + u, \, y + v, \, z + w). \qquad (2\text{-}16.2)$$

Then, strains propagate through the material points to displace the other end of the material length segment onto

$$Q(x + \xi, y + \eta, z + \zeta), \qquad \rightarrow \qquad Q' \, (x + \xi + \delta \, u, \, y + \eta + \delta \, v, \, z + \zeta + \delta \, w) \quad (2\text{-}16.3)$$

Therefore, before deformation, the position vector is

$$\overline{PQ}\langle \xi, \eta, \zeta \rangle . \qquad (2\text{-}16.3.1)$$

After deformation of the material rigid body containing the two point P and Q, the new position vector becomes

$$\overline{P'Q'}\langle \xi + \delta u, \eta + \delta v, \zeta + \delta w \rangle \qquad (2\text{-}16.3.2)$$

As shown in Figure 2-2, each of the three Cartesian projections of elongations receives three contributions. For example, the unit elongation δu, receives the following elongation contributions:

(i)   Extension strain along the x-direction given by $\dfrac{\partial u}{\partial x}\xi$ (since the derivative is unit elongation, multiplied by x-length of the segment; ξ)

96

(ii)   Shearing strain from the y-displacement, along the x-direction, given by $\dfrac{\partial \mathbf{u}}{\partial y}\eta$ (since the derivative is unit elongation, multiplied by y-length of the segment; η)

(iii)   Shearing strain from the z-displacement, along the x-direction, given by $\dfrac{\partial \mathbf{u}}{\partial z}\zeta$ (since the derivative is unit elongation, multiplied by z-length of the segment; ζ)

Therefore, the next infinitesimal deformation of the three projections of

$$\langle \delta u, \delta v, \delta w \rangle = \langle \delta\xi, \delta\eta, \delta\zeta \rangle = \overline{P'Q'}\langle \xi + \delta u, \eta + \delta v, \zeta + \delta w \rangle - \overline{PQ}\langle \xi, \eta, \zeta \rangle \qquad (2\text{-}16.4)$$

Are as follows

$$\delta\xi = \delta u = \frac{\partial \mathbf{u}}{\partial x}\xi + \frac{\partial \mathbf{u}}{\partial y}\eta + \frac{\partial \mathbf{u}}{\partial z}\zeta \qquad (2\text{-}16.4.1)$$

$$\delta\eta = \delta v = \frac{\partial \mathbf{v}}{\partial x}\xi + \frac{\partial \mathbf{v}}{\partial y}\eta + \frac{\partial \mathbf{v}}{\partial z}\zeta \qquad (2\text{-}16.4.2)$$

$$\delta\zeta = \delta w = \frac{\partial \mathbf{w}}{\partial x}\xi + \frac{\partial \mathbf{w}}{\partial y}\eta + \frac{\partial \mathbf{w}}{\partial z}\zeta \qquad (2\text{-}16.4.3)$$

From equation (1-18), we have

$$\xi = \mathbf{p}l,\ \eta = \mathbf{p}m,\ \zeta = \mathbf{p}n$$

Therefore, equations (2-16.4) bear great resemblance of equations (24), except here, strain is asymmetric, by virtue of equations (2-5). i.e.,

$$\frac{\partial \mathbf{w}(x,y,z)}{\partial y} \neq \frac{\partial \mathbf{v}(x,y,z)}{\partial z} \quad \text{if } |\omega_x| > 0 \qquad (2\text{-}17.1)$$

$$\frac{\partial \mathbf{u}(x,y,z)}{\partial z} \neq \frac{\partial \mathbf{w}(x,y,z)}{\partial x} \quad \text{if } |\omega_y| > 0 \qquad (2\text{-}17.2)$$

$$\frac{\partial \mathbf{v}(x,y,z)}{\partial x} \neq \frac{\partial \mathbf{u}(x,y,z)}{\partial y} \quad \text{if } |\omega_z| > 0 \qquad (2\text{-}17.3)$$

Thus, we could conclude the **general strain ellipsoidal equation** in the same fashion we followed in deriving equation (1-19), as follows

$$\mathbf{p}^2\varepsilon_p = \varepsilon_{xx}\xi^2 + \varepsilon_{yy}\eta^2 + \varepsilon_{zz}\zeta^2 + 2\left(\alpha_{xy}\xi\eta + \alpha_{zy}\eta\zeta + \alpha_{xz}\xi\zeta\right) \qquad (2\text{-}18.1)$$

Or

$$\varepsilon_p = \frac{\partial \boldsymbol{u}}{\partial x}l^2 + \frac{\partial \boldsymbol{v}}{\partial y}m^2 + \frac{\partial \boldsymbol{w}}{\partial z}n^2 + \left(\frac{\partial \boldsymbol{u}}{\partial y} + \frac{\partial \boldsymbol{v}}{\partial x}\right)lm + \left(\frac{\partial \boldsymbol{v}}{\partial z} + \frac{\partial \boldsymbol{w}}{\partial y}\right)mn + \left(\frac{\partial \boldsymbol{u}}{\partial z} + \frac{\partial \boldsymbol{w}}{\partial x}\right)nl \qquad (2\text{-}18.2)$$

The **strain surface** in equation (2-18) determines the nine components of strain at the tip of vector ***p*** drawn for a point of reference to an arbitrary point in the rigid body, in same fashion as the stress equation of Cauchy determines the stress, Figure 1-19.

Example 14

(i)   Find the equation of the plane passing through the points P(1,-1,3), Q(2-4,1,-2) and R(-1,-1,1) in an elastic body.

(ii)   Find the second plane that is perpendicular to the above plane and intersects it at the line PQ.

(iii)   Assume the following infinitesimal deformation apply to dimension of the given region

$$u\,(x,y,z) = 0.1\ x + 0.2\ y + 0.3\ z \qquad\qquad (2\text{-}19.1)$$
$$v(x,y,z) =\ 0.05\ x + 0.1\ y + 0.15\ z \qquad\qquad (2\text{-}19.2)$$
$$w(x,y,z) = 0.1\ x\ \text{-}\ 0.1\ y \qquad\qquad (2\text{-}19.3)$$

(iii.a) Find the equations of the two deformed planes
(iii.b) Discuss the changes in the normals to the planes after deformation.

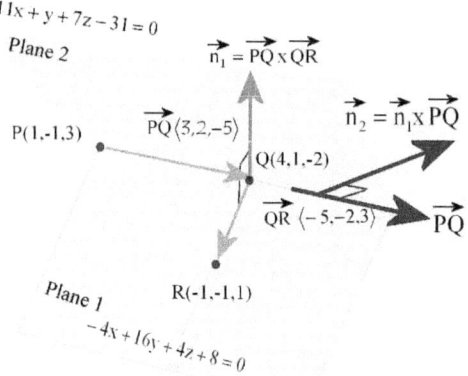

Figure 2-5. Analytic geometry of plane equations. Determining the equation of a plane, normal vector, and perpendicular plane, line of intersection, before and after homogeneous deformation.

Solution

(i) The **position vectors** for the lines connecting points P and Q and Q and R, before deformation, are obtained by from the differences of their projections along the three axes as follows

98

$$\overline{PQ} = \langle 4-1,1+1,-2-3\rangle = \langle 3,2,-5\rangle, \qquad \left|\overline{PQ}\right| = \sqrt{3^2+2^2+5^2} = 6.164414 \tag{2-19.4}$$

$$\overline{QR} = \langle -1-4,-1-1,1+2\rangle = \langle -5,-2,3\rangle, \qquad \left|\overline{QR}\right| = \sqrt{5^2+2^2+3^2} = 6.164414 \tag{2-19.5}$$

Where, the **angled brackets** denote vector with three projections along the three Cartesian coordinates, and signs of projections refer to the direction on each axis.

The vectors normal on the position vectors $\overline{PQ}$ and $\overline{QR}$ are obtained as follows:

$$\overline{n}_1 = \overline{PQ} \times \overline{QR} = \begin{vmatrix} \hat{i} & \hat{j} & \hat{k} \\ 3 & 2 & -5 \\ -5 & -2 & 3 \end{vmatrix} = \langle -4,16,4\rangle = \langle a,b,c\rangle$$

Or, $\qquad \overline{n}_1 = \langle -4,16,4\rangle \tag{2-19.6}$

Thus, the equation of the plane that passes through the three points P, Q, and R is constructed by any of the three points P, Q, or R and the projections of the normal to the plane, (2-19.6) as follows

$$ax + by + cz + d = 0 \tag{2-19.7.1}$$

Let us use P(1,-1,3) and $\overline{n}_1 = \langle -4,16,4\rangle$ to determine the constant d. Thus

$$-4(1)+16(-1)+4(3)+d = 0, \text{ or } d = 8$$

Thus, the equation of the plane is

$$-4x +16y + 4z + 8 = 0 \tag{2-19.7.2}$$

(ii) We now turn to the plane perpendicular to that of (2-19.7.2), passing through PQ.

The vector normal to the line $\overline{PQ}$ is given by

$$\overline{n}_2 = \overline{n}_1 \times \overline{PQ} = \begin{vmatrix} \hat{i} & \hat{j} & \hat{k} \\ -4 & 16 & 4 \\ 3 & 2 & -5 \end{vmatrix} = \langle -80-8,-(20-12),-8-48\rangle = \langle -11,-8,-56\rangle = -8\langle 11,1,7\rangle$$

Or, $\bar{n}_2 = -8\langle 11,1,7\rangle$ (2-19.8)

Thus, the plane passing through PQ and normal to (2-19.7.2) is given by the point P(1,-1,3) and the vector normal $\bar{n}_2 = -8\langle 11,1,7\rangle$

$$11(x-1)+(y+1)+7(z-3)=0$$
$$11x+y+7z-11+1-21=0$$
Or, $\quad 11x+y+7z-31=0$ (2-19.9)

To verify our formulas, we should prove that the **dot product** of the two normals of the two planes vanishes and that the two planes pass through the given points. Thus

$$\overline{n_\uparrow} \bullet \overline{n} = 11(-4)+1(16)+4(7) = -44+16+28 = 0$$ (2-19.9.1)

The line of intersection of plane in (2-19.7.2) and that in (2-19.9) must pass through the two points P and Q. That line is obtained by subtracting the two equations to get

Multiply by 16: $11x+y+7z-31=0$
$$-4x+16y+4z+8=0$$

$$(16X11)x+16y+16X7)z-16x31=0$$
$$-4x+16y+4z+8=0$$
Subtract: $180x+108z-16x31-8=-180x+108z-8(62+1)=0$

Therefore, $5x+3z-14=0$ (2-19.9.2)

Equation (2-19.9.2) must pass by $\overline{PQ}=\langle 4-1,1+1,-2-3\rangle = \langle 3,2,-5\rangle$, which equation is obtained as follows:

$$\frac{x-1}{3}=\frac{y+1}{2}=\frac{z-3}{-5}$$
$$-5(x-1)=3(z-3)$$
$$5x+3z-14=0$$ (2-19.9.3)

From (2-19.9.2) and (2-19.9.3), we are now assured that our obtained solution is consistent with two perpendicular planes (2-19.7.2) and (2-19.9).

(iii) We now determine the **nine components of strain** from equations (2-19.1 through 3) by performing partial differentiation as follows:

$$\frac{\partial u}{\partial x} = 0.1 \qquad \frac{\partial u}{\partial y} = 0.2 \qquad \frac{\partial u}{\partial z} = 0.3$$

$$\frac{\partial v}{\partial x} = 0.05 \qquad \frac{\partial v}{\partial y} = 0.1 \qquad \frac{\partial v}{\partial z} = 0.15 \qquad\qquad (2\text{-}19.10)$$

$$\frac{\partial w}{\partial x} = 0.1 \qquad \frac{\partial w}{\partial y} = -0.1 \qquad \frac{\partial w}{\partial z} = 0$$

(iii.a) Deformations of the line segments $\overline{PQ}$ and $\overline{QR}$

Substituting by the strains from (2-19.10) into (2-19.4) and (2-19.5), recognizing that strains must be integrated over the distances from the original points of deformation, we get

$$P'Q' = \left\langle \left(3 + 3\frac{\partial u}{\partial x} + 2\frac{\partial u}{\partial y} - 5\frac{\partial u}{\partial z}\right), \left(2 + 3\frac{\partial v}{\partial x} + 2\frac{\partial v}{\partial y} - 5\frac{\partial v}{\partial z}\right), \left(-5 + 3\frac{\partial w}{\partial x} + 2\frac{\partial w}{\partial y} - 5\frac{\partial w}{\partial z}\right) \right\rangle$$

$$= \left\langle (3 + 0.3 + 0.4 - 1.5), (2 + 0.15 + 0.2 - 0.75), -5 + 0.3 - 0.2 \right\rangle = \langle 2.2, 1.6, -4.9 \rangle$$

Or,

$$P'Q' = \langle 2.2, 1.6, -4.9 \rangle, \qquad \left| P'Q' \right| = 5.604463 \qquad\qquad (2\text{-}19.11)$$

$$Q'R' = \left\langle \left(-5 - 5\frac{\partial u}{\partial x} - 2\frac{\partial u}{\partial y} + 3\frac{\partial u}{\partial z}\right), \left(-2 - 5\frac{\partial v}{\partial x} - 2\frac{\partial v}{\partial y} + 3\frac{\partial v}{\partial z}\right), \left(3 - 5\frac{\partial w}{\partial x} - 2\frac{\partial w}{\partial y} + 3\frac{\partial w}{\partial z}\right) \right\rangle$$

$$= \left\langle -5 - 0.5 - 0.2 + 0.9, -2 - 0.25 - 0.2 + 0.45, 3 - 0.5 + 0.2 \right\rangle = \langle -4.8, -2, 2.7 \rangle$$

Or,

$$Q'R' = \langle -4.8, -2, 2.7 \rangle, \qquad \left| Q'R' \right| = 5.93043 \qquad\qquad (2\text{-}19.12)$$

(iii.b) Deformations of the normals to the planes

The new normal on the deformed plane through the three given points is obtained similar to (2-19.6), as follows

$$\overline{n}'_1 = \overline{P'Q'} \times \overline{Q'R'} = \begin{vmatrix} \hat{i} & \hat{j} & \hat{k} \\ 2.2 & 1.6 & -4.9 \\ -4.8 & -2 & 2.7 \end{vmatrix} = \left\langle \begin{array}{l} (1.6)(2.7) - (4.9)(2), \\ -(2.2)(2.7) + (4.49)(4.8), \\ -(2.2)(2) + (1.6)(4.8), \end{array} \right\rangle$$

Or,

$$\overline{n}'_1 == \langle -5.48, 15.342, 3.28 \rangle \qquad\qquad (2\text{-}19.13)$$

101

Similarly, the normal to the perpendicular deformed surface is obtained as in (2-19.8) as follows

$$\overline{n}'_2 = \overline{n}'_1 \times \overline{P'Q'} = \begin{vmatrix} \hat{i} & \hat{j} & \hat{k} \\ -5.48 & 15.342 & 3.28 \\ 2.2 & 1.6 & -4.9 \end{vmatrix} = \left\langle \begin{array}{c} -15.342(4.9) - 1.6(3.28), \\ -(5.48(4.9) - 3.28(2.2)), \\ -5.48(1.6) - 15.342(2.2) \end{array} \right\rangle$$

$$\overline{n}'_2 = \langle -80.4238, 19.636, -42.5204 \rangle \tag{2-19.14}$$

The dot product of $\overline{n}'_1 . \overline{n}'_2$ gives the cosine of the angle between the two normals. Therefore,

$$\overline{n}_1 \bullet \overline{n}_2 = \frac{(-5.48)(-80.4238) + (15.342)(19.636) + (3.28)(-42.5204)}{\sqrt{\left[(-5.48)^2 + (15.342)^2 + (3.28)^2\right]\left[(-80.4238)^2 + (19.636)^2 + (-42.5204)^2\right]}} = 0.065709$$

Therefore, the new angle between the two planes =

$$\cos^{-1}(0.065709) = 86.23241° = 1.50504 \text{ radians} \tag{2-19.15.1}$$

Thus, the two perpendicular planes, before deformation, experienced rotation from 90° to the angle 86.23241°.

Net elongations in line segments:

$$\overline{PQ} - \overline{P'Q'} = 6.164414 - 5.604463 = 0.559951 \tag{2-19.15.2}$$

$$\overline{QR} - \overline{Q'R'} = 6.164414 - 5.93043 = 0.233984 \tag{2-19.15.3}$$

2.6. Principal planes and principal surface of extensional strain

We have studied the general stress equation (1-19), which shares great deal of similarity with the relative strain equation (2-18) as follows:

1. Both equations can be reduced into two tensor, deviator and spherical, equations (2-8.2) and (2-31). Thus, we can analyze principal elongation strain and principal stress in separation of shearing strains and shearing stresses.
2. Both equations can degenerate into four different geometrical surfaces based on the signs and balances of the principal elongations or principal stresses.
3. Both equations have three main orthogonal axes, along which shearing strains and shearing stresses vanish, are called principal surfaces of principal stresses and principal strains.
4. Both equations are tools for determining the stress or strain distribution in the interior of the rigid body in terms of given known external surfaces. The two Cauchy equations

enable engineers to visualize the distributions of stresses and strains in crucial design planes.

5. The deviator tensor of the two equations has three characteristic invariants that describe three useful features of the ellipsoid equations. The first invariant of the deviator vanishes under the equilibrium of hydrostatic stress and principal elongation. The second invariant describes the averages of squared deviations from main hydrostatic stress and main elongation strains. The third invariant of the deviator of the two equations describes the average of cubic deviations.

6. The relative strain equation (2-18) differs from (1-19) in its asymmetric tensor due to the rotational strains. Those characterize the nature of the material, which are missing in stress effects.

7. Both stress- and strain- ellipsoidal equations conform to continuity conditions, equations (2-4) and (2-11) and (2-13).

8. The extension strain tensor, equation (2-8.2), comprises an invariant **dilatational strain** as it is tied strictly to the material volume. Since shearing strains reduce material size of infinitesimal volumes to higher order, we could safely conclude that only elongation strains of the spherical tensor ε_{xx}, ε_{yy}, ε_{zz}, (2-8.2), contribute to change of volume.

The relation between the dilatational strain and the volume of the deformed body is obtained as follows.

Let an initial infinitesimal volume, $d\tau = dxdydz$, deform by change in its dimensions as follows

$$d\tau + \delta\,(d\tau) = dx\,(1+\varepsilon_{xx})\,dy\,(1+\varepsilon_{yy})\,dx(1+\varepsilon_{zz}) \qquad (2\text{-}20.1)$$
$$d\tau + \delta\,(d\tau) = dx\,dy\,dy\,(1+\varepsilon_{xx}+\varepsilon_{yy}+\varepsilon_{zz}\,\varepsilon_{yy} +\varepsilon_{zz}\,\varepsilon_{xx} + \varepsilon_{xx}\,\varepsilon_{yy}\,\varepsilon_{zz})$$

Dismissing the multipliers of smaller quantities, and remembering that $d\tau = dxdydz$, we get

$$\delta\,(d\tau) = dx\,dy\,dy\,(\varepsilon_{xx}+\varepsilon_{yy}+\varepsilon_{zz}) \qquad (2\text{-}20.2)$$

Thus, that we can define the **dilatational strain,** Θ_ε, as follows

i.e., $$\Theta_\varepsilon = \frac{\delta(d\tau)}{d\tau} = \varepsilon_{xx} + \varepsilon_{yy} + \varepsilon_{zz} \qquad (2\text{-}20.3)$$

Substitute by the relationships of elongations strains from equations (2-2), we get

$$\frac{\delta(d\tau)}{d\tau} = \frac{\partial u(x,y,z)}{\partial x} + \frac{\partial v(x,y,z)}{\partial z} + \frac{\partial w(x,y,z)}{\partial z} \qquad (2\text{-}20.4)$$

From equation (2-20.3), we observe that the **dilatational strain** Θ_ε takes the form of the triple hydrostatic **stress**, 3 N_0, equation, (2-32.4).

103

2.7. Deviator and spherical strain tensors and invariants

In lieu of repeating the same steps followed in deriving equations (2-33), (2-34), and (2-35) for the invariants of the general stress tensor, we will outline the similar formulas. Start by defining the average dilatation strain by

$$\varepsilon_0 = \frac{1}{3}\left(\varepsilon_{xx} + \varepsilon_{yy} + \varepsilon_{zz}\right) \tag{2-21}$$

2.7.1. Vanishing deviator of the first invariant of the relative strain tensor

(i) Deviator $D_{\Theta\varepsilon}$ of first invariant Θ :

In the case of first invariant of stress, equation (2-33.2) gives

$$D_\Theta = N_1 + N_2 + N_3 - 3N_0 == \left(\begin{array}{l}\text{Vanishing stresses, i.e., no deviation}\\ \text{from the hydrostatic stress } N_0.\end{array}\right) \tag{2-22}$$

Therefore, similar first invariant for strain takes the form

$$D_{\Theta\varepsilon} = \varepsilon_{xx} + \varepsilon_{yy} + \varepsilon_{zz} - 3\varepsilon_0 = 0 = \left(\begin{array}{l}\text{Vanishing strains, i.e., no deviation}\\ \text{from the dilatational strain } \varepsilon_0.\end{array}\right) \tag{2-23}$$

2.7.2. Squared deviations of the second invariant of the relative strain tensor

(ii) Deviator $D_{\Phi\varepsilon}$ of second invariant Φ :

In the case of first invariant of stress, equation (2-34.3) gives

$$D_\Phi = -\frac{1}{6}\left[(N_1 - N_2)^2 + (N_1 - N_3)^2 + (N_2 - N_3)^2\right]$$

And

$$D_\Phi = -\frac{3}{2}\left[\frac{(N_1 - N_0)^2 + (N_2 - N_0)^2 + (N_3 - N_0)^2}{3}\right] = -\frac{3}{2}\left(\begin{array}{l}\text{Average of sqaures of}\\ \text{deviations of } N_1, N_2, \text{and } N_3,\\ \text{from hydrostatè stess, } N_0\end{array}\right)$$

Therefore, similar second invariant for strain takes the form

$$D_{\Phi\varepsilon} = -\frac{1}{6}\left[(\varepsilon_{xx} - \varepsilon_{yy})^2 + (\varepsilon_{xx} - \varepsilon_{zz})^2 + (\varepsilon_{yy} - \varepsilon_{zz})^2\right] \tag{2-24.1}$$

And

$$D_{\Phi x} = -\frac{3}{2}\left[\frac{(\varepsilon_{xx} - \varepsilon_0)^2 + (\varepsilon_{yy} - \varepsilon_0)^2 + (\varepsilon_{zz} - \varepsilon_0)^2}{3}\right] = -\frac{3}{2}\left(\begin{array}{l}\text{Average of sqaures of}\\ \text{deviations of } \varepsilon_{xx}, \varepsilon_{yy}, \varepsilon_{zz}\\ \text{from dilatatiomal straion,} \varepsilon_0\end{array}\right) \tag{2-24.2}$$

2.7.3. Cubic deviations of the third invariant of the relative strain tensor

(iii) Deviator $D_{\Xi\varepsilon}$ of second invariant Ξ :

In the case of third invariant of stress, equation (2-35.2) gives

$$D_{\Sigma} = \frac{1}{3}\left[\left(N_1 - N_0\right)^3 + \left(N_2 - N_0\right)^3 + \left(N_3 - N_0\right)^3 = \begin{pmatrix} \textbf{Average of cubes of} \\ \textbf{deviations of } N_1, N_2, \textbf{and } N_3, \\ \textbf{from hydrostatic stess, } N_0 \end{pmatrix}\right]$$

$$D_{\Sigma\varepsilon} = \frac{1}{3}\left[\left(\varepsilon_{xx} - \varepsilon_0\right)^3 + \left(\varepsilon_{xx} - \varepsilon_0\right)^3 + \left(\varepsilon_{yy} - \varepsilon_0\right)^3 = \begin{pmatrix} \text{Average of cubes of} \\ \text{deviations of } \varepsilon_{xx}, \varepsilon_{xx}, \varepsilon_{yy} \\ \text{from dilatational strain}, \varepsilon_0 \end{pmatrix}\right] \qquad (2\text{-}25)$$

Example 15

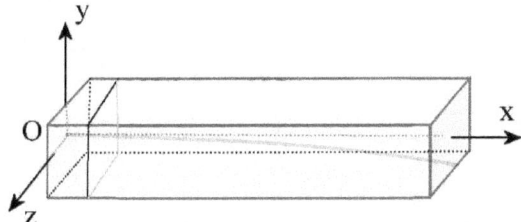

Figure 2-6. Geometry of bending a rod the xOy plane.

In Figure 2-6, a rod is bent in the xOy plane.

    (i)   Write the equation of deflection of the Ox axis.
    (ii)   Determine the condition for built-in end at $x = 0$ when no bending occurs on the z or z axes

Solution

(i) Equation of the Ox line before deflection

$$\frac{x - x_0}{a} = \frac{y - y_0}{b} = \frac{z - z_0}{c} \qquad (2\text{-}26.1)$$

$$\frac{x - x_0}{a} = \frac{y}{b} = \frac{z}{c} = t \qquad (2\text{-}26.2)$$

$$x' = \left(x + \frac{\partial u}{\partial x} x + \frac{\partial u}{\partial y} 0 + \frac{\partial u}{\partial z} 0 \right) = x + \frac{\partial u}{\partial x} x \qquad (2\text{-}26.3)$$

$$y' = 0 + \frac{\partial v}{\partial x} x + \frac{\partial v}{\partial y} 0 + \frac{\partial v}{\partial z} 0 = \frac{\partial v}{\partial x} x \qquad (2\text{-}26.4)$$

The equation of the bent axis is obtained by eliminating x between equations (2-26.3) and (2-26.4), we get

$$y' = \left(\frac{\frac{\partial v}{\partial x}}{1 + \frac{\partial u}{\partial x}} \right) x' \qquad (2\text{-}26.5)$$

Example 16

Given the displacements

$$u = -\frac{xy}{\rho}$$
$$v = \frac{\sigma yz}{\rho} \qquad (2\text{-}27.1)$$
$$w = \frac{x^2 + \sigma\left(y^2 - z^2\right)}{2\rho}$$

Solution

$$\frac{\partial u}{\partial x} = -\frac{y}{\rho}, \quad \frac{\partial u}{\partial y} = -\frac{x}{\rho}, \quad \frac{\partial u}{\partial z} = 0$$
$$\frac{\partial v}{\partial x} = 0, \quad \frac{\partial v}{\partial y} = \frac{\sigma z}{\rho}, \quad \frac{\partial v}{\partial z} = \frac{\sigma y}{\rho}$$
$$\frac{\partial w}{\partial x} = \frac{x}{\rho}, \quad \frac{\partial w}{\partial y} = \frac{\sigma y}{\rho}, \quad \frac{\partial w}{\partial z} = -\frac{\sigma z}{\rho}$$

Dilatation strain

$$\varepsilon_0 = \frac{\partial u}{\partial x} + \frac{\partial v}{\partial y} + \frac{\partial w}{\partial x} = -\frac{y}{\rho} + \frac{\sigma z}{\rho} - \frac{\sigma z}{\rho} = -\frac{y}{\rho}$$

$$\varepsilon_{xx} = \frac{\partial u}{\partial x} = -\frac{y}{\rho},$$
$$\varepsilon_{yy} = \frac{\partial v}{\partial y} = \frac{\sigma z}{\rho}$$
$$\varepsilon_{zz} = \frac{\partial w}{\partial z} = -\frac{\sigma z}{\rho}$$

$$2\alpha_{xy} = \frac{\partial u}{\partial y} + \frac{\partial v}{\partial x} = -\frac{x}{\rho}$$

$$2\alpha_{yz} = \frac{\partial v}{\partial z} + \frac{\partial w}{\partial y} = 2\frac{\sigma y}{\rho}$$

$$2\alpha_{zx} = \frac{\partial u}{\partial z} + \frac{\partial w}{\partial x} = 0 + \frac{x}{\rho} = \frac{x}{\rho}$$

Deflection (2-26.5)

$$y' = \left(\frac{\dfrac{\partial v}{\partial x}}{1 + \dfrac{\partial u}{\partial x}}\right) x'$$

$$x' = x + \frac{\partial u}{\partial x} x = x\left(1 - \frac{y}{\rho}\right)$$

2.8. Reciprocity

We will generalize the notation of geometrical analysis in Example 14 in order to study the reciprocity of shearing stresses and trains between adjacent surfaces.

In an elastic body, consider of the three points

$$\begin{aligned}&P(x_p, y_p, z_p)\\&Q(x_q, y_q, z_q)\\&R(x_r, y_r, z_r)\end{aligned}$$
(2-26)

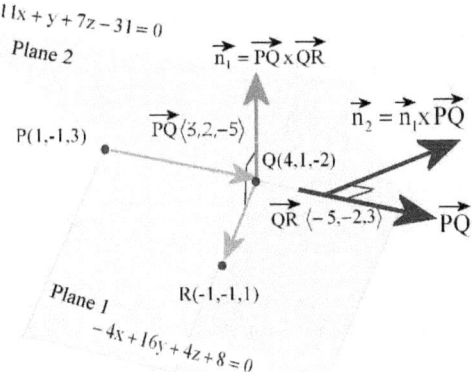

Figure 2-7. Analytic geometry of plane equations. Determining the equation of a plane, normal vector, and perpendicular plane, line of intersection, before and after homogeneous deformation.

(i) The **position vectors** for the lines connecting points P and Q and Q and R, are obtained by from the differences of their projections along the three axes as follows

$$\overline{PQ} = \left\langle x_q - x_p, y_q - y_p, z_q - z_p \right\rangle \tag{2-26.1}$$

$$\left| \overline{PQ} \right| = \sqrt{\left(x_q - x_p\right)^2 + \left(y_q - y_p\right)^2 + \left(z_q - z_p\right)^2} \tag{2-26.2}$$

Similarly, we get

$$\overline{QR} = \left\langle x_r - x_q, y_r - y_q, z_r - z_q \right\rangle \tag{2-26.3}$$

$$\left| \overline{QR} \right| = \sqrt{\left(x_r - x_q\right)^2 + \left(y_r - y_q\right)^2 + \left(z_r - z_q\right)^2} \tag{2-26.4}$$

The **vectors normal** on the position vectors $\overline{PQ}$ and $\overline{QR}$ are obtained as follows:

$$\overline{n}_1 = \overline{PQ} \times \overline{QR} = \begin{vmatrix} \hat{i} & \hat{j} & \hat{k} \\ x_q - x_p & y_q - y_p & z_q - z_p \\ x_p - x_r & y_p - y_r & z_p - z_r \end{vmatrix} \tag{2-26.5.1}$$

In the angled-bracket form, $\overline{n}_1$ takes the form

$$\overline{n}_1 = \left\langle \begin{matrix} \left(y_q - y_p\right)\left(z_p - z_r\right) - \left(y_p - y_r\right)\left(z_q - z_p\right) \\ \left(x_p - x_r\right)\left(z_q - z_p\right) - \left(x_q - x_p\right)\left(z_p - z_r\right) \\ \left(x_q - x_p\right)\left(y_p - y_r\right) - \left(x_p - x_r\right)\left(y_q - y_p\right) \end{matrix} \right\rangle = \left\langle a_1, b_1, c_1 \right\rangle \tag{2-26.5.2}$$

Where

$$\begin{aligned} a_1 &= \left(y_q - y_p\right)\left(z_p - z_r\right) - \left(y_p - y_r\right)\left(z_q - z_p\right) \\ b_1 &= \left(x_p - x_r\right)\left(z_q - z_p\right) - \left(x_q - x_p\right)\left(z_p - z_r\right) \\ c_1 &= \left(x_q - x_p\right)\left(y_p - y_r\right) - \left(x_p - x_r\right)\left(y_q - y_p\right) \end{aligned} \tag{2-26.5.3}$$

Thus, the equation of the plane that passes through the three points P, Q, and R is constructed by any of the three points P, Q, or R and the projections of the normal to the plane, (2-26.5.2) as follows

$$a_1 x + b_1 y + c_1 z + d_1 = 0 \tag{2-26.6}$$

Where, the constant d_1 can be determined by substituting from (2-26) and (2-26.5.3) by the Cartesian projections of point P and the normal vector $\overline{n}_1$.

(ii) We now turn to the <u>plane perpendicular</u> to that of (2-26.6), passing through PQ.

The vector normal to the line $\overline{PQ}$ is given by

$$\overline{n}_2 = \overline{n}_1 \times \overline{PQ} = \begin{vmatrix} \hat{i} & \hat{j} & \hat{k} \\ a_1 & b_1 & c_1 \\ x_q - x_p & y_q - y_p & z_q - z_p \end{vmatrix} \tag{2-26.7.1}$$

In the angled-bracket form, $\overline{n}_2$ takes the form

$$\overline{n}_2 = \left\langle \begin{matrix} b_1(z_q - z_p) - c_1(y_q - y_p) \\ c_1(x_q - x_p) - a_1(z_q - z_p) \\ a_1(y_q - y_p) - b_1(x_q - x_p) \end{matrix} \right\rangle = \langle a_2, b_2, c_2 \rangle \tag{2-26.7.2}$$

Where,
$$\begin{aligned} a_2 &= b_1(z_q - z_p) - c_1(y_q - y_p) \\ b_2 &= c_1(x_q - x_p) - a_1(z_q - z_p) \\ c_2 &= a_1(y_q - y_p) - b_1(x_q - x_p) \end{aligned} \tag{2-26.7.3}$$

Thus, the plane passing through PQ and normal to (2-26.7.3) is given by

$$a_2 x + b_2 y + c_2 z + d_2 = 0 \tag{2-26.8}$$

Where, the constant d_2 can be determined by substituting from (2-26) and (2-26.7.3) by the Cartesian projections of point P and the normal vector $\overline{n}_2$.

(iii) Stresses on planes 1 and 2
Assume that total stresses σ_1 (σ_{1x}, σ_{1y}, σ_{1z}) and σ_2 (σ_{2x}, σ_{2y}, σ_{2z}). Each total stress imposes normal and shearing stresses on the plane it acts upon.

Since planes 1 and 2 contains the same material object, assume that the total stress, $F(F_x, F_y, F_z)$, that affects both surfaces originates from point $O(x_o, y_o, z_o)$.

The transfer of the total stress F from O to P_1 is given by equation (2-4)

$$F_{px} = F_x + \frac{\partial F_x}{\partial x}(x_p - x_o) + \frac{\partial F_x}{\partial y}(y_p - y_o) + \frac{\partial F_x}{\partial z}(z_p - z_o) \tag{2-26.17}$$

$$F_{py} = F_y + \frac{\partial F_y}{\partial x}(x_p - x_o) + \frac{\partial F_y}{\partial y}(y_p - y_o) + \frac{\partial F_y}{\partial z}(z_p - z_o) \tag{2-26.18}$$

$$F_{pz} = F_z + \frac{\partial F_z}{\partial x}(x_p - x_o) + \frac{\partial F_z}{\partial y}(y_p - y_o) + \frac{\partial F_z}{\partial z}(z_p - z_o) \tag{2-26.19}$$

109

Shear stress on plane 1 perpendicular to $\overline{PQ}$ and parallel to $\overline{n}_2$

$$\overline{F}_p * \overline{n}_2 = F_{px}.a_2 + F_{py}.b_2 + F_{pz}.c_2$$

(2-26.23)

Shear stress on plane 2 perpendicular to $\overline{PQ}$ and parallel to $\overline{n}_1$

$$\overline{F}_p * \overline{n}_1 = F_{px}.a_1 + F_{py}.b_1 + F_{pz}.c_1$$

(2-26.24)

$$\overline{F}_1 * \overline{n}_2 = a_2\left(F_x + \frac{\partial F_x}{\partial x}t_1 a_1 + \frac{\partial F_x}{\partial y}t_1 b_1 + \frac{\partial F_x}{\partial z}t_1 c_1 \right) +$$

$$+ b_2\left(F_y + \frac{\partial F_y}{\partial x}t_1 a_1 + \frac{\partial F_y}{\partial y}t_1 b_1 + \frac{\partial F_y}{\partial z}t_1 c_1 \right)$$

$$+ c_2\left(F_z + \frac{\partial F_z}{\partial x}t_1 a_1 + \frac{\partial F_z}{\partial y}t_1 b_1 + \frac{\partial F_z}{\partial z}t_1 c_1 \right)$$

(2-26.25)

$$\overline{F}_2 * \overline{n}_1 = a_1\left(F_x + \frac{\partial F_x}{\partial x}t_2 a_2 + \frac{\partial F_x}{\partial y}t_2 b_2 + \frac{\partial F_x}{\partial z}t_2 c_2 \right) +$$

$$+ b_1\left(F_y + \frac{\partial F_y}{\partial x}t_2 a_2 + \frac{\partial F_y}{\partial y}t_2 b_2 + \frac{\partial F_y}{\partial z}t_2 c_2 \right)$$

$$+ c_1\left(F_z + \frac{\partial F_z}{\partial x}t_2 a_2 + \frac{\partial F_z}{\partial y}t_2 b_2 + \frac{\partial F_z}{\partial z}t_2 c_2 \right)$$

(2-26.26)

The transfer of the total stress F from O to S_2 is given by equation (2-4)

$$F_{2x} = F_x + \frac{\partial F_x}{\partial x}t_2 a_2 + \frac{\partial F_x}{\partial y}t_2 b_2 + \frac{\partial F_x}{\partial z}t_2 c_2$$

(2-26.20)

$$F_{2y} = F_y + \frac{\partial F_y}{\partial x}t_2 a_2 + \frac{\partial F_y}{\partial y}t_2 b_2 + \frac{\partial F_y}{\partial z}t_2 c_2$$

(2-26.21)

$$F_{2z} = F_z + \frac{\partial F_z}{\partial x}t_2 a_2 + \frac{\partial F_z}{\partial y}t_2 b_2 + \frac{\partial F_z}{\partial z}t_2 c_2$$

(2-26.22)

The perpendicular

Therefore, the normal stress on plane 1 by σ_1 (σ_{1x}, σ_{1y}, σ_{1z}) is obtain by its **dot product** with $\bar{n}_1$.

$$\bar{\sigma}_1 * \bar{n}_2 = \bar{\sigma}_2 * \bar{n}_2 \qquad\qquad (2\text{-}26.9)$$

CHAPTER 3

VOLUMETRIC HOOKE'S LAW

3.1. Scope of theoretical relation of cause (stress) and effect (strain)

The theoretical study of the relationship between the distribution of **static stress** and **geometrical strain** of material spatial dimensions is subject to the following assumption:

1. Effects of isolated stresses could be **linearly superimposed** to account for net effect of superimposed stresses.
2. Physical properties of elastic materials are divided into **homogeneous** and **heterogeneous** compositions, in regard to variation of structure at various points in the same body.
3. Homogeneous structures are farther divided into **isotropic** and **anisotropic**. For example, **crystals** show directional variation in their physical properties.
4. The cause and effect, of stress and deformation, is subject to the constraints of:

(i) **Tensile test curve** between the two, which is determined by experimental measurements of the properties of material. For example, the range of elastic behavior of the material before reaching the cut-off proportional limit determines the maximum allowed stress before irregular deformation ensues.

(ii) The **continuity of mass** and **space** of the body. For example, elongation of material in one direction imposes contraction in perpendicular directions, which will be assumed equal in isotropic material, thus rupture could occur despite working within the proportional range of tensile curve. Here, the volume of the body is assumed invariant.

5. Elasticity of materials can be farther classified as **ductile** and **brittle**. The tensile stress-strain curve for ductile materials, such as mild steels, shows linear relationship between stress and strain within the proportionality range. Brittle materials such as hard alloy steels, cast iron, and stone, have very narrow range or none deformation under stress.

3.2. The three components of Hooke's law

Hooke's law expresses stress in terms of elongation strain within the proportionality region of stress-stain curve. It has three components as follows[3]

---

[3] H. M. Ledbetter and R. P. Reed

(i) Young's normal stress-elongation strain:

$$\sigma_x = E\varepsilon_{xx} \tag{3-1.1}$$

Where **E** is **Young's modulus,** in Newton/m^2. This varies from 200 x 10^9 N/m^2 for steal to 3 x 10^9 N/m^2 for wood.

(ii) Poisson's lateral contraction-elongation strain

The elongated material contract laterally in relation to the elongation strain as follows:

$$\varepsilon_{yy} = \varepsilon_{zz} = -v\varepsilon_{xx} \tag{3-1.2}$$

Where v is **Poisson's ratio.**

Elastic Properties of Selected Engineering Materials

Material	Density (kg/m^3)	Young's Modulus 10^9 N/m^2	Ultimate Strength S$_u$ 10^6 N/m^2	Yield Strength S$_y$ 10^6 N/m^2
Steel	7860	200	400	250
Aluminum	2710	70	110	95
Glass	2190	65	50^b	.
Concrete	2320	30	40^b	.
Wood	525	13	50^b	.
Bone	1900	9^b	170^b	.
Polystyrene	1050	3	48	

Material Class	Poisson's Ratio v
Ceramics	0.2
Metals	0.3
Plastics	0.4
Rubber	0.5

(iii) Hooke's shear law:

"Elastic Properties of Metals and Alloys, 1. Iron, Nickel, and Iron. Nickel Alloys" J. Phys. Chem. Ref. Data, Vol. 2, No 3, (531-617) 1973

$$\tau_{xy} = G\varepsilon_{xy}$$
$$\tau_{yz} = G\varepsilon_{yz} \tag{3-1.3}$$
$$\tau_{zx} = G\varepsilon_{zx}$$

Where **G** is the **shear modulus of elasticity**.

3.3. Elastic properties of material

(i) Aluminum[4]

Aluminum is a perfect **ductile metal** that can be studied theoretically with great rapport between experimental observations and theoretical modeling by Hooke's law[5] for one-dimensional stress-strain relation, equation (3-1.1). The **tensile stress strain** curve is obtained from testing thin long rods. Figure 3-1. Thus, the effects of shear strains and lateral contraction arising from two or three-dimensional bodies are greatly reduced.

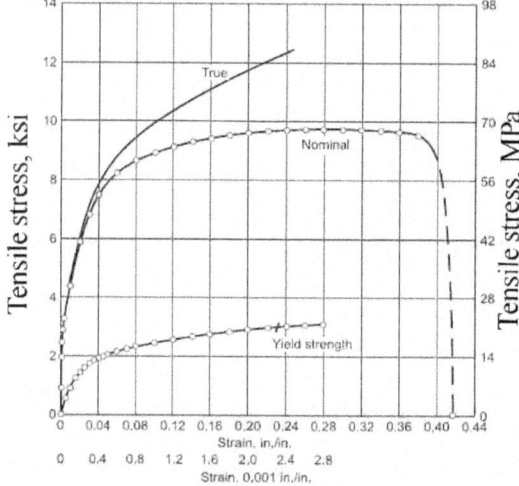

[4] Y. Tamarin, Atlas of Stress-Strain Curves, ASM International; 2 edition (November 1, 2002), 2nd Edition, ISBN-10: 087170739X.

[5] Aluminum and tungsten are the only cubic metals with known elastic coefficients, where the elastic anisotropy factor A≈ 1, since all shear moduli are equal. (source: H. M. Ledbetter and R. P. Reed)

114

Figure 3-1. Tensile stress-strain curve for aluminum. Test specimen diameter, 12.7 mm (0.5 in.). Gage length: 203.2 mm (8 in.). For this aluminum alloy (WA.002 1060-O) rod, the nominal tensile strength, 67.2 MPa (9.75 ksi). True tensile strength, 86.2 MPa (12.5 k Figure 3-1. Tensile stress-strain curve for aluminum. Test specimen diameter, 12.7 mm (0.5 in.). Gage length: 203.2 mm (8 in.). For this aluminum alloy (WA.002 1060-O) rod, the nominal tensile strength, 67.2 MPa (9.75 ksi). True tensile strength, 86.2 MPa (12.5 ksi). Nominal yield strength (0.2% offset), 21 MPa (3.0 ksi). Elongation (in 50.8 mm, or 2 in.), 42.7%. Reduction of area, 91%.

(Source: Alcoa, Aluminum Research Laboratory, New Kensington, PA, Oct 1951)

(ii) **Concrete** [6]

In contrast to ductile materials, concrete does not permit tensile testing of thin long rods due to its **brittle structure**. In Figure 3-1, relative **compressive stress-strain** curve for concrete columns was obtained in terms of surface displacement (not strain), which shows the dependence of the five curves on the geometry of the stressed concrete columns. Nevertheless, the three zones of behavior of concrete under compressive stress are clearly delineated: **elastic proportionality** region, **cut-off peak-stress** region, and **post-peak softening**, which vary from steep to gradual slopes. .

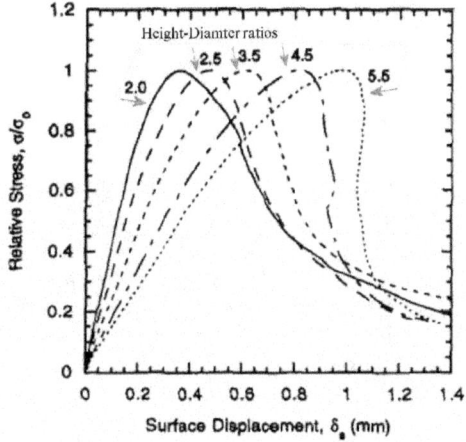

---

[6] Effect of Length on Compressive Strain Softening of Concrete,

Journal of Engineering Mechanics, Vol. 123, No. 1, January 1997, pp. 25-35, Figure 7.

Figure 3-2. Relative compressive stress plotted against surface displacements of concrete columns of various height-to-diameter ratios. (Effect of Length on Compressive Strain Softening of Concrete, Journal of Engineering Mechanics, Vol. 123, No. 1, January 1997, pp. 25-35, Figure 7.)

(iii) **Young's modulus, Poisson's ratio, and shear modulus**

For the sake of acquaintance and completeness, the dependence of the three elastic constants, **E**, **ν**, and **G**, of materials on their chemical properties is stated briefly in terms of the following three graphs.

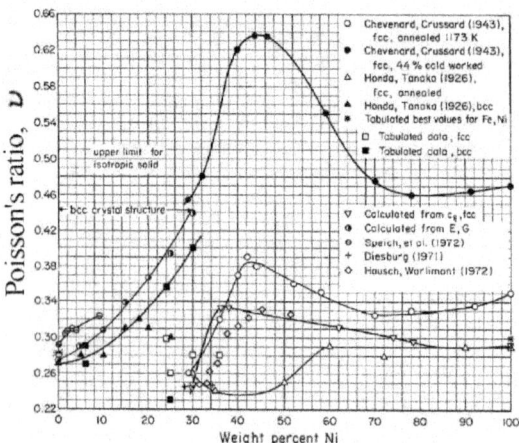

Figure 3-3. Compositional variation of Poisson ratio ν of iron-nickel alloys. (source FIGURE 8. H. M. Ledbetter and R. P. Reed, "Elastic Properties of Metals and Alloys, 1. Iron, Nickel, and Iron. Nickel Alloys" J. Phys. Chem. Ref. Data, Vol. 2, No 3 19 Figure 3-8. Compositional variation of Poisson ratio ν of iron-nickel alloys. (Source FIGURE 8. H. M. Ledbetter and R. P. Reed, "Elastic Properties of Metals and Alloys, 1. Iron, Nickel, and Iron. Nickel Alloys" J. Phys. Chem. Ref. Data, Vol. 2, No 3 1973)

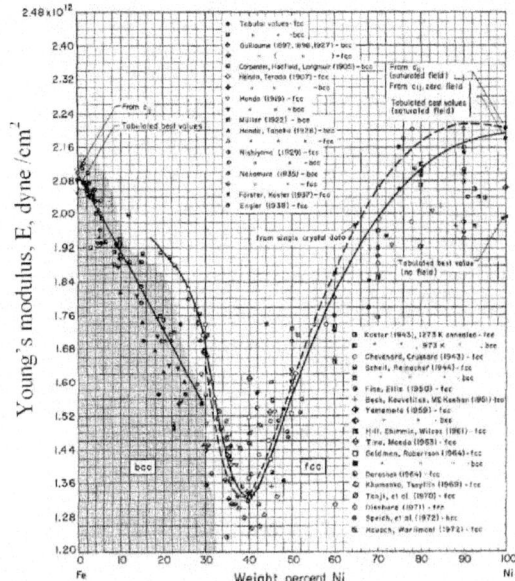

Figure 3-4. Compositional variation of Young's modulus E o f Iron-Nickel alloys (source FIGURE 5. H. M. Ledbetter and R. P. Reed, "Elastic Properties of Metals and Alloys, 1. Iron, Nickel, and Iron. Nickel Alloys" J. Phys. Chem. Ref. Data, Vol. 2, No 3 1973)

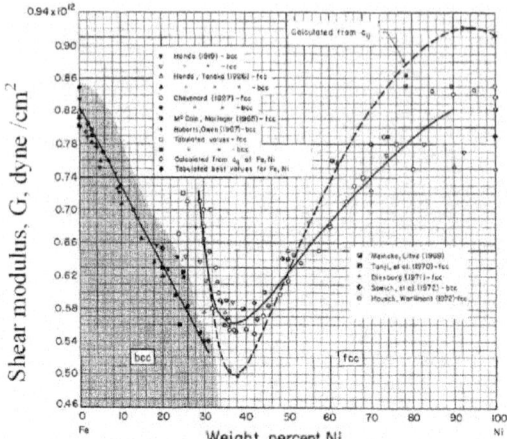

Figure 3-5. Compositional variation of shear modulus G of iron-nickel alloys. (source FIGURE 6. H. M. Ledbetter and R. P. Reed, "Elastic Properties of Metals and Alloys, 1. Iron, Nickel, and Iron. Nickel Alloys" J. Phys. Chem. Ref. Data, Vol. 2, No 3 1973)

3.4 Volumetric Hooke's law

Our assumption of superposition of effects of isolated stresses helps express Hooke's law in terms of the superposition of the three coordinate effects (strains) as follows.

Equations (3-1) are written for the three Cartesian directions of an infinitesimal element of volume as follows:

(i) Expressing stresses in terms of strains

The three normal stresses and elongation strains, in the three Cartesian directions, are described by equation (3-1.1) and (3-1.2). Along each Cartesian coordinate, the length of the infinitesimal element elongates by Young's equation (3-1.1) and contracts twice by Poisson's equation (3-1.2). Therefore, we get

$$\begin{matrix} \text{Young's} & \text{Poisson's two contractions} \\ \text{elongations} & \end{matrix}$$

$$
\begin{aligned}
\varepsilon'_{xx} &= \sigma_x E^{-1} & \varepsilon''_{xx} &= -\nu\varepsilon'_{yy} & \varepsilon'''_{xx} &= -\nu\varepsilon'_{zz} \\
\varepsilon'_{yy} &= \sigma_y E^{-1} & \varepsilon''_{yy} &= -\nu\varepsilon'_{xx} & \varepsilon'''_{yy} &= -\nu\varepsilon'_{zz} \\
\varepsilon'_{zz} &= \sigma_z E^{-1} & \varepsilon''_{zz} &= -\nu\varepsilon'_{xx} & \varepsilon'''_{yy} &= -\nu\varepsilon'_{yy}
\end{aligned}
\tag{3-2}
$$

118

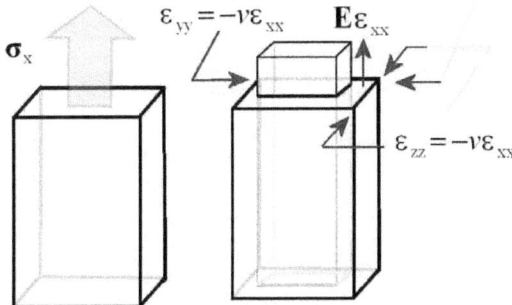

Figure 3-6. Young's elongation and Poisson's contraction

The net superposition of elongations and contractions in the three Cartesian directions, shown in Figure 3-1, is obtained from (3-2) as follows

$$\varepsilon_{xx} = \varepsilon'_{xx} + \varepsilon''_{xx} + \varepsilon'''_{xx} = E^{-1}\left[\sigma_x - \nu(\sigma_y + \sigma_z)\right]$$
$$\varepsilon_{yy} = \varepsilon'_{yy} + \varepsilon''_{yy} + \varepsilon'''_{yy} = E^{-1}\left[\sigma_y - \nu(\sigma_x + \sigma_z)\right] \quad (3\text{-}3.1)$$
$$\varepsilon_{zz} = \varepsilon'_{zz} + \varepsilon''_{zz} + \varepsilon'''_{zz} = E^{-1}\left[\sigma_z - \nu(\sigma_y + \sigma_x)\right]$$

And from (3-1.3), we get

$$\alpha_{xy} = G^{-1}\tau_{xy}$$
$$\alpha_{yz} = G^{-1}\tau_{yz} \quad (3\text{-}3.2)$$
$$\alpha_{zx} = G^{-1}\tau_{zx}$$

Equations (3-3) comprise the **generalized Hooke's volumetric law** of elasticity. It is written in the matrix form as follows

$$\begin{bmatrix} \varepsilon_{xx} \\ \varepsilon_{yy} \\ \varepsilon_{zz} \\ \alpha_{xy} \\ \alpha_{yz} \\ \alpha_{xz} \end{bmatrix} = \begin{bmatrix} E^{-1} & -\nu E^{-1} & -\nu E^{-1} & 0 & 0 & 0 \\ -\nu E^{-1} & E^{-1} & -\nu E^{-1} & 0 & 0 & 0 \\ -\nu E^{-1} & -\nu E^{-1} & E^{-1} & 0 & 0 & 0 \\ 0 & 0 & 0 & G^{-1} & 0 & 0 \\ 0 & 0 & 0 & 0 & G^{-1} & 0 \\ 0 & 0 & 0 & 0 & 0 & G^{-1} \end{bmatrix} \times \begin{bmatrix} \sigma_x \\ \sigma_y \\ \sigma_z \\ \tau_{xy} \\ \tau_{yz} \\ \tau_{xz} \end{bmatrix} \quad (3\text{-}3.3)$$

It is conveniently expressed in terms of the **sum of the normal stresses, Ξ,** and **sum of the three elongations** or **dilatation, ε,** defined as

119

$$\Xi = \sigma_x + \sigma_y + \sigma_z \tag{3-4.1}$$

$$\varepsilon = \varepsilon_{xx} + \varepsilon_{yy} + \varepsilon_{zz} \tag{3-4.2}$$

Before substituting Ξ, and ε, we will first add the three equations (3-3.1) to get

$$\varepsilon_{xx} + \varepsilon_{yy} + \varepsilon_{zz} = E^{-1}\left[\sigma_x + \sigma_y + \sigma_z - 2v\left(\sigma_x + \sigma_y + \sigma_z\right)\right] \tag{3-5.1}$$

Therefore, from (3-4), equation (3-2.4) becomes

$$\varepsilon = \frac{1-2v}{E}\Xi \tag{3-5.2}$$

$$\Xi = \frac{\varepsilon E}{(1-2v)} \tag{3-5.3}$$

Thus, Hooke's law for homogeneous material concludes that **the dilatation is an invariant independent of the inclination**, and only function of the physical constants of the materials, v and **E** and the sum of normal stresses, Ξ.

To emphasize, equation (3-5.2) is written in words as follows:

$$\text{sum of 3 coordinate elongations} = \frac{\text{sum of 3 nomal stresses}(1-2 \ \text{Poisson's ratio})}{\text{Young's modulus}}$$

Equation (3-5.3) imposes an upper limit on Poisson's ratio in the special case when all normal stresses are positive (i.e., $\Xi > 0$)) and body elongates along the three axes of coordinates (i.e., $\varepsilon > 0$), which implies that the denominator $(1-2v)$ must be greater than zero. Thus,

$$v \le \frac{1}{2} \tag{3-5.4}$$

(ii) Expressing strains in terms of stresses

Equation (3-3.1) is written in terms of ε, from (3-4.1), as follows

$$E\varepsilon_{xx} = (1+v)\sigma_x - \frac{vE}{(1-2v)}\varepsilon \tag{3-6.1}$$

Or,
$$\sigma_x = \frac{E}{(1+v)}\varepsilon_{xx} + \frac{vE}{(1-2v)(1+v)}\varepsilon \tag{3-6.2}$$

The constant multipliers of strains in equation (3-6.2) are called **Lamé's coefficients**, and used to simplify notation of equation (3-6.2) and the remaining five equations of stresses as follows.

$$\sigma_x = 2\mu\varepsilon_{xx} + \lambda\varepsilon \tag{3-7.1}$$

$$\sigma_y = 2\mu\varepsilon_{yy} + \lambda\varepsilon \tag{3-7.2}$$

$$\sigma_z = 2\mu\varepsilon_{zz} + \lambda\varepsilon \tag{3-7.3}$$

$$\tau_{xy} = \mu\alpha_{xy} \tag{3-7.4}$$

$$\tau_{yz} = \mu\alpha_{yz} \tag{3-7.5}$$

$$\tau_{zx} = \mu\alpha_{zx} \tag{3-7.6}$$

Where, the **Lamé's coefficients**, are defined by

$$\mu = \frac{E}{2(1+v)} \tag{3-8.1}$$

$$\lambda = \frac{vE}{(1-2v)(1+v)} \tag{3-8.2}$$

$$\mu = G \tag{3-8.3}$$

Since $v \leq \frac{1}{2}$, the two Lamé's coefficients are positive constants of materials.

Equations (3-7) are another representation of generalized volumetric Hooke's law, where stresses are expressed in terms of strains. It is written in the matrix form as follows

$$\begin{bmatrix} \sigma_x \\ \sigma_y \\ \sigma_z \\ \tau_{xy} \\ \tau_{yz} \\ \tau_{xz} \end{bmatrix} = \begin{bmatrix} 2\mu+\lambda & \lambda & \lambda & 0 & 0 & 0 \\ \lambda & 2\mu+\lambda & \lambda & 0 & 0 & 0 \\ \lambda & \lambda & 2\mu+\lambda & 0 & 0 & 0 \\ 0 & 0 & 0 & \mu & 0 & 0 \\ 0 & 0 & 0 & 0 & \mu & 0 \\ 0 & 0 & 0 & 0 & 0 & \mu \end{bmatrix} \times \begin{bmatrix} \varepsilon_{xx} \\ \varepsilon_{yy} \\ \varepsilon_{zz} \\ \alpha_{xy} \\ \alpha_{yz} \\ \alpha_{xz} \end{bmatrix} \tag{3-8.4}$$

Also, equations (3-8) is used to give equation (3-5) different outlook, as follows

Add the three normal stresses from equations (3-7), we get

$$\sigma_x + \sigma_y + \sigma_z = 2\mu(\varepsilon_{xx} + \varepsilon_{yy} + \varepsilon_{zz}) + 3\lambda\lambda \tag{3-9.1}$$

$$\Xi = (2\mu + 3\lambda)\varepsilon \tag{3-9.2}$$

Equation (3-9.2) conveys a simple meaning when interpreted by the use of equation (1-32.4), i.e., $N_0 = \frac{1}{3}(N_1 + N_2 + N_3)$, and equation (2-21), $\varepsilon_0 = \frac{1}{3}(\varepsilon_{xx} + \varepsilon_{yy} + \varepsilon_{zz})$, in case of **principal surfaces** of **principal stresses** and **principal elongation**. Thus, equation (3-9.2) takes the form

$$\Xi = 3N_0 = (2\mu + 3\lambda)3\varepsilon_0$$

121

i.e.,
$$N_0 = (2\mu + 3\lambda)\varepsilon_0 \qquad (3\text{-}10.1)$$

Equation (3-10.1) is an alternative formulation of (3-5.3). It related the **principal dilatation, ε_0,** and **hydrostatic stress, N_0,** in terms of **Lamé's coefficients**. Equation (3-10.1) is conveniently expressed in terms of the **bulk modulus**, χ, defined in terms of Lamé's coefficients, as follows

.

$$\chi = \lambda + \frac{2}{3}\mu \qquad (3\text{-}10.2)$$

i.e.,
$$N_0 = (3\chi)\varepsilon_0 = \chi\varepsilon \qquad (3\text{-}10.3)$$

To be noted that ε is the sum of the three elongations, equation (3-4.2), while ε_0 is the average of the three principal elongations, (1/3), one third of the sum.

3.5. Relationships between Young's modulus, Poisson's ratio, and Lamé's coefficients

Equations (3-8) provide the ratio between the two Lamé's coefficients, as follows

$$\frac{\lambda}{\mu} = \frac{vE}{(1-2v)(1+v)} \Big/ \frac{E}{2(1+v)} = \frac{2v}{(1-2v)} \qquad (3\text{-}11.1)$$

With simple algebraic manipulation of (3-11.1), we obtain expression of Poisson's ration in terms of Lamé's coefficients alone as follows

$$v = \frac{\lambda}{2(\mu + \lambda)} \qquad (3\text{-}11.2)$$

Equation (3-11.2) can be used to replace the Poisson's ratio in (3-8.1) such that we could express Young's modulus in terms of Lamé's coefficients alone. First, (3-11.2) is written as

$$1 + v = \frac{2\mu + 3\lambda}{2(\mu + \lambda)} \qquad (3\text{-}11.2a)$$

Substituting from (3-11.2a) into (3-8.1), we get

$$E = \mu \frac{2\mu + 3\lambda}{\mu + \lambda} \qquad (3\text{-}11.3)$$

3.6. Elastic potential energy

122

The energy delivered to an elastic body cause deformation, which represents work done under the field of inter-molecular forces. Like all **harmonic oscillators**, central forces restore original steady state configuration according to equations of equilibrium.

To obtain the equation for stored elastic potential energy, we start by equations (1-3) and Figure 3-7. We then consider the work done in each of the three Cartesian directions of an infinitesimal element of volume by the three components of stresses along that direction. For example, σ_x, τ_{xz}, τ_{zx}, along the x-axis. The **superposition of effects** from the three axes comprises the net stored energy.

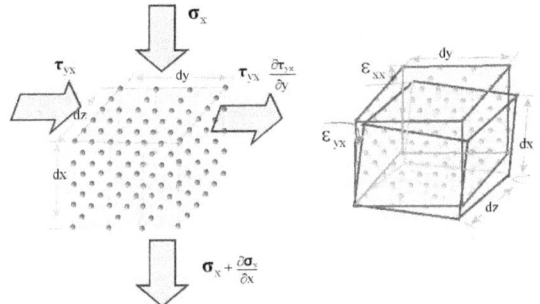

Figure 3-7. Geometry of volumetric deformation of infinitesimal strains and stresses that allow determining the stored elastic potential energy.

(i) On the x-axis, shown in Figure 3-7 and from equation (1-1.1), the normal stress σ_x imparts the force $\sigma_x$dydz on upper surface area; ΔA_{yz}=dy.dz. The produced elongation, du, along the x-axis, equation (2-2.1), is obtained by the product of **volumetric unit strain** ε_{xx}, equation (3-3.1), and the average distance ½ dx. Thus, du = ½ ε_{xx} dx. Therefore, the work done by the normal stress along the x-axis is given by

$$dW_x = \text{force x displacement} = \tfrac{1}{2}\sigma_x\varepsilon_{xx}\,dxdydz \qquad (3\text{-}12.1)$$

The superposition of effects allows us to add the three contributions of compression (tension) energies dW_x, dW_y, and dW_z, such that

$$dW_n = dW_x + dW_y + dW_z = \tfrac{1}{2}\left(\sigma_x\varepsilon_{xx} + \sigma_y\varepsilon_{yy} + \sigma_z\varepsilon_{zz}\right)dxdydz \qquad (3\text{-}12.2)$$

(ii) The shearing stresses, τ_{xz} and τ_{xy}, impose angular shear or rotation, or both. The force imparted by τ_{xy} on dxdz in the direction of y-axis, from equation (1-1.1), is τ_{xy}dxdz, which imposes a couple the tends to torque the infinitesimal volume an angle ε_{yx}, Figure 3-7, along the y-axis. Thus, the average volumetric displacement dv = ½ ε_{yx}dy. Therefore, the work done by the shearing stress along the y-axis is given by

$$dW_{xy} = \text{force x displacement} = \tfrac{1}{2}\tau_{xy}\alpha_{xy}\,dxdydz \tag{3-13.1}$$

The net work done on shearing displacements along the three axes is therefore

$$dW_{s} = \tfrac{1}{2}\big[\tau_{xy}\alpha_{xy} + \tau_{yz}\alpha_{yx} + \tau_{zx}\alpha_{zx}\big]dxdydz \tag{3-13.2}$$

(iii) The net work done is obtained from (3-12.2) and (3-13.2) as the sum of work done by normal and shearing stresses.

$$dW = dW_{n} + dW_{s} = \tfrac{1}{2}\big((\sigma_{x}\varepsilon_{xx} + \sigma_{y}\varepsilon_{yy} + \sigma_{z}\varepsilon_{zz}) + \tau_{xy}\alpha_{xy} + \tau_{yz}\alpha_{yx} + \tau_{zx}\alpha_{zx}\big)dxdydz \tag{3-14.1}$$

Substituting from equations (3-3.1) by the volumetric Hooke's relations for strains, in terms of Young's modulus **E** and Poisson's ratio v, and dividing both sides by the volume element **dxdydz**, we get **elastic energy per unit volume** as follows

$$T = \frac{dW}{dxdydz} = \frac{1}{2}\begin{pmatrix} \sigma_{x}E^{-1}\big[\sigma_{x} - v(\sigma_{y} + \sigma_{z})\big] \\ + \sigma_{y}E^{-1}\big[\sigma_{y} - v(\sigma_{x} + \sigma_{z})\big] \\ + \sigma_{z}E^{-1}\big[\sigma_{z} - v(\sigma_{x} + \sigma_{y})\big] \\ + G^{-1}\tau_{xy}^{2} + G^{-1}\tau_{yz}^{2} + G^{-1}\tau_{zx}^{2} \end{pmatrix} \tag{3-14.2}$$

Replacing **G** by **μ**, from (3-8.1), we get

$$T = \frac{2}{E}\big(\sigma_{x}^{2} + \sigma_{y}^{2} + \sigma_{z}^{2} - 2v(\sigma_{y}\sigma_{x} + \sigma_{y}\sigma_{z} + \sigma_{z}\sigma_{x}) + 2(1+v)(\tau_{xy}^{2} + \tau_{yz}^{2} + \tau_{zx}^{2})\big) \tag{3-14.3}$$

Equation (3-14.3) unit gives the **unit potential energy,** T(x,y,z), in terms of Young's modulus and Poisson's ratio and normal and shearing stresses.

(iv) To express the **unit potential energy,** T(x,y,z), in terms of strains, equations (3-7) are used to replace the stresses in equation (3-14.1) as follows

$$T = \tfrac{1}{2}\begin{pmatrix} (2\mu\varepsilon_{xx} + \lambda\varepsilon)\varepsilon_{xx} + (2\mu\varepsilon_{yy} + \lambda\varepsilon)\varepsilon_{yy} + (2\mu\varepsilon_{zz} + \lambda\varepsilon)\varepsilon_{zz} \\ + \mu(\alpha_{xy}^{2} + \alpha_{yz}^{2} + \alpha_{zx}^{2}) \end{pmatrix} \tag{3-15.1}$$

Equation (3-4.2) is used to replace ε in (3-15.1), thus

$$T = \tfrac{1}{2}\begin{pmatrix} 2\mu(\varepsilon_{xx}^{2} + \varepsilon_{yy}^{2} + \varepsilon_{zz}^{2}) \\ + \lambda(\varepsilon_{xx} + \varepsilon_{yy} + \varepsilon_{zz})(\varepsilon_{xx} + \varepsilon_{yy} + \varepsilon_{zz}) \\ + \mu(\alpha_{xy}^{2} + \alpha_{yz}^{2} + \alpha_{zx}^{2}) \end{pmatrix} \tag{3-15.2}$$

Example 17

A plane surface placed in the xy-plane is subjected to tension along the x-axis. No perpendicular stresses act in the xz- or yz- planes. Consider that elastic deformation if homogeneous, find the following:

(i) The formula for surface strains in terms of the magnitude of the tensile stress and the physical properties of the material.

(ii) Find the shear stress and strains in a plane oriented at angle α with the xOz plane.

(iii) Find the angles of rotation between various sides of a square on the plane after deformation

Solution

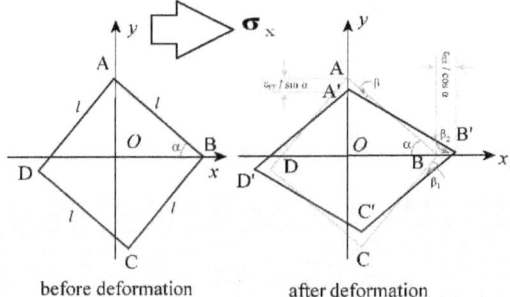

before deformation   after deformation

Figure 3-8. Homogeneous elastic deformation of a plane by tensile stress acting along the x-axis.

First, let us determine the controlling formula of homogeneous elastic deformation under the conditions given in the problem.

From (3-3.1 and 2), we substitute by

$$\alpha_{yz} = \alpha_{zx} = \varepsilon_{zz} \tag{3-16.1}$$

From (3-7), we get

$$\sigma_x = 2\mu\mu_{xx} + \lambda\left(\varepsilon_{xx} + \varepsilon_{yy}\right) \tag{3-16.2}$$

$$\sigma_y = 2\mu\mu_{yy} + \lambda\left(\varepsilon_{xx} + \varepsilon_{yy}\right) \tag{3-16.3}$$

$$\tau_{xy} = \mu\alpha_{xy} \tag{3-16.4}$$

(i) Tension along the x-axis implies that $\sigma_y = 2\mu\varepsilon_{yy} + \lambda\left(\varepsilon_{xx} + \varepsilon_{yy}\right) = 0$ and $\tau_{xy} = \mu\alpha_{xy} = 0$.

125

$$\varepsilon_{yy} = -\frac{\lambda}{\lambda + 2\mu} \varepsilon_{xx} \qquad (3\text{-}16.5)$$

Substituting by ε_{yy} from (3-16.5) into equation (3-16.2) we get

$$\sigma_x = 2\mu\mu_{xx} + \lambda\left(\varepsilon_{xx} - \frac{\lambda}{\lambda + 2\mu}\varepsilon_{xx}\right)$$
$$= 4\left(\frac{\mu + \lambda}{\lambda + 2\mu}\right)\mu\varepsilon_{xx} \qquad (3\text{-}16.6)$$

(ii) Shear stress on a plane inclined on the xOz plane by angle α.

From equations (1-14.6), we get

$$\tau_{uv} = \sigma_x \sin\alpha \cos\alpha = 4\left(\frac{\mu + \lambda}{\lambda + 2\mu}\right)\mu\varepsilon_{xx} \sin\alpha \cos\alpha$$
$$= 2\mu\varepsilon_{xx}\left(\frac{\mu + \lambda}{\lambda + 2\mu}\right)\sin2\alpha \qquad (3\text{-}16.7)$$

Comparing equation (3-16.7) with equation (3-7.4, 5, and 6) we conclude that

$$\tau_{uv} = 2\mu\varepsilon_{xx}\left(\frac{\mu + \lambda}{\lambda + 2\mu}\right)\sin2\alpha = \mu\alpha_{uv} \qquad (3\text{-}16.7a)$$

Therefore, the strain along the plane uv is given by

$$\alpha_{uv} = 2\varepsilon_{xx}\left(\frac{\mu + \lambda}{\lambda + 2\mu}\right)\sin2\alpha \qquad (3\text{-}16.7b)$$

Substituting by ε_{xx} from (3-16.6) into equation (3-16.7b) we get

$$\alpha_{uv} = 2\frac{\sigma_x}{4\mu}\left(\frac{\lambda + 2\mu}{\mu + \lambda}\right)\left(\frac{\mu + \lambda}{\lambda + 2\mu}\right)\sin2\alpha$$
$$= \frac{\sigma_x}{2\mu}\sin2\alpha \qquad (3\text{-}16.7c)$$

(iii) Changes in lengths of the sides of the square ABCD

In Figure 3-8, assume that the ABCD is a given square with side length l laying in the xOy-plane, before deformation.

After deformation, by the stress directed solely along the x-axis, the new shape is A'B'C'D'. The angle α is assumed between the uv-plane and the plane zOx. For simplicity, is taken between the side AD of the square and the x-axis, as shown in Figure 3-8.

Before Deformation, the length

$$OA = l \sin \alpha \qquad (3\text{-}16.8a)$$
$$OB = l \cos \alpha \qquad (3\text{-}16.8b)$$

After deformation, every unit length, of the xOy-plane, along the y-axis shrinks ε_{yy}. Thus, the point A moves to A', a distance equal to the length OA times the unit strain ε_{xx}. Similarly, the point B moves to B', a distance equal to the length OB times the strain ε_{xx}. Therefore,

$$AA' = -\varepsilon_{yy}\, l \sin \alpha \qquad (3\text{-}16.9a)$$
$$BB' = \varepsilon_{xx}\, l \cos \alpha \qquad (3\text{-}16.9b)$$

The negative sign, in equation (3-16.9a), accounts for the perpendicular contraction in the y-direction on the direction of stress in the x-direction according to Hooke's law, equation (3-1.2).

(iv) Angles of rotations of the sides of the square ABCD

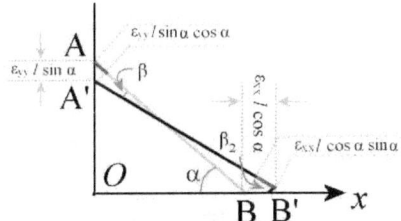

Figure 3-9. Homogeneous elastic deformation of a plane by tensile stress acting along the x-axis.

As shown in Figure 31, in order to calculate the angle of rotation of AB to A'B', we first determine the elongation

$$A'B'\text{-}AB = AA' \cos \alpha + BB' \sin \alpha$$
$$= \varepsilon_{xx}\, l \cos \alpha \sin \alpha - \varepsilon_{yy}\, l \sin \alpha \cos \alpha \qquad (3\text{-}16.9c)$$

The angle β between the lines AB and A' B' is calculated from equations, by the usual assumption of small angle approximation, as follows

127

$$\tan \hat{\beta} \approx \hat{\beta} = \frac{A'B'-AB}{AB}$$

$$= \frac{\varepsilon_{xx} l \cos\alpha\sin\alpha - \varepsilon_{yy} l \cos\alpha\sin\alpha}{l} \qquad (3\text{-}16.10)$$

$$= \cos\alpha\sin\alpha\left(\varepsilon_{xx} - \varepsilon_{yy}\right)$$

Substituting by ε_{yy} from (3-16.5) and by ε_{xx} from (3-16.6) into equation (3-16.10) we get

$$\varepsilon_{yy} = \frac{\sigma_x}{4\mu}\left(\frac{\lambda}{\mu+\lambda}\right) \qquad (3\text{-}16.11)$$

$$\varepsilon_{xx} = \frac{\sigma_x}{4\mu}\left(\frac{\lambda+2\mu}{\mu+\lambda}\right) \qquad (3\text{-}16.12)$$

Therefore, from (3-16.11 and 12), equation (3-16.10) becomes

$$\hat{\beta} = \cos\alpha\sin\alpha\left[\frac{\sigma_x}{4\mu}\left(\frac{\lambda+2\mu}{\mu+\lambda}\right) - \frac{\sigma_x}{4\mu}\left(\frac{\lambda}{\mu+\lambda}\right)\right]$$

$$= \frac{1}{2}\left(\frac{\sigma_x \cos\alpha\sin\alpha}{\mu+\lambda}\right) \qquad (3\text{-}16.13)$$

Substituting by (sin 2α = 2 cos α sin α) in (3-16.13), we finally get,

$$\hat{\beta} = \frac{1}{4}\left(\frac{\sigma_x \sin 2\alpha}{\mu+\lambda}\right) \qquad (3\text{-}16.14)$$

In conclusion, in the case of isotropic materials, the shear stress in equation (3-16.7a), the strain in equation (3-16.7c), and the rotation equation (3-16.14) of the body depend solely on the two Lamé' s coefficients μ and λ.

CHAPTER 4

LAMÉ'S EQUATIONS OF CONTINUITY

4.1. Combining Cauchy's equations of displacements and Hooke's equations of stress

The combined effects of static equilibrium of normal and shear stresses (**Navier's equations**), relations between displacements and strains (**Cauchy's equations**), and physical properties of materials (**Hooke's laws**) are integrated into a system of the differential equations of **Lamé.** Those equations govern the causes and effects of elastic deformation by external forces, with the aid of **surface conditions** of arial projections of stresses from the surface of the rigid body to interior planes of the rigid body.

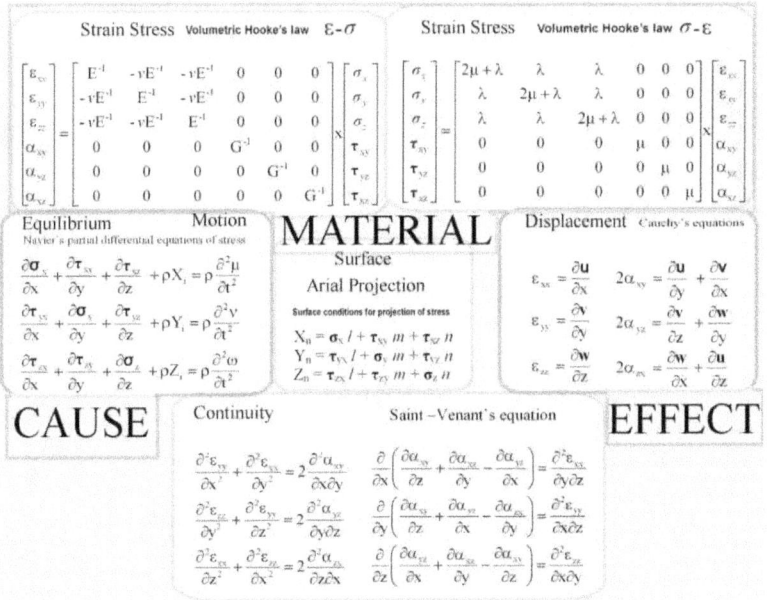

Figure 4-1. The fundamental governing laws of the theory of elasticity.

The derivation of Lamé's equations of the fundamental equations of elasticity is as follows.

129

Cauchy's equations of displacement-strain, equation (2-2.4), are used to replace the strain terms in Hooke's law, equations (3-7) as follows.
Starting with the stresses along the x-axis, we get

$$\sigma_x = 2\mu \frac{\partial u}{\partial x} + \lambda\varepsilon \tag{4-1.1}$$

$$\tau_{xy} = \mu\left(\frac{\partial u}{\partial y} + \frac{\partial v}{\partial x}\right) \tag{4-1.2}$$

$$\tau_{zx} = \mu\left(\frac{\partial u}{\partial z} + \frac{\partial w}{\partial x}\right) \tag{4-1.3}$$

4.2. Combining equations of strain and stress with Navier's equations of static equilibrium

Differentiating (4-1) with respect to the three respective coordinates allows us to combine **Hooke's law** with **Navier's equations** (1-4). Thus,

$$\frac{\partial \sigma_x}{\partial x} = 2\mu \frac{\partial^2 u}{\partial x^2} + \lambda \frac{\partial\varepsilon}{\partial x} \tag{4-2.1}$$

$$\frac{\partial \tau_{xy}}{\partial y} = \mu\left(\frac{\partial^2 u}{\partial y^2} + \frac{\partial^2 v}{\partial x^2}\right) \tag{4-2.2}$$

$$\frac{\partial \tau_{zx}}{\partial z} = \mu\left(\frac{\partial^2 u}{\partial z^2} + \frac{\partial^2 w}{\partial x^2}\right) \tag{4-2.3}$$

From the equation (4-2), which relates displacements to stress and strain, the static equilibrium equations of Navier's (1-4) become

$$\left(2\mu \frac{\partial^2 u}{\partial x^2} + \lambda \frac{\partial\varepsilon}{\partial x}\right) + \mu\left(\frac{\partial^2 u}{\partial y^2} + \frac{\partial^2 v}{\partial x^2}\right) + \mu\left(\frac{\partial^2 u}{\partial z^2} + \frac{\partial^2 w}{\partial x^2}\right) + \rho X_i = \rho \frac{\partial^2 u}{\partial t^2} \tag{4-3.1}$$

Forming Laplacian operation from Navier's equation (4-3.1), it is written as follows

$$\mu\left(\frac{\partial^2 u}{\partial x^2} + \frac{\partial^2 u}{\partial y^2} + \frac{\partial^2 u}{\partial z^2}\right) + \mu\left(\frac{\partial^2 u}{\partial x^2} + \frac{\partial^2 v}{\partial x^2} + \frac{\partial^2 w}{\partial x^2}\right) + \lambda \frac{\partial\varepsilon}{\partial x} + \rho X_i = \rho \frac{\partial^2 u}{\partial t^2} \tag{4-3.2}$$

i.e.,

$$\mu\nabla^2 u + \mu\left(\frac{\partial^2 u}{\partial x^2} + \frac{\partial^2 v}{\partial x^2} + \frac{\partial^2 w}{\partial x^2}\right) + \lambda \frac{\partial\varepsilon}{\partial x} + \rho X_i = \rho \frac{\partial^2 u}{\partial t^2} \tag{4-3.3}$$

Where, the Laplacian operator is defined as

$$\nabla^2 = \frac{\partial^2}{\partial x^2} + \frac{\partial^2}{\partial y^2} + \frac{\partial^2}{\partial z^2} \tag{4-3.4}$$

4.3. Combining principal strain with Navier's equations of static equilibrium

(iv) Substituting by the derivatives of Cauchy's equations (2-2.4 through 2.6), with respect to x, we get

$$\frac{\partial \varepsilon}{\partial x} = \frac{\partial}{\partial x}\left(\varepsilon_{xx} + \varepsilon_{yy} + \varepsilon_{zz}\right) = \frac{\partial}{\partial x}\left(\frac{\partial u}{\partial x} + \frac{\partial v}{\partial x} + \frac{\partial w}{\partial x}\right) \tag{4-3.5}$$

Thus, from (4-3.4) and (4-3.5), equation (4-3.3), and similar equations for the y- and z- directions, we finally obtain the **Lamé's equations** as follows

$$\mu \nabla^2 \mathbf{u} + (\mu + \lambda)\frac{\partial \varepsilon}{\partial x} + \rho X_i = \rho \frac{\partial^2 \mathbf{u}}{\partial t^2} \tag{4-4.1}$$

$$\mu \nabla^2 \mathbf{v} + (\mu + \lambda)\frac{\partial \varepsilon}{\partial y} + \rho Y_i = \rho \frac{\partial^2 \mathbf{v}}{\partial t^2} \tag{4-4.2}$$

$$\mu \nabla^2 \mathbf{w} + (\mu + \lambda)\frac{\partial \varepsilon}{\partial z} + \rho Z_i = \rho \frac{\partial^2 \mathbf{w}}{\partial t^2} \tag{4-4.3}$$

Equations (4-4) describe the **internal body displacements** u, v, and w and **strains** ε, in terms of the material constants μ, λ, ρ, and internal stresses X_i, Y_i, and Z_i.

4.4. Surface conditions for external forces

Equations (4-4) are solved under the **boundary conditions** obtained from the surface conditions equations (1-8). We will use equations (4-1), and similar equations for the y- and z- directions, to replace the stresses in equations (1-8), we get

$$X_n = l\left(2\mu\frac{\partial u}{\partial x} + \lambda\varepsilon\right) + m\mu\left(\frac{\partial u}{\partial y} + \frac{\partial v}{\partial x}\right) + n\mu\left(\frac{\partial u}{\partial z} + \frac{\partial w}{\partial x}\right) \tag{4-5.1}$$

$$Y_n = l\left(\frac{\partial u}{\partial y} + \frac{\partial v}{\partial x}\right) + m\left(2\mu\frac{\partial v}{\partial y} + \lambda\varepsilon\right) + n\mu\left(\frac{\partial v}{\partial z} + \frac{\partial w}{\partial y}\right) \tag{4-5.2}$$

$$Z_n = l\left(\frac{\partial u}{\partial z} + \frac{\partial w}{\partial x}\right) + m\mu\left(\frac{\partial w}{\partial y} + \frac{\partial v}{\partial z}\right) + n\left(2\mu\frac{\partial w}{\partial z} + \lambda\varepsilon\right) \tag{4-5.3}$$

Equations (4-5) can be arranged to simplify notation as follows

$$X_n = l\lambda\varepsilon + \mu\left(l\frac{\partial u}{\partial x} + m\frac{\partial v}{\partial x} + n\frac{\partial w}{\partial x}\right) + \mu\left(l\frac{\partial u}{\partial x} + m\frac{\partial u}{\partial y} + n\frac{\partial u}{\partial z}\right) \qquad (4\text{-}6.1)$$

$$Y_n = m\lambda\varepsilon + \mu\left(l\frac{\partial u}{\partial y} + m\frac{\partial v}{\partial y} + n\frac{\partial w}{\partial y}\right) + \mu\left(l\frac{\partial v}{\partial x} + m\frac{\partial v}{\partial y} + n\frac{\partial v}{\partial z}\right) \qquad (4\text{-}6.2)$$

$$Z_n = n\lambda\varepsilon + \mu\left(l\frac{\partial u}{\partial z} + m\frac{\partial v}{\partial z} + n\frac{\partial w}{\partial z}\right) + \mu\left(l\frac{\partial w}{\partial x} + m\frac{\partial w}{\partial y} + n\frac{\partial w}{\partial z}\right) \qquad (4\text{-}6.3)$$

From Figure 34, we discern that the last brackets on the right side of (4-6) are derivatives of the displacements along the normal. Thus, equations (4-6) can be written as follows.

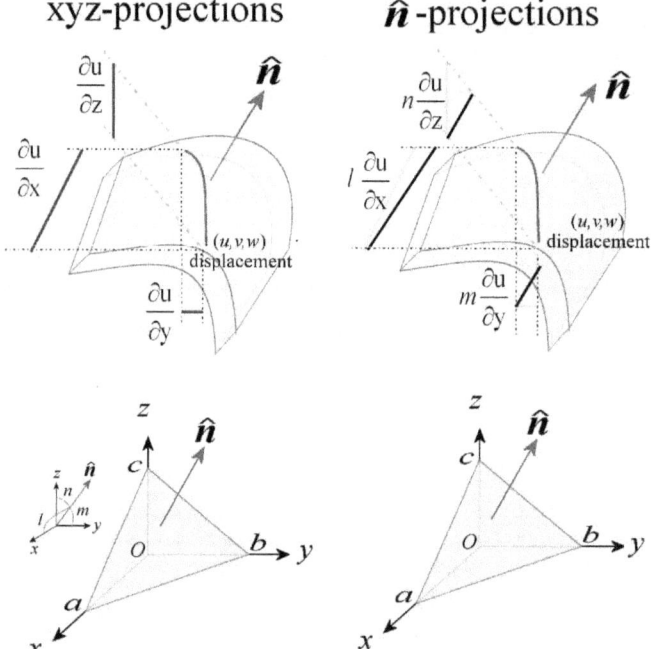

Figure 4-2. Surface projection of components of displacements on an arbitrarily oriented plane with normal (l, m, n).

$$X_n = l\lambda\varepsilon + \mu\left(l\frac{\partial u}{\partial x} + m\frac{\partial v}{\partial x} + n\frac{\partial w}{\partial x}\right) + \mu\frac{\partial u}{\partial n} \qquad (4\text{-}7.1)$$

132

$$Y_n = m\lambda\varepsilon + \mu\left(l\frac{\partial \mathbf{u}}{\partial y} + m\frac{\partial \mathbf{v}}{\partial y} + n\frac{\partial \mathbf{w}}{\partial y} \right) + \mu\frac{\partial \mathbf{v}}{\partial n} \tag{4-7.2}$$

$$Z_n = n\lambda\varepsilon + \mu\left(l\frac{\partial \mathbf{u}}{\partial z} + m\frac{\partial \mathbf{v}}{\partial z} + n\frac{\partial \mathbf{w}}{\partial z} \right) + \mu\frac{\partial \mathbf{w}}{\partial n} \tag{4-7.3}$$

Equations (4-4) and (4-7) comprise the system of equations required to solve internal deformations and external boundary conditions of applied forces.

CHAPTER 5

ELASTIC VIBRATION

5.1. Vibration of unbound surfaces

A. Longitudinal vibration

Example 18

An unbound elastic membrane located in the yz-plane. Assume that the membrane does not have any internal stresses.

(A) Find the possible displacement function of longitudinal vibration along the x-axis.

(B) Find the possible displacement function of transverse vibration along the z-axis.

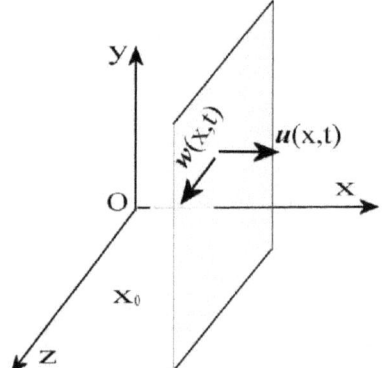

Figure 5-1. Unbound elastic membrane.

Solution

(i) Equations of displacement:

$$u = u(x,t)$$

134

$$v = 0$$
$$w = 0 \qquad (5\text{-}1.1)$$

(ii) Body forces: assumed absent

$$X_i = Y_i = Z_i = 0 \qquad (5\text{-}1.2)$$

(iii) Principal strain from Cauchy's equations

$$\varepsilon = \varepsilon_{xx} + \varepsilon_{yy} + \varepsilon_{zz} = \frac{\partial u}{\partial x} + \frac{\partial v}{\partial y} + \frac{\partial w}{\partial z} = \frac{\partial u}{\partial x} \qquad (5\text{-}1.3)$$

The three derivatives of ε, required Lamé's equations (4-4)

$$\frac{\partial \varepsilon}{\partial x} = \frac{\partial^2 u}{\partial x^2} \qquad \frac{\partial \varepsilon}{\partial y} = \frac{\partial^2 u}{\partial x \partial y} = 0 \qquad \frac{\partial \varepsilon}{\partial z} = \frac{\partial^2 u}{\partial x \partial z} = 0 \qquad (5\text{-}1.4)$$

The Laplacians of equations (5-1.1) are obtained from (5-1.4) as the acceleration terms along the y- and z-axes vanish. Thus

$$\nabla^2 u = \frac{\partial^2 u}{\partial x^2} \qquad \nabla^2 v = 0 \qquad \nabla^2 w = 0 \qquad \frac{\partial^2 v}{\partial t^2} = \frac{\partial^2 w}{\partial t^2} = 0 \qquad (5\text{-}1.5)$$

(iv) Lamé's equations (4-4)

$$\mu \nabla^2 u + (\mu + \lambda)\frac{\partial \varepsilon}{\partial x} + \rho X_i = \rho \frac{\partial^2 u}{\partial t^2} \qquad (5\text{-}1.6a)$$

From (5-1.5), we get

$$\mu \frac{\partial^2 u}{\partial x^2} + (\mu + \lambda)\frac{\partial^2 u}{\partial x^2} = \rho \frac{\partial^2 u}{\partial t^2} \qquad (5\text{-}1.6b)$$

$$(2\mu + \lambda)\frac{\partial^2 u}{\partial x^2} = \rho \frac{\partial^2 u}{\partial t^2} \qquad (5\text{-}1.6c)$$

(v) Wave equation

Equation (5-1.6c) is a wave equation of oscillation usually expressed as follows

$$\ddot{u} - k^2 \frac{\partial^2 u}{\partial x^2} = 0 \qquad (5\text{-}1.7a)$$

$$k^2 = \frac{2\mu + \lambda}{\rho} \qquad (5\text{-}1.7b)$$

$$\ddot{u} = k^2 \frac{\partial^2 u}{\partial x^2} \tag{5-1.7c}$$

Equation (5-1.7) implies that only those forms of the displacement $u(x,t)$, (5-1.1), that satisfy equation (5-1.7) can represent longitudinal oscillation.

B. Transverse vibration

(i) Equations of displacement:

$$u = 0$$
$$v = 0$$
$$w = w(x,t) \tag{5-2.1}$$

(ii) Body forces: assumed absent

$$X_i = Y_i = Z_i = 0 \tag{5-2.2}$$

(iii) Principal strain from Cauchy's equations

Since the displacement w occurs in the plane of the membrane, then all points will move equally due to lack of stress (vanishing area along the x-axis)

$$\varepsilon = \varepsilon_{xx} + \varepsilon_{yy} + \varepsilon_{zz} = \frac{\partial u}{\partial x} + \frac{\partial v}{\partial y} + \frac{\partial w}{\partial z} = 0 \tag{5-2.3}$$

The three derivatives of ε, required Lamé's equations (4-4)

$$\frac{\partial \varepsilon}{\partial x} = \frac{\partial^2 u}{\partial x^2} = 0 \qquad \frac{\partial \varepsilon}{\partial y} = \frac{\partial^2 u}{\partial x \partial y} = 0 \qquad \frac{\partial \varepsilon}{\partial z} = \frac{\partial^2 u}{\partial x \partial z} = 0 \tag{5-2.4}$$

The Laplacians of equations (5-2.1) are obtained from (5-2.4) and the acceleration terms along the y- and z-axes vanish, from (5-2.1). Thus

$$\nabla^2 u = 0 \qquad \nabla^2 v = 0 \qquad \nabla^2 w = \frac{\partial^2 w}{\partial x^2} \qquad \frac{\partial^2 v}{\partial t^2} = \frac{\partial^2 u}{\partial t^2} = 0 \tag{5-2.5}$$

Note that w can only change along the x-direction, which extends outside the membrane.

(iv) Lamé's equations (4-4)

$$\mu \nabla^2 \mathbf{w} + (\mu + \lambda)0 + 0 = \rho \frac{\partial^2 \mathbf{w}}{\partial t^2} \qquad (5\text{-}2.6a)$$

From (5-2.5), we get

$$\mu \frac{\partial^2 \mathbf{w}}{\partial x^2} = \rho \frac{\partial^2 \mathbf{w}}{\partial t^2} \qquad (5\text{-}2.6b)$$

(v) Wave equation

Equation (5-2.6b) is a wave equation of oscillation usually expressed as follows

$$\ddot{\mathbf{w}} - b^2 \frac{\partial^2 \mathbf{w}}{\partial x^2} = 0 \qquad (5\text{-}2.7a)$$

$$b^2 = \frac{\mu}{\rho} \qquad (5\text{-}2.7b)$$

$$\ddot{\mathbf{w}} = b^2 \frac{\partial^2 \mathbf{w}}{\partial x^2} = 0 \qquad (5\text{-}2.7c)$$

Equation (5-2.7) implies that only those forms of the displacement $w(x,t)$, (5-2.1), that satisfy equation (5-2.7) can represent transverse vibration.

Note that longitudinal vibrations (5-1.7) and transverse vibrations (5-2.7) differ in the speeds k and b, but both occur with respect to x-coordinate alone.

C. Harmonic longitudinal vibrations

Example 19

In the planar material surface with homogeneous elastic properties, considered in the previous example, find the ratio between velocities of propagation of longitudinal and transverse harmonic vibration

Solution

Assume a solution to (5-1.7) of the sinusoidal form

$$u = A \sin 2\pi \left(\frac{x}{\lambda_0} - \frac{t}{\tau_0} \right) \qquad (5\text{-}3.1)$$

Where, A, λ_0, and τ_0 are assumed constants, such that

A, the amplitude of the deformation or displacement u.

λ_0, spatial divisor or segments of distances by which distance x (the dividend) is divided along the cycles of the harmonic wave spread over space.

τ_0, similarly temporal divisor or segments of times by which the time t (the dividend) is divided along the cycles of the harmonic wave spread over time.
Substitute from (5-3.1) into (5-1.7), we get

$$\frac{\partial^2}{\partial t^2}\left[A\sin 2\pi\left(\frac{x}{\lambda_0}-\frac{t}{\tau_0}\right)\right] = k^2\frac{\partial^2}{\partial x^2}\left[A\sin 2\pi\left(\frac{x}{\lambda_0}-\frac{t}{\tau_0}\right)\right] \tag{5-3.2}$$

$$k^2 = \frac{\lambda_0^2}{\tau_0^2} = \frac{2\mu+\lambda}{\rho} \tag{5-3.3a}$$

$$\frac{\lambda_0}{\tau_0} = \pm k = \pm\sqrt{\frac{2\mu+\lambda}{\rho}} \tag{5-3.3b}$$

Behavior of a membrane vibrating harmonically:

(i) Strain ε_{xx} along the x-axis

The **unit displacement** of the membrane along the x-axis is given by Cauchy's equation

$$\varepsilon_{xx} = \frac{\partial u}{\partial x} = \frac{2\pi A}{\lambda_0}\cos 2\pi\left(\frac{x}{\lambda_0}-\frac{t}{\tau_0}\right) \tag{5-4}$$

The spatial and temporal behavior of the **principal strain** ε_{xx} determines the **speed of propagation** of vibrations of a material planar body and the **wavelength** and **frequency** of the elastic wave of vibrations.

(ii) Wavelength λ_0

At any point of time (t = C, constant), equation (5-4.1) gives changes of the unit strain ε_{xx} as sinusoidal variations such that, when $x = n\lambda_0$ (where n is integer number, n = 0, 1, 2,.), we have

$$\varepsilon_{xx} = \frac{2\pi A}{\lambda_0}\cos 2\pi\left(\frac{n\lambda_0}{\lambda_0}-\frac{C}{\tau_0}\right)$$
$$= \frac{2\pi A}{\lambda_0}\cos 2\pi(n-C') \tag{5-5}$$
$$= \text{constant}, \qquad n = 0,1,2,..$$

Hence, λ_0 is spatial segment or wavelength by which the vibration waves retain a predicable pattern depending only on t = C.

138

(iii) Wave period τ_0

Again, at any point of space ($x = d$, constant), equation (5-4.1) gives changes of the unit strain ε_{xx} as sinusoidal variations such that, when $t = m\,\tau_0$ (where m is integer number, $n = 0, 1, 2,.$), we have

$$\varepsilon_{xx} = \frac{2\pi A}{\lambda_0}\cos 2\pi\left(\frac{d}{\lambda_0} - \frac{m\tau_0}{\tau_0}\right)$$

$$= \frac{2\pi A}{\lambda_0}\cos 2\pi(D - m) \tag{5-6}$$

$$= \text{constant}, \qquad m = 0,1,2,..$$

Hence, τ_0 is temporal segment or wave period by which the vibration waves retain a predicable pattern depending only on $x = d$.

(iv) Velocity of longitudinal wave propagation v_0

The velocity of propagation of constant amplitude of the unit strain ε_{xx}, equation (5-4), is determined from the wavelength λ_0 and wave period τ_0, and from equation (5-3.3b), as follows

$$v_0 = \frac{\partial x}{\partial t} = \frac{\lambda_0}{\tau_0} = \pm\sqrt{\frac{2\mu + \lambda}{\rho}} \tag{5-7}$$

Equation (5-7) comprises a fundamental conclusion concerning the **speed of propagation** of **mechanical deformation** in homogeneous elastic materials, which depends solely on physical properties: The Lamé's coefficients λ and μ, and material density ρ.

(v) Velocity of transverse wave propagation v_τ

The velocity of propagation of constant amplitude of the unit shear strain ε_{xz}, can be obtained equations similar to (5-3.1) and (5-4) but with speed of transverse propagation of shear, equation (5-2.7b). Thus

$$v_\tau = \frac{\partial z}{\partial t} = \pm\sqrt{\frac{\mu}{\rho}} \tag{5-8}$$

(vi) The ratio (v_0 / v_τ) of longitudinal to transverse velocities of propagation of waves of vibration in matter

From equations (5-7) and (5-8), the ratio (v_0 / v_τ) is given by

139

$$\frac{V_o}{V_\tau} = \pm \sqrt{\frac{2\mu + \lambda}{\mu}}$$

(5-9)

Therefore, the speed of propagation of longitudinal vibrations is greater than the speed of propagation of transverse vibrations, and their ratio depends solely on the elastic properties of the material (i.e., the Lamé's coefficients λ and μ).

Example 20

A steel body of homogeneous elastic properties has the following constants

Young's constant, $E = 2 \times 10^6$ kg/cm^2 (5-10.1)
Poisson's ration, $v = 0.3$ (5-10.2)
Density, $\rho = 7.85$ g/cm^3 (5-10.3)

Find the speed of sound and the ratio of longitudinal to transverse speeds of sound in the given body.

Solution

From equations (3-8), the **Lamé's coefficients**, for the steel, equations (5-10) are calculated as follows

$$\mu = \frac{E}{2(1+v)} = \frac{2(10^6)}{2(1+0.3)} = 7.6923(10^5) \quad \frac{kg}{cm^2}$$

(5-10.4)

$$\lambda = \frac{vE}{(1-2v)(1+v)} = \frac{(0.3)2(10^6)}{(1-0.6)(1+0.3)} = 1.153846(10^6) \quad \frac{kg}{cm^2}$$

(5-10.5)

The steel density, in weight units, is calculated as follows

$$\rho = \frac{7.85 \ \frac{gm}{cm^3}}{981 \ \frac{cm}{sec^2}} = 8.002(10^{-3}) \quad \frac{gm.sec^2}{cm^4} = 8.002(10^{-6}) \quad \frac{kg.sec^2}{cm^4}$$

(5-10.5)

Equation (5-7)

$$V_o = \pm \sqrt{\frac{2\mu + \lambda}{\rho}} = \sqrt{\frac{2(7.6923)(10^5) + 1.153846(10^6) \quad \frac{kg}{cm^2}}{8.002(10^{-6}) \quad \frac{kg.sec^2}{cm^4}}}$$

$$= 579829.3 \quad \frac{cm}{sec}$$

(5-10.6)

$$= 5,798.29 \quad \frac{m}{sec}$$

Equation (5-8) gives

$$v_\tau = \pm\sqrt{\frac{\mu}{\rho}} = \sqrt{\frac{(7.6923)(10^5)\ \dfrac{kg}{cm^2}}{8.002(10^{-6})\ \dfrac{kg.sec^2}{cm^4}}}$$

$$= 309931.8\ \frac{cm}{sec}$$ (5-10.7)

$$= 3,099.32\ \frac{m}{sec}$$

Therefore, equation (5-9) becomes

$$\frac{v_o}{v_\tau} = \frac{5,798.29}{3,099.32} = 1.870829$$ (5-10-8)

5.2. Vibration of bound surfaces

Example 21

Consider an elastic thin long bar exposed to longitudinal force F. Find the equation of vibration of the bar material assuming that transverse strains can be neglected.

Solution

(i) Approximate formulation of Lamé's equations

The Navier's equation for one dimensional equilibrium between normal unit stress σ_x and displacement $u(x,t)$ is obtained from equation (1-4.1) by neglecting the shearing stress derivatives and the internal forces. Thus, we get

$$\frac{\partial \sigma_x(x,y,z)}{\partial x} = \rho\frac{\partial^2 u(x,y,z)}{\partial t^2}$$ (5-11.1)

Where, the normal stress σ_x is replaced by **Cauchy**'s equation (2-2.4) and **Hooke**'s equation (3-1.1), which are re-written here as follows

$$\sigma_x = E\varepsilon_{xx}$$ (5-11.2)

$$\varepsilon_{xx} = \frac{\partial u(x,y,z)}{\partial x}$$ (5-11.3)

Equations (5-11.2) and (5-11.3) gives

$$\sigma_x = E \frac{\partial u(x, y, z)}{\partial x} \tag{5-11.4}$$

Which, upon differentiation with respect to x, gives

$$\frac{\partial \sigma_x}{\partial x} = E \frac{\partial^2 u(x, y, z)}{\partial x^2} \tag{5-11.5}$$

Thus, equations (5-11.1) and (5-11.5) render a partial differential equation of displacement alone as follows

$$\frac{\partial^2 u(x, y, z)}{\partial x^2} = \left(\frac{\rho}{E}\right) \frac{\partial^2 u(x, y, z)}{\partial t^2} \tag{5-11.6}$$

In fact, equation (5-11.6) is a simplified form of Lamé's equation (4-4.1), .i.e.,

$$\mu \nabla^2 \mathbf{u} + (\mu + \lambda) \frac{\partial \varepsilon}{\partial x} + \rho X_i = \rho \frac{\partial^2 \mathbf{u}}{\partial t^2} \tag{5-11.7.1}$$

Since the Laplacian term $\nabla^2 \mathbf{u}$ in equation (5-11.7) is reduced to the single derivative in the x-coordinate. Also, the term $\frac{\partial \varepsilon}{\partial x}$ is also reduced to the derivative of the longitudinal strain on the x-axis, (5-11.3). Thus, the Lamé's equation (5-11.7) becomes

$$\mu \frac{\partial^2 \mathbf{u}}{\partial x^2} + (\mu + \lambda) \frac{\partial^2 \mathbf{u}}{\partial x^2} = \rho \frac{\partial^2 \mathbf{u}}{\partial t^2} \tag{5-11.7.2}$$

Or,

$$(2\mu + \lambda) \frac{\partial^2 \mathbf{u}}{\partial x^2} = \rho \frac{\partial^2 \mathbf{u}}{\partial t^2} \tag{5-11.7.3}$$

Where, from equations (3-8.1) and (3-8.2), we get

$$2\mu + \lambda = \frac{E}{(1 + v)} + \frac{vE}{(1 - 2v)(1 + v)} \tag{5-11.7.4}$$

In the present case of thin and long rod, where harmonic vibration are mostly longitudinal, we could safely put Poisson's ratio, $v = 0$. Thus, equation (5-11.7.4) is reduced to ($2\mu + \lambda = E$). Therefore, the Lamé's equation (5-11.7.3) is reduced to equation (5-11.6).

(ii) Solving Lamé's equation by separation of variables

Our governing equation (5-11.6), which is an approximation of (5-11.7.3), can be solved by assuming two separate functions $X(x)$ and $T(t)$, which product; $u(x,t) = X(x)T(t)$ comprises the solution of the governing equations. Because, there exists many values for each of X or T which do not coincide with respective T or X, which satisfy the governing equation, we will use **Fourier's series** to define the requirements on the products of XT that comprise valid solution.

Thus, $\quad\quad u(x,t) = X(x)T(t)$ $\hfill$ (5-12.1)

Substituting with the derivatives of $u(x,t)$, from (5-12.1) into (5-11.6), we get

$$\frac{\partial^2}{\partial x^2}\big(X(x)T(t)\big) = \left(\frac{\rho}{E}\right)\frac{\partial^2}{\partial t^2}\big(X(x)T(t)\big)$$

$$T\frac{\partial^2 X}{\partial x^2} = \left(\frac{\rho}{E}\right)X\frac{\partial^2 T}{\partial t^2}$$

$\hfill$ (5-12.2.1)

Apparently, equation (5-12.2) can be equated with arbitrary constants $(-\gamma^2)$, which will be defined from the boundary and initial conditions of the problem. First, substitute by

$$a^2 = \frac{E}{\rho}$$

$\hfill$ (5-12.2.2)

Thus, after dividing both sides of equation (5-12.2.1) by $a^2 TX$, we get

$$\frac{1}{X}\cdot\frac{\partial^2 X}{\partial x^2} = \frac{1}{a^2 T}\cdot\frac{\partial^2 T}{\partial t^2} = -\gamma^2$$

$\hfill$ (5-12.3)

Thus, we have two separate equations, one is time-independent, the other coordinate-independent, as follows

$$\frac{\partial^2 X}{\partial x^2} = -\gamma^2 X$$

$\hfill$ (5-12.4)

$$\frac{\partial^2 T}{\partial t^2} = -\gamma^2 a^2 T$$

$\hfill$ (5-12.5)

One possible solution of equations (5-12.4 and 5) is the Fourier's series of cyclical, sinusoidal form as follows

$$X = A\cos\gamma x + B\sin\gamma x$$ $\hfill$ (5-12.6)
$$T = C\cos\gamma at + D\sin\gamma at$$ $\hfill$ (5-12.7)

Thus, from (5-12.1), a possible solution for (5-12.3) is
$$u(x,t) = XT = \big(A\cos\gamma x + B\sin\gamma x\big)\big(C\cos\gamma at + D\sin\gamma at\big)$$ $\hfill$ (5-12.8)

We now have a possible solution for the simplified Lamé's equation with four groups of constants that need be determined from the initial and boundary conditions of the problem.

(iii) Boundary conditions for built-in thin long bar

1. A bar with length l, built-in at x = 0, therefore, $\mathbf{u}(0,t) = 0$

Substitute by x = u = 0 at equation (5-12.8), we get

$$0 = (A + 0)(C \cos \gamma at + D \sin \gamma at)$$

i.e., $\qquad A = 0$ $\hspace{6cm}$ (5-12.9)

Thus, equation (5-12.8) becomes

$$\mathbf{u}(x,t) = XT = (C \cos \gamma at + D \sin \gamma at)\sin \gamma x \hspace{3cm} (5\text{-}12.10)$$

We have included the constant B, equation (5-12.6), in both of C and D, since the three constants are arbitrarily chosen.

2. A bar with length l, built-in at x = l, therefore, $\varepsilon_{xx} = 0$

$$\varepsilon_{xx} = \frac{\partial \mathbf{u}}{\partial x} = (C \cos \gamma at + D \sin \gamma at)\frac{\partial \mathbf{u}}{\partial x}\sin \gamma x$$

$$= (C \cos \gamma at + D \sin \gamma at)\gamma \cos \gamma x \,|_{x=l} \hspace{3cm} (5\text{-}12.11)$$

$$= 0$$

Therefore, $\quad \cos \gamma l = 0 \quad \gamma l = i\dfrac{\pi}{2}, \quad$ where, $\;i = 1,3,5,...$ $\hspace{2cm}$ (5-12.12)

So far, we have determined the constant A, equation (5-12.9) and the constant γ, equation (5-12.12). Thus, equation (5-12.8)

$$\mathbf{u}(x,t) = \sum_{i=1,3,5,...}^{\infty} \sin \frac{i\pi x}{2l}\left(C_i \cos \frac{i\pi at}{2l} + D_i \sin \frac{i\pi at}{2l} \right) \hspace{2cm} (5\text{-}12.13)$$

(iv) Initial conditions for vibrating built –in bar

144

The two remaining groups of constants C_i and D_i are determined from the initial displacement and initial velocity of deformation, where

$$\mathbf{u}(x,0) = f(x)$$

$$\frac{\partial \mathbf{u}(x,0)}{\partial t} = g(x) \tag{5-12.14}$$

Thus, from equation (5-12.13), we get, at $t = 0$,

$$\mathbf{u}(x,0) = \sum_{i=1,3,5...}^{\infty} C_i \sin\frac{i\pi x}{2l} = f(x) \tag{5-12.15}$$

$$\frac{\partial \mathbf{u}(x,0)}{\partial t} = \frac{\pi a}{2l} \sum_{i=1,3,5...}^{\infty} iD_i \sin\frac{i\pi x}{2l} = g(x) \tag{5-12.16}$$

The determination of C_i and D_i are obtained by the using the properties of integration of sinusoidal functions over a full period.

Determining C_i by Euler's formulas:

Thus, multiply both sides of equation (5-12.15) by $\sin\frac{j\pi x}{2l}$, then integrate both sides with respect to x over the period $-l$ to $+l$, we get

$$\int_{-l}^{+l} \sum_{i=1,3,5...}^{\infty} C_i \sin\frac{i\pi x}{2l} \sin\frac{j\pi x}{2l}\, dx = \int_{-l}^{+l} f(x)\sin\frac{i\pi x}{2l}\, dx \tag{5-12.17a}$$

We will now evaluate the following integral in the two cases: $i = j$ and $i \neq j$:

First, let us substitute the product of two sine-functions by the sum of two cosines, integrate, and substitute by the integration limits, as follows

$$\int_{-l}^{+l} C_i \sin\frac{i\pi x}{2l} \sin\frac{j\pi x}{2l}\, dx = \frac{1}{2}\int_{-l}^{+l} C_i\left(\cos\frac{i-j}{2l}\pi x - \cos\frac{i+j}{2l}\pi x\right)dx$$

$$= \frac{C_i}{2}\left[\frac{\sin\left(\dfrac{i-j}{2l}\pi x\right)}{\dfrac{i-j}{2l}\pi} - \frac{\sin\left(\dfrac{i+j}{2l}\pi x\right)}{\dfrac{i+j}{2l}\pi}\right]_{-l}^{l}$$

145

$$= \frac{C_i}{2} \left(\frac{\sin\left(\frac{i-j}{2}\pi\right)}{\frac{i-j}{2l}\pi} - \frac{\sin\left(\frac{i+j}{2}\pi\right)}{\frac{i+j}{2l}\pi} + \frac{\sin\left(\frac{i-j}{2}\pi\right)}{\frac{i-j}{2l}\pi} - \frac{\sin\left(\frac{i+j}{2}\pi\right)}{\frac{i+j}{2l}\pi} \right)$$

$$= lC_i \left(\frac{\sin\left(\frac{i-j}{2}\pi\right)}{\frac{i-j}{2}\pi} - \frac{\sin\left(\frac{i+j}{2}\pi\right)}{\frac{i+j}{2}\pi} \right) \qquad (5\text{-}12.17b)$$

Case 1: $i = j$

In equation (5-12.17b), it is easy to prove the following two relations

$$\lim_{i=j} \left(\frac{\sin\left(\frac{i-j}{2}\pi\right)}{\frac{i-j}{2}\pi} \right) = 1$$

(5-12.18a)

$$\lim_{i=j} \left(\frac{\sin\left(\frac{i+j}{2}\pi\right)}{\frac{i+j}{2}\pi} \right) = \left(\frac{\sin(i\pi)}{i\pi} \right) = 0, \qquad i = 1,3,5,\ldots$$

(5-12.18b)

Thus, from (5-12.18a and b) and (5-12.17b), we get the first Euler's formula for the constants C_i's, as follows.

$$C_i = \frac{1}{l} \int_{-l}^{+l} f(x)\sin\frac{i\pi x}{2l}dx \qquad (5\text{-}12.18c)$$

Case 2: $i \neq j$

In equation (5-12.17b), it is easy to prove the following two relations

146

$$\lim_{i \neq j} \left(\frac{\sin\left(\frac{i-j}{2}\pi\right)}{\frac{i-j}{2}\pi} \right) = 0, \qquad i, j = 1,3,5,.. \qquad i \neq j$$

(5-12.19a)

$$\lim_{i \neq j} \left(\frac{\sin\left(\frac{i+j}{2}\pi\right)}{\frac{i+j}{2}\pi} \right) = \left(\frac{\sin(i\pi)}{i\pi} \right) = 0, \qquad i = 1,3,5,...$$

(5-12.18b)

Determining D_i by Euler's formulas:

Thus, multiply both sides of equation (5-12.16) by $\sin\dfrac{j\pi x}{2l}$, then integrate both sides with respect to x over the period $-l$ to $+l$, we get

$$\frac{i\pi a}{2l} \int_{-l}^{+l} \sum_{i=1,3,5,..}^{\infty} D_i \sin\frac{i\pi x}{2l} \sin\frac{j\pi x}{2l} dx = \int_{-l}^{+l} g(x)\sin\frac{i\pi x}{2l} dx \qquad (5\text{-}12.19a)$$

Since we have already proved in equation (5-12.18a) that only at $i = j$, the integral on the left hand of equation (5-12.19a) is equal to $l i_d$, then we have the second Euler's formula for the constants D_i's, as follows.

$$D_i = \frac{2}{i\pi a} \int_{-l}^{+l} g(x)\sin\frac{i\pi x}{2l} dx \qquad (5\text{-}12.19b)$$

Final solution of Lamé's equation for thin long bound elastic bar is obtained from (5-12.13), (5-12.18c), and (5-12.19b)

$$u(x,t) = \sum_{i=1,3,5,..}^{\infty} \sin\frac{i\pi x}{2l}\left[\left(\frac{1}{l}\int_{-l}^{+l} f(x)\sin\frac{i\pi x}{2l} dx\right)\cos\frac{i\pi at}{2l} + \left(\frac{2}{i\pi a}\int_{-l}^{+l} g(x)\sin\frac{i\pi x}{2l} dx\right)\sin\frac{i\pi at}{2l}\right]$$

(5-12.20)

Where, the initial displacement f(x) and initial speed g(x) are postulated in equation (5-12.14).

CHAPTER 6

TORSION, BENDING, AND SUSPENSION OF A BAR

6.1. Pure shear stress

A. Torsion of a circular bar

Example 22

Find the deformation of a circular cylinder, Figure 6-1, torqued by twisting moment comprised of two opposite couples that produce pure shear stresses

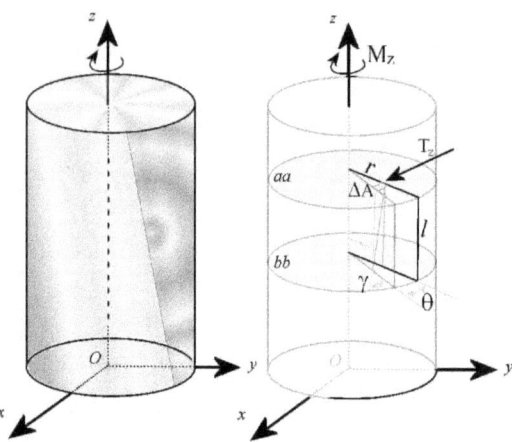

Figure 6-1. Sectional analysis of azimuthal torque of a cylindrical bar assuming spatially linear stresses.

Solution

(i) Torque

Consider two arbitrary cross sections *aa* and *bb* located in the xy-plane and separated by unit length *l* along the z-axis.

Based on the assumption that the torqued bar is only being sheared, therefore, the relative twisting angle between *aa* and *bb*, denoted by θ (measured in the xy-plane), refers to uniform rotation of all points of one cross section with respect to the other.

In order to estimate the internal stresses caused by twisting moment (torque) on section *aa*, we will integrate, over the radial zr-plane, the infinitesimal torque generated at radial point *r* by shearing stress T_z applied at infinitesimal area ΔA as follows.

$$M_z = \int_F T_z r \, dA = \int_F T_z r \left(r dr d\theta \right) \tag{6-1.1}$$

As the infinitesimal torque $T_a$ra rotates the radius *r* an angle θ in the plane *aa*, the unit separation between cross sections *aa* and *bb* rotates an angle γ relative it initial position along the z-direction. Therefore, from, we get

$$\gamma = r\,\theta \tag{6-1.2}$$

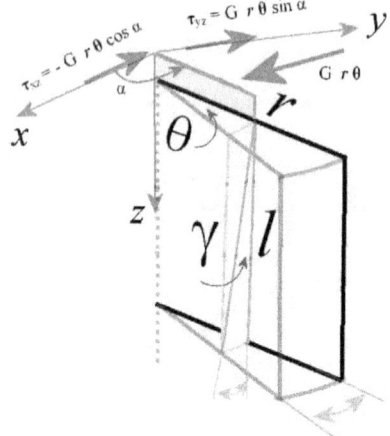

Figure 6-2. Sectional analysis of infinitesimal shear angle γ in relation of cross sectional twisting angle θ.

Thus, from **Hook's shear law**, equation (3-1.3) the shear stress T_z is given by

$$T_z = G\,\gamma = G\ r\,\theta \qquad\qquad (6\text{-}1.3)$$

(ii) Stresses

The nine stresses applying on a 3D solid body are obtained by decomposing T_z into its projections, in addition to our imposed conditions that the body experiences pure shearing stresses. Therefore,

Table 6-1: Nine stresses on cylindrical rod torqued at its base by spatially linear stresses acting in the xy-plane and inducing pure twisting of cross-sections perpendicular to z-axis.

	x-axis stresses	y-axis stresses	z-axis stresses
Normal stress	$\sigma_x = 0$	$\sigma_y = 0$	$\sigma_z = 0$
Shear stress	$\tau_{xz} = \tau_{zx} = -\,G\ r\,\theta \sin\alpha$	$\tau_{yz} = \tau_{zy} = G\ r\,\theta \cos\alpha$	$\tau_{yx} = \tau_{xy} = 0$

Substituting by $\sin\alpha = y / r$ and $\cos\alpha = x / r$, Table 6-1 becomes

Table 6-2: Nine stresses in Table 6-1 represented in terms of x and y coordinates.

	x-axis stresses	y-axis stresses	z-axis stresses
Normal stress	$\sigma_x = 0$	$\sigma_y = 0$	$\sigma_z = 0$

Shear stress $\quad \tau_{xz} = \tau_{xz} = - G\,y\,\theta \qquad \tau_{yz} = \tau_{zy} = G\,x\,\theta \qquad \tau_{yx} = \tau_{xy} = 0$

(iii) Strains

Since all strains are assumed shear strains caused by angular twisting of cross sections perpendicular to the oz-axis, the results of Hooke's shear law are tabulated below.

Table 6-3: Nine strains corresponding to stress in Table 6-2.

	x-axis strains	y-axis strains	z-axis strains
Elongation strain	$\varepsilon_{xx} = 0$	$\varepsilon_{yy} = \alpha_y = 0$	$\varepsilon_{zz} = 0$
Shear strain	$\alpha_{xz} = -\,y\,\theta$	$\alpha_{yz} = \alpha_{zy} = x\,\theta$	$\alpha_{yx} = 0$

From the table, we could determine the displacements by integrating the strains over spatial lengths, using **Cauchy's equations** (2-2) and (2-3), as follows:

$$\varepsilon_{xx} = \frac{\partial u}{\partial x} = 0 \tag{6-1.4a}$$

$$\varepsilon_{yy} = \frac{\partial v}{\partial y} = 0 \tag{6-1.4b}$$

$$\varepsilon_{zz} = \frac{\partial w}{\partial z} = 0 \tag{6-1.4c}$$

$$2\alpha_{xy} = \frac{\partial u}{\partial y} + \frac{\partial v}{\partial x} = 0 \tag{6-1.4d}$$

$$2\alpha_{yz} = \frac{\partial v}{\partial z} + \frac{\partial w}{\partial y} = x\theta \tag{6-1.4e}$$

$$2\alpha_{zx} = \frac{\partial w}{\partial x} + \frac{\partial u}{\partial z} = -y\theta \tag{6-1.4f}$$

(iv) Displacements

Integrating (6-1.4a through 4c), we get the following three arbitrary functions for displacement **u**, **v**, and **w**, of the form

$$\mathbf{u} = \mathbf{u}\,(y,z) \tag{6-1.5a}$$
$$\mathbf{v} = \mathbf{v}\,(x,z) \tag{6-1.5b}$$
$$\mathbf{w} = \mathbf{w}(x,y) \tag{6-1.5c}$$

In order to determine the exact function that suit our assumptions of spatially linear stresses causing pure shearing, we differentiate equations (6-1.5) with respect to x, y, and z, with the constraints imposed by equations (6.1.5) that: **u** is only a function in y and z, **v** is only a function in x and z, and **w** is only a function in x and y.

Therefore, the nine second-derivatives are obtained from (6-1.4) and (6-1.5) as follows.

Differentiating (6-1.5) w.r.t. x, we get

$$\frac{\partial^2 \mathbf{v}(x,z)}{\partial x^2} = 0 \tag{6-1.6a}$$

$$\frac{\partial^2 \mathbf{v}(x,z)}{\partial x \partial z} + \frac{\partial^2 \mathbf{w}(x,y)}{\partial x \partial y} = \theta \tag{6-1.6b}$$

$$\frac{\partial^2 \mathbf{w}(x,y)}{\partial x^2} = 0 \tag{6-1.6c}$$

Differentiating (6-1.5) w.r.t. y, we get

$$\frac{\partial^2 \mathbf{u}(y,z)}{\partial y^2} = 0 \tag{6-1.6d}$$

$$\frac{\partial^2 \mathbf{w}(x,y)}{\partial y^2} = 0 \tag{6-1.6e}$$

$$\frac{\partial^2 \mathbf{w}(x,y)}{\partial x \partial y} + \frac{\partial^2 \mathbf{u}(y,z)}{\partial z \partial y} = -\theta \tag{6-1.6f}$$

Differentiating (6-1.5) w.r.t. z, we get

$$\frac{\partial^2 \mathbf{u}(y,z)}{\partial y \partial z} + \frac{\partial \mathbf{v}^2(z,y)}{\partial x \partial z} = 0 \tag{6-1.6g}$$

$$\frac{\partial^2 \mathbf{v}(x,z)}{\partial z^2} = 0 \tag{6-1.6h}$$

$$\frac{\partial^2 \mathbf{u}(y,z)}{\partial z^2} = 0 \tag{6-1.6i}$$

Our remaining job in obtaining isolated derivatives for u, v, and w is to algebraically perform additions and subtractions on the mixed equations (6-1.6b), (6-1.6f), and (6-1.6g).

Subtract terms of equations (6-1.6f) from terms of equations (6-1.6b), we get

$$\frac{\partial^2 \mathbf{v}(x,z)}{\partial x \partial z} - \frac{\partial^2 \mathbf{u}(y,z)}{\partial z \partial y} = 2\theta \tag{6-1.6j}$$

Add terms of equation (6-1.6j) to terms of equation (6-1.6g)

$$\frac{\partial^2 \mathbf{v}(x,z)}{\partial x \partial z} = 0 \tag{6-1.6k}$$

Substituting from (6-1.6k) into (6-1.6j), we get

$$\frac{\partial^2 \mathbf{u}(y,z)}{\partial z \partial y} = -\theta \tag{6-1.6l}$$

Finally, substituting from (6-1.6k) into (6-1.6b), we get

$$\frac{\partial^2 \mathbf{w}(x,y)}{\partial x \partial y} = 0 \tag{6-1.6m}$$

Hence, to sum up, we obtained isolated second-derivatives for $\mathbf{u}$, $\mathbf{v}$, and $\mathbf{w}$, that are tabulated as follows

Table 6-4. Second-derivatives of displacements u, v, and w for the simplified problem of twisted cylindrical bar under torque applied at its base.

$\dfrac{\partial^2 \mathbf{u}(y,z)}{\partial z \partial y} = -\theta$	$\dfrac{\partial^2 \mathbf{v}(x,z)}{\partial x^2} = 0$	$\dfrac{\partial^2 \mathbf{w}(x,y)}{\partial x^2} = 0$
$\dfrac{\partial^2 \mathbf{u}(y,z)}{\partial y^2} = 0$	$\dfrac{\partial^2 \mathbf{v}(x,z)}{\partial x \partial z} = 0$	$\dfrac{\partial^2 \mathbf{w}(x,y)}{\partial y^2} = 0$
$\dfrac{\partial^2 \mathbf{u}(y,z)}{\partial z^2} = 0$	$\dfrac{\partial^2 \mathbf{v}(x,z)}{\partial z^2} = 0$	$\dfrac{\partial^2 \mathbf{w}(x,y)}{\partial x \partial y} = 0$

(iv.a) Arbitrary displacement functions

Integrating the nine derivatives in Table 6-4, we get,

$$\mathbf{u}(y,z) = -\theta yz + C_2(y) + C_3$$
$$\tag{6-1.7a}$$
$$\mathbf{u}(y,z) = yC_4(z) + C_6$$
$$\tag{6-1.7b}$$
$$\mathbf{u}(y,z) = zC_7(y) + C_9$$
$$\tag{6-1.7c}$$

$$\mathbf{v}(x,z) = xA_1(z) + A_3 \tag{6-1.7d}$$
$$\mathbf{v}(x,z) = \theta xz + A_5(x) + A_6 \tag{6-1.7e}$$
$$\mathbf{v}(x,z) = zA_7(x) + A_9 \tag{6-1.7f}$$

$$\mathbf{W}(x,y) = xB_1(y) + B_3 \tag{6-1.7g}$$
$$\mathbf{W}(x,y) = yB_4(x) + B_6 \tag{6-1.7h}$$
$$\mathbf{W}(x,y) = B_1(y) + B_2(x) + B_3 \tag{6-1.7i}$$

From equations (6-1.7), we conclude that the following forms of displacements are the general solutions of the equations in Table 6-4.

$$u(y,z) = -\theta yz + ay + bz + c \tag{6-1.8a}$$
$$v(x,z) = \theta xz + dx + ez + f \tag{6-1.8b}$$
$$\mathbf{W}(x,y) = gx + hy + k \tag{6-1.8c}$$

(iv.b) Boundary conditions
(A) Proper strains

In order to determine the constants a through k in equations (6-1.8), we start by substituting with those general solutions in equations (6-1.4d through 4f) which relate strains to first derivatives of displacements, as follows

$$\frac{\partial u}{\partial y} + \frac{\partial v}{\partial x} = \frac{\partial}{\partial y}\left(-\theta yz + ay + bz + c\right) + \frac{\partial}{\partial x}\left(\theta xz + dx + ez + f\right)$$
$$-\theta z + a + \theta z + d = 0$$

i.e., $\quad\quad a = -d \tag{6-1.9a}$

$$\frac{\partial v}{\partial z} + \frac{\partial w}{\partial y} = \frac{\partial}{\partial z}\left(\theta xz + dx + ez + f\right) + \frac{\partial}{\partial y}\left(gx + hy + k\right) = x\theta$$
$$\theta x + e + h = x\theta$$

i.e., $\quad\quad e = -h \tag{6-1.9b}$

$$\frac{\partial w}{\partial x} + \frac{\partial u}{\partial z} = \frac{\partial}{\partial z}\left(-\theta yz + ay + bz + c\right) + \frac{\partial}{\partial x}\left(gx + hy + k\right) = -y\theta$$
$$= -\theta y + b + g = -y\theta$$

i.e., $\quad\quad b = -g \tag{6-1.9c}$

(B) Conditions for immobility

At $\quad\quad x = y = z = 0 \rightarrow \quad$ deformation vanish $\rightarrow$ $u = v = w = 0 \tag{6-1.10}$

Substitute by (6-1.9) in (6-1.8) and observing the conditions in (6-1.10), we get

$$c = f = k = 0 \tag{6-1.11}$$

154

Therefore, equations (6-1.8), we get (a = -d, e = -h, b = -g)

$$u(y,z) = -\theta yz + ay + bz \qquad (6-1.12a)$$
$$v(x,z) = \theta xz - ax + ez \qquad (6-1.12b)$$
$$w(x,y) = -bx - ey \qquad (6-1.12c)$$

(C) Conditions for immobility of initial elements at origin

At $\qquad$ $x = y = z = 0 \rightarrow$ $\quad$ deformation vanish $\rightarrow \dfrac{\partial u}{\partial z} = \dfrac{\partial u}{\partial y} = \dfrac{\partial v}{\partial z} = 0$ $\qquad (6-1.13)$

Then from (6-1.12), we get the remaining three constants

$$\frac{\partial u}{\partial z} = -0 + b = 0, \qquad b = 0$$
$$\frac{\partial u}{\partial y} = -0 + a = 0, \qquad a = 0 \qquad (6-1.14)$$
$$\frac{\partial v}{\partial z} = 0 + e = 0, \qquad e = 0$$

Thus, we reach our desired solution

$$u(y,z) = -\theta yz \qquad (6-1.15a)$$
$$v(x,z) = \theta xz \qquad (6-1.15b)$$
$$w(x,y) = 0 \qquad (6-1.15c)$$

(v) Compliance of stresses with equations of internal continuity and equilibrium

In order to ascertain that the stresses in Table 6-2 and strains in Table 6-4 represent acceptable solutions to the following fundamental governing equations of the theory of elasticity:

(A) **Navier's** equations of equilibrium of stresses, (1-4.1)
(B) **Saint –Venant's** equation continuity, (2-13.1)
(C) Volumetric **Hooke's law**, equation (3-3.1)
(D) **Hooke's shear law**, equation (3-3.2), and
(E) **Surface condition**, equations (1-8)

Re-written here for one dimensional stress on the x-axis as follows

Navier's equations $\qquad \dfrac{\partial \sigma_x}{\partial x} + \dfrac{\partial \tau_{xy}}{\partial y} + \dfrac{\partial \tau_{xz}}{\partial z} + \rho X_i = \rho \dfrac{\partial^2 u}{\partial t^2}$ $\qquad (6-1.16)$

Saint –Venant's equation $\qquad \dfrac{\partial}{\partial x}\left(\dfrac{\partial \alpha_{xy}}{\partial z} + \dfrac{\partial \alpha_{xz}}{\partial y} - \dfrac{\partial \alpha_{yz}}{\partial x} \right) = \dfrac{\partial^2 \varepsilon_{xx}}{\partial y \partial z}$ $\qquad (6-1.17)$

155

Volumetric Hooke's law	$\varepsilon_{xx} = E^{-1}\left[\sigma_x - v\left(\sigma_y + \sigma_z\right)\right]$	(6-1.18)
Hooke's shear law	$\varepsilon_{xy} = G^{-1}\tau_{xy}$	(6-1.19)

Surface conditions	$X_n = \sigma_x\, l + \tau_{xy}\, m + \tau_{xz}\, n$	(6-1.20)
	$Y_n = \tau_{yx}\, l + \sigma_y\, m + \tau_{yz}\, n$	(6-1.21)
	$Z_n = \tau_{zx}\, l + \tau_{zy}\, m + \sigma_z\, n$	(6-1.22)

(a) **Navier's** equation, (6-1.16), is satisfied by assuming that $X_i = 0$ and $\dfrac{\partial^2 u}{\partial t^2} = 0$, as follows

$$\frac{\partial}{\partial x}0 + \frac{\partial}{\partial y}0 + \frac{\partial(-Gy\theta)}{\partial z} + 0 = 0$$

Where the twisting angle θ and shear modulus G are assumed constant.

(b) **Saint –Venant**'s equation (6-1.17) and Tables (6-2) and (6-3), we infer that the second derivatives of stresses and strains in Tables (6-2) and (6-3) comply with both education.

$$\frac{\partial}{\partial x}\left(\frac{\partial}{\partial z}0 + \frac{\partial(-y\theta)}{\partial y} - \frac{\partial(x\theta)}{\partial x}\right) = \frac{\partial^2}{\partial y \partial z}0$$

$$\frac{\partial}{\partial x}(0 - \theta - \theta) = 0$$

(c) In both tables, all normal stresses and elongation strains vanish, thus satisfying **Hooke's law**, equation (6-1.18).

(d) Surface equations (6-1.20) though 10) are applied to the lateral walls and bases of the cylinders as shown in Figure 6-2, as follows

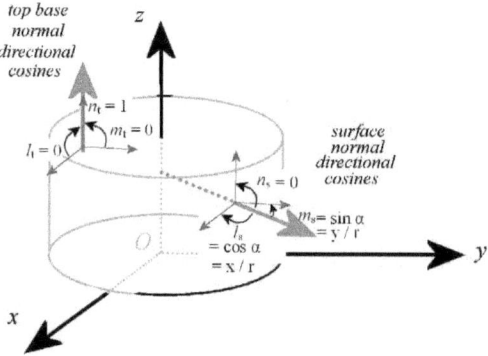

Figure 6-3. Directional cosines of normals on lateral wall and base of cylindrical rod.

The directional cosines of the normal to the lateral surface of the cylinder are:

$$l_s = \cos \alpha = x\,/\,r$$
$$m_s = \sin \alpha = y\,/\,r$$
$$n_s = 0 \tag{6-1.23}$$

Therefore, from Tables 6-2 and 6-3 and surface equations we get

$$X_n = (0)(x/r) + (0)(y/r) - G\,y\,\theta(0) = 0 \tag{6-1.24}$$
$$Y_n = (0)(x/r) + (0)(y/r) + G\,x\,\theta\,(0) = 0 \tag{6-1.25}$$
$$Z_n = =- G\,y\,\theta(x/r) + G\,x\,\theta(y/r) + (0)\,(0) = 0 \tag{6-1.26}$$

The directional cosines of the normal to the base surface of the cylinder are:

$$l_t = 0$$
$$m_t = 0$$
$$n_t = 1 \tag{6-1.27}$$

Therefore, from Tables 6-2 and 6-3 and surface equations we get

$$X_n = (0)(0) + (0)(0) - G\,y\,\theta(1) = - G\,y\,\theta \tag{6-1.28}$$
$$Y_n = (0)(0) + (0)(0) + G\,x\,\theta\,(1) = G\,x\,\theta \tag{6-1.29}$$
$$Z_n = - G\,y\,\theta(0) + G\,x\,\theta(0) + (0)(1) = 0 \tag{6-1.30}$$

Thus, all our equations are properly satisfied by the stresses and strains in Tables 6-2 and 3, the lateral surfaces of the cylindrical rod are devoid of stresses, and the top base is subjected to the shear stresses $X_n = - Gy\theta$ and $Y_n = Gx\theta$.

(vi) Resultant of stresses

The x-component force:

$$\int_0^r \int_0^{2\pi} \left(- Gy\theta\right) r\,dr\,d\phi = -G\theta \int_0^r \int_0^{2\pi} r^2 \sin\phi\,dr\,d\phi$$

$$= G\theta \frac{R^3}{3} \left[\cos\phi\right]_0^{2\pi} \tag{6-1.31}$$

$$= G\theta \frac{R^3}{3} \left[1 - 1\right] = 0$$

The y-component force:

157

$$\int_0^r \int_0^{2\pi} (Gx\theta)rdrd\phi = G\theta \int_0^r \int_0^{2\pi} r^2 \cos\phi drd\phi$$

$$= -G\theta \frac{R^3}{3} [\sin\phi]_0^{2\pi} \qquad (6\text{-}1.32)$$

$$= G\theta \frac{R^3}{3} [1-1] = 0$$

Thus, resultant of the force on the base of the cylinder vanishes.

(vii) Resultant of moment of torsional couple

$$\int_0^r \int_0^{2\pi} T_z r dA = \int_0^r \int_0^{2\pi} (Gr\theta)r(rdrd\phi)$$

$$= G\theta \int_0^r \int_0^{2\pi} r^3 drd\phi \qquad (6\text{-}1.33)$$

$$= 2\pi G\theta \frac{R^4}{4} = G\theta \frac{\pi R^4}{2}$$

$$= G\theta I_p$$

Where, $I_p = \dfrac{\pi R^4}{2}$ is the **polar moment of inertia** around the oz-axis,

6.2. Pure bending stress

Example 23

A prismatic bar is placed along the oz-axis, bent by two opposing couples operating in the xOz-plane and applied at its terminal. Assume that the cross sections of the bar retain their planar form and angulation on the neutrally curved centroid axis of the bar.

Find the strains, stresses, and displacements that conform to the equations of elasticity discussed above.

Solution

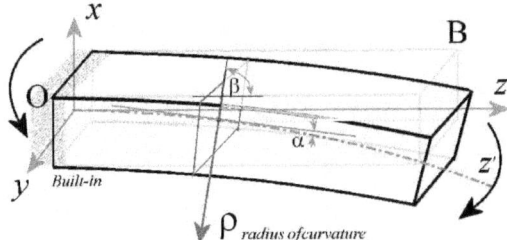

Figure 6-4. A prismatic bar along oz-axis is bent by two equal and opposite couples in the x0z plane such that all cross-section perpendicular on the oz-axis retain their angular and planar orientation on the curved oz' axis.

(i) Curvature strain

In pure bending of the prismatic bar in Figure 6-3, the fibers along the oz-axis experience elongation in proportion to their distance x from the neutral axis oz as illustrated in Figure 6-5

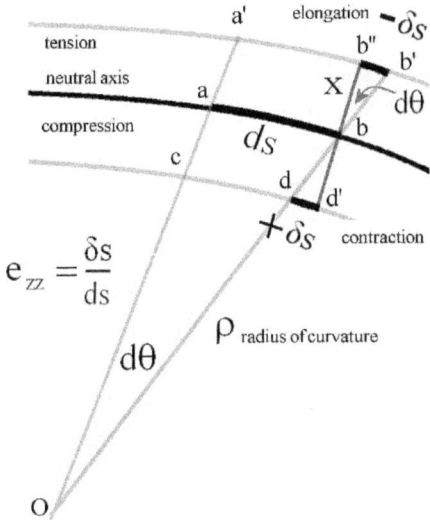

Figure 6-5. Relationship between curvature and elongation of a bar.

In Figure 6-5

ds: The arc distance ab on the neutral axis.
ρ: The distance Oa = Ob. and is the radius of curvature at point a and point b
$d\theta$ The infinitesimal angle of curvature corresponding to the arc ds
x: The distance bb' measured from the neutral axis to an arbitrary point on the plane Ocaa'.
δs: The elongation b"b' on contraction dd' due to the shift z from the neutral axis.

The **unit elongation** at distance x from the neutral axis is given by the ratio δs/ds. Therefore,

$$e_{zz} = -\frac{\delta s}{ds} = -\frac{x d\theta}{\rho d\theta} = -\frac{x}{\rho} \tag{6-2.1}$$

Where the negative sign changes with the direction of measuring x from the neutral axis to account for elongation of contraction, tension or compression.

(ii) Stress

The bent bar is stretched (tension) on the side of the neutral line away from the center of curvature, compressed on the side close to the center of curvature. From Hooke's law, we get

$$\sigma_z = E e_{zz} = -\frac{xE}{\rho} \tag{6-2.2}$$

The remaining five stresses vanish by virtue of our assumption that the cross sections of the bar retain their shapes and perpendicular angulation on the neutral axis, without shearing stresses or lateral stresses perpendicular on the xOz-plane.

(A) Thus, substituting in the surface condition, equation (1-8), by the following directional cosines of the normal to the surface:

$$l_s = 1$$
$$m_s = 0$$
$$n_s = 0 \tag{6-2.3}$$

Therefore, the **surface stresses** are

$$X_n = 0(1) + (0)(0) + (0)(0) = 0 \tag{6-2.4a}$$
$$Y_n = (0)(1) + (0)(0) + (0)(0) = 0 \tag{6-2.4b}$$
$$Z_n = (0)(1) + (0)(0) + (-Ex/\rho)(0) = 0 \tag{6-2.4c}$$

(B) Moment of couple at the bar end B, Figure 6-5

$$M_y = \int_{-a}^{a} \int_{-b}^{b} \sigma_z x\, dA = -\int_{-a}^{a} \int_{-b}^{b} E\frac{x^2}{\rho}\, dA$$

$$= -\frac{E}{\rho}\int_{-a}^{a} \int_{-b}^{b} x^2\, dA = -\frac{E}{\rho}I_p$$

(6-2.5)

Where, I_p is the **moment of inertia** around the oz-axis. It should be noted that the torque M_y bends the bar in the xOz-plane, the direction of the couple is the y-axis that normal on the plane of rotation, hence the subscript "y".

(iii) Strains

Cauchy's equations (2-2) and Hooke's equations (3-3.1)

$$\frac{\partial u}{\partial x} = \varepsilon_{xx} = E^{-1}[\sigma_x - v(\sigma_y + \sigma_z)] = E^{-1}\left[0 - v\left(0 - \frac{Ex}{\rho}\right)\right] = \frac{vx}{\rho}$$

(6-2.6a)

$$\frac{\partial v}{\partial y} = \varepsilon_{yy} = E^{-1}[\sigma_x - v(\sigma_y + \sigma_z)] = E^{-1}\left[0 - v\left(0 - \frac{Ex}{\rho}\right)\right] = \frac{vx}{\rho}$$

(6-2.6b)

$$\frac{\partial w}{\partial z} = \varepsilon_{zz} = E^{-1}[\sigma_x - v(\sigma_y + \sigma_z)] = E^{-1}\left[-\frac{Ex}{\rho} - v(0 - 0)\right] = -\frac{x}{\rho}$$

(6-2.6c)

$$2\alpha_{xy} = \frac{\partial u}{\partial y} + \frac{\partial v}{\partial x} = 0$$

(6-2.6d)

$$2\alpha_{yz} = \frac{\partial v}{\partial z} + \frac{\partial w}{\partial y} = 0$$

(6-2.6e)

$$2\alpha_{zx} = \frac{\partial w}{\partial x} + \frac{\partial u}{\partial z} = 0$$

(6-2.6f)

Table 6-5: Nine strains expressed in terms of stresses on prismatic bar exposed to pure bending

	x-axis strains	y-axis strains	z-axis strains
Elongation strain	$\frac{\partial u}{\partial x} = \varepsilon_{xx} = \frac{vx}{\rho}$	$\frac{\partial v}{\partial y} = \varepsilon_{yy} = \frac{vx}{\rho}$	$\frac{\partial w}{\partial z} = \varepsilon_{zz} = -\frac{x}{\rho}$
Shear strain	$\alpha_{xz} = 0$	$\alpha_{yz} = \alpha_{zy} = 0$	$\alpha_{yx} = 0$

(iv) Displacements

Integrating the partial derivatives in Table 6-5, we get the following arbitrary general functions for displacements.

Table 6-6: General arbitrary functions for displacements.

	x-axis	y-axis	z-axis

Displacements

$$\mathbf{u} = \frac{vx^2}{2\rho} + f(y,z) \qquad \mathbf{v} = \frac{vxy}{\rho} + g(x,z) \qquad \mathbf{w} = -\frac{xz}{\rho} + h(x,y)$$

The unknown functions f, g, h, in Table 6-6, are determined by substituting the derivatives of the displacements in **Cauchy's equations** (2-3), as follows:

$$\frac{\partial f(y,z)}{\partial y} + \frac{\partial g(x,z)}{\partial x} = -\frac{vy}{\rho} \tag{6-2.7a}$$

$$\frac{\partial g(x,z)}{\partial z} + \frac{\partial h(x,y)}{\partial y} = 0 \tag{6-2.7b}$$

$$\frac{\partial h(x,y)}{\partial x} + \frac{\partial f(y,z)}{\partial z} = \frac{z}{\rho} \tag{6-2.7c}$$

Therefore, the nine second-derivatives are obtained from (6-2.7 as follows.

Differentiating (6-2.7) w.r.t. x, we get

$$\frac{\partial^2 f(y,z)}{\partial y \partial x} + \frac{\partial^2 g(x,z)}{\partial x^2} = \frac{\partial^2 g(x,z)}{\partial x^2} = 0 \tag{6-2.8a}$$

$$\frac{\partial^2 g(x,z)}{\partial z \partial x} + \frac{\partial^2 h(x,y)}{\partial y \partial x} = 0 \tag{6-2.8b}$$

$$\frac{\partial^2 h(x,y)}{\partial x^2} + \frac{\partial^2 f(y,z)}{\partial z \partial x} = \frac{\partial^2 h(x,y)}{\partial x^2} = 0 \tag{6-2.8c}$$

Differentiating (6-2.7) w.r.t. y, we get

$$\frac{\partial^2 f(y,z)}{\partial y^2} + \frac{\partial^2 g(x,z)}{\partial x \partial y} = \frac{\partial^2 f(y,z)}{\partial y^2} = -\frac{v}{\rho} \tag{6-2.8d}$$

$$\frac{\partial^2 g(x,z)}{\partial z \partial y} + \frac{\partial h^2(x,y)}{\partial y^2} = \frac{\partial h^2(x,y)}{\partial y^2} = 0 \tag{6-2.8e}$$

$$\frac{\partial^2 h(x,y)}{\partial x \partial y} + \frac{\partial^2 f(y,z)}{\partial z \partial y} = 0 \tag{6-2.8f}$$

Differentiating (6-2.7) w.r.t. z, we get

$$\frac{\partial^2 f(y,z)}{\partial y \partial z} + \frac{\partial^2 g(x,z)}{\partial x \partial z} = 0 \tag{6-2.8g}$$

$$\frac{\partial^2 g(x,z)}{\partial z^2} + \frac{\partial^2 h(x,y)}{\partial y \partial z} = \frac{\partial^2 g(x,z)}{\partial z^2} = 0 \tag{6-2.8h}$$

$$\frac{\partial^2 h(x,y)}{\partial x \partial z} + \frac{\partial^2 f(y,z)}{\partial z^2} = \frac{\partial^2 f(y,z)}{\partial z^2} = \frac{1}{\rho} \tag{6-2.8i}$$

Our next job is to separate the mixed derivatives in equations (6-2.8b), (6-2.8f), and (6-2.8g), as follows:

Subtract equations (6-2.8f) from (6-2.8b), we get

$$\frac{\partial^2 g(x,z)}{\partial z \partial x} - \frac{\partial^2 f(y,z)}{\partial z \partial y} = 0 \tag{6-2.9a}$$

Add equation (6-2.9a) to (6-2.8g), we get

$$\frac{\partial^2 g(x,z)}{\partial x \partial z} = 0 \tag{6-2.9b}$$

Substitute from (6-2.9b) to (6-2.9a), we get

$$\frac{\partial^2 f(y,z)}{\partial z \partial y} = 0 \tag{6-2.9c}$$

Substitute from (6-2.9b) to (6-2.8b), we get

$$\frac{\partial^2 h(x,y)}{\partial y \partial x} = 0 \tag{6-2.9d}$$

Finally, the nine second-derivatives of f, g, and h are obtained in isolation and are listed in Table 6-7.

Table 6-7. Second-derivatives of displacements f, g, h for the simplified problem of pure bending of a prismatic bar under torque applied at its ends.

$\dfrac{\partial^2 f(y,z)}{\partial z \partial y} = 0$	$\dfrac{\partial^2 g(x,z)}{\partial x^2} = 0$	$\dfrac{\partial^2 h(x,y)}{\partial x^2} = 0$
$\dfrac{\partial^2 f(y,z)}{\partial y^2} = -\dfrac{v}{\rho}$	$\dfrac{\partial^2 g(x,z)}{\partial x \partial z} = 0$	$\dfrac{\partial h^2(x,y)}{\partial y^2} = 0$
$\dfrac{\partial^2 f(y,z)}{\partial z^2} = \dfrac{1}{\rho}$	$\dfrac{\partial^2 g(x,z)}{\partial z^2} = 0$	$\dfrac{\partial^2 h(x,y)}{\partial y \partial x} = 0$

Integrating the nine derivatives in Table 6-7, we get,

$$\mathbf{f}(y,z) = C_1(z) + C_2(y) + C_3$$

(6-2.10a)
$$\mathbf{f}(y,z) = -\frac{vy^2}{2\rho} + yC_4(z) + C_6$$

(6-2.10b) $\mathbf{f}(y,z) = \dfrac{z^2}{2\rho} + zC_7(y) + C_9$

(6-2.10c)

$$\mathbf{g}(x,z) = xA_1(z) + A_3 \qquad\qquad\qquad\qquad\qquad (6\text{-}2.10\text{d})$$
$$\mathbf{g}(x,z) = A_4(z) + A_5(x) + A_6 \qquad\qquad\qquad (6\text{-}2.10\text{e})$$
$$\mathbf{g}(x,z) = zA_7(x) + A_9 \qquad\qquad\qquad\qquad\quad (6\text{-}2.10\text{f})$$

$$\mathbf{h}(x,y) = xB_1(y) + B_3 \qquad\qquad\qquad\qquad\quad (6\text{-}2.10\text{g})$$
$$\mathbf{h}(x,y) = yB_4(x) + B_6 \qquad\qquad\qquad\qquad\quad (6\text{-}2.10\text{h})$$
$$\mathbf{h}(x,y) = B_1(y) + B_2(x) + B_3 \qquad\qquad\qquad (6\text{-}2.10\text{i})$$

From equations (6-2.10), we conclude that the following forms of functions f, g, and h are the complementary displacement solutions in Table 6-6.

$$\mathbf{f}(y,z) = \frac{1}{2\rho}\left(z^2 - vy^2\right) + a_1 y + b_1 z + c_1 \qquad\qquad (6\text{-}2.11\text{a})$$
$$\mathbf{g}(x,z) = a_2 z + b_2 x + c_2 \qquad\qquad\qquad\qquad\quad (6\text{-}2.11\text{b})$$
$$\mathbf{h}(x,y) = a_3 x + b_3 y + c_3 \qquad\qquad\qquad\qquad\quad (6\text{-}2.11\text{c})$$

(iv.b) Boundary conditions

(A) Proper strains

In order to determine the constants a through c in equations (6-2.11), we start by substituting with those general solutions in equations (6-2.7a through 7c) which relate strains to first derivatives of displacements, as follows

$$-\frac{vy}{\rho} + a_1 + b_2 = -\frac{vy}{\rho}$$

i.e., $\qquad a_1 = -b_2 \qquad\qquad\qquad\qquad\qquad\qquad\qquad (6\text{-}2.11\text{d})$

$$a_2 + b_3 = 0$$

i.e., $\qquad a_2 = -b_3 \qquad\qquad\qquad\qquad\qquad\qquad\qquad (6\text{-}2.11\text{e})$

164

$$a_3 + \frac{z}{\rho} + b_1 = \frac{z}{\rho}$$

i.e., $\qquad a_3 = -b_1$ $\qquad\qquad$ (6-2.11f)

Equation (6-2.11) and Table 6-6 provide the general functions for displacement of pure bending of a prismatic bar as follows

$$\mathbf{u} = \frac{vx^2}{2\rho} + \frac{1}{2\rho}\left(z^2 - vy^2\right) + a_1 y + b_1 z + c_1 \qquad\qquad (6\text{-}2.12\text{a})$$

$$\mathbf{v} = \frac{vxy}{\rho} + a_2 z - a_1 x + c_2 \qquad\qquad (6\text{-}2.12\text{b})$$

$$\mathbf{w} = -\frac{xz}{\rho} - b_1 x - a_2 y + c_3 \qquad\qquad (6\text{-}2.12\text{c})$$

(B) Conditions for immobility

At $\qquad x = y = z = 0 \rightarrow$ deformation vanish $\rightarrow$ $\mathbf{u} = \mathbf{v} = \mathbf{w} = 0$ $\qquad$ (6-2.13a)

Substitute by (6-2.13a) in (6-2.12), we get

$$c_1 = c_2 = c_3 = 0 \qquad\qquad (6\text{-}2.13\text{b})$$

(C) Conditions for immobility of initial elements at origin

At $\qquad x = y = z = 0 \rightarrow$ deformation vanish $\rightarrow \dfrac{\partial u}{\partial z} = \dfrac{\partial v}{\partial x} = \dfrac{\partial v}{\partial z} = 0$ $\qquad$ (6-2.14a)

Then differentiating equations (6-2.12) and substituting by (6-2.14a), we get the remaining three constants

$$\frac{\partial \mathbf{u}}{\partial z} = 0 + b_1 = 0$$

$$\frac{\partial \mathbf{v}}{\partial x} = 0 - a_1 = 0 \qquad\qquad (6\text{-}2.14\text{b})$$

$$\frac{\partial \mathbf{v}}{\partial z} = a_2 = 0$$

Thus, (6-2.13b) ($c_1 = c_2 = c_3 = 0$) and (6-2.14b) ($a_1 = a_2 = b_1 = 0$), we reach our desired solution from (6-2.12) as follows

$$u = \frac{1}{2\rho}\left[z^2 + v\left(x^2 - y^2\right)\right]$$ (6-2.15a)

$$v = \frac{vxy}{\rho}$$ (6-2.15b)

$$w = -\frac{xz}{\rho}$$ (6-2.15c)

The radius of curvature, ρ, of material of linear elastic properties is obtained from the differential equation for the **deflection of a beam** with the following known constants:

M   is the **bending moment** at a point x of the beam,
E   is its **Young's modulus** and
I   the **moment of inertia** of the cross section of the beam about its neutral axis.

Without proof, we write relation between M, E, I and ρ, as follows.

$$\rho = \frac{EI}{M}$$ (6-2.16)

(v) Deformations

(a) Points

After deformation, the coordinates of the new points are related to the old points by adding the displacements from equations to the coordinates of the old points as follows

$$x_1 = x_0 + u = x_0 + \frac{1}{2\rho}\left[z_0^2 + v\left(x_0^2 - y_0^2\right)\right]$$ (6-2.16a)

$$y_1 = y_0 + v = y_0 + \frac{vx_0y_0}{\rho}$$ (6-2.16b)

$$z_1 = z_0 + w = z_0 - \frac{x_0z_0}{\rho}$$ (6-2.16c)

(b) Lines

The equation of the neutral axis before deformation is $x = y = 0$.
After deformation, the equation of the axis is obtained from equation (6-2.16)

$$x_1 = \frac{z^2}{2\rho} \tag{6-2.17a}$$

$$y_1 = 0 \tag{6-2.17b}$$

$$z_1 = z_0 \tag{6-2.17c}$$

Differentiating equation (6-2.17a) gives the slope of the tangent to the deformed neutral line as

$$\tan\alpha = \frac{\partial x_1}{\partial z} = \frac{\partial}{\partial z}\left(\frac{z^2}{2\rho}\right) = \frac{z_0}{\rho} \tag{6-2.17d}$$

(c) Planes

Before deformation, any cross sectional plane has the equation $z = z_0$.
After deformation, the equation of the same cross section becomes

$$z_1 = z_0 + \mathbf{w} = z_0 - \frac{xz_0}{\rho}$$
$$= z_0\left(1 - \frac{x}{\rho}\right) \tag{6-2.18}$$

Equation (6-2.18) represents a plane inclined in the x and z axes, but perpendicular to the y-axis.

The slope of the cross-sectional plane on the z-axis is obtained by differentiating (6-2.18) as follows

$$\tan\beta = \frac{\partial x}{\partial z} = \frac{1}{z_0 \dfrac{\partial}{\partial x}\left(1 - \dfrac{x}{\rho}\right)} = \frac{1}{-\dfrac{z_0}{\rho}} = -\frac{\rho}{z_0} \tag{6-2.18a}$$

From equations (6-2.17d) and (6-2.18), we note that

$$\tan\alpha\,\tan\beta = \frac{z_0}{\rho}\left(-\frac{\rho}{z_0}\right) = -1 \tag{6-2.18b}$$

Thus, as we have assumed, the cross sections across the deformed neutral axis are normal to it and remain planar.

(d) Rectangular cross section

167

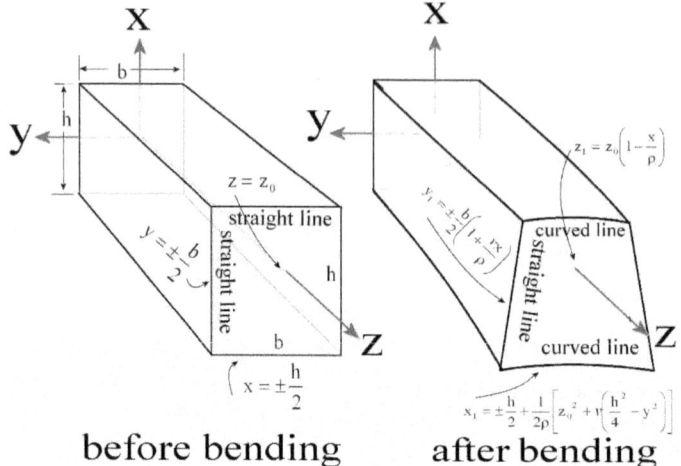

Figure 6-6. Deformation of cross sections of a prismatic bar upon pure bending

As shown in Figure 6-6, before deformation, a bar with rectangular cross section has the following equations

$$z = z_0$$
$$y = \pm\frac{b}{2}$$
$$x = \pm\frac{h}{2}$$

(6-2.19)

Where, the constants b and h are the width and height of the rectangle.

After deformation, equations (6-2.16b) and (6-2.16c) contribute displacements to (6-2.19a) as follows

$$z_1 = z_0 - \frac{xz}{\rho}$$

(6-2.20a)

$$y_1 = \pm\frac{b}{2} + \frac{vxy}{\rho}$$

(6-2.20b)

$$x_1 = \pm\frac{h}{2} + \frac{1}{2\rho}\left[z^2 + v(x^2 - y^2)\right]$$

(6-2.20c)

Replace

z by z_0 in (6-2.20a),

168

y by ± b/2, in (6-2.20b), and
x by ± h/2, in (6-2.20c)

We get

$$z_1 = z_0\left(1 - \frac{x}{\rho}\right)$$ (6-2.21a)

$$y_1 = \pm\frac{b}{2}\left(1 + \frac{\nu x}{\rho}\right)$$ (6-2.21b)

$$x_1 = \pm\frac{h}{2} + \frac{1}{2\rho}\left[z_0^2 + \nu\left(\frac{h^2}{4} - y^2\right)\right]$$ (6-2.21c)

Equation (6-2.21a) has been discussed under (6-2.18), and represents the plane inclined on the z and x axes.

Equation (6-2.21b) represents the linear sides of the deformed cross section, which are first degree in x.

Equation (6-2.21c) represents the curved top and base of the deformed rectangle, which are second degree in y.

6.3. Suspension of a bar

A cylindrical prism of arbitrary cross section suspended from its upper end and stretched by its own weight.

(i) Stresses

We will place the origin of coordinates at the free end of bar, $\sigma_z = 0$ at $z = 0$. Thus, the stresses applied to the body take the following forms:

	x-axis stresses	y-axis stresses	z-axis stresses
Normal stress	$\sigma_x = 0$	$\sigma_y = 0$	$\sigma_z - \rho g z$
Shear stress	$\tau_{xz} = \tau_{zx} = 0$	$\tau_{yz} = \tau_{zy} = 0$	$\tau_{yx} = \tau_{xy} = 0$

(ii) Strains

Cauchy's equations (2-2) and Hooke's equations (3-3.1)

$$\frac{\partial u}{\partial x} = \varepsilon_{xx} = E^{-1}\left[\sigma_x - \nu\left(\sigma_y + \sigma_z\right)\right] = E^{-1}\left[0 - \nu(0 + \rho g z)\right] = -\frac{\nu\rho g}{E}z$$ (6-3.1a)

169

$$\frac{\partial v}{\partial y} = \varepsilon_{yy} = E^{-1}\left[\sigma_x - v\left(\sigma_y + \sigma_z\right)\right] = E^{-1}\left[0 - v(0 + \rho g z)\right] = -\frac{v\rho g}{E} z \qquad (6\text{-}3.1b)$$

$$\frac{\partial w}{\partial z} = \varepsilon_{zz} = E^{-1}\left[\sigma_x - v\left(\sigma_y + \sigma_z\right)\right] = E^{-1}\left[\rho g z - v(0 - 0)\right] = \frac{\rho g}{E} z \qquad (6\text{-}3.1c)$$

$$2\alpha_{xy} = \frac{\partial u}{\partial y} + \frac{\partial v}{\partial x} = 0 \qquad (6\text{-}3.1d)$$

$$2\alpha_{yz} = \frac{\partial v}{\partial z} + \frac{\partial w}{\partial y} = 0 \qquad (6\text{-}3.1e)$$

$$2\alpha_{zx} = \frac{\partial w}{\partial x} + \frac{\partial u}{\partial z} = 0 \qquad (6\text{-}3.1f)$$

Table 6-8: Nine strains expressed in terms of stresses on prismatic bar suspended from its top

	x-axis strains	y-axis strains	z-axis strains
Elongation strain	$\dfrac{\partial u}{\partial x} = \varepsilon_{xx} = -\dfrac{v\rho g}{E} z$	$\dfrac{\partial v}{\partial y} = \varepsilon_{yy} = -\dfrac{v\rho g}{E} z$	$\dfrac{\partial w}{\partial z} = \varepsilon_{zz} = \dfrac{\rho g}{E} z$
Shear strain	$\alpha_{xz} = 0$	$\alpha_{yz} = \alpha_{zy} = 0$	$\alpha_{yx} = 0$

(iv) Displacements

Integrating the partial derivatives in Table 6-8, we get the following arbitrary general functions for displacements.

Table 6-9: General arbitrary functions for displacements.

	x-axis	y-axis	z-axis
Displacements	$u = -\dfrac{v\rho g}{E} xz + f(y,z)$	$v = -\dfrac{v\rho g}{E} yz + g(x,z)$	$w = \dfrac{\rho g}{2E} z^2 + h(x,y)$

The unknown functions f, g, h, in Table 6-9, are determined by substituting the derivatives of the displacements in **Cauchy's equations** (2-3), as follows:

$$\frac{\partial f(y,z)}{\partial y} + \frac{\partial g(x,z)}{\partial x} = 0 \qquad (6\text{-}3.2a)$$

$$\frac{\partial g(x,z)}{\partial z} + \frac{\partial h(x,y)}{\partial y} = \frac{v\rho g}{E} y \qquad (6\text{-}3.2b)$$

$$\frac{\partial h(x,y)}{\partial x} + \frac{\partial f(y,z)}{\partial z} = \frac{v\rho g}{E} x \qquad (6\text{-}3.2c)$$

Therefore, the nine second-derivatives are obtained from (6-3.2) as follows.

Differentiating (6-3.2) w.r.t. x, we get

$$\frac{\partial^2 f(y,z)}{\partial y \partial x} + \frac{\partial^2 g(x,z)}{\partial x^2} = \frac{\partial^2 g(x,z)}{\partial x^2} = 0 \qquad (6\text{-}3.3\text{a})$$

$$\frac{\partial^2 g(x,z)}{\partial z \partial x} + \frac{\partial^2 h(x,y)}{\partial y \partial x} = 0 \qquad (6\text{-}3.3\text{b})$$

$$\frac{\partial^2 h(x,y)}{\partial x^2} + \frac{\partial^2 f(y,z)}{\partial z \partial x} = \frac{\partial^2 h(x,y)}{\partial x^2} = \frac{\nu \rho g}{E} \qquad (6\text{-}3.3\text{c})$$

Differentiating (6-3.2) w.r.t. y, we get

$$\frac{\partial^2 f(y,z)}{\partial y^2} + \frac{\partial^2 g(x,z)}{\partial x \partial y} = \frac{\partial^2 f(y,z)}{\partial y^2} = 0 \qquad (6\text{-}3.3\text{d})$$

$$\frac{\partial^2 g(x,z)}{\partial z \partial y} + \frac{\partial h^2(x,y)}{\partial y^2} = \frac{\partial h^2(x,y)}{\partial y^2} = \frac{\nu \rho g}{E} \qquad (6\text{-}3.3\text{e})$$

$$\frac{\partial^2 h(x,y)}{\partial x \partial y} + \frac{\partial^2 f(y,z)}{\partial z \partial y} = 0 \qquad (6\text{-}3.3\text{f})$$

Differentiating (6-3.2) w.r.t. z, we get

$$\frac{\partial^2 f(y,z)}{\partial y \partial z} + \frac{\partial^2 g(x,z)}{\partial x \partial z} = 0 \qquad (6\text{-}3.3\text{g})$$

$$\frac{\partial^2 g(x,z)}{\partial z^2} + \frac{\partial^2 h(x,y)}{\partial y \partial z} = \frac{\partial^2 g(x,z)}{\partial z^2} = 0 \qquad (6\text{-}3.3\text{h})$$

$$\frac{\partial^2 h(x,y)}{\partial x \partial z} + \frac{\partial^2 f(y,z)}{\partial z^2} = \frac{\partial^2 f(y,z)}{\partial z^2} = 0 \qquad (6\text{-}3.3\text{i})$$

Our next job is to separate the mixed derivatives in equations (6-3.3b), (6-3.3f), and (6-3.3g), as follows:

Subtract equations (6-3.3f) from (6-3.3b), we get

$$\frac{\partial^2 g(x,z)}{\partial z \partial x} - \frac{\partial^2 f(y,z)}{\partial z \partial y} = 0 \qquad (6\text{-}3.4\text{a})$$

Add equation (6-3.4a) to (6-3.3g), we get

$$\frac{\partial^2 g(x,z)}{\partial x \partial z} = 0 \qquad (6\text{-}3.4\text{b})$$

Substitute from (6-3.4b) to (6-3.4a), we get

$$\frac{\partial^2 f(y,z)}{\partial z \partial y} = 0 \qquad\qquad (6\text{-}3.4c)$$

Substitute from (6-3.4b) to (6-3.3b), we get

$$\frac{\partial^2 h(x,y)}{\partial y \partial x} = 0 \qquad\qquad (6\text{-}3.4d)$$

Finally, the nine second-derivatives of f, g, and h are obtained in isolation and are listed in Table 6-10.

Table 6-10. Second-derivatives of displacements f, g, h for the simplified problem of pure bending of a prismatic bar under torque applied at its ends.

$$\frac{\partial^2 f(y,z)}{\partial z \partial y} = 0 \qquad\qquad \frac{\partial^2 g(x,z)}{\partial x^2} = 0 \qquad\qquad \frac{\partial^2 h(x,y)}{\partial x^2} = \frac{v\rho g}{E}$$

$$\frac{\partial^2 f(y,z)}{\partial y^2} = 0 \qquad\qquad \frac{\partial^2 g(x,z)}{\partial x \partial z} = 0 \qquad\qquad \frac{\partial h^2(x,y)}{\partial y^2} = \frac{v\rho g}{E}$$

$$\frac{\partial^2 f(y,z)}{\partial z^2} = 0 \qquad\qquad \frac{\partial^2 g(x,z)}{\partial z^2} = 0 \qquad\qquad \frac{\partial^2 h(x,y)}{\partial y \partial x} = 0$$

Integrating the nine derivatives in Table 6-10, we get,

$$\mathbf{f}(y,z) = C_1(z) + C_2(y) + C_3$$
$$(6\text{-}3.5a) \qquad\qquad \mathbf{f}(y,z) = yC_4(z) + C_6$$
$$(6\text{-}3.5b) \qquad\qquad \mathbf{f}(y,z) = zC_7(y) + C_9$$

$$(6\text{-}3.5c)$$

$$\mathbf{g}(x,z) = xA_1(z) + A_3 \qquad\qquad (6\text{-}3.5d)$$
$$\mathbf{g}(x,z) = A_4(z) + A_5(x) + A_6 \qquad\qquad (6\text{-}3.5e)$$
$$\mathbf{g}(x,z) = zA_7(x) + A_9 \qquad\qquad (6\text{-}3.5f)$$

$$\mathbf{h}(x,y) = \frac{v\rho g}{2E} x^2 + xB_1(y) + B_3 \qquad\qquad (6\text{-}3.5g)$$

$$\mathbf{h}(x,y) = \frac{v\rho g}{2E} y^2 + yB_4(x) + B_6 \qquad\qquad (6\text{-}3.5h)$$

$$\mathbf{h}(x,y) = B_1(y) + B_2(x) + B_3 \qquad\qquad (6\text{-}3.5i)$$

172

From equations (6-3.5), we conclude that the following forms of functions f, g, and h are the complementary displacement solutions in Table 6-9.

$$f(y,z) = a_1 y + b_1 z + c_1 \qquad\qquad (6\text{-}3.6a)$$

$$g(x,z) = a_2 z + b_2 x + c_2 \qquad\qquad (6\text{-}3.6b)$$

$$h(x,y) = \frac{v\rho g}{2E}\left(x^2 + y^2\right) + a_3 x + b_3 y + c_3 \qquad\qquad (6\text{-}3.6c)$$

(iv.b) Boundary conditions

(A) Proper strains

In order to determine the constants a through c in equations (6-3.6), we start by substituting with those general solutions in equations (6-3.2a through 2c) which relate strains to first derivatives of displacements, as follows

$$a_1 + b_2 = 0$$

i.e., $\qquad a_1 = -b_2 \qquad\qquad\qquad\qquad (6\text{-}3.7a)$

$$a_2 + b_3 = 0$$

i.e., $\qquad a_2 = -b_3 \qquad\qquad\qquad\qquad (6\text{-}3.7b)$

$$a_3 + b_1 = 0$$

i.e., $\qquad a_3 = -b_1 \qquad\qquad\qquad\qquad (6\text{-}3.7c)$

Equation (6-3.7) and Table 6-9 provide the general functions for displacement of pure suspension of a prismatic bar as follows

$$u = -\frac{v\rho g}{E}xz + a_1 y + b_1 z + c_1 \qquad\qquad (6\text{-}3.8a)$$

$$v = -\frac{v\rho g}{E}yz + a_2 z - a_1 x + c_2 \qquad\qquad (6\text{-}3.8b)$$

$$w = \frac{\rho g}{2E}z^2 + \frac{v\rho g}{2E}\left(x^2 + y^2\right) - b_1 x - a_2 y + c_3 \qquad\qquad (6\text{-}3.8c)$$

(B) Conditions for immobility

At $\qquad x = y = 0$ and $z = l \rightarrow \quad$ deformation vanish $\rightarrow \; u = v = w = 0 \qquad\qquad (6\text{-}3.9a)$

Notice that here $z = l$, which differs from the previous example.

173

Substitute by (6-3.9) in (6-3.8), we get

$$0 = b_1 l + c_1 \qquad\qquad (6\text{-}3.10a)$$
$$0 = a_2 l + c_2 \qquad\qquad (6\text{-}3.10b)$$
$$0 = \frac{\rho g}{2E} l^2 + c_3 \qquad\qquad (6\text{-}3.10c)$$

Therefore,

$$c_1 = - b_1 l$$
$$c_2 = - a_2 l$$
$$c_3 = -\frac{\rho g}{2E} l^2$$

(C) Conditions for immobility of initial elements at origin

At $\qquad$ $x = y = 0$ and $z = l$ $\rightarrow$ $\quad$ deformation vanish $\rightarrow$ $\dfrac{\partial u}{\partial z} = \dfrac{\partial v}{\partial x} = \dfrac{\partial v}{\partial z} = 0$ $\qquad$ (6-3.11a)

Then differentiating equations (6-3.8) and substituting by (6-3.10), we get the remaining three constants

$$\frac{\partial u}{\partial z} = 0 + b_1 = 0$$
$$\frac{\partial v}{\partial x} = 0 - a_1 = 0 \qquad\qquad (6\text{-}3.11b)$$
$$\frac{\partial v}{\partial z} = -a_2 = 0$$

Thus, (6-3.10a) ($c_1 = c_2 = 0$ and $c_3 = -\dfrac{\rho g}{2E} l^2$) and (6-3.10b) ($a_1 = a_2 = b_1 = 0$), we reach our desired solution from (6-3.8a) as follows

$$u = -\frac{\nu \rho g}{E} xz \qquad\qquad (6\text{-}3.12a)$$
$$v = -\frac{\nu \rho g}{E} yz \qquad\qquad (6\text{-}3.12b)$$
$$w = \frac{\rho g}{2E} z^2 + \frac{\nu \rho g}{2E}(x^2 + y^2) - \frac{\rho g}{2E} l^2 \qquad\qquad (6\text{-}3.12c)$$

(v) Deformations

174

(a) Points

After deformation, the coordinates of the new points are related to the old points by adding the displacements from equations to the coordinates of the old points as follows

$$x_1 = x_0 + \mathbf{u} = x_0\left(1 - \frac{v\rho g}{E} z_0\right) \qquad (6\text{-}3.13\text{a})$$

$$y_1 = y_0 + \mathbf{v} = y_0\left(1 - \frac{v\rho g}{E} z_0\right) \qquad (6\text{-}3.13\text{b})$$

$$z_1 = z_0 + \mathbf{w} = z_0 + \frac{\rho g}{2E}\left[z_0^2 - l^2 + v\left(x_0^2 + y_0^2\right)\right] \qquad (6\text{-}3.13\text{c})$$

Equations (6-3.13) show the displacements of the new points of the prismatic bar are all functions of the weight (ρg), the coordinates of the points (x_0, y_0, z_0), and Young's and Poisson's constants; E and v, respectively.

(b) Lines

Equations (6-3.13) show that points on the central axis of the bar, which equation prior to deformation is $x = y = 0$, experience elongation alone along the z-axis, given by putting $x_0 = y_0 = 0$ in (6-3.13).

Therefore, the axis of the bar has the new equation

$$x_1 = 0\left(1 - \frac{v\rho g}{E} z_0\right) = 0 \qquad (6\text{-}3.14\text{a})$$

$$y_1 = 0\left(1 - \frac{v\rho g}{E} z_0\right) = 0 \qquad (6\text{-}3.14\text{b})$$

$$z_1 = z_0 + \frac{\rho g}{2E}\left[z_0^2 - l^2 + 0\right] \qquad (6\text{-}3.14\text{c})$$

Clearly, at $z_0 = l$, elongation vanishes.

The total elongation of the prismatic bar at it axis is $\frac{\rho g}{2E} l^2$.

(c) Planes

Before deformation, any cross sectional plane has the equation $z = z_0$.
After deformation, the equation of the same cross section is obtained by equations (6-3.13) by substituting by x_0 and y_0 into z_1, to get.

$$z_1 = z_0 + \frac{\rho g}{2E}\left[z_0^2 - l^2 + \nu\left(\left(\frac{x_1}{1 - \frac{\nu \rho g}{E}z_0} \right)^2 + \left(\frac{y_1}{1 - \frac{\nu \rho g}{E}z_0} \right)^2 \right) \right]$$

$$= z_0 + \frac{\rho g}{2E}\left[z_0^2 - l^2 + \frac{\nu}{\left(1 - \frac{\nu \rho g}{E}z_0\right)^2}\left(x_1^2 + y_1^2\right) \right]$$

(6-3.15)

Equation (6-3.15) describes a paraboloid of axis of revolution along the z-axis.

(d) Circular cross section

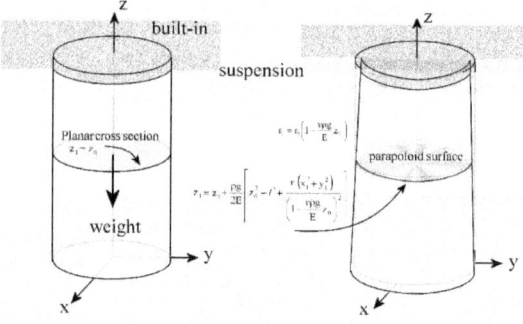

before bending   after bending

Figure 6-7. Deformation of cross sections of a prismatic bar by its own weight

As shown in Figure 6-7, before deformation, a bar with circular cross section has the following equations

$$r_0^2 = x_0^2 + y_0^2$$

(6-3.16a)

Where, r_0 is the radius of the cross section prior to deformation,

After deformation, equations (6-3.13a and 13b) give the new circle as

176

$$r_1{}^2 = x_0{}^2\left(1 - \frac{v\rho g}{E} z_0\right)^2 + y_0{}^2\left(1 - \frac{v\rho g}{E} z_0\right)^2$$

$$= r_0{}^2\left(1 - \frac{v\rho g}{E} z_0\right)^2$$

(6-3.16b)

$$r_1 = r_0\left(1 - \frac{v\rho g}{E} z_0\right)$$

(6-3.16c)

CHAPTER 7

PLANE ELASTICITY PROBLEMS

7.1. Plane strain approximations

(i) Planar displacements

In problems when one coordinate can be disregarded, the **equations of displacements** are simplified into two as follows

$$\mathbf{u} = f(x.y) \tag{7-1.1}$$
$$\mathbf{v} = g(x,y) \tag{7-1.2}$$

where, the z-dependent displacement vanishes

$$\mathbf{w} = 0$$

Similarly, the equations of strains are also simplified into three equations as follows

$$\varepsilon_{xx} = \frac{\partial f(x, y)}{\partial x} \tag{7-2.1}$$

$$\varepsilon_{yy} = \frac{\partial g(x, y)}{\partial y} \tag{7-2.2}$$

$$2\alpha_{xy} = \frac{\partial f(x, y)}{\partial y} + \frac{\partial g(x, y)}{\partial x} \tag{7-2.3}$$

Where, the z-dependent strains vanish

$$\varepsilon_{zz} = \frac{\partial \mathbf{w}}{\partial z} = 0$$

$$2\alpha_{yz} = \frac{\partial \mathbf{v}}{\partial z} + \frac{\partial \mathbf{w}}{\partial y} = 0$$

$$2\alpha_{zx} = \frac{\partial \mathbf{w}}{\partial x} + \frac{\partial \mathbf{u}}{\partial z} = 0$$

(ii) Modified Hooke's law for planar strains

Cauchy's equations (2-2) and **Hooke's equations** (3-3.1) are reduced into two

Since

$$\frac{\partial \mathbf{w}}{\partial z} = \varepsilon_{zz} = E^{-1}\left[\sigma_x - v\left(\sigma_y + \sigma_z\right)\right] = 0$$

Therefore,

$$\sigma_z = v\left(\sigma_x + \sigma_y\right) \tag{7-3.1}$$

$$\frac{\partial \mathbf{u}}{\partial x} = \varepsilon_{xx} = E^{-1}\left[\sigma_x - v\left(\sigma_y + \sigma_z\right)\right]$$

$$= E^{-1}\left[\sigma_x - v\left(\sigma_y + v\left(\sigma_y + \sigma_x\right)\right)\right]$$

$$= E^{-1}\left[\left(1 - v^2\right)\sigma_x - v\sigma_y\left(1 + v\right)\right]$$

$$= E^{-1}\left(1 - v^2\right)\left[\sigma_x - v\sigma_y\frac{\left(1 + v\right)}{\left(1 - v^2\right)}\right] \tag{7-3.2}$$

$$= E^{-1}\left(1 - v^2\right)\left[\sigma_x - \sigma_y\frac{v}{\left(1 - v\right)}\right]$$

$$= E_1^{-1}\left[\sigma_x - v_1\sigma_y\right]$$

Similarly,

$$\frac{\partial \mathbf{v}}{\partial y} = \varepsilon_{yy} = E_1^{-1}\left(\sigma_y - v_1\sigma_x\right) \tag{7-3.3}$$

Where, the **modified Young's modulus** E_1 and **modified Poisson's ratio** v_1 are defined by (7-3.2) as

$$E_1 = \frac{E}{\left(1 - v^2\right)} \tag{7-4.1}$$

$$v_1 = \frac{v}{\left(1 - v\right)} \tag{7-4.2}$$

Hooke's shear law, equation (3-7.4) and **Lamé's coefficient**, equation (3-8.1) give

$$\tau_{xy} = \mu\alpha_{xy} \tag{7-5.1}$$

$$\mu = \frac{E}{2\left(1 + v\right)} \tag{7-5.2}$$

Therefore,

$$\alpha_{xy} = \frac{2\left(1 + v\right)}{E}\tau_{xy} \tag{7-5.3}$$

179

From (7-4), we can prove that the fraction $\dfrac{2(1+v)}{E}$ is equal to the modified fraction $\dfrac{2(1+v_1)}{E_1}$ as follows. Add 1 to both sides of (7-4.2) then divide by members the two equations (7-4.1) and (7-4.2), we get

$$\frac{E_1}{1+v_1} = \frac{\dfrac{E}{(1-v^2)}}{1+\dfrac{v}{(1-v)}} = \frac{E}{(1+v)} \qquad (7\text{-}5.4)$$

Thus, from (7-5.4), equation (7-5.3), becomes

$$\alpha_{xy} = \frac{2(1+v_1)}{E_1}\tau_{xy} \qquad (7\text{-}5.5)$$

(iii) Planar stress functions

$$\sigma_x = \sigma_x(x,y) \qquad (7\text{-}6.1)$$
$$\sigma_y = \sigma_y(x,y) \qquad (7\text{-}6.2)$$
$$\sigma_z = v(\sigma_y + \sigma_x) \qquad (7\text{-}6.3)$$
$$\tau_{xy} = \tau_{xy}(x,y) \qquad (7\text{-}6.4)$$
$$\tau_{yz} = 0 \qquad (7\text{-}6.5)$$
$$\tau_{zx} = 0 \qquad (7\text{-}6.6)$$

(iv) Navier's equations of equilibrium of stresses

Equations (1-4) become

$$\frac{\partial\sigma_x}{\partial x} + \frac{\partial\tau_{xy}}{\partial y} + \rho X_i = 0 \qquad (7\text{-}7.1)$$

$$\frac{\partial\tau_{xy}}{\partial x} + \frac{\partial\sigma_y}{\partial y} + \rho Y_i = 0 \qquad (7\text{-}7.2)$$

(v) Surface condition

Equations (1-8) are reduced into two equations

$$X_n = \sigma_x\, l + \tau_{xy}\, m \qquad (7\text{-}8.1)$$
$$Y_n = \tau_{yx}\, l + \sigma_y\, m \qquad (7\text{-}8.2)$$

(vi) Saint –Venant's equation continuity equations

Under the planar approximation, equations (2-11) are reduced to a single continuity equation

$$\frac{\partial^2 \varepsilon_{yy}}{\partial x^2} + \frac{\partial^2 \varepsilon_{xx}}{\partial y^2} = 2\frac{\partial^2 \alpha_{xy}}{\partial x \partial y} \qquad (7\text{-}9)$$

7.2. Planar stress approximations

(i) Hooke's law for planar stress

All stresses in the z-direction vanish

$$\sigma_z = \tau_{xz} = \tau_{yz} = 0 \qquad (7\text{-}10.1)$$

Equation (7-6.3) changes such that $\sigma_z = 0$. Therefore, equations (7-5) become

$$\varepsilon_{xx} = E^{-1}\left(\sigma_x - v\sigma_y\right) \qquad (7\text{-}10.2)$$

$$\varepsilon_{yy} = E^{-1}\left(\sigma_y - v\sigma_x\right) \qquad (7\text{-}10.3)$$

$$\alpha_{xy} = \frac{2(1+v)}{E}\tau_{xy} \qquad (7\text{-}10.4)$$

(ii) Navier-Hooke's equations to Maurice Lévy's equation

In order to obtain a unified differential equation for stresses and strains in static equilibrium we need to tied Hooke's equation (7-10) (yet with modified Young's modulus and modified Poisson's ratio) with Navier's equation (7-7), as follows.

A. Differentiating equations (7-10) twice, we get

$$\frac{\partial^2 \varepsilon_{xx}}{\partial y^2} = E_1^{-1}\left(\frac{\partial^2 \sigma_x}{\partial y^2} - v_1\frac{\partial^2 \sigma_y}{\partial y^2}\right) \qquad (7\text{-}11.1)$$

$$\frac{\partial^2 \varepsilon_{yy}}{\partial x^2} = E_1^{-1}\left(\frac{\partial^2 \sigma_y}{\partial x^2} - v_1\frac{\partial^2 \sigma_x}{\partial x^2}\right) \qquad (7\text{-}11.2)$$

$$\frac{\partial^2 \alpha_{xy}}{\partial x \partial y} = \frac{2(1+v_1)}{E_1}\frac{\partial^2 \tau_{xy}}{\partial x \partial y} \qquad (7\text{-}11.3)$$

181

(B) Differentiating Navier's equations (7-7) gives

$$\frac{\partial^2 \sigma_x}{\partial x^2} + \frac{\partial^2 \tau_{xy}}{\partial x \partial y} + \rho \frac{\partial X_i}{\partial x} = 0 \qquad (7\text{-}12.1)$$

$$\frac{\partial^2 \tau_{xy}}{\partial x \partial y} + \frac{\partial^2 \sigma_y}{\partial y^2} + \rho \frac{\partial Y_i}{\partial y} = 0 \qquad (7\text{-}12.2)$$

Adding the two equations by members, we obtain the mixed derivative of the shear stress, which will enter (7-11). First, the addition gives

$$\frac{\partial^2 \sigma_x}{\partial x^2} + \frac{\partial^2 \tau_{xy}}{\partial x \partial y} + \rho \frac{\partial X_i}{\partial x} = 0 \qquad (7\text{-}12.1)$$

$$2 \frac{\partial^2 \tau_{xy}}{\partial x \partial y} = -\frac{\partial^2 \sigma_x}{\partial x^2} - \frac{\partial^2 \sigma_y}{\partial y^2} - \rho \left(\frac{\partial X_i}{\partial x} + \frac{\partial Y_i}{\partial y} \right) \qquad (7\text{-}12.2)$$

Equation (7-12.2) is used to remove the shear stress derivative from (7-11.3). Thus, we get

$$-\frac{\partial^2 \alpha_{xy}}{\partial x \partial y} = \frac{(1+v_1)}{E_1} \left[\frac{\partial^2 \sigma_x}{\partial x^2} + \frac{\partial^2 \sigma_y}{\partial y^2} + \rho \left(\frac{\partial X_i}{\partial x} + \frac{\partial Y_i}{\partial y} \right) \right] \qquad (7\text{-}12.3)$$

(C) Adding equations (7-12.3) and (7-11.1 and 11.2), by members, we get

$$\frac{\partial^2 \varepsilon_{xx}}{\partial y^2} + \frac{\partial^2 \varepsilon_{yy}}{\partial x^2} - \frac{\partial^2 \alpha_{xy}}{\partial x \partial y} = E_1^{-1} (1+v_1) \left[\frac{\partial^2 \sigma_x}{\partial x^2} + \frac{\partial^2 \sigma_y}{\partial y^2} + \rho \left(\frac{\partial X_i}{\partial x} + \frac{\partial Y_i}{\partial y} \right) \right]$$

$$+ E_1^{-1} \left(\frac{\partial^2 \sigma_y}{\partial x^2} - v_1 \frac{\partial^2 \sigma_x}{\partial x^2} \right) + E_1^{-1} \left(\frac{\partial^2 \sigma_x}{\partial y^2} - v_1 \frac{\partial^2 \sigma_y}{\partial y^2} \right)$$

$$= E_1^{-1} \left[\left(\frac{\partial^2}{\partial x^2} + \frac{\partial^2}{\partial y^2} \right) \left(\sigma_y + \sigma_x \right) + (1+v_1) \rho \left(\frac{\partial X_i}{\partial x} + \frac{\partial Y_i}{\partial y} \right) \right]$$

$$= E_1^{-1} \left[\nabla^2 \left(\sigma_y + \sigma_x \right) + (1+v_1) \rho \left(\frac{\partial X_i}{\partial x} + \frac{\partial Y_i}{\partial y} \right) \right] \qquad (7\text{-}13)$$

From Saint-Venant's equation of continuity, equation (7-9) and (7-13), we get

$$\nabla^2 \left(\sigma_y + \sigma_x \right) = -(1+v_1) \rho \left(\frac{\partial X_i}{\partial x} + \frac{\partial Y_i}{\partial y} \right) \qquad (7\text{-}14)$$

Where, ∇^2 is the **Laplacian operator** defined in the two dimensional coordinates by

$$\nabla^2 = \frac{\partial^2}{\partial x^2} + \frac{\partial^2}{\partial y^2} \qquad (7.14.1)$$

In cylindrical polar coordinate, the Laplacian takes the form

$$\nabla^2 = \frac{1}{r}\frac{\partial}{\partial r} + \frac{1}{r^2}\frac{\partial^2}{\partial \theta^2} + \frac{\partial^2}{\partial r^2} \qquad (7.14.2)$$

(iii) Interpretation of Maurice Lévy's equation

Maurice Lévy's equation (7-14), which is obtained by combing **Navier's equation** for static balance of internal stresses and **Hooke's law** that govern geometrical deformation in relation to stresses via material properties, offers the following explanations about the theory of elasticity:

1. When **internal forces are constants**, such as gravitational forces ($\rho Y_i = \rho g = p$ and $X_i = 0$), then the derivatives of internal forces vanish

$$\frac{\partial X_i}{\partial x} = \frac{\partial 0}{\partial x} = 0$$
$$\frac{\partial}{\partial y}\rho g = 0 \qquad (7-15.1)$$

Therefore, **Maurice Levy's equation** (7-14) becomes

$$\nabla^2\left(\sigma_y + \sigma_x\right) = 0 \qquad (7-15.2)$$

Hence, in the absence of internal forces the stresses σ_x and σ_y are related to spatial coordinates in a harmonic relationship where the changes in coordinates cancel each other on the second derivatives.

2. Substituting by the principal stress, $\Theta = \sigma_x + \sigma_y + \sigma_z$, from equation (1-32.1), **Maurice Levy's equation** (7-14) becomes

$$\nabla^2\left(\sigma_y + \sigma_x + \sigma_z\right) = \nabla^2\Theta = 0 \qquad (7-15.3)$$

But, since the approximation of **planar stress** implies that $\sigma_x = 0$, then (7-15.3) applies to non-planar stresses as well.

Farther, **planar strain** in the xy-plane, which implies that $\sigma_z = v\left(\sigma_y + \sigma_x\right)$ equation (7-6.3), also conform to **Maurice Lévy's equation** (7-15.3), because

$$\nabla^2\left[\sigma_y + \sigma_x + v\left(\sigma_y + \sigma_x\right)\right] = (1 + v)\nabla^2\left(\sigma_y + \sigma_x\right) = 0$$
$$\nabla^2\left(\sigma_y + \sigma_x\right) = 0 \tag{7-15.4}$$

Similarly, the equations (7-15.4), written in cylindrical polar coordinates, reads as follows

$$\nabla^2\left(\sigma_r + \sigma_\theta\right) = 0 \tag{7-15.5}$$

(iv) Solution of problems of plane stress

The three xy-planar stresses σ_x, σ_y, and τ_{xy}, presenting the solution of planar stress problems are obtain in the following fashion:

1. **Navier's equations** (7-7) tie normal and shearing stresses to internal forces on the balance of gradients. (first spatial derivatives of stresses)
2. **Maurice Levy's equation** ties the second derivatives of the normal stresses in a condition for static balance of stresses and strains.
3. The **surface conditions**, equations (7-8) link external forces to internal stresses.
4. The three obtained stresses σ_x, σ_y, and τ_{xy} enter **Hooke's law**, equations (7-10) to determine the required strains ε_{xx}, ε_{yy}, and α_{xy}.
5. The obtained three strains enter **Cauchy's equations** (7-3) to determine the displacements, **u** and **v**.

7.3. Polynomial stress function

George Biddell Airy's stress function ties together the following system of five equations that govern problems of elasticity:

1- Navier's static balance.
2- Cauchy's geometrical relations of strains and displacements.
3- Saint-Venant's conditions of continuity.
4- Hooke's law of stress-strain, and
5- Surface condition between external forces and internal stresses.

To derive **Airy's stress equation** we start by Navier's equations (7-7) in the **homogeneous** form $(X_i = Y_i = 0)$ as follows

$$\frac{\partial \sigma_x}{\partial x} + \frac{\partial \tau_{xy}}{\partial y} = 0 \tag{7-16.1}$$

$$\frac{\partial \tau_{xy}}{\partial x} + \frac{\partial \sigma_y}{\partial y} = 0 \tag{7-16.2}$$

We then search for **general solution** of the homogeneous Navier's equation in the form of derivatives of a **scalar stress function** φ(x,y). Since stresses are pressure forces affecting areas, we propose the solutions for the three planar stresses as the **second derivatives** of φ(x,y).

Therefore,

$$\sigma_x = \frac{\partial^2 \varphi(x,y)}{\partial y^2} \tag{7-17.1}$$

$$\sigma_y = \frac{\partial^2 \varphi(x,y)}{\partial x^2} \tag{7-17.2}$$

$$\tau_{xy} = -\frac{\partial^2 \varphi(x,y)}{\partial x \partial y} \tag{7-17.3}$$

The cylindrical polar form takes the following form

$$\sigma_r = \frac{1}{r}\frac{\partial \varphi(r,\theta)}{\partial r} + \frac{1}{r^2} \cdot \frac{\partial^2 \varphi(r,\theta)}{\partial \theta^2} \tag{7-17.4}$$

$$\sigma_\theta = \frac{\partial^2 \varphi(r,\theta)}{\partial r^2} \tag{7-17.5}$$

$$\tau_{r\theta} = -\frac{\partial}{\partial r}\left(\frac{1}{r}\frac{\partial \varphi(r,\theta)}{\partial \theta} \right)$$
$$= -\frac{1}{r}\frac{\partial^2 \varphi(r,\theta)}{\partial r \partial \theta} + \frac{1}{r^2} \cdot \frac{\partial \varphi(r,\theta)}{\partial \theta} \tag{7-17.6}$$

It is to be noted that the normal stresses are represented by derivatives across the perpendicular plane on which the stresses act, the shearing stresses by derivatives across the planes of shear.

The inhomogeneous Navier's equations would then require a **particular solution** to add to its general solution of homogeneous form (7-17), as follows:

$$\sigma_x = \sigma_y = 0$$
$$\tau_{xy} = -px \tag{7-18}$$

Substituting by the expressions for normal stresses from equation (7-17) into **Lévy's equation** (7-15.3), we get

$$\nabla^2 \left(\frac{\partial^2 \varphi(x,y)}{\partial x^2} + \frac{\partial^2 \varphi(x,y)}{\partial y^2} \right) = 0$$

$$\nabla^2 \left(\nabla^2 \varphi(x,y) \right) = 0 \qquad (7\text{-}19.1)$$

$$\frac{\partial^4 \varphi(x,y)}{\partial x^4} + 2\frac{\partial^4 \varphi(x,y)}{\partial y^2 \partial x^2} + \frac{\partial^4 \varphi(x,y)}{\partial y^4} = 0$$

Similarly, the cylindrical polar form is

$$\nabla^2 \left(\nabla^2 \varphi(r,\theta) \right) = 0$$

$$\left(\frac{1}{r}\frac{\partial}{\partial r} + \frac{1}{r^2}\frac{\partial^2}{\partial \theta^2} + \frac{\partial^2}{\partial r^2} \right)\left(\frac{1}{r}\frac{\partial}{\partial r} + \frac{1}{r^2}\frac{\partial^2}{\partial \theta^2} + \frac{\partial^2}{\partial r^2} \right)\varphi(r,\theta) = 0 \qquad (7\text{-}19.2)$$

$$\left(+\frac{1}{r^3}\frac{\partial}{\partial r} - \frac{1}{r^2}\frac{\partial^2}{\partial r^2} + \frac{2}{r}\frac{\partial^3}{\partial r^3} + \frac{5}{r^4}\frac{\partial^4}{\partial \theta^4} + \frac{\partial^4}{\partial r^4} \right)\varphi(r,\theta) = 0$$

Integration of the fourth order derivative of **Airy's stress equation** (7-19) give stress distribution at any point in the body, which are tied to the boundary condition with the surface equations (7-8) such that

$$X_n = \frac{\partial^2 \varphi(x,y)}{\partial y^2} l(x,y) - \left(\frac{\partial^2 \varphi(x,y)}{\partial x \partial y} + px \right) m(x,y)$$

$$Y_n = -\left(\frac{\partial^2 \varphi(x,y)}{\partial x \partial y} + px \right) l(x,y) + \frac{\partial^2 \varphi(x,y)}{\partial x^2} m(x,y) \qquad (7\text{-}20)$$

(i) Stress function $\varphi(x,y)$ and moment of force M(x,y)

The meaning of the stress function $\varphi(x,y)$ becomes apparent after integrating equations (7-20)

a. Force acting on the surface

$$\frac{\partial \varphi(x,y)}{\partial y} = -\int_{s_0}^{s} X_n ds - A(x)$$

$$\frac{\partial \varphi(x,y)}{\partial x} = -\int_{s_0}^{s} Y_n ds - B(y) \qquad (7\text{-}20)$$

Thus, the gradient of the stress function $\varphi(x,y)$ represents the force acting on the surface.

b. Moment of force acting on the surface

$$\frac{\partial \varphi(x, y)}{\partial s} = \frac{\partial M}{\partial s} = Q$$

$$\varphi(x, y) = M(x, y)$$

(7-21)

Thus, the stress function $\varphi(x,y)$ represents the moment of the force acting on the surface.

c. Power order of stress function polynomial

1. Second order stress function polynomial

Apparently, the second order polynomial is the lowest possible order acceptable that could generate constant stresses, as follows:

$$\varphi(x, y) = ax^2 + bxy + cy^2$$

(7-22.1)

From equation (7-17), a stress function of second order leads to **constant stress,** independent of the spatial coordinates. In the case of (7-22.1), the coefficients of coordinates represent the three planar stresses as follows

$$\sigma_x = \frac{\partial^2 \varphi(x, y)}{\partial y^2} = 2c$$

(7-22.2)

$$\sigma_y = \frac{\partial^2 \varphi(x, y)}{\partial x^2} = 2a$$

(7-22.3)

$$\tau_{xy} = -\frac{\partial^2 \varphi(x, y)}{\partial x \partial y} = -b$$

(7-22.4)

2. Third order polynomial

$$\varphi(x, y) = ax^3 + bx^2y + cxy^2 + dy^3 + ex^2 + fxy + gy^2$$

(7-23.1)

From equation (7-17), a stress function of second order leads to spatially linear **stress,** dependent of the spatial coordinates.

$$\sigma_x = \frac{\partial^2 \varphi(x, y)}{\partial y^2} = 2cx + 6dy + 2g$$

(7-23.2)

$$\sigma_y = \frac{\partial^2 \varphi(x, y)}{\partial x^2} = 6ax + 2by + 2e$$

(7-23.3)

$$\tau_{xy} = -\frac{\partial^2 \varphi(x, y)}{\partial x \partial y} = -2bx - 2cy - f - px$$

(7-23.4)

Apparently, also, the fourth order derivative, of a third order polynomial, satisfies **Lévy's equation** (7-19) which obviate the need for manipulating the coefficients of the polynomials.

3. Fourth order polynomial

Here, the fourth order derivative, of a fourth order polynomial, does not satisfy **Lévy's equation** (7-19) unless the coefficients of the stress functions are chosen carefully.

d. Limitations of algebraic polynomial stress function

Choosing the proper algebraic polynomial that fits the stresses, permitted by the equations of the theory of elasticity, is arbitrary and limited to few problems where the power of the polynomial could be discerned from the linearity of variables of the problem.

Clearly, the methods pf algebraic polynomial stress functions are limited to planar stresses and strains where only two coordinates are concerned.

(ii) Pure bending of cantilever

Example 24

Find the stresses and forces acting on a cantilever when its stress function is given by

$$\varphi(x, y) = \frac{k}{6} y^3 \qquad (7\text{-}24.1)$$

Solution

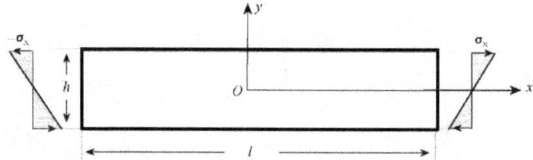

Figure 7-1. Pure bending of a cantilever.

1. Stresses

$$\sigma_x = \frac{\partial^2 \varphi(x, y)}{\partial y^2} = ky \qquad (7\text{-}24.2)$$

$$\sigma_y = \frac{\partial^2 \varphi(x,y)}{\partial x^2} = 0 \qquad (7\text{-}24.3)$$

$$\tau_{xy} = -\frac{\partial^2 \varphi(x,y)}{\partial x \partial y} - \rho g = -\rho g \qquad (7\text{-}24.4)$$

The force acting in the x-direction is determined as follows

$$F = \int_{-h/2}^{h/2} \sigma_x \, dy = \int_{-h/2}^{h/2} ky \, dy = k \left(\frac{y^2}{2} \right)_{-h/2}^{h/2}$$
$$= k \frac{h^2}{4} \qquad (7\text{-}24.5)$$

The shearing force is determined as follows

$$\int_{-h/2}^{h/2} \tau_{xy} \, dy = -\rho g \int_{-h/2}^{h/2} dy = -\rho g (y)_{-h/2}^{h/2}$$
$$= -\rho g h \qquad (7\text{-}24.6)$$

2. Displacements

Hooke's law and Cauchy's equation (7-3) link displacement to the stresses in equations (7-25) as follows.

Normal stress of (7-24.2), with Hooke's volumetric law and Cauchy's equation give:

$$\frac{\partial u}{\partial x} = \varepsilon_{xx} = E^{-1} \left[\sigma_x - v(\sigma_y + \sigma_z) \right]$$
$$= \frac{k}{E} y \qquad (7\text{-}24.7)$$

$$\frac{\partial v}{\partial y} = \varepsilon_{yy} = E^{-1} (\sigma_y - v\sigma_x)$$
$$= -\frac{vk}{E} y \qquad (7\text{-}24.8)$$

Shearing stress of (7-24.4), with Hooke's volumetric shear law (7-5.5) and Cauchy's equation (7-2.3) give:

$$\frac{\partial v}{\partial x} + \frac{\partial u}{\partial y} = -\rho g \frac{2(1+v)}{E} \qquad (7\text{-}24.9)$$

Integrating the two equations (7-24.7 and (7-24.8) with respect to x and y, respectively we get

$$u = \frac{k}{E} yx + f(y) \tag{7-24.10}$$

$$v = -\frac{vk}{2E} y^2 + g(x) \tag{7-24.11}$$

Where the unknown functions f(y) and g(x) are to be determined by substituting by the derivatives of (7-24.10) and (7-24.11) in (7-24.9).

Differentiating equation (7-24.10) with respect to y, we get

$$\frac{\partial u}{\partial y} = \frac{k}{E} x + \frac{\partial f(y)}{\partial y} \tag{7-24.12}$$

Differentiating equation (7-24.11) with respect to x, we get

$$\frac{\partial v}{\partial x} = \frac{\partial g(x)}{\partial x} \tag{7-24.13}$$

Adding (7-24.12) and (7-24.13) and equating the sum of members with those of (7-24.9), we get

$$\frac{\partial g(x)}{\partial x} + \frac{k}{E} x + \frac{\partial f(y)}{\partial y} = -\rho g \frac{(1+v)}{E} \tag{7-24.14}$$

We notice that the terms in equation (7-24.14) can be sorted out by separation of variables into x-dependent terms and y-dependent terms. Thus, we could simplify equation (7-24.14) as follows

$$\frac{\partial g(x)}{\partial x} + \frac{k}{E} x = C_1 \tag{7-24.15}$$

$$\frac{\partial f(y)}{\partial y} = C_2 \tag{7-24.16}$$

Where,

$$C_1 + C_2 = -\rho g \frac{(1+v)}{E} \tag{7-24.17}$$

Integrating the last two equations, each with respect to its divisor differential we get

$$g(x) = -\frac{kx^2}{2E} + xC_1 + C_3 \tag{7-24.18}$$

$$f(y) = yC_2 + C_4 \tag{7-24.19}$$

From equations (7-24.18 and 24.19), the equations of displacements (7-24.10 and 24.11) become

190

$$u = \frac{k}{E} yx + yC_2 + C_4 \tag{7-24.20}$$

$$v = -\frac{vk}{2E} y^2 - \frac{k}{2E} x^2 + xC_1 + C_3 \tag{7-24.21}$$

Where the four constants C's are determined from the boundary conditions as follows.

3. Boundary conditions

Let us assume the following boundary conditions: $x = y = 0 \rightarrow u = v = 0$.
Therefore, $C_3 = C_4 = 0$.

Second, the immobility condition on tangents at $x = y = 0$ will eliminate either C_2 or C_1 depending on whether we adopt $\frac{\partial v}{\partial x} = 0$ or $\frac{\partial u}{\partial y} = 0$.

Let us consider the option of $C_2 = 0$. Then from equation (7-24.17), $C_1 = -\rho g \frac{(1+v)}{E}$.
Therefore, equations (7-24.20 and 24.21) become

$$u_1 = \frac{k}{E} yx \tag{7-24.22}$$

$$v_1 = -\frac{vk}{2E} y^2 - \frac{k}{2E} x^2 - x\rho g \frac{(1+v)}{E} \tag{7-24.23}$$

Let us consider the other option of $C_1 = 0$. Then from equation (7-24.17), $C_2 = -\rho g \frac{(1+v)}{E}$.
Therefore, equations (7-24.20 and 24.21) become

$$u_2 = \frac{k}{E} yx - y\rho g \frac{(1+v)}{E} \tag{7-24.24}$$

$$v_2 = -\frac{vk}{2E} y^2 - \frac{k}{2E} x^2 \tag{7-24.25}$$

From equations (7-24.22) and (7-24.24), we conclude that the cross sectional planes (at $x =$ constant) remain plane after deformation in the case of pure bending.

From equations (7-24.23) and (7-24.25), we also conclude that the beam axis curves to the second power of y.

4. Axial deflection

At $y = 0$, the equations of the axis of the cantilever are

First fixation option:

$$u_1(x,0) = 0 \tag{7-24.26}$$

191

$$V_1(x,0) = -\frac{k}{2E}x^2 - x\rho g\frac{(1+v)}{E} \qquad (7\text{-}24.27)$$

Second fixation option:

$$u_2(x,0) = 0 \qquad (7\text{-}24.28)$$

$$V_2(x,0) = -\frac{k}{2E}x^2 \qquad (7\text{-}24.29)$$

In the case of first optional fixing of the end of the cantilever, the coordinates of the new points, after deformation, on the deformed axis are related to the old points by adding the displacements from equations (7-24.22 and 24.23) to the coordinates of the old points as follows

$$x_1 = x_0 + u = x_0 + \frac{k}{E}y_0 x_0 \qquad (7\text{-}24.30)$$

$$y_1 = y_0 + V = y_0 - \frac{vk}{2E}y_0^2 - \frac{k}{2E}x_0^2 - x_0\rho g\frac{(1+v)}{E} \qquad (7\text{-}24.31)$$

The angle of the tangent to the cross section is obtained from (7-24.30) as

$$\tan\alpha = \frac{\partial x_1}{\partial y_0} = \frac{k}{E}x_0 \qquad (7\text{-}24.32)$$

The angle of the tangent to the axis is obtained from (7-24.31) as

$$\tan\beta = \frac{1}{\dfrac{\partial y_1}{\partial x_0}} = \frac{-1}{\dfrac{k}{E}x_0 - \rho g\dfrac{(1+v)}{E}} \qquad (7\text{-}24.33)$$

The dot product of the tangent to the cross sectional plane (7-24.32) and the tangent to the beam axis (7-24.33) is

$$\tan\alpha\,\tan\beta = \frac{-\dfrac{k}{E}x_0}{\dfrac{k}{E}x_0 - \rho g\dfrac{(1+v)}{E}} \qquad (7\text{-}24.34)$$

If we neglect the internal forces, the above expression is reduced to

$$\tan\alpha\,\tan\beta = \frac{-\dfrac{k}{E}x_0}{\dfrac{k}{E}x_0 - 0} = -1 \qquad (7\text{-}24.35)$$

192

Thus, the cross sections planes of cantilever remain perpendicular to the beam axis when internal forces (beam weight) are omitted. (i.e., shearing stresses omitted).

(iii) Forced bending of cantilever

Example 25

Given the following three stresses on a cantilever built-in on one end and loaded on the other end as shown in Figure 7-2.

$$\sigma_x = \frac{M}{J}y = -\frac{Q(l-x)}{J}y \qquad (7\text{-}25.1)$$

$$\sigma_y = 0 \qquad (7\text{-}25.2)$$

$$\tau_{xy} = \frac{Q}{2J}\left(\frac{h^2}{4} - y^2\right) \qquad (7\text{-}25.3)$$

Determine Airy's stress function and prove that it satisfies the equations of elasticity.

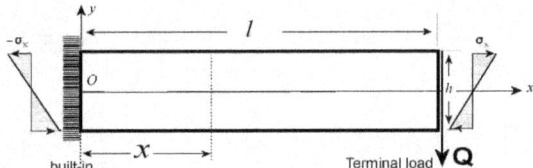

Figure 7-2. Loaded bending of a cantilever.

Solution

1. Stresses

Let us start by equations (7-17) and (7-25.1 through 25.3) in order to find the stress function of the loaded cantilever as follows

$$\frac{\partial^2 \varphi}{\partial y^2} = \sigma_x = -\frac{Q(l-x)}{J}y \qquad (7\text{-}25.4)$$

$$\frac{\partial^2 \varphi}{\partial x^2} = \sigma_y = 0 \qquad (7\text{-}25.5)$$

$$-\frac{\partial^2\varphi}{\partial x\partial y} = \tau_{xy} = \frac{Q}{2J}\left(\frac{h^2}{4} - y^2\right) \tag{7-25.6}$$

2. Stress function

Integrating equations (7-25.4) twice we get

$$\varphi(x,y) = -\frac{Q(l-x)}{6J}y^3 + yf(x) + g(x) \tag{7-25.7}$$

The two unknown functions f(x) and g(x) are determined in two steps of differentiation of (7-25.7) as follows.

Differentiating (7-25.7) twice with respect to x and equating with zero, from equation (7-25.5), we get

$$\frac{\partial^2\varphi}{\partial x^2} = y\frac{\partial^2 f(x)}{\partial x^2} + \frac{\partial^2 g(x)}{\partial x^2} = 0$$
$$\frac{\partial^2 g(x)}{\partial x^2} = -y\frac{\partial^2 f(x)}{\partial x^2} \tag{7-25.8}$$

Differentiating (7-25.7) twice, once with respect to x then with respect to y and using equation (7-25.6), we get

$$-\frac{\partial^2\varphi}{\partial x\partial y} = \frac{-Q}{2J}y^2 - \frac{\partial f(x)}{\partial x} = \frac{Q}{2J}\left(\frac{h^2}{4} - y^2\right)$$
$$\frac{\partial f(x)}{\partial x} = \frac{-Q}{2J}y^2 - \frac{Q}{2J}\left(\frac{h^2}{4} - y^2\right) = -\frac{Qh^2}{8J} \tag{7-25.9}$$

Integrating (7-27.9), we get

$$f(x) = -\frac{Qh^2}{8J}x + C_1 \tag{7-25.10}$$

From (7-25.9) and (7-25.8), we have

$$\frac{\partial^2 g(x)}{\partial x^2} = 0 \tag{7-25.11}$$

Integrating (7-25.11) twice we get

$$g(x) = xC_2 + C_3 \tag{7-25.12}$$

194

From (7-25.7), (7-25.10), and (7-25.12), the stress function become

$$\varphi(x, y) = -\frac{Q(l-x)}{6J} y^3 - \frac{Qh^2}{8J} xy + yC_1 + xC_2 + C_3 \qquad (7-25.13)$$

Therefore, the stresses given in (7-25.1 through 25.3) constitute viable solution of the equations of elasticity.

3. Surface conditions

At $y = \pm h/2$, equation (7-25.3) shows that $\tau_{xy} = 0$ and equation (7-25.3) gives $\sigma_y = 0$.

At $x = l$, equation (7-25.1) gives $\sigma_x = 0$.

At $x = 0$, equation (7-25.1) gives $\sigma_x = - Q\,l\,y\,/\,J$.

4. Terminal force

Integrating the shear stress on the cross section at $x = l$, and noting that moment of inertia $J = \delta h^3/12$, (where δ is taken $= 1$, in the planar case) we get the shearing force; Q, as follows:

$$
\begin{aligned}
\int_{-h/2}^{h/2} \tau_{xy}\,dy &= \int_{-h/2}^{h/2} \frac{Q}{2J}\left(\frac{h^2}{4} - y^2\right)dy = \frac{Q}{2J}\left(\frac{h^2}{4}y - \frac{y^3}{3}\right)\Bigg|_{-h/2}^{h/2} \\
&= \frac{Q}{2J}\left(\frac{h^3}{8} - \frac{h^3}{24} + \frac{h^3}{8} - \frac{h^3}{24}\right) \\
&= \frac{Qh^3}{12J} \\
&= Q
\end{aligned}
\qquad (7-25.14)
$$

5. Displacements

Hooke's law and Cauchy's equation (7-3) link displacement to the stresses in equations (7-25) as follows.

Normal stress of (7-25.4), with Hooke's volumetric law and Cauchy's equation give:

$$
\begin{aligned}
\frac{\partial u}{\partial x} = \varepsilon_{xx} &= E^{-1}\left[\sigma_x - \nu\left(\sigma_y + \sigma_z\right)\right] \\
&= -\frac{Q(l-x)}{EJ} y
\end{aligned}
\qquad (7-25.15)
$$

195

$$\frac{\partial v}{\partial y} = \varepsilon_{yy} = E^{-1}\left(\sigma_y - v_1\sigma_x\right)$$

$$= \frac{vQ(l-x)}{EJ}y \tag{7-25.16}$$

Shearing stress of (7-25.6), with Hooke's volumetric shear law (7-5.5) and Cauchy's equation (7-2.3) give:

$$\frac{\partial v}{\partial x} + \frac{\partial u}{\partial y} = 2\alpha_{zx} = \frac{2(1+v)Q}{EJ}\left(\frac{h^2}{4} - y^2\right) \tag{7-25.17}$$

Integrating the two equations with respect to x and y, respectively we get

$$u = -\frac{Q}{EJ}y\left(lx - \frac{x^2}{2}\right) + f(y) \tag{7-25.18}$$

$$v = \frac{vQ}{2EJ}y^2(l-x) + g(x) \tag{7-25.19}$$

Where the unknown functions f(y) and g(x) are to be determined by substituting by the derivatives of (7-25.18) and (7-25.19) in (7-25.17).

Differentiating equation (7-25.18) with respect to y, we get

$$\frac{\partial u}{\partial y} = -\frac{Q}{EJ}\left(lx - \frac{x^2}{2}\right) + \frac{\partial f(y)}{\partial y} \tag{7-25.20}$$

Differentiating equation (7-25.19) with respect to x, we get

$$\frac{\partial v}{\partial x} = -\frac{vQ}{2EJ}y^2 + \frac{\partial g(x)}{\partial x} \tag{7-25.21}$$

Adding (7-25.20) and (7-25.21) and equating the sum of members with those of (7-25.17), we get

$$-\frac{vQ}{2EJ}y^2 + \frac{\partial g(x)}{\partial x} - \frac{Q}{EJ}\left(lx - \frac{x^2}{2}\right) + \frac{\partial f(y)}{\partial y} = \frac{(1+v)Q}{EJ}\left(\frac{h^2}{4} - y^2\right) \tag{7-25.22}$$

We notice that the terms in equation (7-25.22) can be sorted out by separation of variables into x-dependent terms and y-dependent terms. Farther, since g and f are arbitrary functions, we could assumed that the include the constant factor (Q/EJ). Thus, we could simplify equation (7-25.22) as follows

196

$$\left[\frac{\partial g(x)}{\partial x} - lx + \frac{x^2}{2}\right] + \left[\frac{\partial f(y)}{\partial y} - \frac{v}{2}y^2 + (1+v)y^2\right] = (1+v)\frac{h^2}{4} \tag{7-25.23}$$

Or

$$\frac{\partial g(x)}{\partial x} - lx + \frac{x^2}{2} = C_1 \tag{7-25.23a}$$

$$\frac{\partial f(y)}{\partial y} + \left(1 + \frac{v}{2}\right)y^2 = C_2 \tag{7-25.23b}$$

Where,

$$C_1 + C_2 = (1+v)\frac{h^2}{4} \tag{7-25.23c}$$

Integrating the last two equations, each with respect to its divisor differential we get

$$g(x) = l\frac{x^2}{2} - \frac{x^3}{6} + xC_1 + C_3 \tag{7-25.24}$$

$$f(y) = -\left(1 + \frac{v}{2}\right)\frac{y^3}{3} + yC_2 + C_4 \tag{7-25.25}$$

From equations (7-25.24 and 25.25), the equations of displacements (7-25.18 and 25.19) become

$$u = \frac{Q}{EJ}\left(-lxy + \frac{x^2 y}{2} - \left(\frac{2+v}{6}\right)y^3 + yC_2 + C_4\right) \tag{7-25.26}$$

$$v = \frac{Q}{EJ}\left(\frac{v}{2}y^2 l - \frac{v}{2}y^2 x + l\frac{x^2}{2} - \frac{x^3}{6} + xC_1 + C_3\right) \tag{7-25.27}$$

Where the four constants C's are determined from the boundary conditions as follows.

6. Boundary conditions

a. Initial immobility implies $u = v = 0$ at $x = y = 0$. Therefore,

$$C_3 = C_4 = 0 \tag{7-25.28}$$

b. Initial lack of rotation or constant tangent, implies $\dfrac{\partial v}{\partial x} = 0$. Therefore,

$$C_1 = 0 \tag{7-25.29}$$

Then, from (7-25.23c), we get

$$C_2 = (1+v)\frac{h^2}{4} \tag{7-25.30}$$

197

Finally, the equations of displacements (7-25.26 and 25.27 become

$$u_1 = \frac{Q}{EJ}\left(-lxy + \frac{x^2 y}{2} - \left(\frac{2+v}{6}\right)y^3 + (1+v)\frac{h^2}{4}y\right)$$
(7-25.31)

$$v_1 = \frac{Q}{EJ}\left(\frac{v}{2}y^2 l - \frac{v}{2}y^2 x + l\frac{x^2}{2} - \frac{x^3}{6}\right)$$
(7-25.32)

c. Initial lack of rotation or constant tangent, implies $\dfrac{\partial u}{\partial y} = 0$. Therefore,

$$C_2 = 0$$
(7-25.33)

Then, from (7-25.23c), we get

$$C_1 = (1+v)\frac{h^2}{4}$$
(7-25.34)

Finally, the equations of displacements (7-25.26 and 25.27 become

$$u_2 = \frac{Q}{EJ}\left(-lxy + \frac{x^2 y}{2} - \left(\frac{2+v}{6}\right)y^3\right)$$
(7-25.35)

$$v_2 = \frac{Q}{EJ}\left(\frac{v}{2}y^2 l - \frac{v}{2}y^2 x + l\frac{x^2}{2} - \frac{x^3}{6} + (1+v)\frac{h^2}{4}x\right)$$
(7-25.36)

From the two sets of equations of displacements (7-25.31 and 32) and (7-25.35 and 36), it is clear that the fixing condition of the cantilever at $x = y = 0$, affects the equations of displacements.

7. Axial deflection

The cantilever axis Ox, Figure 7-2, is defined by $y = 0$. Thus, from equations (7-25.31 and 32), we get

$$u_1(x,0) = 0$$
(7-25.37a)

$$v_1(x,0) = \frac{x^2 Q}{2EJ}\left(l - \frac{x}{3}\right)$$
(7-25.37b)

Maximal deflection is given by putting $x = l$ in the above equation, thus

$$v_1(l,0) = \frac{l^3 Q}{3EJ}$$
(7-25.37c)

From equations (7-25.35 and 36), we get

198

$$\mathbf{u}_2(x,0) = 0 \tag{7-25.37d}$$

$$\mathbf{V}_2(x,0) = \frac{xQ}{2EJ}\left(xl - \frac{x^2}{3} + (1+v)\frac{h^2}{2}\right) \tag{7-25.37e}$$

Maximal deflection is given by putting $x = l$ in the above equation, thus

$$\mathbf{V}_2(l,0) = \frac{lQ}{EJ}\left(\frac{l^2}{3} + (1+v)\frac{h^2}{4}\right) \tag{7-25.37f}$$

And, from (3-8.1),

$$G = \mu = \frac{E}{2(1+v)}$$

The moment of inertia $\qquad\qquad J = \delta h^3/12$

And, from (3-8.1), $\qquad\qquad\qquad \mu = \dfrac{E}{2(1+v)}$

Equation (7-25.37f) can be expressed as follows

$$\mathbf{V}_2(l,0) = \frac{l^3Q}{3EJ} + \frac{3lQ}{2Gh} \tag{7-25.37g}$$

Therefore, under such scheme of immobilization, maximal deflection increased over equation (7-25.37c) by the **shear** component ($3lQ/2Gh$).

8. Cross sectional deformation

A cross sectional plane in the cantilever in Figure 7-2 has the following equations

Before deformation: $\qquad\qquad x = x_0$

After deformation, and on the assumption of the <u>immobilization adopted in equation</u> (7-25.31), we get

$$x = x_0 + u_0$$
$$= x_0 + \frac{Q}{EJ}\left(-lx_0 y + \frac{x_0^2 y}{2} - \left(\frac{2+v}{6}\right)y^3 + (1+v)\frac{h^2}{4}y\right) \tag{7-25.38}$$

Therefore, the cross-sectional plane of a loaded cantilever deforms to a **parabola**. Even at $x = 0$, the cross-sectional plane has the equation (substitute by $x_0 = 0$, in (7-25.38))

$$x = \frac{Q}{EJ}\left[-\left(\frac{2+v}{6}\right)y^3 + (1+v)\frac{h^2}{4}y\right] \tag{7-25.39}$$

199

The angle between the normal to the cross-section of the cantilever at x = 0 is given by the derivative of x with respect to y, in equation (7-25.39).

$$\frac{\partial x}{\partial y} = \frac{Q}{EJ}\left[-\left(\frac{2+v}{2}\right)y^2 + (1+v)\frac{h^2}{4}\right]_{\substack{y=0\\x=0}}$$

$$= \frac{Q(1+v)}{EJ}\frac{h^2}{4}$$

(7-25.40)

Since,

The moment of inertia $\qquad\qquad J = \delta h^3/12$

And, from (3-8.1), $\qquad\qquad \mu = \dfrac{E}{2(1+v)}$

i.e., $\qquad\qquad \dfrac{\partial x}{\partial y} = \dfrac{3Q}{2Gh} > 0$

(7-25.41)

Therefore, the deformed cross section, at x = 0 and y = 0, does not align its normal with the axis of the cantilever.

Of course, if we adopted the <u>alternative immobilization in equation</u> (7-25.35), then equation (7-25.38) becomes

$$x = x_0 + u_0$$

$$= x_0 + \frac{Q}{EJ}\left(-lx_0y + \frac{x_0^2 y}{2} - \left(\frac{2+v}{6}\right)y^3\right)$$

(7-25.42)

Therefore, under the alternative immobilization conditions, the cross-sectional plane of a loaded cantilever deforms also to a parabola. Put x = 0, in (7-25.42), the cross-sectional plane has the equation

$$x = -\frac{Q}{6EJ}(2+v)y^3$$

(7-25.43)

We note that, despite changing the fixing condition at the end of the cantilever, equation (7-25.41) remains valid.

(iv) Uniformly loaded beam supported at both ends

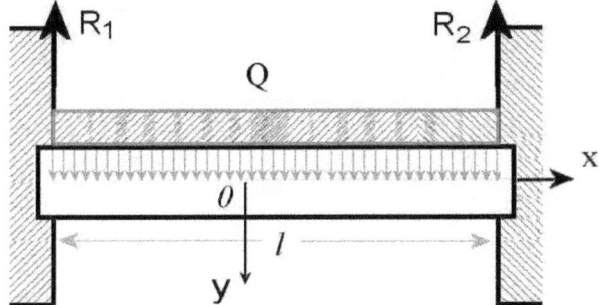

Figure 7-3. Uniformly loaded beam built-in at both ends

Example 26

Starting with the following provisional equations for stresses on the beam in Figure 7-3:

$$\frac{\partial^2 \varphi}{\partial y^2} = \sigma_x = \frac{Q}{2J}\left(\frac{l^2}{4} - x^2\right)y \tag{7-26.1}$$

$$-\frac{\partial^2 \varphi}{\partial x \partial y} = \tau_{xy} = -\frac{Qx}{2J}\left(\frac{h^2}{4} - y^2\right) \tag{7-26.2}$$

(i) Determine the resulting stress function $\varphi(x,y)$ and determine the requirements for its compliance with Lévy's equations (7-19)

(ii) Determine the shearing forces and torques at both ends.

(iii) Determine the conditions for vanishing couples at the ends of the beam.

(iv) Find the equation of deflection of the beam.

Solution

1. Stress function

Integrating (7-26.1) twice with respect to y, we get

$$\frac{\partial \varphi(x,y)}{\partial y} = \frac{Q}{4J}\left(\frac{l^2}{4} - x^2\right)y^2 + f_1(x) \tag{7-26.3a}$$

$$\varphi(x,y) = \frac{Q}{12J}\left(\frac{l^2}{4} - x^2\right)y^3 + yf_1(x) + f_2(x) \tag{7-26.3b}$$

From equation (7-26.2) and (7-26.3a) we can determine one among the unknown function $f_1(x)$. Differentiate (7-26.3b) with respect to x to get

$$\frac{\partial^2 \varphi(x,y)}{\partial y \partial x} = \frac{-Qx}{2J} y^2 + \frac{\partial f_1(x)}{\partial x} \qquad (7\text{-}26.4\text{a})$$

Therefore, from (7-26.2) and (7-26.4), we get

$$\frac{\partial f_1(x)}{\partial x} = \frac{Qh^2}{8J} x \qquad (7\text{-}26.4\text{b})$$

Integrating with respect to x, we get

$$f_1(x) = \frac{Qh^2}{16J} x^2 + C_1 \qquad (7\text{-}26.4\text{c})$$

From (7-26.3b) and (7-26.4c) we reach the stress function of the form

$$\varphi(x,y) = \frac{Q}{12J}\left(\frac{l^2}{4} - x^2\right)y^3 + y\left(\frac{Qh^2}{16J}x^2 + C_1\right) + f_2(x) \qquad (7\text{-}26.5)$$

2. Compliant stress function

In order for stress function (7-26.5) to satisfy Lévy's equation (7-19), the mixed fourth derivatives of $\varphi(x,y)$ must vanish. For that purpose, we must add term that cancels the remainder of the mixed fourth derivative of (7-26.5) as follows

Since the fourth mixed derivatives of (7-26.5) gives

$$\frac{\partial^4 \varphi(x,y)}{\partial x^2 \partial y^2} = \frac{-Q}{J} y \qquad (7\text{-}27.1)$$

Therefore, we need a term function that cancels the twice of the term $-Qy/J$ in order for the stress function $\varphi(x,y)$ to satisfy the Lévy's equation. Therefore, the desired function should have the form

$$\frac{\partial^4 \psi(x,y)}{\partial x^4} + 2\frac{\partial^4 \psi(x,y)}{\partial x^2 \partial y^2} + \frac{\partial^4 \psi(x,y)}{\partial y^4} = \frac{2Q}{J} y \qquad (7\text{-}27.2)$$

We conclude that required term function $\psi(x,y)$ should have the form

$$\psi(x,y) = y\left(Ay^4 + Bx^4 + Cx^2y^2\right) \qquad (7\text{-}27.3)$$

202

Our choice of $\psi(x,y)$, equation (7-27.3), is based on the order of differentiation in the Lévy's equation, (7-27.2). We could dismiss the mixed term in (7-27.3) since it already exist in (7-26.5).

Thus, the two constants A and B, in equation (7-27.3) are determined as follows.

Performing the differentiations in (7-27.2) on (7-27.3), we get

$$4.3.2.By + 5.4.3.2.Ay = \frac{2Q}{J}y$$

$$24B + 120A = \frac{2Q}{J} \qquad (7\text{-}27.4a)$$

i.e.,

$$A = \frac{Q}{60J} + \frac{B}{5} \qquad (7\text{-}27.4b)$$

Thus, the term function $\psi(x,y)$ in equation (7-27.3)

$$\psi(x,y) = \left[\frac{Q}{60J} + \frac{B}{5}\right]y^5 + Bx^4 y \qquad (7\text{-}27.5)$$

Equation (7-27.5) has only the unknown B. Hence, the Airy's stress function (7-26.5) becomes

$$\varphi(x,y) = \frac{Q}{12J}\left(\frac{l^2}{4} - x^2\right)y^3 + \frac{Qh^2}{16J}yx^2 + f_2(x) + \left[\frac{Q}{60J} + \frac{B}{5}\right]y^5 + Bx^4 y \qquad (7\text{-}27.6)$$

We have omitted any term that has isolated first order power coordinate since the second derivative of the stress function eliminate those terms.

We thus has the stress function (7-27.6) with two unknowns, B and $f_2(x)$.which will be determined from the boundary conditions.

3. Compliant stresses

The newly modified stress function (7-27.6) renders the following modifications in the acceptable stresses in the beam.

$$\frac{\partial^2 \varphi}{\partial y^2} = \sigma_x = \frac{Q}{2J}\left(\frac{l^2}{4} - x^2\right)y + \left[\frac{Q}{3J} + 4B\right]y^3 \qquad (7\text{-}28.1)$$

$$\frac{\partial^2 \varphi}{\partial x^2} = \sigma_y = -\frac{Q}{6J}y^3 + \frac{Qh^2}{8J}y + 12Byx^2 + \frac{\partial^2 f_2(x)}{\partial x^2} \qquad (7\text{-}28.2)$$

$$-\frac{\partial^2 \varphi}{\partial x \partial y} = \tau_{xy} = \frac{Q}{2J}xy^2 - \frac{Qh^2}{8J}x - 4Bx^3 \qquad (7\text{-}28.3)$$

4. Boundary conditions

a. Lower surface of beam, where there is no load and y = h/2, we get

$$y = \frac{h}{2}$$
$$\sigma_y = 0 \qquad\qquad (7\text{-}29.1)$$
$$\tau_{xy} = 0$$

Substituting in equations (7-28.2) and (7-28.3), we get

$$\sigma_y = -\frac{Qh^3}{48J} + \frac{Qh^3}{16J} + 6Bhx^2 + \frac{\partial^2 f_2(x)}{\partial x^2} = 0 \qquad (7\text{-}29.2)$$

$$\tau_{xy} = \frac{Q}{8J}xh^2 - \frac{Qh^2}{8J}x - 4Bx^3 = 0 \qquad (7\text{-}29.3)$$

Therefore,

$$B = 0 \qquad\qquad (7\text{-}29.4)$$

$$\frac{\partial^2 f_2(x)}{\partial x^2} = \frac{2Qh^3}{48J} = -\frac{2Qh^3}{48h^3}12 = -\frac{Q}{2} \qquad (7\text{-}29.5)$$

b. Upper surface of beam, where the load Q is uniform and y = -h/2, we get

$$y = -\frac{h}{2}$$
$$-\sigma_y = \sigma_{-y} = Q \qquad\qquad (7\text{-}30.1)$$

The negative sign assigned to the normal stress σy accounts for the opposite action per Newton's third law of action and counter-action.

Substituting in equations (7-28.1) and (7-28.3), we get

$$\sigma_y = \frac{Qh^3}{48J} - \frac{Qh^3}{16J} + \frac{\partial^2 f_2(x)}{\partial x^2} = -Q \qquad (7\text{-}30.2)$$

Which gives

$$\frac{\partial^2 f_2(x)}{\partial x^2} = -Q + \frac{2Qh^3}{48J} = -Q + \frac{2Q}{48}12 = -Q + \frac{2Q}{48}12 = -\frac{Q}{2}$$ (7-30.3)

Therefore, the three internal stresses which conform to **Lévy's equations** (7-19) are

$$\sigma_x = \frac{Q}{2J}\left(\frac{l^2}{4} - x^2 + \frac{2}{3}y^2\right)y$$ (7-31.1)

$$\sigma_y = \frac{Q}{2J}\left(\frac{h^2}{4}y - \frac{y^3}{3} - J\right)$$ (7-31.2)

$$\tau_{xy} = \frac{Q}{2J}x\left(-y^2 + \frac{h^2}{4}\right)$$ (7-31.3)

5. Beam ends conditions

a. Shear stress at beam ends

At both ends, $x = \pm l/2$, the shearing force is determined by integrating the shearing stress over the thickness of the beam.

$$\begin{aligned}
\int_{-h/2}^{h/2} \tau_{xy}dy &= \int_{-h/2}^{h/2} \frac{Q}{2J}x\left(-y^2 + \frac{h^2}{4}\right)dy \\
&= \frac{Q}{2J}\left(\pm\frac{l}{2}\right)\left[-\frac{y^3}{3} + \frac{h^2}{4}y\right]_{-h/2}^{h/2} \\
&= \frac{Q}{2J}\left(\pm\frac{l}{2}\right)\left[\left(-\frac{h^3}{24} + \frac{h^3}{8}\right) - \left(\frac{h^3}{24} - \frac{h^3}{8}\right)\right] \\
&= \frac{Q}{2J}\left(\pm\frac{l}{2}\right)\frac{h^3}{6} = \pm\frac{Ql}{2}
\end{aligned}$$ (7-32.1)

Thus, the shearing force on each end is half of the total load applied on the beam.

b. Normal stresses at ends.

From equation (7-31.1), normal stress vanished only on the axis, $y = 0$. At the upper and lower surfaces, near the edges, we get

$$\sigma_x\left(\pm\frac{l}{2}, y\right) = \frac{Q}{2J}\left(\frac{l^2}{4} - \frac{l^2}{4} + \frac{2}{3}y^2\right)y = \frac{Q}{3J}y^3 = 4Q\left(\frac{y}{h}\right)^3$$ (7-32.2)

The net force of the x-direction vanishes as follows

205

$$\int_{-h/2}^{h/2} \sigma_x dy = 4Q\left(\frac{y^4}{4h^3}\right)_{-h/2}^{h/2} = 0 \tag{7-32.3}$$

c. Torque at beam ends

The moment of couple of the normal stress at the ends of the beam is determined as follows

$$M = \int_{-h/2}^{h/2} \sigma_x y \, dy = 4Q\left(\frac{y^5}{5h^3}\right)_{-h/2}^{h/2}$$

$$= 4Q\left(2\frac{h^5}{5h^3(2^5)}\right) =$$

$$= Q\frac{h^2}{20} \tag{7-32.3}$$

6. Manipulating the momenta of end couple

In order to eliminate the moment of couple, $M = Q\dfrac{h^2}{20}$ in equation (7-32.3),   we   could   add

canceling stress $\dfrac{Q}{2J}\chi y$ in equation (7-31.1) such that

$$\sigma_x = \frac{Q}{2J}\left(\frac{l^2}{4} - x^2 + \frac{2}{3}y^2 + \chi\right)y$$

$$\tag{7-33.1}$$

The moment of couple of the normal stress at the ends of the beam is determined as follows

$$M = \int_{-h/2}^{h/2} \sigma_x y \, dy = \frac{Q}{2J}\int_{-h/2}^{h/2}\left(\frac{l^2}{4} - x^2 + \frac{2}{3}y^2 + \chi\right)_{x=l} y^2 dy$$

$$= \frac{Q}{2J}\int_{-h/2}^{h/2}\left(\frac{2}{3}y^4 + \chi y^2\right)dy = \frac{Q}{2J}\left(\frac{2}{15}y^5 + \frac{\chi y^3}{3}\right)_{-h/2}^{h/2}$$

$$= \frac{Q}{2J}\left(\frac{4}{15}\left(\frac{h}{2}\right)^5 + \frac{2\chi}{3}\left(\frac{h}{2}\right)^3\right) = \frac{Q}{2J}\left(\frac{h^2}{5} + 2\chi\right)\frac{h^3}{24} \tag{7-33.2}$$

$$= \frac{Q}{4}\left(\frac{h^2}{5} + 2\chi\right)$$

Thus, the value of χ that cancels the moment M is obtained by equating (7-33.2) with zero. Thus, we get

206

$$\frac{Q}{2J}\left(\frac{h^2}{5}+2\chi\right)\frac{h^3}{24}=0$$

$$\chi=-\frac{h^2}{10}$$

(7-33.2)

Thus, the equation for **couple-free beam** ends, (7-33.1) takes the form

$$\sigma_x=\frac{Q}{2J}\left(\frac{l^2}{4}-\frac{h^2}{10}-x^2+\frac{2}{3}y^2\right)y$$

(7-33.3)

7. Displacements

Cauchy-Hooke's equations, together with the equations of stresses (7-31.2) and (7-33.3) give:

$$\frac{\partial u}{\partial x}=\varepsilon_{xx}=E^{-1}\left[\sigma_x-\nu(\sigma_y+\sigma_z)\right]$$

$$=\frac{Q}{2J}E^{-1}\left[\left(\frac{l^2}{4}-\frac{h^2}{10}-x^2+\frac{2}{3}y^2\right)y-\nu\left(\frac{h^2}{4}y-\frac{y^3}{3}-J\right)\right]$$

(7-34.1)

$$=\frac{Q}{2J}E^{-1}\left[\nu J+\left(\frac{l^2}{4}-\left(\frac{2+5\nu}{4}\right)h^2-x^2+\frac{2+\nu}{3}y^2\right)y\right]$$

Integrating, we get

$$u(x,y)=\frac{Q}{2J}E^{-1}\left[\nu Jx+\left(\frac{l^2}{4}x-\left(\frac{2+5\nu}{4}\right)h^2x-\frac{x^3}{3}+\frac{2+\nu}{3}y^2x\right)y\right]+f_1(y)$$ (7-34.2)

Similarly,

$$\frac{\partial v}{\partial y}=\varepsilon_{yy}=E^{-1}\left(\sigma_y-\nu_1\sigma_x\right)$$

$$=\frac{Q}{2J}E^{-1}\left(\left(\frac{h^2}{4}y-\frac{y^3}{3}-J\right)-\nu\left(\frac{l^2}{4}-\frac{h^2}{10}-x^2+\frac{2}{3}y^2\right)y\right)$$

(7-34.3)

$$=\frac{Q}{2J}E^{-1}\left(-J+\left(\frac{2\nu+5}{20}h^2-\nu\left(\frac{l^2-x^2}{4}\right)\right)y-\frac{(2\nu+1)y^3}{3}\right)$$

Integrating, we get

$$v(x,y)=\frac{Q}{2J}E^{-1}\left(-Jy+\left(\frac{2\nu+5}{20}h^2-\nu\left(\frac{l^2-x^2}{4}\right)\right)\frac{y^2}{2}-\frac{(2\nu+1)y^4}{12}\right)+f_2(x)$$ (7-34.4)

207

Where the two function $f_1(y)$ and $f_2(x)$ are determine from the fixation conditions at the ends of the beam.

8. Summary of calculations

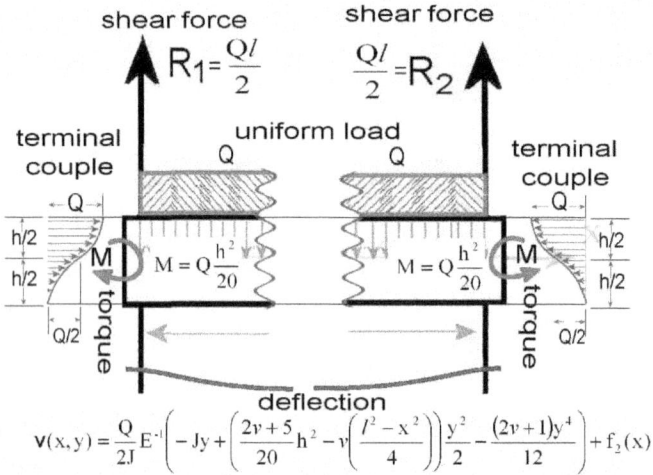

$$v(x,y) = \frac{Q}{2J}E^{-1}\left(-Jy + \left(\frac{2v+5}{20}h^2 - v\left(\frac{l^2-x^2}{4}\right)\right)\frac{y^2}{2} - \frac{(2v+1)y^4}{12}\right) + f_2(x)$$

Figure 7-4. Summary of calculations of deflection, shear stresses, terminal couples, and proper stress distributions on uniformly loaded beam.

(v) Vertically loaded triangular dam

Example 27

A dam cross section yOS is exposed to water pressure on the side yOz (z is perpendicular to the plane of the paper).

i. Find the normal stress and shearing stress on the opposite side. SOz in terms of the angle θ between the two faces yOz and SOz and the profile of water pressure.

ii. Find the stresses at the sections *aa* and *bb*

208

iii. Find the condition when the face SOz is free of load

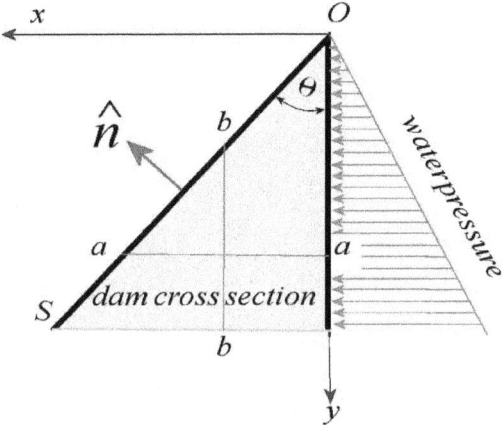

Figure 7-5. Triangular cross-section in a dam resisting pressure water in a river oriented in the x-direction, with depth along the y-axis on the vertical side.

Solution

(1) Stresses

Since all forces, external and internal, are linear functions of coordinates and the known boundary conditions are limited to 4, we will adopt equations (7-23.2) without the three constant g, e, and f, for the three internal strains

$$\sigma_x = 2cx + 6dy \tag{7-35.1a}$$
$$\sigma_y = 6ax + 2by \tag{7-35.1b}$$
$$\tau_{xy} = -2bx - 2cy - \rho gx \tag{7-35.1c}$$

(2) Boundary conditions

a. Surface yOz at x = 0, the pressure on the surface yOz = P_oy

Therefore, boundary conditions on the yOz surface are:

$$\sigma_{-x} = -\sigma_x = P_o y \tag{7-35.2a}$$
$$\tau_{-xy} = -\tau_{xy} = 0 \tag{7-35.2b}$$

209

Thus, equations (7-35.1) become

$$\sigma_x = 2c\ 0 + 6d\ y = -P_o y \qquad (7\text{-}35.2c)$$
$$\tau_{xy} = -2\ b\ 0 - 2cy = 0 \qquad (7\text{-}35.2d)$$

Thus,

$$6d = -P_o \qquad (7\text{-}35.2e)$$
$$c = 0 \qquad (7\text{-}35.2f)$$

Thus, the equations of stress after applying the boundary conditions at yOz are

$$\sigma_x = -P_o y \qquad (7\text{-}35.2g)$$
$$\sigma_y = 6ax + 2by \qquad (7\text{-}35.2h)$$
$$\tau_{xy} = -2bx - \rho gx \qquad (7\text{-}35.2i)$$

b. Surface SOz is devoid of external forces, therefore the boundary conditions on the SOz side of the dam are

$$X_n = 0 \qquad (7\text{-}35.3a)$$
$$Y_n = 0 \qquad (7\text{-}35.3b)$$

The directional cosines of the normal to SOz are

$$l_1 = \cos\theta,$$
$$m_1 = -\sin\theta \quad \text{(directed in the negative direction of y-axis)}$$
$$n_1 = \cos\pi/2 \qquad (7\text{-}35.3c)$$

Equations (1-8.1 through 3) help transfer the stresses from the surface yOz of equation (7-35.1) onto the SOz surface as follows.

$$x - y\tan\theta \qquad (7\text{-}35.4a)$$
$$X_n = \sigma_x\, l_1 + \tau_{xy}\, m_1 + \tau_{xz}\, n_1 \qquad (7\text{-}35.4b)$$

Substituting from equations (7-35.2g through 2i) into (7-35.4b), we get

$$X_n = \sigma_x \cos\theta - \tau_{xy} \sin\theta = 0$$
$$= (-P_o y)\cos\theta - (-2bx - px)\sin\theta = 0$$
$$= (-P_o y)\cos\theta - (-2b-p)\, y \tan\theta \sin\theta = 0$$

$$2b + \rho g = P_o \cot^2\theta \qquad (7\text{-}35.4c)$$

And

$$Y_n = \tau_{yx}\, l + \sigma_y\, m + \tau_{yz}\, n$$
$$= (-2b-p)\, y \tan\theta \cos\theta - (6a\, y \tan\theta + 2by)\sin\theta = 0$$

210

$$4b + \rho g = -6a \tan \theta \qquad (7\text{-}35.4d)$$

Equations (7-35.4c) and (7-35.4d) give the two constants b and a, as follows

$$a = (\rho g - 2 P_o \cot^2 \theta)/ (6 \tan \theta) \qquad (7\text{-}35.4e)$$
$$b = (- \rho g + P_o \cot^2 \theta) / 2 \qquad (7\text{-}35.4f)$$

Thus, the equations of stresses on the dam which satisfy Lévy's equation (7-19) are

$$\sigma_x = -P_o y \cdot \qquad (7\text{-}35.5a)$$
$$\sigma_y = \cot\theta\left(\rho g - 2P_o \cot^2 \theta\right)x + \left(- \rho g + P_o \cot^2 \theta\right)y \qquad (7\text{-}35.5b)$$
$$\tau_{xy} = -P_0 x \cot^2 \theta \qquad (7\text{-}35.5c)$$

The terms which include P_o are contributions from the normal stress σ_x, equations (7-35.2a) due to water pressure. The terms which include ρg are contributions from internal forces (7-14).

(3) Surface *aa*

$$y = \text{constant} = C \qquad (7\text{-}35.6a)$$
$$\sigma_x = -P_o C = C_1 \qquad (7\text{-}35.6b)$$
$$\sigma_y = \cot\theta\left(\rho g - 2P_o \cot^2 \theta\right)x + \left(- \rho g + P_o \cot^2 \theta\right)C$$
$$\qquad (7\text{-}35.6c)$$
$$= C_2 x + C_3$$
$$\tau_{xy} = -P_0 x \cot^2 \theta$$
$$\qquad (7\text{-}35.6d)$$
$$= C_4 x$$

(4) Surface *bb*

$$x = \text{constant} = C \qquad (7\text{-}35.7a)$$
$$\sigma_x = -P_o y \qquad (7\text{-}35.7b)$$
$$\sigma_y = \cot\theta\left(\rho g - 2P_0 \cot^2 \theta\right)C + \left(- \rho g + P_0 \cot^2 \theta\right)y$$
$$\qquad (7\text{-}35.7c)$$
$$= C_2 + C_3 y$$
$$\tau_{xy} = -P_0 C \cot^2 \theta = C_4 \qquad (7\text{-}35.7d)$$

Summary

Stresses	Surface *aa*	Surface *bb*
$\sigma_x = -P_o y$	$y = \text{constant} = C$	$x = \text{constant} = C$
$\sigma_y = \cot\theta\left(\rho g - 2P_0 \cot^2 \theta\right)x$	$\sigma_x = -P_o C = C_1$	$\sigma_x = -P_o y$
$+\left(- \rho g + P_0 \cot^2 \theta\right)y$	$\sigma_y = C_2 x + C_3$	$\sigma_y = C_2 + C_3 y$
	$\tau_{xy} = C_4 x$	$\tau_{xy} = C_4$

$$\tau_{xy} = -P_0 x \cot^2 \theta$$

7.4. Separation of variables or geometrical polynomials

(i) Separation of coordinate functions

The approach of choosing algebraic polynomial for the stress function $\varphi(x,y)$ entails the task of guessing proper fit for two coordinates x and y. Here, the separation of variables reduces the choice to single coordinates via the products of polynomials, or trigonometric polynomials.

Starting from the Airy-Lévy's equation

$$\frac{\partial^4 \varphi(x,y)}{\partial x^4} + 2\frac{\partial^4 \varphi(x,y)}{\partial y^2 \partial x^2} + \frac{\partial^4 \varphi(x,y)}{\partial y^4} = 0 \qquad (7\text{-}36.1)$$

And the solution

$$\varphi(x,y) = X(x)Y(y) \qquad (7\text{-}36.2)$$

Therefore, (7-36.1) becomes

$$Y(y)\frac{d^4 X(x)}{dx^4} + 2\frac{d^2 X(x)}{dx^2} \cdot \frac{d^2 Y(y)}{dy^2} + X(x)\frac{d^4 Y(y)}{dy^4} = 0 \qquad (7\text{-}36.3)$$

(ii) Choice of periodic polynomial functions

In order to disentangle the X and Y in equation (7-36), we will choose periodic functions that allow us to assume the following solutions

$$\frac{d^4 X(x)}{dx^4} = k^4 X(x) \qquad (7\text{-}37.1)$$

$$\frac{d^2 X(x)}{dx^2} = -k^2 X(x) \qquad (7\text{-}37.2)$$

The periodicity of X is proven by differentiating (7-37.2) twice to get

$$\frac{d^4 X(x)}{dx^4} = -k^2 \frac{d^2 X(x)}{dx^2} \qquad (7\text{-}37.3)$$

From (7-37.2) and (7-37.3), we discern the periodicity since we have

212

$$\frac{d^4X(x)}{dx^4} = -k^2 \frac{d^2X(x)}{dx^2} = -k^2\left(-k^2X(x)\right) = k^4X(x) \qquad (7\text{-}37.4)$$

Thus, the choice of the negative sign after every two differentiations guaranteed the periodicity of $X(x)$.

Now, substituting from equations (7-37.1) and (7-37.2) into (7-36.3) we get

$$Y(y)k^4X(x) - 2k^2X(x).\frac{d^2Y(y)}{dy^2} + X(x)\frac{d^4Y(y)}{dy^4} = 0 \qquad (7\text{-}38.1)$$

$$\left(Y(y)k^4 - 2k^2.\frac{d^2Y(y)}{dy^2} + \frac{d^4Y(y)}{dy^4}\right)X(x) = 0 \qquad (7\text{-}38.2)$$

Thus, from equations (7-38.2) and (7-37.2) we can write two separate differential equations for $X(x)$ and $Y(y)$ as follows:

$$\frac{d^2X(x)}{dx^2} + k^2X(x) = 0 \qquad (7\text{-}39.1)$$

$$Y(y)k^4 - 2k^2.\frac{d^2Y(y)}{dy^2} + \frac{d^4Y(y)}{dy^4} = 0 \qquad (7\text{-}39.2)$$

(iii) Laplace solution of differential equations

a. The Lapalce transform of equation (7-39.1) is

$$L\left\{\frac{d^2X(x)}{dx^2} + k^2X(x)\right\} = L\{0\} \qquad (7\text{-}40.1)$$

We will use the property of Laplace transform that equates multiplication by s with integration as follows.

$$s^2 x(s) - s\ X(0) - X'(0) + k^2 x(s) = 0 \qquad (7\text{-}40.2)$$

For the purpose of immediate solution, we will assume that the boundary constants $X(0)$ and $X'(0)$ amount to a constant C, such that the inverse Laplace transform and required solution of (7-39.1) is

$$X(x) = L^{-1}\left\{\frac{C}{s^2 + k^2}\right\} = \frac{C}{k}\sin kx \qquad (7\text{-}40.3)$$

Since k and C are arbitrary constants, we could express $X(x)$ in terms of sum of sines and cosines as follows:

213

$$X(x) = C_1 \sin kx + C_2 \cos kx \qquad\qquad (7\text{-}40.4)$$

b. The Lapalce transform of equation (7-39.2) is

$$L\left\{Y(y)k^4 - 2k^2 \cdot \frac{d^2Y(y)}{dy^2} + \frac{d^4Y(y)}{dy^4}\right\} = L\{0\} \qquad\qquad (7\text{-}41.1)$$

The Laplace transform is

$$k^4 \, y(s) - 2 \, k^2 \, s^2 \, y(s) + s^4 \, y(s) = C \qquad\qquad (7\text{-}41.2)$$
$$y(s) \, (k^4 - 2 \, k^2 \, s^2 + s^4) = C$$

i.e., $\qquad y(s) = \dfrac{C}{\left(s^2 - k^2\right)^2} \qquad\qquad (7\text{-}41.3)$

The inverse Laplace transform of (7-41.3)

$$Y(y) = L^{-1}\left\{\frac{C}{\left(s^2 - k^2\right)\left(s^2 - k^2\right)}\right\} \qquad\qquad (7\text{-}41.4)$$

Equation (7-41.4) can be written in terms of the product of two transforms, which are determined by the product theorem of **convolution integrals** as follows

$$Y(y) = L^{-1}\left\{\frac{C}{\left(s^2 - k^2\right)\left(s^2 - k^2\right)}\right\} = C \sinh(ky) * \frac{1}{k}\sinh(ky) \qquad\qquad (7\text{-}41.4)$$

Where, $\qquad \sinh(ky) * \sinh(ky) = \displaystyle\int_0^t \sinh k(y - \lambda) \, \sinh(k\lambda) \, d\lambda$

$$= \int_0^y \left(\frac{e^{k(y-\lambda)} - e^{-k(y-\lambda)}}{2} \right) \left(\frac{e^{k\lambda} - e^{-k\lambda}}{2} \right) \ d\lambda$$

$$= \frac{1}{4} \int_0^y \left(e^{ky} - e^{k(y-2\lambda)} - e^{-k(y-2\lambda)} + e^{-ky} \right) \ d\lambda$$

$$= \frac{1}{4} \left[\left(e^{ky} + e^{-ky} \right) \lambda + e^{ky} \frac{e^{-2\lambda k}}{2k} - e^{-ky} \frac{e^{2\lambda k}}{2k} \right]_0^y$$

$$= \frac{1}{4} \left[\left(e^{ky} + e^{-ky} \right) y + \frac{1}{2k} \left(e^{ky} e^{-2yk} - e^{-ky} e^{2yk} \right) - \frac{1}{2k} \left(e^{ky} - e^{-ky} \right) \right]$$

$$= \frac{1}{2} \left[y \cosh ky - \frac{1}{k} \sinh ky \right]$$

i.e., $\qquad Y(y) = \dfrac{C}{2k} \left[y \cosh ky - \dfrac{1}{k} \sinh ky \right]$ $\qquad\qquad$ (7-41.5)

Similar to equation (7-40.3), we can write equation (7-41.5) in more general form as follows

$$Y(y) = C_3 y \cosh ky + C_4 y \sinh ky + C_5 \cosh ky + C_6 \sinh ky \qquad\qquad (7\text{-}41.5)$$

(iv) General solution of Airy-Lévy's equation

The desired solution of Airy-Lévy's equation (7-36.2) is thus.
2s

$$\varphi(x, y) = X(x) Y(y)$$

$$= \left(C_1 \sin kx + C_2 \cos kx \right) \left[\begin{matrix} C_3 y \cosh ky + C_4 y \sinh ky \\ + C_5 \cosh ky + C_6 \sinh ky \end{matrix} \right] \qquad (7\text{-}42)$$

7.5. Determination of coefficients of Fourier's series from trigonometric integrals

(i) Terminal axial stresses at x = 0 and x = l

The Fourier's expansion stress function, equation (7-42), is subjected to vanishing terminal normal stresses along the beam long axis Ox, Figure 7-6. From equation (7-17.1), we get

215

$$\sigma_x\big|_{\substack{x=0 \\ x=l}} = \frac{\partial^2 \varphi(x,y)}{\partial y^2}\bigg|_{\substack{x=0 \\ x=l}} = 0$$

$$= X(x)\frac{d^2 Y(y)}{dy^2}$$

$$= \left(C_1 \sin kx + C_2 \cos kx\right)_{\substack{x=0 \\ x=l}}\left[\begin{array}{l}C_3 y \cosh ky + C_4 y \sinh ky \\ + C_5 \cosh ky + C_6 \sinh ky\end{array}\right]$$

$$= 0$$

(7-43.1)

Clearly, the y-terms in equation (7-43.1) do not vanish at the ends of the beam. Thus, the only vanishing term is the x-dependent function as follows:

$$\left(C_1 \sin kx + C_2 \cos kx\right)_{\substack{x=0 \\ x=l}} = 0 \tag{7-43.2}$$

$$C_2 \cos kx = 0 \tag{7-43.3}$$

$$C_1 \sin(kl) + C_2 \cos(kl) = 0 \tag{7-43.4}$$

Equation (7-43.3) is satisfied at the zeros of the sine by

$$k = \frac{n\pi}{l}, \quad n = 1, 2, 3 \ldots \infty \tag{7-43.5}$$

We omitted the value of n = 0 as it will defeat our assumption of periodicity in equations (7-37.1) and (7-37.2). Such periodicity allows us summing many terms of sinusoidal functions that should approximate the exact solution of non-periodic polynomials. (Closeness of solution).

Equation (7-43.4) is satisfied at the zeros of the sine term and

$$C_2 = 0 \tag{7-43.6}$$

We could also include the constant C_1 in the coefficients of y-dependent terms.

Therefore, after applying the terminal boundary conditions of vanishing terminal axial normal stresses, equation (7-42) becomes

$$\varphi(x,y) = \sum_{n=1}^{\infty} \sin\left(n\pi\frac{x}{l}\right)\left[\begin{array}{l}A_n y \cosh\left(n\pi\frac{y}{l}\right) + B_n y \sinh\left(n\pi\frac{y}{l}\right) \\ + C_n \cosh\left(n\pi\frac{y}{l}\right) + D_n \sinh\left(n\pi\frac{y}{l}\right)\end{array}\right] \tag{7-44}$$

216

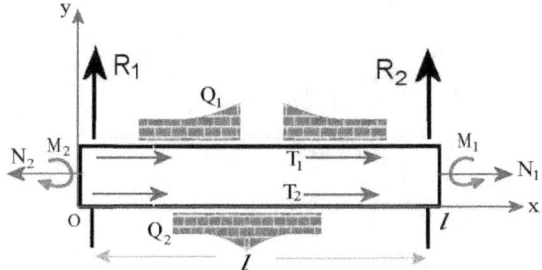

Figure 7-6. Stresses and couples used in determining the coefficients of the Fourier's series of the stress function, equation (7-42).

(ii) Boundary conditions at the upper surface of the beam; y = h

Equation (7-17.2) is satisfied at the upper surface by the externally applied normal load $Q_1(x)$ as follows

$$\sigma_y\Big|_{y=h} = \frac{\partial^2 \varphi(x,y)}{\partial x^2}\Big|_{y=h} = Q_1(x) \tag{7-45.1}$$

Equation (7-17.3) is satisfied by the externally applied tangential stress $T_1(x)$

$$\tau_{xy}\Big|_{y=h} = -\frac{\partial^2 \varphi(x,y)}{\partial x \partial y} = -T_1(x) \tag{7-45.2}$$

Applying the differentiation and boundary values of equation (7-45.1) on the stress function (7-44), we get

$$Q_1(x) = -\left(\frac{\pi}{l}\right)^2 \sum_{n=1}^{\infty} n^2 \sin\left(n\pi\frac{x}{l}\right) \left[\begin{array}{l} A_n h\cosh\left(n\pi\frac{h}{l}\right) + B_n h\sinh\left(n\pi\frac{h}{l}\right) \\ + C_n \cosh\left(n\pi\frac{h}{l}\right) + D_n \sinh\left(n\pi\frac{h}{l}\right) \end{array}\right] \tag{7-45.3}$$

Similarly, applying the differentiation and boundary values of equation (7-45.2) on the stress function (7-44), we get

217

$$T_1(x) = \left(\frac{\pi}{l}\right)\sum_{n=1}^{\infty} n \cos\left(n\pi\frac{x}{l}\right)\begin{bmatrix} \left[A_n + D_n\left(\frac{n\pi}{l}\right)\right]\cosh\left(n\pi\frac{h}{l}\right) \\ +\left[B_n + C_n\left(\frac{n\pi}{l}\right)\right]\sinh\left(n\pi\frac{h}{l}\right) \\ +A_n\left(n\pi\frac{h}{l}\right)\sinh\left(n\pi\frac{h}{l}\right) \\ +B_n\left(n\pi\frac{h}{l}\right)\cosh\left(n\pi\frac{h}{l}\right) \end{bmatrix}$$ (7-45.4)

(iii) Boundary conditions at the lower surface of the beam; y = 0

Equation (7-17.2) is satisfied at the lower surface by the externally applied normal load $Q_2(x)$ as follows

$$-\sigma_y\big|_{y=0} = -\frac{\partial^2\varphi(x,y)}{\partial x^2}\bigg|_{y=0} = Q_2(x)$$ (7-46.1)

Equation (7-17.3) is satisfied by the externally applied tangential stress $T_2(x)$

$$-\tau_{xy}\big|_{y=0} = +\frac{\partial^2\varphi(x,y)}{\partial x \partial y} = T_2(x)$$ (7-46.2)

Applying the differentiation and boundary values of equation (7-46.1) on the stress function (7-44), we get

$$Q_2(x) = \left(\frac{\pi}{l}\right)^2 \sum_{n=1}^{\infty} n^2 \sin\left(n\pi\frac{x}{l}\right)[C_m]$$ (7-46.3)

Similarly, applying the differentiation and boundary values of equation (7-46.2) on the stress function (7-44), we get

$$T_2(x) = +\left(\frac{\pi}{l}\right)\sum_{n=1}^{\infty} n \cos\left(n\pi\frac{x}{l}\right)\left[A_m + D_m\left(\frac{n\pi}{l}\right)\right]$$ (7-46.4)

(iv) Nth coefficient of Fourier series

218

The four coefficients A_n, B_n, C_n, and D_n are determined by the usual means used in Fourier series analysis, by multiplying by $\sin(k\pi x/l)$ and $\cos(k\pi x/l)$ then integrating over the period $x = [0, l]$. Such process eliminates all terms other than $n = k$.

a. Integral of products of two sine functions

We will take advantage of the following property of integration of product of two sine functions

$$\int_0^l \sin\left(n\pi\frac{x}{l}\right)\sin\left(k\pi\frac{x}{l}\right)dx = \frac{1}{2}\int_0^l \left[\cos\left(n-k\right)\pi\frac{x}{l}\right) - \cos\left(n+k\right)\pi\frac{x}{l}\right) + \right]dx$$

$$= \frac{l}{2}\left[\frac{\sin\left(n-k\right)\pi\frac{x}{l}\right)}{\left(n-k\right)\pi} - \frac{\sin\left(n+k\right)\pi\frac{x}{l}\right)}{\left(n+k\right)\pi}\right]_0^l \qquad (7\text{-}47.1)$$

$$= \frac{l}{2}\left[\frac{\sin(n-k)\pi}{\left(n-k\right)\pi} - \frac{\sin(n+k)\pi}{\left(n+k\right)\pi}\right]$$

When $n = k$, we can substitute by

$$\lim_{n\to l}\frac{\sin\left[(n-k)\pi\right]}{\left(n-k\right)\pi} = 1 \qquad (7\text{-}47.2)$$

$$\lim_{n\to l}\frac{\sin\left[(n+k)\pi\right]}{\left(n+k\right)\pi} = 0 \qquad (7\text{-}47.3)$$

Without proof, the equalities (7-47.2) and (7-47.2) can be justified on the basis that the sine of angle approaches the value of the angle, in radians, as the angle get smaller. Thus, the ratio in (7-47.2) approaches unity as the angle is getting smaller. In contrast, in (7-47.3), the angle is getting larger while the sine vanishes for integer values of $(n + k)$.

Therefore, for $n = k$, we get

$$\int_0^l \sin\left(n\pi\frac{x}{l}\right)\sin\left(k\pi\frac{x}{l}\right)dx = \frac{l}{2} \qquad (7\text{-}47.4)$$

Similarly,

$$\int_0^l \cos\left(n\pi\frac{x}{l}\right)\cos\left(k\pi\frac{x}{l}\right)dx = \frac{1}{2}\int_0^l \left[\cos\left((n-k)\pi\frac{x}{l}\right) + \cos\left((n+k)\pi\frac{x}{l}\right) + \right]dx$$

$$= \frac{l}{2}\left[\frac{\sin\left((n-k)\pi\frac{x}{l}\right)}{(n-k)\pi} + \frac{\sin\left((n+k)\pi\frac{x}{l}\right)}{(n+k)\pi}\right]_0^l \qquad (7\text{-}47.5)$$

$$= \frac{l}{2}\left[\frac{\sin(n-k)\pi}{(n-k)\pi} + \frac{\sin(n+k)\pi}{(n+k)\pi}\right]$$

Similarly, for n = k, we can get

$$\int_0^l \cos\left(n\pi\frac{x}{l}\right)\cos\left(k\pi\frac{x}{l}\right)dx = \frac{l}{2} \qquad (7\text{-}47.6)$$

b. From equation (7-45.3), after multiplication by sin (kπx/l) and integration over the period x = [0, l], we get

$$\int_0^l Q_1(x)\sin\frac{k\pi x}{l}dx = -\left(\frac{\pi}{l}\right)^2\sum_{n=1}^{\infty}n^2\begin{bmatrix}A_n h\cosh\dfrac{n\pi h}{l}\\[4pt]+B_k h\sinh\dfrac{n\pi h}{l}\\[4pt]+C_k\cosh\dfrac{n\pi h}{l}\\[4pt]+D_k\cosh\dfrac{n\pi h}{l}\end{bmatrix}\int_0^l \sin\frac{k\pi x}{l}\sin\frac{n\pi x}{l}dx \qquad (7\text{-}48.1)$$

Where the bracketed four-terms are lumped for convenience. Therefore, when n = k, we get

$$\int_0^l Q_1(x)\sin\frac{k\pi x}{l}dx = -\frac{2\pi^2 k^2}{l}\begin{bmatrix}A_k h\cosh\dfrac{k\pi h}{l} + B_k h\sinh\dfrac{k\pi h}{l}\\[4pt]+C_k\cosh\dfrac{k\pi h}{l} + D_k\cosh\dfrac{k\pi h}{l}\end{bmatrix} \qquad (7\text{-}48.2)$$

Equation (7-48.2) connects the four nth coefficients with the upper surface load. We will need three more equations to determine the all four coefficients.

c. From equation (7-45.4), after multiplication by cos (kπx/l) and integration over the period x = [0, l], we get

220

$$\int_0^l T_1(x)\cos\frac{k\pi x}{l}\,dx = +\left(\frac{\pi k}{2}\right)\begin{bmatrix} A_k\left(\cosh\frac{k\pi h}{l} + \frac{k\pi h}{l}\sinh\frac{k\pi h}{l}\right) \\[2mm] + B_k\left(\sinh\frac{k\pi h}{l} + \frac{k\pi h}{l}\cosh\frac{k\pi h}{l}\right) \\[2mm] + C_k\frac{k\pi}{l}\sinh\frac{k\pi h}{l} \\[2mm] + D_k\frac{k\pi}{l}\cosh\frac{k\pi h}{l} \end{bmatrix} \qquad (7\text{-}49)$$

Equation (7-49) is the second equation that connects the four nth coefficients to the surface load, $T_1(x)$.

d. From equation (7-46.3), after multiplication by sin $(k\pi x/l)$ and integration over the period x = [0, l], we get

$$\int_0^l Q_2(x)\sin\frac{k\pi x}{l}\,dx = \left(\frac{\pi^2 k^2}{2l}\right)C_k \qquad (7\text{-}50)$$

e. From equation (7-46.4), after multiplication by cos $(k\pi x/l)$ and integration over the period x = [0, l], we get

$$\int_0^l T_2(x)\cos\frac{k\pi x}{l}\,dx = \frac{\pi k}{2}\left[A_k + D_k\left(\frac{k\pi}{l}\right)\right] \qquad (7\text{-}51)$$

(v) Summary of Fourier series and coefficients of stress function

Table 7-1: Fourier series and coefficients of stress function in planar stress

$$\varphi(x,y) = \sum_{n=1}^{\infty}\sin\left(n\pi\frac{x}{l}\right)\begin{bmatrix} A_n y\cosh\left(n\pi\frac{y}{l}\right) + B_n y\sinh\left(n\pi\frac{y}{l}\right) \\[2mm] + C_n\cosh\left(n\pi\frac{y}{l}\right) + D_n\sinh\left(n\pi\frac{y}{l}\right) \end{bmatrix} \qquad (7\text{-}44)$$

$$\begin{bmatrix} A_k h\cosh\frac{k\pi h}{l} + B_k h\sinh\frac{k\pi h}{l} \\[2mm] + C_k\cosh\frac{k\pi h}{l} + D_k\cosh\frac{k\pi h}{l} \end{bmatrix} = -\frac{l}{2\pi^2 k^2}\int_0^l Q_1(x)\sin\frac{k\pi x}{l}\,dx \qquad (7\text{-}48.2)$$

$$
\left[\begin{array}{l} A_k\left(\cosh\dfrac{k\pi h}{l} + \dfrac{k\pi h}{l}\sinh\dfrac{k\pi h}{l} \right) \\[2ex] +B_k\left(\sinh\dfrac{k\pi h}{l} + \dfrac{k\pi h}{l}\cosh\dfrac{k\pi h}{l} \right) \\[2ex] +C_k\dfrac{k\pi}{l}\sinh\dfrac{k\pi h}{l} + D_k\dfrac{k\pi}{l}\cosh\dfrac{k\pi h}{l} \end{array} \right] = \dfrac{2}{\pi k}\int_0^l T_1(x)\cos\dfrac{k\pi x}{l}\,dx
$$

$$\tag{7-49}$$

$$
C_k = \frac{2l}{\pi^2 k^2}\int_0^l Q_2(x)\sin\frac{k\pi x}{l}\,dx
\tag{7-50}
$$

$$
A_k + D_k\left(\frac{k\pi}{l}\right) = \frac{2}{\pi k}\int_0^l T_2(x)\cos\frac{k\pi x}{l}\,dx
\tag{7-51}
$$

(vi) Utility of geometrical polynomials

1. Variable loads

The great numbers of coefficients in Fourier series permit the calculation of diverse configurations of loads on all surfaces of a two-dimensional body. Table 7-1 enables us to change the distributions of $Q_1(x)$, $Q_2(x)$, $T_1(x)$, and $T_2(x)$ which alter the values of the coefficients of the sinusoidal functions. The number of terms under the summation sign determines the accuracy of solution.

2. Non-periodic solution

In equation (7-43.5), we omitted the value of $k = 0$, which corresponds to non-periodic $X(x)$, equations (7-37.1) and (7-37.2). Let us examine the stress function, equation (7-36.2), for which $X(x)$ is not periodic. Thus, substituting by $k = 0$ and integrating (7-37.2) twice, we get

$$
X(x) = ax + b
\tag{7-52.1}
$$

Similarly, substituting by $k = 0$ in equation (7-39.2) and integrating four times we get

$$
Y(y) = A\frac{y^3}{6} + B\frac{y^2}{2} + Cy + D
\tag{7-52.2}
$$

Thus, the stress function (7-36.2) become

$$
\begin{aligned}
\varphi(x,y) &= X(x)Y(y) \\
&= (ax+b)\left(A\frac{y^3}{6} + B\frac{y^2}{2} + Cy \right)
\end{aligned}
\tag{7-52.3}
$$

Equation (7-52.3) has only five constants which, with the help of equations (7-17), offer limited solutions of elasticity problems with complex boundary conditions.

7.6. Determination of coefficients of Fourier's series from hyperbolic linear polynomials

We have dwelled on the determinations of the nth coefficients of **hyperbolic functions**, in the stress function, equation (7-44), by using the two integrals of sines, equation (7-47.1), and cosines, equations (7-47.5). The Fourier's coefficients obtained in such manner entail the integrals of external loads, Table 7-1, in such manner that complicates the computation of the Fourier's coefficients. Equations (7-48.2), (7-49), (7-50), and (7-51).

(i) Linearizing the hyperbolic Fourier functions

We will now approach the determination of the **Fourier coefficients** in equation (7-44) by **linearizing the hyperbolic** y-terms such that we could separate the geometric criteria of the beam from the parameters of the loads. Since we assume to have two surfaces (upper and lower) and two stresses (normal and shear), therefore we need four constants to satisfy the four boundary conditions as follows.

$$\varphi(x, y) = \sum_{n=1}^{\infty} \sin\left(\frac{n\pi x}{l}\right) \sum_{m-1}^{4} A_{mn} \left(\begin{array}{l} a_{mn} \cosh y_1 + b_{mn} \sinh y_1 \\ + c_{mn} y_1 \cosh y_1 + d_{mn} y_1 \sinh y_1 \end{array} \right) \qquad (7\text{-}53.1)$$

Where,

$$y_1 = \frac{n\pi y}{l} \qquad (7\text{-}53.2)$$

Where the coefficients a_{mn}, b_{mn}, c_{mn}, and d_{mn} will be tailored to geometry of the beam only, independent of the load, while the A_{mn} are load-dependent.

Consequently, equations (7-17.2) and (7-17.3) become

1. Normal stress at upper and lower surfaces of the beam

$$\begin{aligned} \sigma_y &= \frac{\partial^2 \varphi(x, y)}{\partial x^2} \\ &= -\sum_{n=1}^{\infty} \left(\frac{n\pi}{l}\right)^2 \sin\left(\frac{n\pi x}{l}\right) \sum_{m-1}^{4} A_{mn} \left(\begin{array}{l} a_{mn} \cosh y_1 + b_{mn} \sinh y_1 \\ + c_{mn} y_1 \cosh y_1 + d_{mn} y_1 \sinh y_1 \end{array} \right) \end{aligned} \qquad (7\text{-}54.1)$$

2. Shear stress at upper and lower surfaces of the beam

223

$$\tau_{xy} = -\frac{\partial^2 \varphi(x,y)}{\partial x \partial y}$$

$$= \sum_{n=1}^{\infty} \left(\frac{n\pi}{l}\right) \cos\left(\frac{n\pi x}{l}\right) \sum_{m-1}^{4} A_{mn} \begin{pmatrix} a_{mn} \sinh y_1 + b_{mn} \cosh y_1 \\ + c_{mn}(\cosh y_1 + y_1 \sinh y_1) \\ + d_{mn}(\sinh y_1 + y_1 \cosh y_1) \end{pmatrix} \qquad (7\text{-}54.2)$$

Boundary conditions at the upper and lower surfaces of the beam

(ii) Fourier series expansion of external loads

Before we settle on the values of the coefficients a_{mn}, b_{mn}, c_{mn}, and d_{mn} of the linear expansions, we will first expand the external loads in terms of Fourier's series such that the four A_{mn} represent external loads alone.

$$Q_1(x) = \sum_{n=1}^{\infty} B_{1n} \sin\left(\frac{n\pi x}{l}\right) \qquad (7\text{-}55.1)$$

$$Q_2(x) = \sum_{n=1}^{\infty} B_{2n} \sin\left(\frac{n\pi x}{l}\right) \qquad (7\text{-}55.2)$$

$$T_1(x) = \sum_{n=1}^{\infty} C_{1n} \cos\left(\frac{n\pi x}{l}\right) \qquad (7\text{-}55.3)$$

$$T_2(x) = \sum_{n=1}^{\infty} C_{2n} \cos\left(\frac{n\pi x}{l}\right) \qquad (7\text{-}55.4)$$

Where, the **load coefficients** B_{1n}, B_{2n}, C_{1n}, and C_{2n} are given coefficients describing the external load distribution.

From the two equations (7-54) and the four equations (7-55) and the boundary conditions at the two surfaces $y = 0$ and $y = h$, we realize the rationale of settling on four linear functions in (7-54) for each constant A_{mn}.

(iii) Boundary conditions at upper and lower surfaces

The linear expansions in (7-54) are first denoted as follows:

$$Y_{mn}(y_1) = a_{mn} \cosh y_1 + b_{mn} \sinh y_1 + c_{mn} y_1 \cosh y_1 + d_{mn} y_1 \sinh y_1 \qquad (7\text{-}56.1)$$

$$\left(\frac{l}{n\pi}\right) Y'_{mn}(y_1) = a_{mn} \sinh y_1 + b_{mn} \cosh y_1 + c_{mn}\begin{pmatrix} \cosh y_1 \\ + y_1 \sinh y_1 \end{pmatrix} + d_{mn}\begin{pmatrix} \sinh y_1 \\ + y_1 \cosh y_1 \end{pmatrix} \qquad (7\text{-}56.2)$$

224

The sixteen constants a_{mn}, b_{mn}, c_{mn}, and d_{mn} (m =1, 2, 3, 4) will be defined such that each A_{mn} has single values at the two surfaces and with the two stresses (normal and shear).

1. Lower surface y = 0

At y = 0, we will set

$$Y_{1n}(0) = 1 \quad\text{and}\quad Y_{mn}(0) = 0 \quad\text{when } m \neq 1 \qquad (7\text{-}57.1)$$
$$(l/n\pi)Y'_{2n}(0) = 1 \quad\text{and}\quad (l/n\pi)Y'_{mn}(0) = 0 \quad\text{when } m \neq 2 \qquad (7\text{-}57.2)$$

Thus, equations (7-56) give

$$Y_{1n}(0) = a_{1n} = 1 \qquad (7\text{-}58.1)$$
$$Y_{2n}(0) = a_{2n} = 0 \qquad (7\text{-}58.2)$$
$$Y_{3n}(0) = a_{3n} = 0 \qquad (7\text{-}58.3)$$
$$Y_{4n}(0) = a_{4n} = 0 \qquad (7\text{-}58.4)$$

$$Y'_{1n}(0) = b_{1n} + c_{1n} = 0 \qquad (7\text{-}58.5)$$
$$Y'_{2n}(0) = b_{2n} + c_{2n} = 1 \qquad (7\text{-}58.6)$$
$$Y'_{3n}(0) = b_{3n} + c_{3n} = 0 \qquad (7\text{-}58.7)$$
$$Y'_{4n}(0) = b_{4n} + c_{4n} = 0 \qquad (7\text{-}58.8)$$

2. Upper surface y = h

At y = h, we will denote

$$y_h = \frac{n\pi h}{l} \qquad (7\text{-}59.1)$$

And will choose the following values

$$Y_{3n}(y_h) = 1 \quad\text{and}\quad Y_{mn}(y_h) = 0 \quad\text{when } m \neq 3 \qquad (7\text{-}59.2)$$
$$(l/n\pi)Y'_{4n}(y_h) = 1 \quad\text{and}\quad (l/n\pi)Y'_{mn}(y_h) = 0 \quad\text{when } m \neq 4 \qquad (7\text{-}59.3)$$

Thus, equations (7-56) give

$$Y_{1n}(y_h) = a_{1n}\cosh y_h + b_{1n}\sinh y_h + c_{1n}y_h\cosh y_h + d_{1n}y_h\sinh y_h = 0 \qquad (7\text{-}60.1)$$
$$Y_{2n}(y_h) = a_{2n}\cosh y_h + b_{2n}\sinh y_h + c_{2n}y_h\cosh y_h + d_{2n}y_h\sinh y_h = 0 \qquad (7\text{-}60.2)$$
$$Y_{3n}(y_h) = a_{3n}\cosh y_h + b_{3n}\sinh y_h + c_{3n}y_h\cosh y_h + d_{3n}y_h\sinh y_h = 1 \qquad (7\text{-}60.3)$$
$$Y_{4n}(y_h) = a_{4n}\cosh y_h + b_{4n}\sinh y_h + c_{4n}y_h\cosh y_h + d_{4n}y_h\sinh y_h = 0 \qquad (7\text{-}60.4)$$

$$\left(\frac{l}{n\pi}\right)Y'_{1n}(y_h) = a_{1n}\sinh y_h + b_{1n}\cosh y_h + c_{1n}\left(\begin{array}{c}\cosh y_h \\ + y_h\sinh y_h\end{array}\right) + d_{1n}\left(\begin{array}{c}\sinh y_h \\ + y_h\cosh y_h\end{array}\right) = 0 \qquad (7\text{-}60.5)$$

$$\left(\frac{l}{n\pi}\right)Y_{2n}'(y_h) = a_{2n}\sinh y_h + b_{2n}\cosh y_h + c_{2n}\left(\begin{array}{c}\cosh y_h \\ + y_h\sinh y_h\end{array}\right) + d_{2n}\left(\begin{array}{c}\sinh y_h \\ + y_h\cosh y_h\end{array}\right) = 0 \qquad (7\text{-}60.6)$$

$$\left(\frac{l}{n\pi}\right)Y_{3n}'(y_h) = a_{3n}\sinh y_h + b_{3n}\cosh y_h + c_{3n}\left(\begin{array}{c}\cosh y_h \\ + y_h\sinh y_h\end{array}\right) + d_{3n}\left(\begin{array}{c}\sinh y_h \\ + y_h\cosh y_h\end{array}\right) = 0 \qquad (7\text{-}60.7)$$

$$\left(\frac{l}{n\pi}\right)Y_{4n}'(y_h) = a_{4n}\sinh y_h + b_{4n}\cosh y_h + c_{4n}\left(\begin{array}{c}\cosh y_h \\ + y_h\sinh y_h\end{array}\right) + d_{4n}\left(\begin{array}{c}\sinh y_h \\ + y_h\cosh y_h\end{array}\right) = 1 \qquad (7\text{-}60.8)$$

3. Solving instantaneous linear equations (7-58) and (7-58)

a. Equations (7-58.1 and 5) and (7-60.1 and 5), we get

$$b_{1n} = -\frac{\left(\sinh y_h \cosh y_h + y_h\right)}{\sinh^2 y_h - y^2{}_h} \qquad (7\text{-}61.1)$$

$$c_{1n} = \frac{\left(\sinh y_h \cosh y_h + y_h\right)}{\sinh^2 y_h - y^2{}_h} \qquad (7\text{-}61.2)$$

$$d_{1n} = -\frac{\sinh^2 y_h}{\sinh^2 y_h - y^2{}_h}$$

$$(7\text{-}61.3)$$

b. Equations (7-58.2 and 6) and (7-60.2 and 6), we get

$$b_{2n} = -\frac{y^2{}_h}{\sinh^2 y_h - y^2{}_h} \qquad (7\text{-}62.1)$$

$$c_{2n} = -\frac{\sinh^2 y_h}{\sinh^2 y_h - y^2{}_h} \qquad (7\text{-}62.2)$$

$$d_{2n} = -\frac{\cosh y_h \sinh y_h - y_h}{\sinh^2 y_h - y^2{}_h} \qquad (7\text{-}62.3)$$

c. Equations (7-58.3 and 7) and (7-60.3 and 7), we get

$$b_{3n} = \frac{\left(\sinh y_h + y_h\cosh y_h\right)}{\left(\sinh^2 y_h - y^2{}_h\right)} \qquad (7\text{-}63.1)$$

$$c_{3n} = -\frac{\left(\sinh y_h + y_h\cosh y_h\right)}{\left(\sinh^2 y_h - y^2{}_h\right)} \qquad (7\text{-}63.2)$$

$$d_{3n} = \frac{y_h\sinh y_h}{\left(\sinh^2 y_h - y^2{}_h\right)} \qquad (7\text{-}63.3)$$

d. Equations (7-58.4 and 8) and (7-60.4 and 8), we get

$$b_{4n} = -\frac{y_h \sinh y_h}{\sinh^2 y_h - y^2_h} \qquad (7\text{-}64.1)$$

$$c_{4n} = \frac{y_h \sinh y_h}{\sinh^2 y_h - y^2_h} \qquad (7\text{-}64.2)$$

$$d_{4n} = \frac{\sinh y_h - y_h \cosh y_h}{\sinh^2 y_h - y^2_h} \qquad (7\text{-}64.3)$$

7.7. Boundary load conditions with hyperbolic Fourier's coefficients

1. At the lower surface ($y = 0$), equations (7-57) allow us to simplify equations (7-54) as follows

Substitute by $Y_{1n}(0) = 1$ and $Y_{mn}(0) = 0$, when $m \neq 1$, from equations (7-57) in equation (7-54.1), we get

$$Q_2(x) = -\sum_{n=1}^{\infty} \left(\frac{n\pi}{l}\right)^2 \sin\left(\frac{n\pi x}{l}\right) A_{1n} \qquad (7\text{-}65.1)$$

Similarly, substitute by $Y'_{2n}(0) = 1$ and $Y'_{mn}(0) = 0$, when $m \neq 2$, from equations (7-57) in equation (7-54.2)

$$T_2(x) = \sum_{n=1}^{\infty} \left(\frac{n\pi}{l}\right) \cos\left(\frac{n\pi x}{l}\right) A_{2n} \qquad (7\text{-}65.2)$$

Thus, the **hyperbolic linearization** and choices of the coefficients of the linear functions rendered the series expansions of the stress function simple enough to allow us separating the beam-dependent coefficients a_{mn}, b_{mn}, c_{mn}, and d_{mn} from the load-dependent coefficients A_{mn}.
Replacing the load function $Q_2(x)$ and $T_2(x)$ from equations (7-55.2) and (7-55.4), equations (7-65.1 and 2) become

$$\sum_{n=1}^{\infty} B_{2n} \sin\left(\frac{n\pi x}{l}\right) = -\sum_{n=1}^{\infty} \left(\frac{n\pi}{l}\right)^2 \sin\left(\frac{n\pi x}{l}\right) A_{1n} \qquad (7\text{-}65.3)$$

$$\sum_{n=1}^{\infty} C_{2n} \cos\left(\frac{n\pi x}{l}\right) = \sum_{n=1}^{\infty} \left(\frac{n\pi}{l}\right) \cos\left(\frac{n\pi x}{l}\right) A_{2n} \qquad (7\text{-}65.4)$$

Therefore, our linearization of the hyperbolic functions of the Fourier series concluded in the simple relationships between the **internal stresses load coefficients** A_{1n} and A_{2n} and the **external load coefficients** B_{2n} and C_{2n} as follows

227

$$B_{2n} = -\left(\frac{n\pi}{l}\right)^2 A_{1n} \qquad (7\text{-}66.1)$$

$$C_{2n} = \left(\frac{n\pi}{l}\right) A_{2n} \qquad (7\text{-}66.2)$$

2 At the upper surface (y = h), equations (7-59) allow us to simplify equations (7-54) as follows.

Substitute by $Y_{3n}(y_h) = 1$ and $Y_{mn}(y_h) = 0$, when $m \neq 3$, from equations (7-59.2) in equation (7-54.1), we get

$$Q_1(x) = -\sum_{n=1}^{\infty} \left(\frac{n\pi}{l}\right)^2 \sin\left(\frac{n\pi x}{l}\right) A_{3n} \qquad (7\text{-}67.1)$$

Substitute by $Y'_{4n}(y_h) = 1$ and $Y'_{mn}(y_h) = 0$, when $m \neq 4$, from equations (7-59.3) in equation (7-54.2), we get

$$T_1(x) = \sum_{n=1}^{\infty} \left(\frac{n\pi}{l}\right) \cos\left(\frac{n\pi x}{l}\right) A_{4n} \qquad (7\text{-}67.2)$$

Replacing the load function $Q_1(x)$ and $T_1(x)$ from equations (7-55.1) and (7-55.3), equations (7-67.1 and 2) become

$$\sum_{n=1}^{\infty} B_{1n} \sin\left(\frac{n\pi x}{l}\right) = -\sum_{n=1}^{\infty} \left(\frac{n\pi}{l}\right)^2 \sin\left(\frac{n\pi x}{l}\right) A_{3n} \qquad (7\text{-}67.3)$$

$$\sum_{n=1}^{\infty} C_{1n} \cos\left(\frac{n\pi x}{l}\right) = \sum_{n=1}^{\infty} \left(\frac{n\pi}{l}\right) \cos\left(\frac{n\pi x}{l}\right) A_{4n} \qquad (7\text{-}67.4)$$

Similarly, our linearization of the hyperbolic functions of the Fourier series concluded in the simple relationships between the **internal stresses load coefficients** A_{3n} and A_{4n} and the **external load coefficients** B_{1n} and C_{1n} as follows

$$B_{1n} = -\left(\frac{n\pi}{l}\right)^2 A_{3n} \qquad (7\text{-}68.1)$$

$$C_{1n} = \left(\frac{n\pi}{l}\right) A_{4n} \qquad (7\text{-}68.2)$$

Summary of coefficients

Table 7-2. Fourier series expansions of hyperbolic linear functions and load-dependent and beam-dependent coefficients.

Substitutions:

$$y_1 = \frac{n\pi y}{l}, \quad y_h = \frac{n\pi h}{l}, \quad s = \sinh y_h, \quad c = \cosh y_h, \quad g = \sinh^2 y_h - y_h^2$$

$$\varphi(x,y) = \sum_{n=1}^{\infty} \sin\left(\frac{n\pi x}{l}\right) \sum_{m=1}^{4} A_{mn} \left(\begin{array}{l} a_{mn}\cosh y_1 + b_{mn}\sinh y_1 \\ + c_{mn}y_1\cosh y_1 + d_{mn}y_1\sinh y_1 \end{array} \right)$$

$a_{1n} = 1$	$b_{1n} = -(s.c + y_h)/g$	$c_{1n} = (s.c + y_h)/g$	$d_{1n} = -s^2/g$
$a_{2n} = 0$	$b_{2n} = -y^2h/g$	$c_{2n} = -s^2/g$	$d_{2n} = -(s.c - y_h)/g$
$a_{3n} = 0$	$b_{3n} = (s + y_h.c)/g$	$c_{3n} = -(s + y_h.c)/g$	$d_{3n} = y_h.s/g$
$a_{4n} = 0$	$b_{4n} = -y_h s/g$	$c_{4n} = y_h.s/g$	$d_{4n} = (s - y_h.c)/g$

$$A_{1n} = -\left(\frac{l}{n\pi}\right)^2 B_{2n} \qquad A_{2n} = \left(\frac{l}{n\pi}\right)C_{2n} \qquad A_{3n} = -\left(\frac{l}{n\pi}\right)^2 B_{1n} \qquad A_{4n} = \left(\frac{l}{n\pi}\right)C_{1n}$$

$$Q_2(x) = \sum_{n=1}^{\infty} B_{2n}\sin\left(\frac{n\pi x}{l}\right) \quad T_2(x) = \sum_{n=1}^{\infty} C_{2n}\cos\left(\frac{n\pi x}{l}\right) \quad Q_1(x) = \sum_{n=1}^{\infty} B_{1n}\sin\left(\frac{n\pi x}{l}\right) \quad T_1(x) = \sum_{n=1}^{\infty} C_{1n}\cos\left(\frac{n\pi x}{l}\right)$$

Stresses

The stress function, equation (7-53.1), is simplified in terms of the linear hyperbolic functions $Y_{mn}(y)$ as follows

$$\varphi(x,y) = \sum_{n=1}^{\infty} \sin\left(\frac{n\pi x}{l}\right) \sum_{m=1}^{4} A_{mn} Y_{mn}(y_1) \qquad (7\text{-}69.1)$$

Where,

$$Y_{mn}(y_1) = a_{mn}\cosh y_1 + b_{mn}\sinh y_1 + c_{mn}y_1\cosh y_1 + d_{mn}y_1\sinh y_1 \qquad (7\text{-}69.2)$$

Similarly, the three stresses obtained from (7-69) are expressed as follows

$$\sigma_y = \frac{\partial^2 \varphi(x,y)}{\partial x^2} = -\sum_{n=1}^{\infty}\left(\frac{n\pi}{l}\right)^2 \sin\left(\frac{n\pi x}{l}\right) \sum_{m=1}^{4} A_{mn} Y_{mn}(y_1) \qquad (7\text{-}70.1)$$

$$\sigma_x = \frac{\partial^2 \varphi(x,y)}{\partial y^2} = \sum_{n=1}^{\infty} \sin\left(\frac{n\pi x}{l}\right) \sum_{m=1}^{4} A_{mn} Y''_{mn}(y_1) \qquad (7\text{-}70.2)$$

$$\tau_{xy} = -\frac{\partial^2 \varphi(x,y)}{\partial x \partial y} = \sum_{n=1}^{\infty}\left(\frac{n\pi}{l}\right) \cos\left(\frac{n\pi x}{l}\right) \sum_{m=1}^{4} A_{mn} Y'_{mn}(y_1) \qquad (7\text{-}70.3)$$

Where,

$$Y'_{mn}(y_1) = \left(\frac{n\pi}{l}\right)\left[a_{mn}\sinh y_1 + b_{mn}\cosh y_1 + c_{mn}\left(\begin{array}{c}\cosh y_1 \\ + y_1\sinh y_1\end{array}\right) + d_{mn}\left(\begin{array}{c}\sinh y_1 \\ + y_1\cosh y_1\end{array}\right)\right] \quad (7\text{-}70.4)$$

$$Y''_{mn}(y_1) = \left(\frac{n\pi}{l}\right)^2\left[\begin{array}{c}a_{mn}\cosh y_1 + b_{mn}\sinh y_1 + c_{mn}(2\sinh y_1 + y_1\cosh y_1) \\ + d_{mn}(2\cosh y_1 + y_1\sinh y_1)\end{array}\right] \quad (7\text{-}70.5)$$

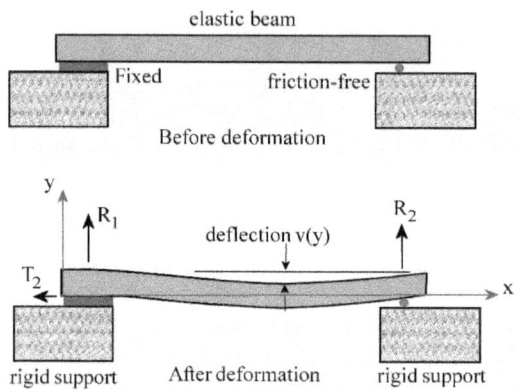

Figure 7-7. Deflection of elastic beam supported on rigid bodies with variable friction contacts

Displacements

Cauchy-Hooke's equation (7-3.3) gives the differential relation between the normal stresses σ_x and σ_y, equations (7-70.1) and (7-70.2) and the vertical displacement $v(x,y)$ in terms of the load-coefficients in Table 7-2.

The contact configuration depicted in Figure 7-7 is accounted for through the load-dependent coefficients in Table 7-2.

The vertical displacement of the beam in Figure 7-7 is obtained through equations (7-3.3) and equations (7-70.1) and (7-70.2) as follows

$$\frac{\partial v}{\partial y} = \varepsilon_{yy} = E_1^{-1}\left(\sigma_y - v_1\sigma_x\right) \quad (7\text{-}71.1)$$

$$v(x,y) = \int E_1^{-1}\left(\sigma_y - v_1\sigma_x\right)dy + f(x)$$

230

$$\mathbf{v}(x,y) = \int \left. E_1^{-1} \left[\begin{array}{l} -\sum\limits_{n=1}^{\infty}\left(\dfrac{n\pi}{l}\right)^2 \sin\left(\dfrac{n\pi x}{l}\right)\sum\limits_{m-1}^{4} A_{mn} Y_{mn}(y_1) \\ -v_1 \sum\limits_{n=1}^{\infty}\sin\left(\dfrac{n\pi x}{l}\right)\sum\limits_{m-1}^{4} A_{mn} Y_{mn}''(y_1) \end{array} \right] dy + f(x) \right.$$

$$\mathbf{v}(x,y) = -E_1^{-1}\sum_{n=1}^{\infty}\sin\left(\frac{n\pi x}{l}\right)\sum_{m-1}^{4} A_{mn} \int \left[\left(\frac{n\pi}{l}\right)^2 Y_{mn}(y_1) + v_1 Y_{mn}''(y_1) \right] dy + f(x) \qquad (7\text{-}71.2)$$

From equations (7-69.2) and (7-70.5), the integral in equation (7-71.2) becomes

$$\int \left[\begin{array}{l} \left(\begin{array}{l} a_{mn}\cosh y_1 + b_{mn}\sinh y_1 \\ + c_{mn} y_1 \cosh y_1 + d_{mn} y_1 \sinh y_1 \end{array}\right) \\ + v_1 \left(\begin{array}{l} a_{mn}\cosh y_1 + b_{mn}\sinh y_1 \\ + c_{mn}\left(2\sinh y_1 + y_1 \cosh y_1\right) \\ + d_{mn}\left(2\cosh y_1 + y_1 \sinh y_1\right) \end{array}\right) \end{array} \right] dy \qquad (7\text{-}71.3)$$

Upon integration and organization of terms we get

$$\mathbf{v}(x,y) = -E_1^{-1}\sum_{n=1}^{\infty}\left(\frac{n\pi}{l}\right)\sin\left(\frac{n\pi x}{l}\right)\sum_{m-1}^{4} A_{mn} \left[\begin{array}{l} a_{mn}\left(1+v_1\right)\sinh y_1 \\ + b_{mn}\left(1+v_1\right)\cosh y_1 \\ + c_{mn}\left[\begin{array}{l}\left(1+v_1\right)y_1\sinh y_1 \\ +\left(v_1-1\right)\cosh y_1\end{array}\right] \\ + d_{mn}\left[\begin{array}{l}\left(1+v_1\right)y_1\cosh y_1 \\ +\left(v_1-1\right)\sinh y_1\end{array}\right] \end{array} \right] + f(x) \qquad (7\text{-}71.4)$$

Special loading cases

1. Upper surface loading alone (with support at both ends

From Table 7-2, we get

$$Q_2(x) = T_1(x) = T_2(x) = 0 \;\rightarrow\; B_{2n} = C_{2n} = C_{1n} = 0 \;\rightarrow\; A_{1n} = A_{2n} = A_{4n} = 0 \qquad (7\text{-}72.1)$$

Thus, equations (7-70) give the following stresses:

$$\sigma_y = -\sum_{n=1}^{\infty}\left(\frac{n\pi}{l}\right)^2 \sin\left(\frac{n\pi x}{l}\right) A_{3n} Y_{3n}(y_1) \qquad (7\text{-}72.2)$$

$$\sigma_x = \sum_{n=1}^{\infty}\sin\left(\frac{n\pi x}{l}\right) A_{3n} Y_{3n}''(y_1) \qquad (7\text{-}72.3)$$

231

$$\tau_{xy} = \sum_{n=1}^{\infty} \left(\frac{n\pi}{l}\right) \cos\left(\frac{n\pi x}{l}\right) A_{3n} Y'_{3n}(y_1) \qquad (7\text{-}72.4)$$

Similarly, equation (7-71.3) gives the following **vertical displacement**

$$V(x,y) = -E_1^{-1} \sum_{n=1}^{\infty} \left(\frac{n\pi}{l}\right) \sin\left(\frac{n\pi x}{l}\right) A_{3n} \left[\begin{pmatrix} b_{3n}(1+v_1)\cosh y_1 \\ +c_{3n}\begin{bmatrix}(1+v_1)y_1\sinh y_1 \\ +(v_1-1)\cosh y_1\end{bmatrix} \\ +d_{3n}\begin{bmatrix}(1+v_1)y_1\cosh y_1 \\ +(v_1-1)\sinh y_1\end{bmatrix}\end{pmatrix}\right] \qquad (7\text{-}72.5)$$

Where, $f(x) = 0$ since $v(x,y) = 0$, when $x = 0$ or l (condition for rigid support).

The **angle of rotation** of the beam is obtained by differentiating equation (7-27.5) and substituting by $x = 0$ or l, and $y = 0$, as follows.

$$\frac{\partial V(x,y)}{\partial x} = -E_1^{-1} \sum_{n=1}^{\infty} \left(\frac{n\pi}{l}\right)^2 A_{3n}\left[(b_{3n}(1+v_1)+c_{3n}(v_1-1))\right] \qquad (7\text{-}72.5)$$

Substituting by b_{3n} and c_{3n} from Table 7-2, we get

$$\frac{\partial V(x,y)}{\partial x} = -E_1^{-1} \sum_{n=1}^{\infty} \left(\frac{n\pi}{l}\right)^2 A_{3n}(s+y_h.c)((1+v_1)-(v_1-1))/g$$

$$\tan\alpha = -2E_1^{-1}\sum_{n=1}^{\infty}\left(\frac{n\pi}{l}\right)^2 A_{3n}\left(\frac{\sinh\left(\frac{n\pi h}{l}\right)+\left(\frac{n\pi h}{l}\right).\cosh\left(\frac{n\pi h}{l}\right)}{\sinh^2\left(\frac{n\pi h}{l}\right)-\left(\frac{n\pi h}{l}\right)^2}\right) \qquad (7\text{-}72.6)$$

Where is the angle of rotation of the beam axis ($x = 0$, $y = 0$) at the end of beam and with the x-axis.

7.8. Beam with infinite span

The Fourier series expansion of the stress function in (7-42) dealt with a beam with finite span l on which a sum of sinusoidal functions of wavelength, $k = n\pi/l$, could be tailored to present precise solution of internal stresses, external loads, and spatial deformations.

However, as the span l increases, the number of wavelets ($\sin kx$, $\cos kx$, $\sinh ky$ and $\cosh ky$) represent continuous distribution, such that the summation sign turns to integration as follows.

Stress function

232

$$\varphi(x,y) = \int_{k=-\infty}^{k=\infty} \left(C_1(k)\sin kx + C_2(k)\cos kx\right)\left[\begin{array}{l} C_3(k)ky\cosh ky + C_4(k)ky\sinh ky \\ + C_5(k)\cosh ky + C_6(k)\sinh ky \end{array}\right]dk \qquad (7\text{-}73)$$

Where the six coefficients $C(k)$'s are determined from the loading boundary conditions as before.

The **internal stresses** are obtained from the stress function, equation (7-73), as follows

$$\sigma_y = \frac{\partial^2 \varphi(x,y)}{\partial x^2} = -\int_{k=-\infty}^{k=\infty} k^2 \left(\begin{array}{l} C_1(k)\sin kx \\ + C_2(k)\cos kx \end{array}\right)\left[\begin{array}{l} C_3(k)ky\cosh ky + C_4(k)ky\sinh ky \\ + C_5(k)\cosh ky + C_6(k)\sinh ky \end{array}\right]dk \qquad (7\text{-}74.1)$$

$$\sigma_x = \frac{\partial^2 \varphi(x,y)}{\partial y^2} = \int_{k=-\infty}^{k=\infty} k^2 \left(\begin{array}{l} C_1(k)\sin kx \\ + C_2(k)\cos kx \end{array}\right)\left[\begin{array}{l} C_3(k)(ky\cosh ky + 2\sinh ky) \\ + C_4(k)(ky\sinh ky + 2\cosh ky) \\ + C_5(k)\cosh ky + C_6(k)\sinh ky \end{array}\right]dk \qquad (7\text{-}74.2)$$

$$\tau_{xy} = -\frac{\partial^2 \varphi(x,y)}{\partial x \partial y} = \int_{k=-\infty}^{k=\infty} k^2 \left(\begin{array}{l} C_1(k)\cos kx \\ - C_2(k)\sin kx \end{array}\right)\left[\begin{array}{l} C_3(k)(ky\sinh ky + \cosh ky) \\ + C_4(k)(ky\cosh ky + \sinh ky) \\ + C_5(k)\sinh ky + C_6(k)\cosh ky \end{array}\right]dk \qquad (7\text{-}74.3)$$

The Cauchy-Hooke's law becomes

$$\frac{\partial v}{\partial y} = \varepsilon_{yy} = E_1^{-1}\left(\begin{array}{l} -\int_{k=-\infty}^{k=\infty} k^2 \left(\begin{array}{l} C_1(k)\sin kx \\ + C_2(k)\cos kx \end{array}\right)\left[\begin{array}{l} C_3(k)ky\cosh ky + C_4(k)ky\sinh ky \\ + C_5(k)\cosh ky + C_6(k)\sinh ky \end{array}\right]dk \\ -v_1\int_{k=-\infty}^{k=\infty} k^2 \left(\begin{array}{l} C_1(k)\sin kx \\ + C_2(k)\cos kx \end{array}\right)\left[\begin{array}{l} C_3(k)(ky\cosh ky + 2\sinh ky) \\ + C_4(k)(ky\sinh ky + 2\cosh ky) \\ + C_5(k)\cosh ky + C_6(k)\sinh ky \end{array}\right]dk \end{array}\right)$$

i.e.,

$$\frac{\partial v}{\partial y} = \varepsilon_{yy} = -E_1^{-1}\int_{k=-\infty}^{k=\infty} k^2 \left(\begin{array}{l} C_1(k)\sin kx \\ + C_2(k)\cos kx \end{array}\right)\left[\begin{array}{l} C_3(k)\left[\begin{array}{l}(1+v_1)ky\cosh ky \\ + 2v_1\sinh ky\end{array}\right] \\ + C_4(k)\left[\begin{array}{l}(1+v_1)ky\sinh ky \\ + 2v_1\cosh ky\end{array}\right] \\ + C_5(k)(1+v_1)\cosh ky \\ + C_6(k)(1+v_1)\sinh ky \end{array}\right]dk \qquad (7\text{-}74.4)$$

7.9. Cylindrical tube with infinite length

Radial stresses, independent of polar angle are encountered in cylindrical tubes with circular cross section.

1. Cylindrical polar radial Lévy's stress function

a. The radial form of equation (7-19.2), independent of θ, is written in the cylindrical polar form as follows

$$\nabla^2\left(\nabla^2\varphi(r,\theta)\right)=0$$

$$\left(\frac{1}{r}\frac{\partial}{\partial r}+\frac{\partial^2}{\partial r^2}\right)\left(\frac{1}{r}\frac{\partial}{\partial r}+\frac{\partial^2}{\partial r^2}\right)\varphi(r,\theta)=0 \qquad (7\text{-}75.1)$$

$$\left(\frac{1}{r^3}\frac{\partial}{\partial r}-\frac{1}{r^2}\frac{\partial^2}{\partial r^2}+\frac{2}{r}\frac{\partial^3}{\partial r^3}+\frac{\partial^4}{\partial r^4}\right)\varphi(r,\theta)=0$$

Or,

$$\nabla^2\left(\nabla^2\varphi(r,\theta)\right)=\frac{1}{r}\frac{\partial}{\partial r}\left[r\frac{\partial}{\partial r}\left[\frac{1}{r}\frac{\partial}{\partial r}\left(r\frac{\partial}{\partial r}\right)\right]\right]\varphi(r,\theta)=0 \qquad (7\text{-}75.2)$$

b. Stresses

Equations (7-17.4 and 5) give

$$\sigma_r=\frac{1}{r}\frac{\partial\varphi(r,\theta)}{\partial r} \qquad (7\text{-}75.3)$$

$$\sigma_\theta=\frac{\partial^2\varphi(r,\theta)}{\partial r^2} \qquad (7\text{-}75.4)$$

$$\tau_{r\theta}=-\frac{\partial}{\partial r}\left(\frac{1}{r}\frac{\partial\varphi(r,\theta)}{\partial\theta}\right)=0 \qquad (7\text{-}75.5)$$

Integrating equation (7-75.2) four times with respect to r gives

First integration

$$r\frac{\partial}{\partial r}\left[\frac{1}{r}\frac{\partial}{\partial r}\left(r\frac{\partial}{\partial r}\right)\right]\varphi(r,\theta)=C_1$$

i.e.,

$$\frac{\partial}{\partial r}\left[\frac{1}{r}\frac{\partial}{\partial r}\left(r\frac{\partial}{\partial r}\right)\right]\varphi(r,\theta)=\frac{C_1}{r} \qquad (7\text{-}75.6)$$

Second integration gives

$$\frac{1}{r}\frac{\partial}{\partial r}\left(r\frac{\partial}{\partial r}\right)\varphi(r,\theta) = C_1 \ln r + C_2 \qquad (7\text{-}75.7)$$

i.e.,
$$\frac{\partial}{\partial r}\left(r\frac{\partial}{\partial r}\right)\varphi(r,\theta) = C_1 r \ln r + rC_2 \qquad (7\text{-}75.8)$$

Let us integrate the (r lnr) term by parts

$$\begin{aligned}
\int r \ln r\, dr &= \frac{1}{2}\int \ln r\, dr^2 \\
&= \frac{1}{2}\left[r^2 \ln r - \int r^2 d(\ln r)\right] \\
&= \frac{1}{2}\left[r^2 \ln r - \int r^2 \frac{1}{r} dr\right] \\
&= \frac{1}{2}\left[r^2 \ln r - \frac{1}{2}r^2\right] \\
&= \frac{1}{4}r^2[2\ln r - 1]
\end{aligned} \qquad (7\text{-}75.8a)$$

i.e.,

Using equation (7-75.8a), the third integration gives

$$r\frac{\partial}{\partial r}\varphi(r,\theta) = C_1\left(\frac{1}{4}r^2(2\ln r - 1)\right) + \frac{1}{2}r^2 C_2 + C_3$$

i.e.,
$$\frac{\partial}{\partial r}\varphi(r,\theta) = C_1\left(\frac{1}{4}r(2\ln r - 1)\right) + \frac{1}{2}rC_2 + \frac{1}{r}C_3 \qquad (7\text{-}75.9)$$

Similarly, using equation (7-75.8a), the fourth integration gives

$$\begin{aligned}
\varphi(r,\theta) &= C_1\left(\frac{1}{2}\left(\frac{1}{4}r^2(2\ln r - 1)\right) - \frac{1}{8}r^2\right) + \frac{1}{4}r^2 C_2 + C_3 \ln r + C_4 \\
&= C_1 r^2 \ln r + C_3 \ln r + C_2 r^2 + C_4
\end{aligned} \qquad (7\text{-}75.10)$$

Where, we have fused constants into four general constants, without altering the general solution for a stress function of radial stresses independent of angular changes.

Therefore, equation (7-75.10) renders the stresses in equations (7-75.3 through 5) read

$$\sigma_r = \frac{1}{r}\frac{\partial\varphi(r,\theta)}{\partial r} = \frac{1}{r}\frac{\partial}{\partial r}\left(C_1 r^2 \ln r + C_3 \ln r + C_2 r^2 + C_4\right)$$

$$= \frac{C_3}{r^2} + 2C_1 \ln r + C_1 + 2C_2 \qquad (7\text{-}75.11)$$

$$\sigma_\theta = \frac{\partial^2 \varphi(r,\theta)}{\partial r^2} = \frac{\partial^2}{\partial r^2}\left(C_1 r^2 \ln r + C_3 \ln r + C_2 r^2 + C_4\right)$$

$$= -\frac{C_3}{r^2} + 2C_1 \ln r + 3C_1 + 2C_2 \qquad (7\text{-}75.12)$$

$$\tau_{r\theta} = \tau_{\theta r} = 0 \qquad (7\text{-}75.13)$$

2. Lamé's circular cylindrical tube

Uniform external and internal pressures

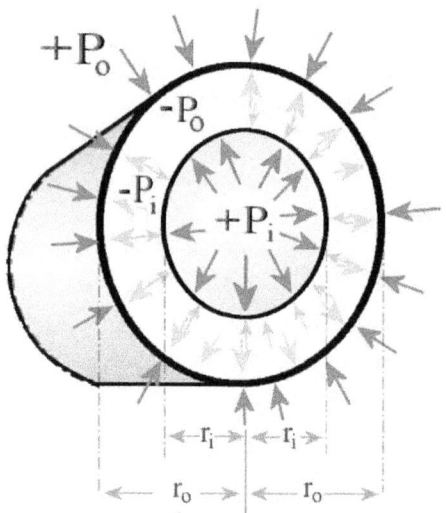

Figure 7-8. Boundary conditions on circular cylindrical double walled connected tubes.

Boundary conditions:

$$
\begin{aligned}
r = r_i \qquad & \sigma_{-r} = +\,p_1 \\
& \sigma_{+r} = -\,p_1 \\
r = r_o \qquad & \sigma_{+r} = -\,p_2 \qquad (7\text{-}75.14)
\end{aligned}
$$

Symmetry condition

$$\varepsilon_{rr} = E^{-1}(\sigma_r - v\sigma_\theta) = \frac{\partial \mathbf{u}(r,\theta,z)}{\partial r} \qquad (7\text{-}75.15)$$

$$\varepsilon_{\theta\theta} = E^{-1}(\sigma_\theta - v\sigma_r) = \frac{1}{r}\frac{\partial \mathbf{v}(r,\theta,z)}{\partial \theta} + \frac{\mathbf{u}}{r} \qquad (7\text{-}75.16)$$

Substituting from (7-75.11 and 12) in equations (7-75.15 and 16), we get

$$\frac{C_3}{r^2}(1+v) + 2(1-v)C_1 \ln r + (1-3v)C_1 + 2(1-v)C_2 = E\frac{\partial \mathbf{u}(r,\theta,z)}{\partial r} \qquad (7\text{-}75.17)$$

and

$$-(1+v)\frac{C_3}{r^2} + 2(1-v)C_1 \ln r + (3-v)C_1 + (1-v)2C_2 = E\frac{\mathbf{u}}{r}$$

i.e.,

$$-(1+v)\frac{C_3}{r} + 2(1-v)C_1 r \ln r + r(3-v)C_1 + r(1-v)2C_2 = E\mathbf{u} \qquad (7\text{-}75.18)$$

Differentiating we get

$$(1+v)\frac{C_3}{r^2} + 2(1-v)C_1[1 - \ln r] + (3-v)C_1 + (1-v)2C_2 = E\frac{\partial \mathbf{u}(r,\theta,z)}{\partial r}$$

Subtracting, we get

$$2 \ln r(1-v)C_1 = 0$$

Therefore,

$$C_1 = 0 \qquad (7\text{-}75.19)$$

Solving equations (7-75.11 and 12) under the boundary conditions (7-75.14) we get

$$-p_i = \frac{C_3}{r_i^2} + 2C_2 \qquad (7\text{-}75.20)$$

$$-p_o = \frac{C_3}{r_o^2} + 2C_2 \qquad (7\text{-}75.21)$$

Subtracting

$$C_3 = (p_o - p_i)\frac{r_i^2 r_o^2}{r_o^2 - r_i^2} \qquad (7\text{-}75.22)$$

$$C_2 = -\frac{1}{2}\left(\frac{p_o r_o^2 - p_i r_i^2}{r_o^2 - r_i^2}\right) \qquad (7\text{-}75.23)$$

Stresses

237

$$\sigma_r = \frac{(p_o - p_i)r_i^2 r_o^2}{r^2(r_o^2 - r_i^2)} - \left(\frac{p_o r_o^2 - p_i r_i^2}{r_o^2 - r_i^2}\right) \qquad (7\text{-}75.24)$$

$$\sigma_\theta = -\frac{(p_o - p_i)r_i^2 r_o^2}{r^2(r_o^2 - r_i^2)} - \left(\frac{p_o r_o^2 - p_i r_i^2}{r_o^2 - r_i^2}\right) \qquad (7\text{-}75.25)$$

Stresses at $r = r_o$

$$\sigma_r = -\frac{(p_o)r_i^2}{(r_o^2 - r_i^2)} + \left(\frac{p_o r_o^2}{r_o^2 - r_i^2}\right) = -p_o \qquad (7\text{-}75.26)$$

$$\sigma_\theta = \left(\frac{2 p_i r_i^2 - p_o(r_o^2 + r_i^2)}{r_o^2 - r_i^2}\right) \qquad (7\text{-}75.27)$$

Stresses at $r = r_i$

$$\sigma_r = \frac{(-p_i)r_o^2}{(r_o^2 - r_i^2)} - \left(\frac{-p_i r_i^2}{r_o^2 - r_i^2}\right) = -p_i \qquad (7\text{-}75.28)$$

$$\sigma_\theta = \frac{2 p_o r_o^2 + p_i(r_i^2 + r_o^2)}{r_o^2 - r_i^2}$$

Displacement

Substituting from equations (7-75.24 and 25) into equation (7-75.16), we get

$$
\begin{aligned}
u(r) &= rE^{-1}(\sigma_\theta - v\sigma_r) \\
&= \frac{1}{E(r_o^2 - r_i^2)}\left[-(1+v)(p_o - p_i)\frac{r_i^2 r_o^2}{r} - (1-v)r(p_o r_o^2 - p_i r_i^2)\right] \\
&= \frac{1}{Er(r_o^2 - r_i^2)}\left(\begin{array}{l} p_o\left(-(1+v)r_i^2 r_o^2 - (1-v)r^2 r_o^2\right) \\ + p_i\left((1+v)r_i^2 r_o^2 + (1-v)r^2 r_i^2\right) \end{array}\right) \qquad (7\text{-}75.29) \\
&= \frac{1}{E}\left[\frac{r_i^2 r_o^2}{r(r_o^2 - r_i^2)}(1+v)(p_i - p_o) + \frac{r}{(r_o^2 - r_i^2)}(1-v)(r_i^2 p_i - r_o^2 p_o)\right] \\
&= \frac{A}{r} + Br
\end{aligned}
$$

Where

238

$$A = \frac{r_i^2 r_o^2}{E(r_o^2 - r_i^2)}(1 + v)(p_i - p_o)$$

$$B = \frac{1}{E(r_o^2 - r_i^2)}(1 - v)(r_i^2 p_i - r_o^2 p_o)$$

(7-75.30)

Total bending force

$$
\begin{aligned}
\int_{r_i}^{r_o} \sigma_\theta dr &= -\frac{(p_o - p_i)r_i^2 r_o^2}{(r_o^2 - r_i^2)} \int_{r_i}^{r_o} \frac{dr}{r^2} - \left(\frac{p_o r_o^2 - p_i r_i^2}{r_o^2 - r_i^2}\right)\int_{r_i}^{r_o} dr \\
&= -\frac{(p_o - p_i)r_i^2 r_o^2}{(r_o^2 - r_i^2)}\left[\frac{1}{r_i} - \frac{1}{r_o}\right] - \left(\frac{p_o r_o^2 - p_i r_i^2}{r_o^2 - r_i^2}\right)(r_o - r_i) \\
&= -\frac{(p_o - p_i)r_i r_o}{(r_o + r_i)} - \left(\frac{p_o r_o^2 - p_i r_i^2}{r_o + r_i}\right) \\
&= -\frac{1}{(r_o + r_i)}\left[(p_o - p_i)r_i r_o + p_o r_o^2 - p_i r_i^2\right]
\end{aligned}
$$

(7-75.31)

At $r_i = 0$

$$\int_0^{r_o} \sigma_\theta dr = -p_o r_o$$

(7-75.32)

Total bending couple

$$
\begin{aligned}
M = \int_{r_i}^{r_o} \sigma_\theta r dr &= -\frac{(p_o - p_i)r_i^2 r_o^2}{(r_o^2 - r_i^2)}\int_{r_i}^{r_o}\frac{dr}{r} - \left(\frac{p_o r_o^2 - p_i r_i^2}{r_o^2 - r_i^2}\right)\int_{r_i}^{r_o} r dr \\
&= -\frac{(p_o - p_i)r_i^2 r_o^2}{(r_o^2 - r_i^2)}\left[\ln\frac{r_o}{r_i}\right] - \left(\frac{p_o r_o^2 - p_i r_i^2}{r_o^2 - r_i^2}\right)\frac{(r_o^2 - r_i^2)}{2} \\
&= -\frac{(p_o - p_i)r_i^2 r_o^2}{(r_o^2 - r_i^2)}\left[\ln\frac{r_o}{r_i}\right] - \frac{1}{2}(p_o r_o^2 - p_i r_i^2)
\end{aligned}
$$

(7-75.33)

3. Bending a circular ring

Boundary conditions:

$$
\begin{aligned}
r = r_i \qquad & \sigma_r = 0 \\
r = r_o \qquad & \sigma_r = 0
\end{aligned}
$$

$$\int_{r_i}^{r_o} \sigma_\theta r\, dr = M$$

$$\int_{r_i}^{r_o} \sigma_\theta\, dr = 0 \tag{7-76.1}$$

Applying the boundary conditions in equations (7-76.1) to the equation of radial and azimuthal stresses, (7-75.11 and 12), we get

Stresses

$$\frac{C_3}{r_i^2} + 2C_1 \ln r_i + C_1 + 2C_2 = 0 \tag{7-76.2}$$

$$\frac{C_3}{r_o^2} + 2C_1 \ln r_o + C_1 + 2C_2 = 0 \tag{7-76.3}$$

Force

$$\int_{r_i}^{r_o} \left(-\frac{C_3}{r^2} + 2C_1 \ln r + 3C_1 + 2C_2 \right) dr$$

$$= \left[\frac{C_3}{r} + 2C_1 r(\ln(r) - 1) + 3rC_1 + 2rC_2 \right]_{r_i}^{r_o} \tag{7-76.4}$$

$$= -C_3 \left(\frac{r_o - r_i}{r_o r_i} \right) + C_1 \left[2(r_o \ln(r_o) - r_i \ln(r_i)) + (r_o - r_i) \right] + 2C_2(r_o - r_i) = 0$$

Eliminations of constants

Subtracting (7-76.2) from (7-76.3), we get

$$C_3 = 2C_1 \left(\frac{r_o^2 r_i^2}{r_o^2 - r_i^2} \right) \ln \frac{r_o}{r_i} \tag{7-76.5a}$$

Then, substituting in (7-76.2), we get

$$C_2 = -\frac{C_3}{2r_o^2} - C_1 \left(\ln r_o + \frac{1}{2} \right)$$

$$= -C_1 \left[\left(\frac{r_i^2}{r_o^2 - r_i^2} \right) \ln \frac{r_o}{r_i} + \left(\ln r_o + \frac{1}{2} \right) \right] \tag{7-76.5b}$$

$$= -C_1 \frac{1}{\left(r_o^2 - r_i^2 \right)} \left[r_o^2 \ln r_o - r_i^2 \ln r_i + \frac{1}{2} \left(r_o^2 - r_i^2 \right) \right]$$

Couple

$$M = \int_{r_i}^{r_o} \sigma_\theta r\, dr$$

$$= \int_{r_i}^{r_o} \left(-\frac{C_3}{r} + 2C_1 r \ln r + 3rC_1 + 2rC_2 \right) dr$$

$$= -C_3 \ln \frac{r_o}{r_i} + C_1 \frac{1}{2}\left(\begin{array}{c} r_o^2[2\ln(r_o)-1] \\ -r_i^2[2\ln(r_i)-1] \end{array} \right) + \frac{3}{2}\left(r_o^2 - r_i^2\right)C_1 + \left(r_o^2 - r_i^2\right)C_2 \qquad (7\text{-}76.6)$$

$$= -C_3 \ln \frac{r_o}{r_i} + C_1\left(r_o^2 \ln(r_o) - r_i^2 \ln(r_i) + \left(r_o^2 - r_i^2\right)\right) + \left(r_o^2 - r_i^2\right)C_2$$

Eliminating C_2 and C_3 between (7-76.5), (7-76.6), and (7-76.7), we get

$$M = -C_3 \ln \frac{r_o}{r_i} + C_1\left(r_o^2 \ln(r_o) - r_i^2 \ln(r_i) + \left(r_o^2 - r_i^2\right)\right) + \left(r_o^2 - r_i^2\right)C_2$$

$$C_1\left(r_o^2 \ln(r_o) - r_i^2 \ln(r_i) + \left(r_o^2 - r_i^2\right)\right) = M + C_3 \ln \frac{r_o}{r_i} - \left(r_o^2 - r_i^2\right)C_2 \qquad (7\text{-}76.7a)$$

i.e.,

$$C_1\left(r_o^2 \ln(r_o) - r_i^2 \ln(r_i) + \left(r_o^2 - r_i^2\right)\right) = M + C_1\left(\begin{array}{c} 2\left(\dfrac{r_o^2 r_i^2}{r_o^2 - r_i^2}\right)\left(\ln \dfrac{r_o}{r_i}\right)^2 \\ + \left(r_o^2 \ln r_o - r_i^2 \ln r_i\right) \\ + \dfrac{1}{2}\left(r_o^2 - r_i^2\right) \end{array} \right) \qquad (7\text{-}76.7b)$$

Or

$$C_1\left(\begin{array}{c} r_o^2 \ln(r_o) - r_i^2 \ln(r_i) + \left(r_o^2 - r_i^2\right) - 2\left(\dfrac{r_o^2 r_i^2}{r_o^2 - r_i^2}\right)\left(\ln \dfrac{r_o}{r_i}\right)^2 \\ - \left(r_o^2 \ln r_o - r_i^2 \ln r_i\right) - \dfrac{1}{2}\left(r_o^2 - r_i^2\right) \end{array} \right) = M \qquad (7\text{-}76.7c)$$

$$C_1 = \frac{2M\left(r_o^2 - r_i^2\right)}{\left[\left(r_o^2 - r_i^2\right)^2 - 4r_o^2 r_i^2\left(\ln \dfrac{r_o}{r_i}\right)^2\right]} \qquad (7\text{-}76.8)$$

$$C_2 = -C_1 \frac{1}{\left(r_o^2 - r_i^2\right)}\left[r_o^2 \ln r_o - r_i^2 \ln r_i + \frac{1}{2}\left(r_o^2 - r_i^2\right)\right]$$

$$= -\frac{M\left[2\left(r_o^2 \ln r_o - r_i^2 \ln r_i\right) + \left(r_o^2 - r_i^2\right)\right]}{\left(\left(r_o^2 - r_i^2\right)^2 - 4r_o^2 r_i^2\left(\ln \frac{r_o}{r_i}\right)^2\right)} \qquad (7\text{-}76.9)$$

$$C_3 = \frac{4Mr_o^2 r_i^2 \ln \frac{r_o}{r_i}}{\left(\left(r_o^2 - r_i^2\right)^2 - 4r_o^2 r_i^2\left(\ln \frac{r_o}{r_i}\right)^2\right)} \qquad (7\text{-}76.10)$$

Summary of constants

$$A = \left(r_o^2 - r_i^2\right)^2 \qquad\qquad M = \int_{r_i}^{r_o} \sigma_\theta r\,dr \qquad\qquad \int_{r_i}^{r_o} \sigma_\theta dr = 0$$
$$\quad\; - 4r_o^2 r_i^2\left(\ln \frac{r_o}{r_i}\right)^2$$

$$C_1 = \frac{2M}{A}\left(r_o^2 - r_i^2\right) \qquad C_2 = -\frac{M\left[2\left(r_o^2 \ln r_o - r_i^2 \ln r_i\right) + \left(r_o^2 - r_i^2\right)\right]}{\left(\left(r_o^2 - r_i^2\right)^2 - 4r_o^2 r_i^2\left(\ln \frac{r_o}{r_i}\right)^2\right)} \qquad C_3 = \frac{4M}{A} r_o^2 r_i^2 \ln \frac{r_o}{r_i}$$

$$\sigma_r = \frac{C_3}{r^2} + 2C_1 \ln r + C_1 + 2C_2 = \frac{4M}{A}\left(\frac{r_o^2 r_i^2}{r^2}\ln \frac{r_o}{r_i} + r_o^2 \ln \frac{r}{r_o} - r_i^2 \ln \frac{r}{r_i}\right)$$

$$\sigma_\theta = -\frac{C_3}{r^2} + 2C_1 \ln r + 3C_1 + 2C_2 = \frac{4M}{A}\left(-\frac{r_o^2 r_i^2}{r^2}\ln \frac{r_o}{r_i} + r_o^2 \ln \frac{r}{r_o} - r_i^2 \ln \frac{r}{r_i} + \left(r_o^2 - r_i^2\right)\right)$$

$$\tau_{r\theta} = 0$$

$$\varepsilon_{rr} = E^{-1}\left(\sigma_r - v\sigma_\theta\right)$$

$$= \frac{4M}{AE}\left((1+v)\frac{r_o^2 r_i^2}{r^2}\ln \frac{r_o}{r_i} + (1-v)\left(r_o^2 \ln \frac{r}{r_o} - r_i^2 \ln \frac{r}{r_i}\right) - v\left(r_o^2 - r_i^2\right)\right)$$

$$\varepsilon_{\theta\theta} = E^{-1}\left(\sigma_\theta - v\sigma_r\right)$$

$$= \frac{4M}{AE}\left(-(1+v)\frac{r_o^2 r_i^2}{r^2}\ln \frac{r_o}{r_i} + (1-v)\left(r_o^2 \ln \frac{r}{r_o} - r_i^2 \ln \frac{r}{r_i}\right) + \left(r_o^2 - r_i^2\right)\right)$$

Displacement

242

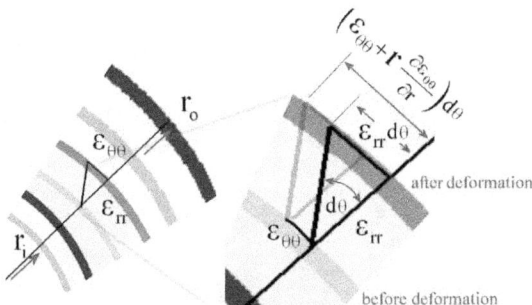

Figure 7-9. Contributions to azimuthal elongation and rotation of radial coordinates as the cylindrical circular tube expands.

1. As circular curves elongate, Figure 7-9, the radial strain ε_{rr} contributes to by negative azimuthal contraction of

$$- \varepsilon_{rr}\, d\theta \qquad\qquad (7\text{-}77.1)$$

2. The azimuthal strain proper $\varepsilon_{\theta\theta}$ contributes to expanding azimuthal strain by ($\varepsilon_{\theta\theta}\, d\theta$), in addition to $\left(r \dfrac{\partial \varepsilon_{\theta\theta}}{\partial r} \right)$, thus netting

$$\left(\varepsilon_{\theta\theta} + r \frac{\partial \varepsilon_{\theta\theta}}{\partial r} \right) d\theta \qquad\qquad (7\text{-}77.2)$$

3. The next azimuthal elongation compounded with radial elongation is

$$\left(\varepsilon_{\theta\theta} + r \frac{\partial \varepsilon_{\theta\theta}}{\partial r} - \varepsilon_{rr} \right) d\theta \qquad\qquad (7\text{-}77.3)$$

Thus, the net angular unit change

$$
\varepsilon_{\theta\theta} + r\frac{\partial \varepsilon_{\theta\theta}}{\partial r} - \varepsilon_{rr} = \frac{4M}{AE}\left(-(1+v)\frac{r_o^2 r_i^2}{r^2}\ln\frac{r_o}{r_i} + (1-v)\left(r_o^2 \ln\frac{r}{r_o} - r_i^2 \ln\frac{r}{r_i} \right) + \left(r_o^2 - r_i^2 \right) \right)
$$
$$
+ r\frac{4M}{AE}\left(+\frac{(1-v)}{r}\left(r_o^2 - r_i^2 \right) - \frac{r_i^2}{r} \right) \qquad\qquad (7\text{-}77.4)
$$
$$
- \frac{4M}{AE}\left((1+v)\frac{r_o^2 r_i^2}{r^2}\ln\frac{r_o}{r_i} + (1-v)\left(r_o^2 \ln\frac{r}{r_o} - r_i^2 \ln\frac{r}{r_i} \right) - v\left(r_o^2 - r_i^2 \right) \right)
$$

Arranging, we get **unit-angular change** after deformation as

$$\varepsilon_{\theta\theta} + r\frac{\partial \varepsilon_{\theta\theta}}{\partial r} - \varepsilon_{rr} = \frac{4M}{AE}\left(r_o^2 - r_i^2\right)(3 - 2v) \tag{7-77.5}$$

Therefore, according to our approximation of **radial stress function**, independent of azimuthal and longitudinal coordinates, the unit-angular change is independent of the radial coordinate r. Therefore, cross-sectional planes, perpendicular to the z-axis, remain **planar** after deformation (i.e., independent of r).

7.10. Finite force applied on half-plane. Flamant-Boussinesg

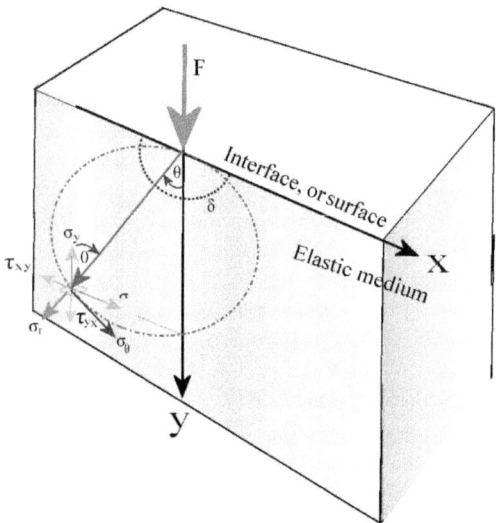

Figure 7-10. Geometry of lumped force to deal with singular solutions.

a. Empirical assumptions for finite force

In order deal with inherent difficulty with singular solution in **inverse square law** when coordinate approaches the center, r = 0, we will adopt **Cauchy-like** approach to isolation of singularities by designating an area of radius, δ, Figure 7-10, where the integration of the stress on such area equal to the external force as follows.

1. Using the rule of surface area, the radial stress σ-r acts on the area cos θ (δdθdz). We will take dz = 1, in order to limit our treatment into the xy-planar geometry. Therefore, the finite force acting on the surface is defined by:

$$\int_{-\pi/2}^{+\pi/2} \sigma_{-r}\left[(1)\delta\cos\theta d\theta\right] = F \qquad (7\text{-}78.1)$$

We will make the following assumptions, based on the physical nature of the problem:

2. Radial stress is proportional to; cos θ, and inversely proportional to the radial distance; r. Keeping in mind that, in cylindrical polar coordinate, the z-dimension is taken as unity.

$$\sigma_r\left(\frac{kg.cm}{sec^2}\cdot\frac{1}{cm^2}\right) = -k\frac{\cos\theta}{r.(1)}\left(\frac{1}{cm^2}\right)$$

i.e.,
$$\sigma_r = -k\frac{\cos\theta}{r} \qquad (7\text{-}78.2)$$

3. Azimuthal normal stresses are absent

$$\sigma_\theta = 0 \qquad (7\text{-}78.3)$$

4. Azimuthal shearing at radius δ = 0

$$\tau_{r\theta} = \tau_{\theta r} = 0 \qquad (7\text{-}78.4)$$

Substituting from (7-78.2) into (7-78.1), we get

$$\int_{-\pi/2}^{+\pi/2} k\frac{\cos\theta}{\delta}\delta\cos\theta d\theta = \frac{k}{2}\left[\theta + \cos\theta\sin\theta\right]_{-\pi/2}^{\pi/2}$$

$$= \frac{k\pi}{2} = F \qquad (7\text{-}78.5)$$

$$k = \frac{2F}{\pi}$$

b. Navier's compatibility of simplified radial stress

The Navier's equation of equilibrium in cylindrical polar coordinate reads:

$$\frac{\partial\sigma_r(r,\theta,z)}{\partial r} + \frac{\partial\tau_{r\theta}(r,\theta,z)}{r\partial\theta} + \frac{\sigma_r(r,\theta,z) - \sigma_\theta(r,\theta,z)}{r} + \frac{\partial\tau_{rz}(r,\theta,z)}{\partial z} + \Pi_i\rho = \rho\frac{\partial^2 u_r(r,\theta,z)}{\partial t^2} \qquad (7\text{-}78.6)$$

In the present case of infinite length cylinder, absence of internal forces, and static motion, the Navier's state of equilibrium becomes

$$\frac{\partial \sigma_r(r,\theta)}{\partial r} + \frac{\partial \tau_{r\theta}(r,\theta)}{r\partial\theta} + \frac{\sigma_r(r,\theta) - \sigma_\theta(r,\theta)}{r} = 0 \tag{7-78.7}$$

Substituting from (7-78.2 through 5) into (7-78.7), we get

$$\frac{\partial}{\partial r}\left(-\frac{2F}{\pi}\frac{\cos\theta}{r}\right) + \frac{1}{r}\left(-\frac{2F}{\pi}\frac{\cos\theta}{r}\right) = 0 \tag{7-78.8}$$

Therefore, our simplified empirical stresses satisfy the Navier's equations.

c. Stress function for simplified radial stresses

The stress function for the simplified empirical stress is obtained from equations (7-17.4 through 6) and (7-78.2) as follow

$$\sigma_r = \frac{1}{r}\frac{\partial\varphi(r,\theta)}{\partial r} + \frac{1}{r^2}\cdot\frac{\partial^2\varphi(r,\theta)}{\partial\theta^2} = -\frac{2F}{\pi}\frac{\cos\theta}{r} \tag{7-79.1}$$

$$\sigma_\theta = \frac{\partial^2\varphi(r,\theta)}{\partial r^2} = 0 \tag{7-79.2}$$

$$\tau_{r\theta} = -\frac{\partial}{\partial r}\left(\frac{1}{r}\frac{\partial\varphi(r,\theta)}{\partial\theta}\right) = 0 \tag{7-79.3}$$

Integrating equation (7-79.2) with respect to r, we get

$$\varphi(r,\theta) = rf(\theta) \tag{7-79.4}$$

Substituting from (7-79.4) into (7-79.1), we get the **ordinary differential equation** in θ as follows

$$f(\theta) + \frac{d^2 f(\theta)}{d\theta^2} = -\frac{2F}{\pi}\cos\theta$$

i.e.,
$$\left(1 + D^2\right)f(\theta) = -\frac{2F}{\pi}\cos\theta \tag{7-79.5}$$

The **complementary solution (c.s.)** of (7-79.5) is

$$c.s = A\sin\theta + B\cos\theta \tag{7-79.6}$$

The **particular solution (p.s.)** takes the form

246

$$p.s = \theta(C\sin\theta + D\cos\theta) \qquad (7\text{-}79.6)$$

Putting the c.s. from (7-79.6) in equation (7-79.5), we get

$$\theta(C\sin\theta + D\cos\theta) + \frac{d^2}{d\theta^2}\left[\theta(C\sin\theta + D\cos\theta)\right] = -\frac{2F}{\pi}\cos\theta$$

$$\theta(C\sin\theta + D\cos\theta) + \frac{d}{d\theta}\left[\begin{array}{l}(C\sin\theta + D\cos\theta) + \\ + \theta C\cos\theta - \theta D\sin\theta\end{array}\right] = -\frac{2F}{\pi}\cos\theta$$

$$\theta(C\sin\theta + D\cos\theta) + \left[\begin{array}{l}(C\cos\theta - D\sin\theta) \\ + C\cos\theta - D\sin\theta \\ - \theta C\sin\theta - \theta D\cos\theta\end{array}\right] = -\frac{2F}{\pi}\cos\theta$$

$$2C\cos\theta - 2D\sin\theta = -\frac{2F}{\pi}\cos\theta$$

Equating the coefficients of $\sin\theta$ and $\cos\theta$, we get

$$\begin{array}{l} C = -\dfrac{F}{\pi} \\ D = 0 \end{array} \qquad (7\text{-}79.7)$$

The general solution, equation (7-79.4), becomes

$$\varphi(r,\theta) = r\left[A\sin\theta + B\cos\theta - \theta\frac{F}{\pi}(\sin\theta)\right] \qquad (7\text{-}79.8)$$

Substituting the stress function from (7-79.8) into equations (7-79.1 through 3), we could easily see that our empirical stresses in equations (7-78.2) satisfy Navier's equation of equilibrium (7-78.7). Noting that the terms $(A\sin\theta + B\cos\theta)$ comprise a linear function which vanish by the operations involved in equations (7-79.1 through 3). Or,

$$\varphi(r,\theta) = -\frac{F}{\pi}\theta r(\sin\theta) \qquad (7\text{-}79.9)$$

d. Normal stresses

According to our assumptions, in equations (7-78.2) and (7-78.3), circles centered around the point of application of force, Figure 7-10, comprise surfaces of normal stresses or **principal surfaces**, since shearing stresses are assumed null. Thus, an arbitrary point (x,y), in the elastic medium, is defined by

$$\cos\theta = \frac{r}{y} \qquad (7\text{-}79.10)$$

247

Where, y is the point of intersection of the tangent to the surface area where the radial stress applies (or the normal to the radial stress). Substituting from equation (7-79.10) into equation (7-78.2), we get

$$\sigma_r = -k\frac{\cos\theta}{r} = -\frac{2F}{\pi y} \qquad (7\text{-}79.11)$$

Thus, normal stress in constant on circles with radius depth y, independent of the θ.

e. Stress distribution

From Figure 7-11, we the stresses at an arbitrary point P(x,y) are

$$\sigma_x = (\sigma_r \sin\theta)\cos\theta \qquad (7\text{-}79.12)$$
$$\sigma_y = (\sigma_r \cos\theta)\cos\theta \qquad (7\text{-}79.13)$$
$$\tau_{xy} = (\sigma_r \sin\theta)\sin\theta \qquad (7\text{-}79.14)$$

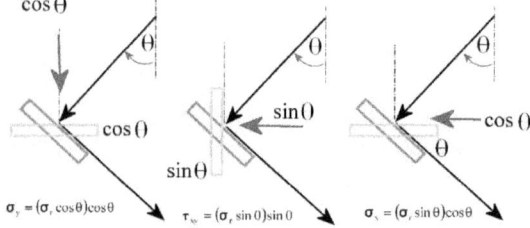

Figure 7-11. Stress distribution in simplified planar stress function.

Also,

$$\sin\theta = \frac{x}{\sqrt{x^2 + y^2}}$$
$$\qquad (7\text{-}79.15)$$
$$\cos\theta = \frac{y}{\sqrt{x^2 + y^2}}$$

Therefore, the stress distribution are obtained from equations (7-79.1) and (7-79.11 through 15), as follows

$$\sigma_x = \left(-\frac{2F}{\pi}\frac{\cos\theta}{r}\sin\theta\right)\cos\theta = -\frac{2F}{\pi}\frac{xy^2}{\left(x^2+y^2\right)^2} \tag{7-79.16}$$

$$\sigma_y = \left(-\frac{2F}{\pi}\frac{\cos\theta}{r}\cos\theta\right)\cos\theta = -\frac{2F}{\pi}\frac{y^3}{\left(x^2+y^2\right)^2} \tag{7-79.17}$$

$$\tau_{xy} = \left(-\frac{2F}{\pi}\frac{\cos\theta}{r}\sin\theta\right)\sin\theta = -\frac{2F}{\pi}\frac{x^2y}{\left(x^2+y^2\right)^2} \tag{7-79.18}$$

f. Representing uniform load in terms of finite load

A uniformly distributed load on the x-axis, Figure 7-11, can be described as a sum of stresses of lumped loads, as follows.

The infinitesimal load is given by

$$F(x)dx = F(x)\frac{rd\theta}{\cos\theta} \tag{7-79.19}$$

Substituting into equation (7-78.2), with the use of equation (7-78.5), we get

i.e.,

$$\sigma_r = -k\frac{\cos\theta}{r} = -\frac{2}{\pi}F(x)\frac{rd\theta}{\cos\theta}\frac{\cos\theta}{r}$$

$$= -\frac{2F(x)}{\pi}d\theta \tag{7-79.20}$$

Figure 7-12. Uniform load representation in terms of lumped load.

Thus, equations (7-79.12 through 14) become

$$d\sigma_x = (d\sigma_r \sin\theta)\cos\theta = -\frac{2F(x)}{\pi}\sin\theta\cos\theta d\theta \qquad (7\text{-}79.21)$$

$$d\sigma_y = (d\sigma_r \cos\theta)\cos\theta = -\frac{2F(x)}{\pi}\cos^2\theta d\theta \qquad (7\text{-}79.22)$$

$$d\tau_{xy} = (d\sigma_r \sin\theta)\sin\theta = -\frac{2F(x)}{\pi}\sin^2\theta d\theta \qquad (7\text{-}79.23)$$

Integrating over θ, from one of the **loaded surface**, θ_1, to the other end, θ_2, we get

$$\sigma_x = -\frac{2}{\pi}\int_{\theta_1}^{\theta_2} F(r\sin\theta)\sin\theta\cos\theta d\theta \qquad (7\text{-}79.24)$$

$$\sigma_y = -\frac{2}{\pi}\int_{\theta_1}^{\theta_2} F(r\sin\theta)\cos^2\theta d\theta \qquad (7\text{-}79.25)$$

$$\tau_{xy} = -\frac{2}{\pi}\int_{\theta_1}^{\theta_2} F(r\sin\theta)\sin^2\theta d\theta \qquad (7\text{-}79.26)$$

In case of **constant load** $F(r\sin\theta) = p$, constant, we get

$$\sigma_x = -\frac{2p}{\pi}\int_{\theta_1}^{\theta_2} \sin\theta\cos\theta d\theta$$

$$= -\frac{p}{\pi}\int_{\theta_1}^{\theta_2} \sin 2\theta d\theta = \frac{p}{2\pi}(\cos 2\theta_2 - \cos 2\theta_1) \qquad (7\text{-}79.27)$$

$$\sigma_y = -\frac{2p}{\pi}\int_{\theta_1}^{\theta_2} \cos^2\theta d\theta$$

$$= -\frac{p}{\pi}\int_{\theta_1}^{\theta_2} (\cos 2\theta + 1)d\theta \qquad (7\text{-}79.28)$$

$$= -\frac{p}{2\pi}(\sin 2\theta_2 - \sin 2\theta_1 + 2(\theta_2 - \theta_1))$$

$$\tau_{xy} = -\frac{2p}{\pi} \int_{\theta_1}^{\theta_2} \sin^2\theta \, d\theta$$

$$= -\frac{p}{\pi} \int_{\theta_1}^{\theta_2} (\cos 2\theta - 1) d\theta \qquad\qquad (7\text{-}79.29)$$

$$= -\frac{p}{2\pi}(\sin 2\theta_2 - \sin 2\theta_1 - 2(\theta_2 - \theta_1))$$

7.11. Finite force applied on the vertex of wedge

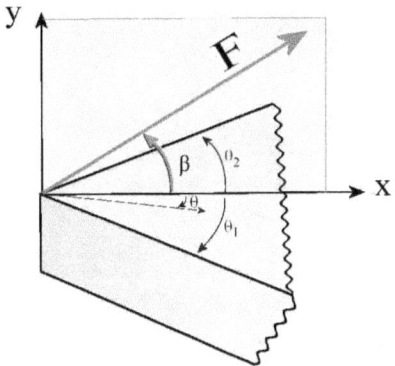

Figure 7-13. Arbitrary load applied to the vertex of a wedge.

a. Stress equation

The finite load applied to the finite geometry of a wedge, **Figure 7-13**, can be accommodated by modifying Flamant-Boussinesg's empirical equation (7-78.2) such that the angle θ is limited by the finite angle of wedge as follows:

$$\sigma_r = -k\frac{\cos(\theta - \theta_o)}{r} \qquad\qquad (7\text{-}80.1)$$

The two constants k and θ_o are determined from the finite nature of the wedge geometry and load as follows

$$F_x = \int_{\theta_1}^{\theta_2} \sigma_{-r} \delta \cos\theta \, d\theta = F \cos\beta$$

$$F_y = \int_{\theta_1}^{\theta_2} \sigma_{-r} \delta \sin\theta \, d\theta = F \sin\beta \qquad (7\text{-}80.2)$$

Where δ is defined in equation (7-78.1) and Figure 7-10, as the radius of Cauchy's circle around the point of application of the force, set empirically as $\pi x/2$ or $1.57x$, where x is width of the area of application of force. Substituting from (7-80.1) into (7-80.2), and noting that $r = \delta$, we get

$$
\begin{aligned}
\int_{\theta_1}^{\theta_2} k \frac{\cos(\theta - \theta_o)}{\delta} \delta \cos\theta \, d\theta &= k \int_{\theta_1}^{\theta_2} \cos(\theta - \theta_o) \cos\theta \, d\theta \\
&= k \left(\cos\theta_o \int_{\theta_1}^{\theta_2} \cos^2\theta \, d\theta + \sin\theta_o \int_{\theta_1}^{\theta_2} \sin\theta \cos\theta \, d\theta \right) \\
&= k \left(\cos\theta_o \left(\frac{1}{2}(2\theta_1 + \sin 2\theta_1) \right) + 0 \right) \\
&= F \cos\beta
\end{aligned}
\qquad (7\text{-}80.3)
$$

And

$$
\begin{aligned}
\int_{\theta_1}^{\theta_2} k \frac{\cos(\theta - \theta_o)}{\delta} \delta \sin\theta \, d\theta &= k \int_{\theta_1}^{\theta_2} \cos(\theta - \theta_o) \sin\theta \, d\theta \\
&= k \left(\cos\theta_o \int_{\theta_1}^{\theta_2} \cos\theta \sin\theta \, d\theta + \sin\theta_o \int_{\theta_1}^{\theta_2} \sin^2\theta \, d\theta \right) \\
&= k \left(\sin\theta_o \left(0 + \frac{1}{2}(2\theta_1 - \sin 2\theta_1) \right) + 0 \right) \\
&= F \sin\beta
\end{aligned}
\qquad (7\text{-}80.4)
$$

Or

$$\frac{1}{2} k \sin\theta_o (2\theta_1 - \sin 2\theta_1) = F \sin\beta \qquad (7\text{-}80.5)$$

$$\frac{1}{2} k \cos\theta_o (2\theta_1 + \sin 2\theta_1) = F \cos\beta \qquad (7\text{-}80.6)$$

Where, $2\theta_1$ is the angle of the vertex of the wedge.

Dividing (7-80.5) by (7-80.6), we get

$$\tan\theta_o = \tan\beta \frac{2\theta_1 + \sin 2\theta_1}{2\theta_1 - \sin 2\theta_1} \qquad (7\text{-}80.7)$$

$$\sin^2\theta_o = \frac{\tan^2\theta_o}{1+\tan^2\theta_o}$$

$$= \frac{(2\theta_1 + \sin 2\theta_1)^2 \tan^2\beta.}{(2\theta_1 - \sin 2\theta_1)^2 + \tan^2\beta(2\theta_1 + \sin 2\theta_1)^2}$$

(7-80.8)

Substituting by $\sin\theta_o$ from (7-80.7) into (7-80.5), we get

$$k = \frac{2F \, \sin\beta}{\sin\theta_o(2\theta_1 - \sin 2\theta_1)}$$

$$= 2F \cdot \sqrt{\frac{\cos^2\beta}{(2\theta_1 + \sin 2\theta_1)^2} + \frac{\sin^2\beta}{(2\theta_1 - \sin 2\theta_1)^2}}$$

(7-80.9)

b. Summary of equations

Table 7-3. Equations of finite stress applied to the vertex of wedge.

Stress equation	$\sigma_r = -k\dfrac{\cos(\theta - \theta_o)}{r}$
Wedge angle parameter	$\tan\theta_o = \tan\beta\dfrac{2\theta_1 + \sin 2\theta_1}{2\theta_1 - \sin 2\theta_1}$
Finite Force parameter	$k = 2F \sqrt{\dfrac{\cos^2\beta}{(2\theta_1 + \sin 2\theta_1)^2} + \dfrac{\sin^2\beta}{(2\theta_1 - \sin 2\theta_1)^2}}$

Note that, putting $\theta_1 = \pi/2$, in equations in the equations in Table 7-3, we get

$$\theta_o = \beta$$
$$k = \frac{2F}{\pi}$$
$$\sin x = \sum_{n=0}^{\infty} \frac{(-1)^n x^{2n+1}}{(2n+1)!} = x - \frac{x^3}{3.2} + \frac{x^5}{5.4.3.2} = x - \frac{x^3}{6} + \frac{x^5}{120}$$

c. Compression of a wedge

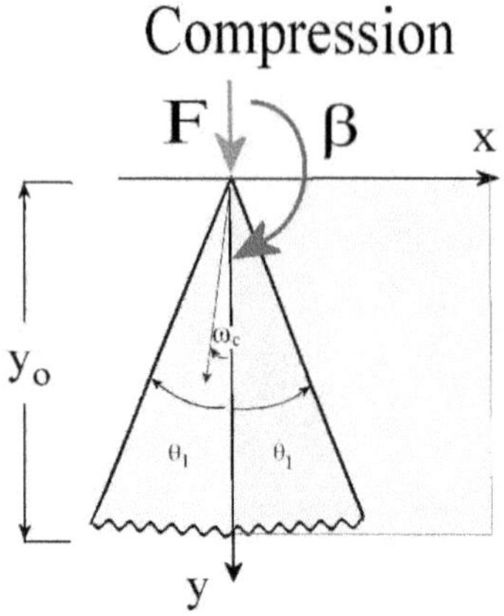

Figure 7-14. Configuration of compression of wedge $\beta = 0$.

As Figure 7-14 shows, putting $\beta = 0$ in the equations in Table 7-3, we get

$$\theta_o = 0 \qquad (7\text{-}81.1)$$

$$\sigma_r = -\frac{2F}{r(2\theta_1 + \sin 2\theta_1)}\cos\theta \qquad (7\text{-}81.2)$$

Where, $\qquad k = \dfrac{2F}{(2\theta_1 + \sin 2\theta_1)} \qquad (7\text{-}81.3)$

At depth $y = y_o$, equations (7-79.16 through 18), become

$$\sigma_x = \left(-k\frac{\cos\theta}{r}\sin\theta\right)\cos\theta = -\frac{2F}{(2\theta_1 + \sin 2\theta_1)}\frac{xy_o^2}{\left(x^2 + y_o^2\right)^2} \qquad (7\text{-}81.4)$$

$$\sigma_y = \left(-k\frac{\cos\theta}{r}\cos\theta\right)\cos\theta = -\frac{2F}{(2\theta_1 + \sin 2\theta_1)}\frac{y_o^3}{\left(x^2 + y_o^2\right)^2} \qquad (7\text{-}81.5)$$

$$\tau_{xy} = \left(-k\frac{\cos\theta}{r}\sin\theta\right)\sin\theta = -\frac{2F}{(2\theta_1 + \sin 2\theta_1)}\frac{x^2 y_o}{\left(x^2 + y_o^2\right)^2} \qquad (7\text{-}81.6)$$

The ratio between the three stresses at depth y_o, is therefore

254

$$\sigma_y : \sigma_x : \tau_{xy} :: y_o^2 : xy_o : x^2 \qquad (7\text{-}81.7)$$

Thus, as the ratio $y_o : x$ increases, the normal stress along the y-axis increases while the shear stress that along x-axis decreases.

Substituting by

$$\tan \omega_c = \frac{x}{y_o} \qquad (7\text{-}81.8)$$

Where ω_c varies from $-\theta_1$ to θ_1.

We get

$$\sigma_x = -\frac{2F}{(2\theta_1 + \sin 2\theta_1)} \frac{1}{(1 + \tan^2 \omega_c)^2} \left(\frac{x}{y_o^2} \right) \qquad (7\text{-}81.9)$$

$$\sigma_y = -\frac{2F}{(2\theta_1 + \sin 2\theta_1)} \frac{1}{(1 + \tan^2 \omega_c)^2} \left(\frac{1}{y_o} \right) \qquad (7\text{-}81.10)$$

$$\tau_{xy} - \frac{2F}{(2\theta_1 + \sin 2\theta_1)} \frac{1}{(1 + \tan^2 \omega_c)^2} \left(\frac{x^2}{y_o^3} \right) \qquad (7\text{-}81.11)$$

Equation (7-81.9) shows that as the width of the wedge decreases (small x), σ_x vanishes. Equation (7-81.10) shows the slight dependence of σ_y on the width of the wedge through $\tan^2 \omega_c$. Equation (7-81.11) shows the greater shear stresses with the increase of width and reduction in height of the wedge.

d. Bending of a wedge

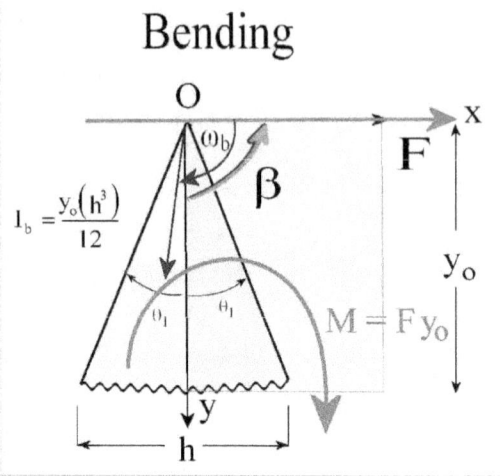

Figure 7-15. Configuration of bending of wedge $\beta = \pi/2$.

As Figure 7-15 shows, putting $\beta = \pi/2$ in the equations in Table 7-3, we get

$$\theta_o = \frac{\pi}{2} \tag{7-82.1}$$

$$\sigma_r = -\frac{2F}{r(2\theta_1 - \sin 2\theta_1)} \cos\left(\theta - \frac{\pi}{2}\right) \tag{7-82.2}$$

Where, $\qquad k = \dfrac{2F}{(2\theta_1 - \sin 2\theta_1)} \tag{7-82.3}$

At depth y_o, the force F induced couple $M = Fy_o$ and the **moment of inertia** of a wedge rotating around its base y_o, with height h, is

$$I_b = \frac{y_o\left(h^3\right)}{12} \tag{7-82.4}$$

It should be noted that the moment of inertia I_b, does not take into account the rotation on the vertex of the base. I_b merely is our own approximation for rotation around the central divider of the wedge.

Also, all angles in equations (7-79.16 through 18) are rotated by $\pi/2$, therefore we have

$$\sigma_x = \left(-k\frac{\sin\theta}{r}\sin\left(\theta - \frac{\pi}{2}\right)\right)\cos\left(\theta - \frac{\pi}{2}\right) = -\frac{2M}{\left(2\theta_1 - \sin 2\theta_1\right)}\frac{xy_o}{\left(x^2 + y_o^2\right)^2} \qquad (7\text{-}82.5)$$

$$\sigma_y = \left(-k\frac{\sin\theta}{r}\cos\left(\theta - \frac{\pi}{2}\right)\right)\cos\left(\theta - \frac{\pi}{2}\right) = -\frac{2M}{\left(2\theta_1 - \sin 2\theta_1\right)}\frac{y_o^2}{\left(x^2 + y_o^2\right)^2} \qquad (7\text{-}82.6)$$

$$\tau_{xy} = \left(-k\frac{\sin\theta}{r}\sin\left(\theta - \frac{\pi}{2}\right)\right)\sin\left(\theta - \frac{\pi}{2}\right) = -\frac{2M}{\left(2\theta_1 - \sin 2\theta_1\right)}\frac{x^2}{\left(x^2 + y_o^2\right)^2} \qquad (7\text{-}82.7)$$

Substituting by

$$\tan\omega_b = \frac{x}{y_o} \qquad (7\text{-}82.8)$$

Where ω_b is measured from the x-axis, and varies from to $(\pi/2) - \theta_1$ to $(\pi/2) + \theta_1$.

And expanding the sine in its power series as follows

$$\sin x = \sum_{n=0}^{\infty} \frac{(-1)^n x^{2n+1}}{(2n+1)!} = x - \frac{x^3}{3.2} + \frac{x^5}{5.4.3.2} = x - \frac{x^3}{6} + \frac{x^5}{120} \qquad (7\text{-}82.9)$$

$$\tan\theta_1 = \tan\theta_2 = \frac{h}{2y_o} \qquad (7\text{-}82.10)$$

For small wedge angle, we get

$$2\theta_1 - \sin 2\theta_1 = 2\theta_1 - \left(2\theta_1 - \frac{(2\theta_1)^3}{6}\right) = \frac{(2\theta_1)^3}{6} \qquad (7\text{-}82.11)$$

$$\sigma_x = -\frac{Mxy_o}{I_b}\left(\frac{\tan(\theta_1)}{\theta_1}\right)^3 \sin^4\omega_b \qquad (7\text{-}82.12)$$

$$\sigma_y = -\frac{My_o^2}{I_b}\left(\frac{\tan(\theta_1)}{\theta_1}\right)^3 \sin^4\omega_b \qquad (7\text{-}82.13)$$

$$\tau_{xy} = -\frac{Mx^2}{I_b}\left(\frac{\tan(\theta_1)}{\theta_1}\right)^3 \sin^4\omega_b \qquad (7\text{-}82.14)$$

The ratio between the three stresses at depth y_o, is also

$$\sigma_y : \sigma_x : \tau_{xy} :: y_o^2 : xy_o : x^2 \qquad (7\text{-}82.15)$$

For small values of θ_1, $\sin\omega_b \rightarrow 1$, thus equations (7-82.12 through 14) become

$$\sigma_x \approx -\frac{Mxy_o}{I_b} \tag{7-82.16}$$

$$\sigma_y \approx -\frac{My_o^2}{I_b} \tag{7-82.17}$$

$$\tau_{xy} \approx -\frac{Mx^2}{I_b} \tag{7-82.18}$$

CHAPTER 8

BI-HARMONIC EQUATION

8.1. Bi-Harmonic equation of plane stress in polar cylindrical coordinates

The plane problem in cylindrical polar coordinate takes the bi-harmonic Laplacian form:

$$\nabla^2\left(\nabla^2\phi(r,\theta)\right) = 0 \tag{8-1.1}$$

$$\nabla^2 = \frac{1}{r}\frac{\partial}{\partial r} + \frac{\partial^2}{\partial r^2} + \frac{1}{r^2}\frac{\partial^2}{\partial\theta^2} \tag{8-1.2}$$

We assume the geometrical product of the polynomials $R(r)$ and $\Theta(\theta)$ as follows

$$\phi(r,\theta) = rR(r)\Theta(\theta) \tag{8-1.3}$$

Performing the first Laplancian operation, we get

$$\begin{aligned}
\nabla^2\left[rR(r)\Theta(\theta)\right] &= \left(\frac{1}{r}\frac{\partial}{\partial r} + \frac{\partial^2}{\partial r^2}\right)rR(r)\Theta(\theta) + \frac{R(r)}{r}\frac{d^2\Theta(\theta)}{d\theta^2} \\
&= \frac{1}{r}\left(r\frac{dR(r)}{dr} + R(r)\right)\Theta(\theta) + \left(r\frac{d^2R(r)}{dr^2} + 2\frac{dR(r)}{dr}\right)\Theta(\theta) + \frac{R(r)}{r}\frac{d^2\Theta(\theta)}{d\theta^2}
\end{aligned} \tag{8-1.4}$$

Which can be written in the abbreviated form as follows

$$\nabla^2\varphi(r,\theta) = r\left(R''(r) + 3\frac{1}{r}R'(r) + \frac{1}{r^2}R(r)\right)\Theta(\theta) + r\left(\frac{1}{r^2}R(r)\Theta''(\theta)\right) \tag{8-1.5}$$

The second Laplacian becomes

$$\nabla^2\left[\nabla^2\varphi(r,\theta)\right] = \left(\frac{1}{r}\frac{\partial}{\partial r} + \frac{\partial^2}{\partial r^2} + \frac{1}{r^2}\frac{\partial^2}{\partial\theta^2}\right)\left[r\left(R'' + \frac{3R'}{r} + \frac{R}{r^2}\right)\Theta + r\left(\frac{1}{r^2}R\Theta''\right)\right]$$

$$= \Theta\left(\frac{1}{r}\frac{d}{dr} + \frac{d^2}{dr^2}\right)r\left(R'' + \frac{3}{r}R' + \frac{1}{r^2}R\right)$$

$$+ \frac{1}{r}\left(R'' + 3\frac{1}{r}R' + \frac{1}{r^2}R\right)\frac{d^2\Theta}{d\theta^2} \qquad (8\text{-}1.6)$$

$$+ \Theta''\left(\frac{1}{r}\frac{d}{dr} + \frac{d^2}{dr^2}\right)\left(\frac{1}{r}R\right)$$

$$+ \frac{1}{r^3}R(r)\frac{d^4\Theta}{d\theta^4}$$

We will perform the differentiation in equation (8-1.5) as follows:

$$\Theta\left(\frac{1}{r}\frac{d}{dr} + \frac{d^2}{dr^2}\right)\left(rR'' + 3R' + \frac{1}{r}R\right) = \frac{1}{r}\left(rR''' + 4R'' + \frac{1}{r}R' - \frac{1}{r^2}R\right)$$

$$+ \left(rR'''' + 5R''' + \frac{1}{r}R'' - \frac{2}{r^2}R' + \frac{2}{r^3}R\right)$$

$$\Theta\left(\frac{1}{r}\frac{d}{dr} + \frac{d^2}{dr^2}\right)\left(rR'' + 3R' + \frac{1}{r}R\right) = \left(rR'''' + 6R''' + \frac{5}{r}R'' - \frac{1}{r^2}R' + \frac{1}{r^3}R\right)\Theta \qquad (8\text{-}1.7)$$

$$\Theta''\left(\frac{1}{r}\frac{d}{dr} + \frac{d^2}{dr^2}\right)\left(\frac{1}{r}R\right) = \left(\frac{1}{r}R'' - \frac{1}{r^2}R' + \frac{1}{r^3}R\right)\Theta'' \qquad (8\text{-}1.8)$$

Substituting from (8-1.7) and (8-1.8) into (8-1.6), we get

$$\nabla^2\left[\nabla^2\varphi(r,\theta)\right] = \left(rR'''' + 6R''' + \frac{5}{r}R'' - \frac{1}{r^2}R' + \frac{1}{r^3}R\right)\Theta$$

$$+ \frac{1}{r}\left(R'' + 3\frac{1}{r}R' + \frac{1}{r^2}R\right)\Theta'' + \left(\frac{1}{r}R'' - \frac{1}{r^2}R' + \frac{1}{r^3}R\right)\Theta'' \qquad (8\text{-}1.9)$$

$$+ \frac{1}{r^3}R\Theta''''$$

$$= 0$$

$$\nabla^2\left[\nabla^2\varphi(r,\theta)\right] = \left(rR'''' + 6R''' + \frac{5}{r}R'' - \frac{1}{r^2}R' + \frac{1}{r^3}R\right)\Theta$$

i.e.,
$$+ 2\left(\frac{1}{r}R'' + \frac{1}{r^2}R' + \frac{1}{r^3}R\right)\Theta'' \qquad (8\text{-}1.10)$$

$$+ \frac{1}{r^3}R\Theta''''$$

$$= 0$$

Multiplying by r^3, we get

$$\left(r^4R'''' + 6r^3R''' + 5r^2R'' - rR' + R\right)\Theta + 2\left(r^2R'' + rR' + R\right)\Theta'' + R\Theta'''' = 0 \qquad (8\text{-}1.11)$$

8.2. Variable separation constant

Equation (8-1.11) represents the final form of the **bi-harmonic equation** for plane stress function with vanishing internal forces and static equilibrium.

In order to separate $R(r)$ from $\Theta(\theta)$, we can divide each member of equation (8-1.11) by $R(r)$ and differentiate with respect to r in order to remove Θ'''' (θ) from our characteristic equation. Thus, we get

$$\frac{d}{dr}\left[\frac{r^4R'''' + 6r^3R''' + 5r^2R'' - rR' + R}{R}\right]\Theta = -2\frac{d}{dr}\left[\frac{r^2R'' + rR' + R}{R}\right]\Theta'' \qquad (8\text{-}1.12)$$

i.e.,
$$\frac{\Theta''}{\Theta} = -\frac{\dfrac{d}{dr}\left[\dfrac{r^4R'''' + 6r^3R''' + 5r^2R'' - rR' + R}{R}\right]}{2\dfrac{d}{dr}\left[\dfrac{r^2R'' + rR' + R}{R}\right]} = -\lambda^2 \qquad (8\text{-}1.13)$$

Where λ is a separation constant to be determined from the boundary conditions of the problem.

8.2.1. Case one: $\dfrac{d}{dr}\left[\dfrac{r^2R'' + rR' + R}{R}\right] \neq 0$

The polar function could therefore be written as

$$\Theta'' + \lambda^2\Theta = 0 \qquad (8\text{-}2.1)$$

Differentiating twice, we also get

$$\Theta'''' = -\lambda^2 \Theta''$$

$$= -\lambda^2\left(-\lambda^2 \Theta\right) \tag{8-2.2}$$

$$= \lambda^4 \Theta$$

With the solution

$$\Theta(\theta) = A_\lambda \cos(\lambda\theta) + B_\lambda \sin(\lambda\theta) \tag{8-2.3}$$

Substitution from equation (8-2.2) in (8-1.11), we get

$$\left(r^4 R'''' + 6r^3 R''' + 5r^2 R'' - rR' + R\right)\Theta - 2\left(r^2 R'' + rR' + R\right)\lambda^2\Theta + \lambda^4\Theta R = 0 \tag{8-2.4}$$

Dividing by $\Theta(\theta)$, we get

$$r^4 R'''' + 6r^3 R''' + \left(5 - 2\lambda^2\right)r^2 R'' - (1 + 2\lambda^2)rR' + \left(1 - 2\lambda^2 + \lambda^4\right)R = 0 \tag{8-2.5}$$

We could adopt the solution obtained previous in equations (7-75.2) and (7-75.10)

$$\nabla^2\left(\nabla^2\varphi(r,\theta)\right) = \frac{1}{r}\frac{\partial}{\partial r}\left(r\frac{\partial}{\partial r}\left[\frac{1}{r}\frac{\partial}{\partial r}\left(r\frac{\partial}{\partial r}\right)\right]\right)\varphi(r,\theta) = 0 \tag{8-2.6a}$$

$$\varphi(r,\theta) = C_1 r^2 \ln r + C_3 \ln r + C_2 r^2 + C_4 \tag{8-2.6b}$$

Yet, we will follows the alternative method by starting with the trial solution to equation (8-2.5)

$$R(r) = r^\beta \tag{8-2.7}$$

Substituting from (8-2.7) in (8-2.5), we get

$$\beta(\beta-1)(\beta-2)(\beta-3)r^\beta + 6\beta(\beta-1)(\beta-2)r^\beta + \left(5 - 2\lambda^2\right)\beta(\beta-1)r^\beta$$
$$- (1+2\lambda^2)\beta r^\beta + \left(1 - 2\lambda^2 + \lambda^4\right)r^\beta = 0 \tag{8-2.8}$$

$$(\beta^4 - 5\beta^3 + 6\beta^2)$$
$$-\beta^3 + 5\beta^2 - 6\beta$$
$$+6\beta^3$$
$$-12\beta^2$$
$$-6\beta^2$$
$$+(5 - 2\lambda^2)\beta^2 \tag{8-2.9}$$
$$+12\beta$$
$$-(5 - 2\lambda^2)\beta$$
$$-(1 + 2\lambda^2)\beta$$
$$+(1 - 2\lambda^2 + \lambda^4) = 0$$

$$\beta^4 - 2(\lambda^2 + 1)\beta^2 + (\lambda^2 - 1)^2 = 0 \tag{8-2.10}$$

With the solution

$$\beta^2 = \frac{2(\lambda^2 + 1) \pm \sqrt{4(\lambda^2 + 1)^2 - 4(\lambda^2 - 1)^2}}{2}$$
$$= \frac{2(\lambda^2 + 1) \pm \sqrt{16\lambda^2}}{2}$$
$$= (\lambda^2 + 1) \pm 2\lambda \tag{8-2.11}$$
$$= (\lambda^2 + 2\lambda + 1), \quad (\lambda^2 - 2\lambda + 1)$$
$$= (\lambda + 1)^2, \quad (\lambda - 1)^2$$

$$\beta : \pm(\lambda + 1), \quad \pm(\lambda - 1) \tag{8-2.12}$$

8.2.2. Case two: λ are integers and $\lambda \neq 0$ or $\lambda \neq 1$

With integer values of λ such that $\lambda \neq 0$ or 1, we avoid repetitive roots. Thus, from equations (8-2.3) and (8-2.7), the proposed solution in equation (8-1.3), becomes

$$\phi(r, \theta) = rR(r)\Theta(\theta)$$
$$= \sum_{\lambda=2}^{n} \left(C_{\lambda 1} r^{\lambda + 2} + C_{\lambda 2} r^{-\lambda} + C_{\lambda 3} r^{\lambda} + C_{\lambda 4} r^{-\lambda + 2} \right) [A_\lambda \cos(\lambda\theta) + B_\lambda \sin(\lambda\theta)] \tag{8-3.1}$$

8.2.3. Case three: $\lambda = 0$

Equation (8-2.1) becomes
$$\Theta'' = 0 \tag{8-4.1}$$

Thus,
$$\Theta(\theta) = A\theta + B \tag{8-4.2}$$

Also, equation (8-2.12) becomes
$$\beta : \pm 1, \quad \pm(-1) \tag{8-4.3}$$

And equation (8-1.13) gives
$$\frac{d}{dr}\left[\frac{r^4 R''''+6r^3 R'''+5r^2 R''-rR'+R}{R}\right] = 0$$

i.e.,
$$r^4 R''''+6r^3 R'''+5r^2 R''-rR'+R = CR \tag{8-4.4}$$

with the solution comprising of four independent terms: r, $\ln(r)$, $(1/r)$ and $(1/r)\ln(r)$.

And the final solution (8-3.1) becomes
$$\phi_0(r,\theta) = \left(C_1 r^2 + C_2 r^2 \ln(r) + C_3 + C_4 \ln(r)\right)(A\theta + B) \tag{8-4.5}$$

8.2.4. Case four: $\lambda = 1$

Equation (8-2.1) becomes
$$\Theta'' = \Theta \tag{8-5.1}$$

Thus,
$$\Theta(\theta) = A\cos\theta + B\sin\theta \tag{8-5.2}$$

Also, equation (8-2.12) becomes
$$\beta : \pm 2, \quad 0 \tag{8-5.3}$$

This, the $R(r)$ solution comprises of four independent terms: r^2, $(1/r^2)$, 1 and $\ln(r)$.

And the final solution (8-3.1) becomes
$$\phi_1(r,\theta) = \left(C_1 r^3 + C_2 r^{-1} + C_3 r + C_4 r \ln(r)\right)(A\cos\theta + B\sin\theta) \tag{8-5.4}$$

8.2.5. Case five: $\dfrac{d}{dr}\left[\dfrac{r^2R''+rR'+R}{R}\right] = 0$

Equation (8-1.12) becomes

$$\frac{d}{dr}\left[\frac{r^4R''''+6r^3R'''+5r^2R''-rR'+R}{R}\right]\Theta = 0 \tag{8-6.1}$$

We therefore have two constant terms

$$\int \frac{d}{dr}\left[\frac{r^2R''+rR'+R}{R}\right]dr = \int d\left[\frac{r^2R''+rR'+R}{R}\right]$$

$$= \frac{r^2R''+rR'+R}{R} \tag{8-6.2}$$

$$= C_1$$

$$\int \frac{d}{dr}\left[\frac{r^4R''''+6r^3R'''+5r^2R''-rR'+R}{R}\right]dr = \int d\left[\frac{r^4R''''+6r^3R'''+5r^2R''-rR'+R}{R}\right]$$

$$= \frac{r^4R''''+6r^3R'''+5r^2R''-rR'+R}{R} \tag{8-6.3}$$

$$= C_2$$

Therefore, equation (8-1.11) becomes

$$\Theta''''+2C_1\Theta''+C_2\Theta = 0 \tag{8-6.4}$$

Thus, equations (8-6.2 through 6.4) tie $R(r)$ and $\Theta(\theta)$ through the two constants C_1 and C_2.

We will use the general solution, equation (8-2.7), in equations (8-6.2 through 6.4) to determine the proper values of C_1 and C_2 such that the three equations yield congruent solution. Therefore, equation (8-6.2)

$$r^2R''+rR'+R = RC_1$$
$$\beta(\beta-1)+\beta+1 = C_1$$

i.e., $\qquad \beta^2 = C_1-1$

Or, $\qquad C_1 = \beta^2+1 \tag{8-6.5}$

Similarly, equation (8-6.3) gives

$$r^4 R''''+6r^3 R'''+5r^2 R''-rR'+R = RC_2$$
$$\beta(\beta-1)(\beta-2)(\beta-3) + 6\beta(\beta-1)(\beta-2) + 5\beta(\beta-1) - \beta + 1 = C_2$$

$$\beta^4 - 2\beta^2 + 1 - C_2 = 0$$

$$\beta^2 = \frac{2 \pm \sqrt{4 - 4(1 - C_2)}}{2} = 1 \pm \sqrt{C_2}$$

Or,
$$C_2 = \pm(\beta^2 - 1)^2 \qquad (8\text{-}6.6)$$

From equations (8-6.5) and (8-6.6), equation (8-6.4) becomes

$$\Theta''''+2(\beta^2 + 1)\Theta''+(\beta^2 - 1)^2 \Theta = 0 \qquad (8\text{-}6.7)$$

In order to determine the values of β which satisfy equation (8-6.7), we propose the solution

$$\Theta(\theta) = e^{\eta\theta} \qquad (8\text{-}6.8)$$

Therefore, substituting from (8-6.8) into (8-6.7) we get

$$\eta^4 + 2(\beta^2 + 1)\eta^2 + (\beta^2 - 1)^2 = 0 \qquad (8\text{-}6.9)$$

$$\eta^2 = \frac{-2(\beta^2 + 1) \pm \sqrt{4(\beta^2 + 1)^2 - 4(\beta^2 - 1)^2}}{2}$$

$$= -(\beta^2 + 1) \pm 2\beta$$

$$\eta^2 = -(\beta+1)^2, -(\beta-1)^2$$

i.e.,
$$\eta = \pm\ i\ (\beta \pm 1) \qquad (8\text{-}6.10a)$$

Where,
$$i = \sqrt{-1} \qquad (8\text{-}6.10b)$$

8.2.6. Case six: $\beta = 0$

Therefore, equations (8-6.5) and (8-6.6), give $C_1 = 1$ and $C_2 = \pm 1$ and equations (8-6.2) yields

$$\frac{r^2R''+rR'+R}{R} = 1$$

$$rR''+R' = 0$$

$$r\frac{d^2R}{dr^2}+\frac{dR}{dr} = 0$$

$$\int\left(r\frac{d^2R}{dr^2}+\frac{dR}{dr}\right)dr = C$$

$$\int r\frac{d^2R}{dr^2}\,dr = C - R$$

$$\int rd\left(\frac{dR}{dr}\right) = C - R$$

$$r\frac{dR}{dr}-\int \frac{dR}{dr}d(r) = C - R$$

$$r\frac{dR}{dr}-\int d(R) = C - R$$

$$r\frac{dR}{dr} = C$$

$$dR = C\frac{dr}{r}$$

$$R = C\ln(r)$$

(8-7.1)

And

$$\Theta''''+2\Theta''+\Theta = 0 \qquad\qquad (8\text{-}7.2)$$

With the roots of the characteristic equation given by (8-6.10a), $\eta = \pm\ i$, with the solution

$$\Theta(\theta) = A\cos\theta + B\sin\theta + C\theta\cos\theta + D\theta\sin\theta$$
(8-7.3)

Thus, the solution to equation (8-1.3) becomes

$$\phi_0(r,\theta) = rR(r)\Theta(\theta)$$

$$= (C_1r + C_2r\ln(r))(A\cos\theta + B\sin\theta + C\theta\cos\theta + D\theta\sin\theta)$$
(8-7.4)

8.2.7 Case seven: $\beta = 1$

Equation (8-6.6) gives $C_2 = \pm(\beta^2 - 1)^2 = 0$, equation (8-6.5) gives $C_1 = 2$, and $\eta = \pm i$ and $\pm 2i$. Thus, equation (8-6.4) becomes

$$\Theta'''' + 4\Theta'' = 0 \tag{8-8.1}$$

And equation (8-6.2) becomes

$$r^2 R'' + rR' - R = 0 \tag{8-8.2}$$

Thus, the general solution becomes

$$\phi_1(r,\theta) = rR(r)\Theta(\theta)$$
$$= (C_1 r^2 + C_2)(A\cos 2\theta + B\sin 2\theta + C\theta + D)$$
$$\tag{8-8.3}$$

8.3. Summary of solution of stress function in polar cylindrical coordinates

$\dfrac{d}{dr}\left[\dfrac{r^2 R'' + rR' + R}{R}\right] \neq 0$	
λ integers $\lambda \neq 0$ $\lambda \neq 1$	$\phi(r,\theta) = \displaystyle\sum_{\lambda=2}^{n} \left(C_{\lambda 1} r^{\lambda+2} + C_{\lambda 2} r^{-\lambda} + C_{\lambda 3} r^{\lambda} + C_{\lambda 4} r^{-\lambda+2}\right)\left[A_\lambda \cos(\lambda\theta) + B_\lambda \sin(\lambda\theta)\right]$
$\lambda = 0$	$\phi_0(r,\theta) = (C_1 r^2 + C_2 r^2 \ln(r) + C_3 + C_4 \ln(r))(A\theta + B)$
$\lambda - 1$	$\phi_1(r,\theta) = (C_1 r^3 + C_2 r^{-1} + C_3 r + C_4 r \ln(r))(A\cos\theta + B\sin\theta)$
$\dfrac{d}{dr}\left[\dfrac{r^2 R'' + rR' + R}{R}\right] = 0$	
$\beta = 0$	$\phi_0(r,\theta) = (C_1 r + C_2 r \ln(r))(A\cos\theta + B\sin\theta + C\theta\cos\theta + D\theta\sin\theta)$
$\beta = 1$	$\phi_1(r,\theta) = (C_1 r^2 + C_2)(A\cos 2\theta + B\sin 2\theta + C\theta + D)$

8.4. Stresses

1. Normal radial stresses

$$\sigma_r = \frac{1}{r}\frac{\partial\varphi(r,\theta)}{\partial r} + \frac{1}{r^2}\cdot\frac{\partial^2\varphi(r,\theta)}{\partial\theta^2}$$

$$\frac{d}{dr}\left[\frac{r^2 R''+rR'+R}{R}\right] \neq 0$$

λ integers $\lambda \neq 0$
$\lambda \neq 1$

$$\sigma_r(r,\theta) = \sum_{\lambda=2}^{n} \begin{pmatrix} C_{\lambda 1}(\lambda+2-\lambda^2)r^\lambda - (\lambda+\lambda^2)C_{\lambda 2}r^{-\lambda-2} \\ + (\lambda-\lambda^2)C_{\lambda 3}r^{\lambda-2} + (-\lambda+2-\lambda^2)C_{\lambda 4}r^{-\lambda} \end{pmatrix}\begin{bmatrix} A_\lambda \cos(\lambda\theta) \\ + B_\lambda \sin(\lambda\theta) \end{bmatrix}$$

$\lambda - 0$

$$\sigma_r(r,\theta) = (2C_1 + 2C_2\ln(r) + C_2 + C_4 r^{-2})(A\theta + B)$$

$\lambda - 1$

$$\sigma_r(r,\theta) = (2C_1 r - 2C_2 r^{-3} + C_4 r^{-1}(1-\ln(r)))(A\cos\theta + B\sin\theta)$$

$$\frac{d}{dr}\left[\frac{r^2 R''+rR'+R}{R}\right] = 0$$

$\beta = 0$

$$\sigma_r(r,\theta) = C_2 r^{-1}(A\cos\theta + B\sin\theta + C\theta\cos\theta + D\theta\sin\theta)$$
$$- 2(C_1 r^{-1} + C_2 r^{-1}\ln(r))(C\sin\theta - D\cos)$$

$\beta = 1$

$$\phi_1(r,\theta) = (C_1 r^2 + C_2)(A\cos 2\theta + B\sin 2\theta + C\theta + D)$$

2. Angular polar stresses

	$\sigma_\theta = \dfrac{\partial^2 \varphi(r,\theta)}{\partial r^2}$
$\dfrac{d}{dr}\left[\dfrac{r^2 R''+rR'+R}{R}\right] \neq 0$	
λ integers $\lambda \neq 0$ $\lambda \neq 1$	$\sigma_\theta(r,\theta) = \displaystyle\sum_{\lambda=2}^{n} \begin{pmatrix} C_{\lambda 1}(\lambda+2)(\lambda+1)r^\lambda + \lambda(\lambda+1)C_{\lambda 2}r^{-\lambda-2} \\ + \lambda(\lambda-1)C_{\lambda 3}r^{\lambda-2} + C_{\lambda 1}(\lambda+2)(\lambda+1)r^\lambda \\ + \lambda(\lambda+1)C_{\lambda 2}r^{-\lambda-2} \\ + (-\lambda+2)(-\lambda+1)C_{\lambda 4}r^{-\lambda} \end{pmatrix}\begin{bmatrix} A_\lambda \cos(\lambda\theta) \\ + B_\lambda \sin(\lambda\theta) \end{bmatrix}$
$\lambda = 0$	$\sigma_\theta(r,\theta) = (2C_1 + C_2(2\ln(r)+3) - C_4 r^{-2})(A\theta + B)$
$\lambda - 1$	$\sigma_\theta(r,\theta) = (6C_1 r + 2C_2 r^{-3} + C_4 r^{-1})(A\cos\theta + B\sin\theta)$
$\dfrac{d}{dr}\left[\dfrac{r^2 R''+rR'+R}{R}\right] = 0$	
$\beta = 0$	$\sigma_\theta(r,\theta) = C_2 r^{-1}(A\cos\theta + B\sin\theta + C\theta\cos\theta + D\theta\sin\theta)$
$\beta = 1$	$\sigma_\theta(r,\theta) = 2C_1(A\cos 2\theta + B\sin 2\theta + C\theta + D)$

3. Shear stresses

	$$\tau_{r\theta} = -\frac{1}{r}\frac{\partial^2\varphi(r,\theta)}{\partial r\partial\theta} + \frac{1}{r^2}\cdot\frac{\partial\varphi(r,\theta)}{\partial\theta}$$
$\dfrac{d}{dr}\left[\dfrac{r^2 R''+rR'+R}{R}\right] \neq 0$	
λ integers $\lambda \neq 0$ $\lambda \neq 1$	$$\tau_{r\theta}(r,\theta) = -\lambda\sum_{\lambda=2}^{n}\begin{pmatrix} C_{\lambda1}(\lambda+2)r^\lambda - \lambda C_{\lambda2}r^{-\lambda-2} \\ + \lambda C_{\lambda3}r^{\lambda-2} + (-\lambda+2)C_{\lambda4}r^{-\lambda} \end{pmatrix}[-A_\lambda\sin(\lambda\theta)+B_\lambda\cos(\lambda\theta)]$$ $$-\lambda^2\sum_{\lambda=2}^{n}\begin{pmatrix} C_{\lambda1}r^\lambda + C_{\lambda2}r^{-\lambda-2} \\ + C_{\lambda3}r^{\lambda-2} + C_{\lambda4}r^{-\lambda} \end{pmatrix}[A_\lambda\cos(\lambda\theta)+B_\lambda\sin(\lambda\theta)]$$
$\lambda = 0$	$$\tau_{r\theta}(r,\theta) = -A\left(2C_1 + C_2 + 2C_2\ln(r) + C_4 r^{-1}\right)$$
$\lambda = 1$	$$\tau_{r\theta}(r,\theta) = -\left(3C_1 r - C_2 r^{-3} + C_3 r^{-1} + C_4 r^{-1}\ln(r) + C_4 r^{-1}\right)(B\cos\theta - A\sin\theta)$$ $$-\left(C_1 r + C_2 r^{-3} + C_3 r^{-1} + C_4 r^{-1}\ln(r)\right)(B\sin\theta + A\cos\theta)$$
$\dfrac{d}{dr}\left[\dfrac{r^2 R''+rR'+R}{R}\right] = 0$	
$\beta = 0$	$$\tau_{r\theta}(r,\theta) = -\left(C_1 r^{-1} + C_2 r^{-1}\ln(r) + r^{-1}C_2\right)\begin{pmatrix} -A\sin\theta + B\cos\theta \\ + C\cos\theta + D\sin\theta \\ -C\theta\sin\theta + D\theta\cos\theta \end{pmatrix}$$ $$+\left(C_1 r^{-1} + C_2 r^{-1}\ln(r)\right)\begin{pmatrix} -A\cos\theta - B\sin\theta - C\sin\theta + D\cos\theta \\ -C\sin\theta + D\cos\theta - C\theta\cos\theta - D\theta\sin\theta \end{pmatrix}$$
$\beta = 1$	$$\tau_{r\theta}(r,\theta) = -\left(2C_1\right)\left(-2A\sin2\theta + 2B\cos2\theta + C\right)$$ $$+\left(C_1 + C_2 r^{-2}\right)\left(-4A\cos2\theta - 4B\sin2\theta\right)$$

270

CHAPTER 9

TORSION OF PRISMATICAL BARS

9-1. Prismatical Circular Cylindrical Bar

Torsion of prismatical bars

A cylindrical tube exposed to torsion couple at its terminal cross sections.

(i) Find the stresses at point P on the upper surface. State the condition for vanishing normal stress

(ii) Find the stresses at point M on the lateral wall. State the condition of vanishing load.

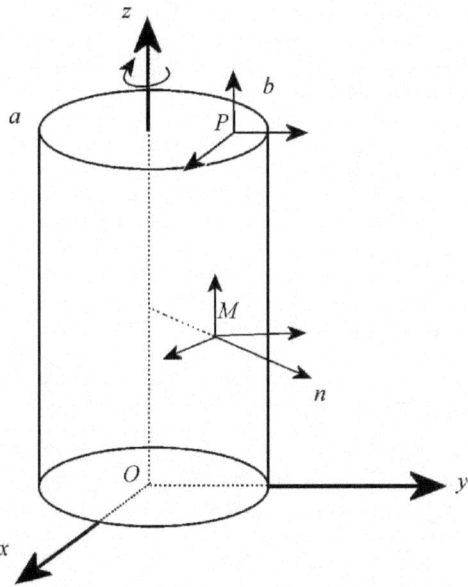

Figure 9-1. Solid circular cylindrical bar twisted at the top surface.

Solution

Neglect internal forces in Navier's equilibrium equation

$$X_i = Y_i = Z_i = 0 \qquad\qquad (9\text{-}1)$$

Displacement

We assume that the xy-Displacements are linear functions of respective perpendicular coordinate (u of y and z, v of x and z) and that the z-displacement is the **torsion function** $\varphi(x,y)$, also dependent on the perpendicular coordinates x and y alone.

$$u(x,y,z) = -\tau\, yz \qquad\qquad (9\text{-}2.1)$$
$$v(x,y,z) = \tau\, xz \qquad\qquad (9\text{-}2.2)$$
$$w(x,y,z) = \tau\, \varphi(x,y) \qquad\qquad (9\text{-}2.3)$$

Where τ is an unknown **angle of twist** (torsional angle) to be determined.

From equations (9-2), we get

$$\varepsilon = \varepsilon_{xx} + \varepsilon_{yy} + \varepsilon_{zz}$$
$$= \frac{\partial u(x,y,z)}{\partial x} + \frac{\partial v(x,y,z)}{\partial y} + \frac{\partial w(x,y,z)}{\partial z} = 0 \qquad\qquad (9\text{-}3.1)$$

Also,

$$\nabla^2 u = -\left(\frac{\partial^2}{\partial x^2} + \frac{\partial^2}{\partial y^2} + \frac{\partial^2}{\partial z^2} \right)\tau yz = 0 \qquad\qquad (9\text{-}3.2)$$

$$\nabla^2 v = \left(\frac{\partial^2}{\partial x^2} + \frac{\partial^2}{\partial y^2} + \frac{\partial^2}{\partial z^2} \right)\tau xz = 0 \qquad\qquad (9\text{-}3.3)$$

$$\nabla^2 w = \left(\frac{\partial^2}{\partial x^2} + \frac{\partial^2}{\partial y^2} + \frac{\partial^2}{\partial z^2} \right)\tau\varphi(x, y)$$
$$= \tau\left(\frac{\partial^2\varphi(x,y)}{\partial x^2} + \frac{\partial^2\varphi(x,y)}{\partial y^2} \right) \qquad\qquad (9\text{-}3.4)$$

Therefore, substituting from equations (9-3) into **Lamé's equations** (4-4)

$$\mu\nabla^2 u + (\mu + \lambda)\frac{\partial\varepsilon}{\partial x} + \rho X_i = \rho\frac{\partial^2 u}{\partial t^2} \qquad \text{satisfied} \qquad (9\text{-}4.1)$$

$$\mu\nabla^2 v + (\mu + \lambda)\frac{\partial\varepsilon}{\partial y} + \rho Y_i = \rho\frac{\partial^2 v}{\partial t^2} \qquad \text{satisfied} \qquad (9\text{-}4.2)$$

$$\mu\nabla^2\mathbf{w} + (\mu + \lambda)\frac{\partial\varepsilon}{\partial z} + \rho Z_i = \rho\frac{\partial^2\mathbf{w}}{\partial t^2} \qquad \text{implies that} \qquad (9\text{-}4.3)$$

i.e.,
$$\frac{\partial^2\varphi(x,y)}{\partial x^2} + \frac{\partial^2\varphi(x,y)}{\partial y^2} = 0 \qquad (9\text{-}5)$$

Stresses

From Cauchy- Hooke's law, equations (4-1), we get

$$\sigma_x = 2\mu\frac{\partial u}{\partial x} + \lambda\varepsilon = 2\mu 0 + \lambda 0 = 0 \qquad (9\text{-}6.1)$$

$$\sigma_y = 2\mu\frac{\partial v}{\partial y} + \lambda\varepsilon = 2\mu 0 + \lambda 0 = 0 \qquad (9\text{-}6.2)$$

$$\sigma_z = 2\mu\frac{\partial w}{\partial z} + \lambda\varepsilon = 2\mu 0 + \lambda 0 = 0 \qquad (9\text{-}6.3)$$

$$\tau_{xy} = \mu\left(\frac{\partial u}{\partial y} + \frac{\partial v}{\partial x}\right) = \mu(0+0) = 0 \qquad (9\text{-}6.4)$$

$$\tau_{zx} = \mu\left(\frac{\partial w}{\partial x} + \frac{\partial u}{\partial z}\right) = \mu\left(\tau\frac{\partial\varphi}{\partial x} - \tau y\right) = \mu\tau\left(\frac{\partial\varphi}{\partial x} - y\right) \qquad (9\text{-}6.5)$$

$$\tau_{zy} = \mu\left(\frac{\partial w}{\partial y} + \frac{\partial v}{\partial z}\right) = \mu\left(\tau\frac{\partial\varphi}{\partial y} + \tau x\right) = \mu\tau\left(\frac{\partial\varphi}{\partial y} + x\right) \qquad (9\text{-}6.6)$$

Therefore, our proposed displacements, equations (9-2), satisfy the equations of elasticity and led to vanishing normal stresses, while allowing shear stresses in the xy-plane.

Surface condition

The Laplacian equation (9-5) represents planar torsion function, which requires the **surface conditions** (1-8) in order to solve $\varphi(x,y)$ in terms of the external stresses

$$X_n = \sigma_x\, l + \tau_{xy}\, m + \tau_{xz}\, n \qquad (9\text{-}7.1)$$
$$Y_n = \tau_{yx}\, l + \sigma_y\, m + \tau_{yz}\, n \qquad (9\text{-}7.2)$$
$$Z_n = \tau_{zx}\, l + \tau_{zy}\, m + \sigma_z\, n \qquad (9\text{-}7.3)$$

On the lateral wall, Figure, $n = \cos(\pi/2) = 0$. Therefore, from equations (9-6), and (9-7), we get

$$X_n = \sigma_x\, l + \tau_{xy}\, m + \tau_{xz}\, n = 0\, l + 0\, m + \tau_{xz}\, 0 = 0 \qquad (9\text{-}7.1)$$
$$Y_n = \tau_{yx}\, l + \sigma_y\, m + \tau_{yz}\, n = 0\, l + 0\, m + \tau_{xz}\, 0 = 0 \qquad (9\text{-}7.2)$$
$$Z_n = \tau_{zx}\, l + \tau_{zy}\, m + \sigma_z\, n = \tau_{zx}\,(x/r) + \tau_{zy}\,(y/r) + \sigma_z\, 0$$

$$= \mu\tau\left(\frac{\partial\varphi}{\partial x} - y\right)l + \mu\tau\left(\frac{\partial\varphi}{\partial y} + x\right)m = 0 \qquad\qquad (9\text{-}7.3)$$

Equation (9-7.3) will be used to determine the **torsion function** $\varphi(x,y)$ from the boundary conditions on the surface, Figure 9-2. Thus,

$$l = \frac{dy}{ds} \qquad\qquad m = -\frac{dx}{ds} \qquad\qquad (9\text{-}7.4)$$

From equations (9-7.3) and (9-7.4), we get

$$\mu\tau\left(\frac{\partial\varphi}{\partial x} - y\right)\frac{dy}{ds} - \mu\tau\left(\frac{\partial\varphi}{\partial y} + x\right)\frac{dx}{ds} = 0 \qquad\qquad (9\text{-}7.5)$$

i.e.,

$$\frac{\partial\varphi}{\partial x}dy - \frac{\partial\varphi}{\partial y}dx = xdx + ydy \qquad\qquad (9\text{-}7.6)$$

Conjugate harmonic function $\psi(x,y)$

Due to the scrambled structure of equation (9-7.6), a function $\psi(x,y)$ **conjugate** to $\varphi(x,y)$ and defined by the following **Cauchy-Riemann condition**

$$\frac{\partial\varphi}{\partial y} = -\frac{\partial\psi}{\partial x} \qquad\qquad \frac{\partial\varphi}{\partial x} = \frac{\partial\psi}{\partial y} \qquad\qquad (9\text{-}7.7)$$

Should render equation (9-7.6) of the following differentiable form

$$\frac{\partial\psi}{\partial y}dy + \frac{\partial\psi}{\partial x}dx = xdx + ydy \qquad\qquad (9\text{-}7.8)$$

Therefore, the alternative differential form of equation (9-7.8) is

$$d\psi = d\left(\frac{x^2 + y^2}{2}\right) \qquad\qquad (9\text{-}7.9)$$

Therefore, integrating we get

$$\psi = \left(\frac{x^2 + y^2}{2}\right) + C$$

$$= \frac{r^2}{2} + C$$

(9-7.10a)

Thus, from equation (9-7.7), we get

$$\frac{\partial \varphi}{\partial y} = -x \qquad \frac{\partial \varphi}{\partial x} = y$$

$$\varphi = -xy + C_1$$

$$\varphi = yx + C_2$$

(9-7.10b)

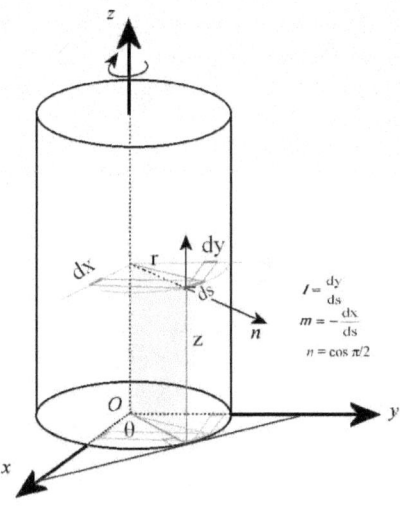

Figure 9-2. Radial and azimuthal planes at the lateral point M.

Torsion shear stress

From equations (9-7.7) and (9-7.10), we obtain the derivatives of $\varphi(x,y)$ needed in equations (9-6.5) and (9-6.6), as follows

$$\tau_{zx} = \mu\tau\left(\frac{\partial \psi}{\partial y} - y\right)$$

(9-8.1)

$$\tau_{zy} = \mu\tau\left(-\frac{\partial\psi}{\partial x} + x\right)$$

(9-8.2)

Comparing our initial assumptions, in equations (9-2.3), we note that both derivatives of ψ must be zeros in order that $u = -y\tau_{zx}$ and $v = x\tau_{zy}$. i.e., $\psi(x,y) = 0$ in the case of plane stresses. Thus, from equation (9-7.10), $x^2+y^2 =$ constant. Thus, the cross sections of the cylinders must remain circular.

Ludwig Prandtl's shear stress function $\Phi(x,y)$

The constant term in equation (9-7.10) defines the constant shear stress at the outer contours $r =$ constant and **L. Prandtl's stress function** as follows

$$\Phi(x,y) = \mu\tau\left(\psi - \frac{x^2 + y^2}{2}\right)$$

(9-9.1)

Thus, the newly introduced shear stress function in (9-9.1), gives upon differentiation the shear stresses in equations (9-8.1) and (9-8.2), as follows

$$\tau_{zx} = \frac{\partial\Phi(x,y)}{\partial y}$$

(9-9.2)

$$\tau_{zy} = -\frac{\partial\Phi(x,y)}{\partial x}$$

(9-9.3)

Resultant shearing forces

$$\tau_x = \iint \tau_{zx}\,dxdy = \iint \frac{\partial\Phi(x,y)}{\partial y}\,dxdy$$

$$= \mu\tau\int_{x_0}^{x_1}\int_{y_0}^{y_1}\left(\frac{\partial\Phi(x,y)}{\partial y}\,dy\right)dx$$

$$= \mu\tau\int_{x_0}^{x_1}[\Phi(y_1) - \Phi(y_0)]dx$$

(9-9.4)

$$= 0$$

Since, from (9-7.10) and (9-9.1), Φ is constant on the lateral wall of the cylinder.

Moment of couple

The couple tangential to wall of the cylinder is the sum of the torques of all stresses on the outer wall (lateral wall) of the cylinder. With all normal stresses vanishing, the only remaining shear stresses give the following sum (integral) couple.

$$\mathbf{M}_t = \iint \left(\tau_{zy} x - \tau_{zx} y \right) dxdy$$

$$= \iint \left(-x \frac{\partial \Phi(x,y)}{\partial x} - y \frac{\partial \Phi(x,y)}{\partial y} \right) dxdy \tag{9-9.5}$$

Equation (9-9.5) can be integrated by parts as follows

$$\mathbf{M}_t = -\int dy \int \left(x \frac{\partial \Phi(x,y)}{\partial x} dx \right) - \int dx \int \left(y \frac{\partial \Phi(x,y)}{\partial y} dy \right)$$

$$= -\int dy \int (x d\Phi) - \int dx \int (y d\Phi)$$

$$= -\int \left[[x\Phi]_{x_0}^{x_1} - \int \Phi dx \right] dy - \int \left[[y\Phi]_{y_0}^{y_1} - \int \Phi dy \right] dx$$

$$= -\int \left([x_1 \Phi(x_1,y) - x_0 \Phi(x_0,y)] - \int \Phi dx \right) dy$$

$$\quad - \int \left([y_1 \Phi(x,y_1) - y_0 \Phi(x,y_0)] - \int \Phi dy \right) dx \tag{9-9.6}$$

Since Φ is constant on the lateral wall of the cylinder, we get

$$\mathbf{M}_t = -\int \left([x_1 \Phi(x_1,y) - x_0 \Phi(x_0,y)] - \int \Phi dx \right) dy$$

$$\quad - \int \left([y_1 \Phi(x,y_1) - y_0 \Phi(x,y_0)] - \int \Phi dy \right) dx$$

$$= -2C[x_1 - x_0](y_1 - y_0) + 2\iint \Phi dydx$$

$$= -2C[x_1 - x_0](y_1 - y_0) + 2\mu\tau \iint \left(\psi - \frac{x^2 + y^2}{2} \right) dydx \tag{9-9.7}$$

The **angle of twist (torsional angle)** τ is thus given by

$$\tau = \frac{\mathbf{M}_t + 2C[x_1 - x_0](y_1 - y_0)}{2\mu \iint \left(\psi - \frac{x^2 + y^2}{2} \right) dydx} \tag{9-9.8a}$$

Our final solution, equation (9-9.8), can be simplified in the form

$$\tau = \frac{\mathbf{M}_t + c}{2\mu K} \tag{9-9.8b}$$

The two constants "a" and "K" are interpreted based on their position in our formulation.

The constant "$c = 2C[x_1 - x_0](y_1 - y_0)$" depends entirely on the radius of the cylinder and therefore could be attributed to the structural material or geometrical composition of the cylinder.

The constant,

$$K = \int \int \left(\psi - \frac{x^2 + y^2}{2} \right) dy dx \qquad (9\text{-}9.8c)$$

μK expresses the deformation of the cylinder or **torsional rigidity**, (μK), as a result of the couple M_t.

Displacement

Setting the arbitrary constant $c = 0$, equations (9-2), (9-7.10), and (9-9.8) give

$$u(x,y) = -\frac{M_t}{2\mu K} yz \qquad (9\text{-}9.9a)$$

$$v(x,y) = -\frac{M_t}{2\mu K} xz \qquad (9\text{-}9.9b)$$

$$w(x,y) = \pm\frac{M_t}{2\mu K} xy \qquad (9\text{-}9.9c)$$

9-2. Prismatical Elliptic Cylindrical Bar

Complex stress and torsion functions

Saint-Venant's method derives its rationale from complex function conjugation, similar to equation (9-7.7) and enables us to drive the **stress function** $\varphi(x,y)$ and its conjugate function $\psi(x,y)$ which could describe deformation of non-circular prismatical bars.

In complex variables, a planar stress problem could be represented by linear variation of the squared variable $z(x,iy)$ as follows

$$-\psi(x,y) + i\varphi(x,y) = \lambda z^2(x,y)$$
$$= \lambda(x + iy)^2 \qquad (9\text{-}10.1)$$
$$= \lambda(x^2 - y^2) + 2ixy$$

278

Thus, the **stress function** is defined as

$$-\psi(x,y) = \lambda(x^2 - y^2)$$

(9-10.2)

And the **torsion function** as

$$\varphi(x,y) = 2\lambda xy$$

(9-10.3)

Where the z-displacement is defined as

$$w(x,y) = \tau\varphi(x,y)$$

(9-10.4)

Where, τ is the torsion **moment couple**.

Thus, equating the expression for ψ from equation (9-10.2) with that from equation (9-7.10), we get the equation of the contour of the cross section of the prismatic rod

$$-\lambda(x^2 - y^2) = \left(\frac{x^2 + y^2}{2}\right) + C$$

i.e.,

$$-\left(\frac{2\lambda + 1}{2C}\right)x^2 + \left(\frac{2\lambda - 1}{2C}\right)y^2 = 1$$

(9-10.5)

Or

$$\frac{x^2}{a^2} + \frac{y^2}{b^2} = 1$$

(9-10.6)

Where,

i.e.,

$$a^2 = \frac{-2C}{2\lambda + 1}$$

(9-10.7a)

i.e.,

$$b^2 = \frac{2C}{2\lambda - 1}$$

(9-10.7a)

Eliminating C and λ between (9-10.5a and 5b) we get

$$\lambda = \frac{b^2 - a^2}{2(a^2 + b^2)}$$

(9-10.7a)

$$C = \frac{-a^2 b^2}{(a^2 + b^2)}$$

(9-10.7a)

Displacement

279

From equations (9-2), (9-10.4) and (9-10.7a), we get

$$u(x,y,z) = -\tau\, yz \tag{9-11.1}$$
$$v(x,y,z) = \tau\, xz \tag{9-11.2}$$
$$w(x,y) = \left(\frac{b^2 - a^2}{a^2 + b^2}\right)\tau xy \tag{9-11.3}$$

Thus, the use of complex function $z^2(x,iy)$, equation (9-10.1), with the assumption that real part $\psi(x,y)$ comprises the **stress function**, equation (9-10.2), and the imaginary part $\varphi(x,y)$ comprises the **torsion function** of z, equation(9-10.3), we were able to modify the equations of a circular cylinder into an elliptic cylinder and thus reach the expression for z-displacement (9-11.3). It remains to find the formula of the moment of couple τ.

Shear stresses

From equations (9-8.1) and (9-8.2), we get

$$\begin{aligned}
\tau_{zx} &= \mu\tau\left(\frac{\partial\psi}{\partial y} - y\right) = \mu\tau\left[-\left(\frac{b^2 - a^2}{2(a^2 + b^2)}\right)\frac{\partial}{\partial y}(x^2 - y^2) - y\right] \\
&= \mu\tau\left[\left(\frac{b^2 - a^2}{a^2 + b^2}\right)y - y\right] \\
&= -\mu\left(\frac{2a^2}{a^2 + b^2}\right)\tau y
\end{aligned} \tag{9-12.1}$$

And

$$\begin{aligned}
\tau_{zy} &= \mu\tau\left(-\frac{\partial\psi}{\partial x} + x\right) = \mu\tau\left(\left(\frac{b^2 - a^2}{2(a^2 + b^2)}\right)\frac{\partial}{\partial x}(x^2 - y^2) + x\right) \\
&= \mu\tau\left(\left(\frac{b^2 - a^2}{(a^2 + b^2)}\right)x + x\right) \\
&= \mu\left(\frac{2b^2}{(a^2 + b^2)}\right)\tau x
\end{aligned} \tag{9-12.2}$$

Moment of couple

As in the case of **circular cylinder**, the couple tangential to wall of the elliptic cylinder is the sum of the torques of all stresses on the outer wall (lateral wall) of the cylinder. With all normal stresses vanishing, the only remaining shear stresses give the following sum (integral) couple.

$$\mathbf{M}_t = -\iint_{cs} \left(\tau_{zy} x - \tau_{zx} y \right) dxdy$$

$$= -\iint_{cs} \left[\mu \left(\frac{2b^2}{a^2+b^2} \right) \tau x^2 + \mu \left(\frac{2a^2}{a^2+b^2} \right) \tau y^2 \right] dxdy \qquad (9\text{-}13.1)$$

$$= -\mu \left(\frac{2\tau}{a^2+b^2} \right) \iint_{cs} \left(b^2 x^2 + a^2 y^2 \right) dxdy$$

$$= -\mu \left(\frac{2\tau b^2 a^2}{a^2+b^2} \right) \iint_{c.s} dxdy$$

Where integration is carried over the cross-section (c.s) of the elliptic bar.

First, integrate with respect to y, we get

$$\mathbf{M}_t = -\mu \left(2\tau \frac{b^2 a^2}{a^2+b^2} \right) \int_{-a}^{a} [y]_{-y}^{+y} dx$$

$$= -\mu \left(4\tau \frac{b^2 a^2}{a^2+b^2} \right) \int_{-a}^{a} y dx \qquad (9\text{-}13.2)$$

Then, substitute by the following parametric equations of ellipse

$$x = a \cos \theta \qquad (9\text{-}13.3)$$
$$y = b \sin \theta \qquad (9\text{-}13.4)$$

Thus, the integral in (9-13.2) is carried out over range 0 to π, with (dx = -a sinθdθ) as follows:

$$\mathbf{M}_t = -\mu \left(4\tau \frac{b^2 a^2}{a^2+b^2} \right) \int_{0}^{\pi} \left(-ba \sin^2 \theta \right) d\theta$$

$$= \mu \left(4\tau \frac{b^3 a^3}{a^2+b^2} \right) \int_{0}^{\pi} \left(\frac{1-\cos 2\theta}{2} \right) d\theta$$

$$= \mu \left(2\tau \frac{b^3 a^3}{a^2+b^2} \right) \left[\theta - \frac{\sin 2\theta}{2} \right]_{0}^{\pi} \qquad (9\text{-}13.5)$$

$$= \mu \left(2\pi \frac{b^3 a^3}{a^2+b^2} \right) \tau$$

Torsional angle or angle of twist

$$\tau = \frac{M_t}{\mu\left(2\pi\dfrac{b^3a^3}{a^2+b^2}\right)}$$

(9-13.6a)

Equation (9-13.6) can be written in terms of the **polar moment of the inertia** of the elliptic cross-section as follows

$$\tau = \frac{M_t}{\mu J_p}$$

(9-13.6b)

Where

$$J_p = 2\pi\frac{b^3a^3}{a^2+b^2}$$

(9-13.6c)

Deformed cross-section contour

The z-dependent displacement, equation (9-11.3), is now written in terms of the angle of twist, equation (9-13.6a), as follows

$$w(x,y) = \left(\frac{b^2-a^2}{a^2+b^2}\right)\frac{M_t}{\mu\left(2\pi\dfrac{b^3a^3}{a^2+b^2}\right)}xy$$

$$= \frac{M_t}{\mu\left(2\pi\dfrac{b^3a^3}{b^2-a^2}\right)}xy$$

(9-14)

Comparing equations (9-9.9c), for z-displacement for circular, and equation (9-14), for elliptic bar, notice the similar structure with the exception that the **torsional rigidity,**

$$\mu K = 2\pi\mu\frac{b^3a^3}{b^2+a^2}$$

(9-15)

is defined in terms of the major and minor radii of the ellipse.

We could now determine the **torsional rigidity** μK for **circular cylinder** in equation (9-9.8c), by making a = b = r, in equation (9-13.6a) that

$$K = \iint \left(\psi - \frac{x^2 + y^2}{2} \right) dy\,dx$$

$$= 2\pi \frac{b^3 a^3}{a^2 + b^2} \bigg|_{a=b=r} = 2\pi \frac{r^3 r^3}{r^2 + r^2} = \pi r^4 \qquad (9\text{-}16)$$

Resultant shear stress

$$\tau = \sqrt{\tau_{zy}^2 + \tau_{zx}^2}$$

$$= \mu\tau \sqrt{\left(\frac{2b^2}{(a^2 + b^2)} \right)^2 x^2 + \left(\frac{2a^2}{(a^2 + b^2)} \right)^2 y^2} \qquad (9\text{-}17.1)$$

$$= \mu\tau \frac{2b^2 a^2}{(a^2 + b^2)} \sqrt{\frac{x^2}{a^4} + \frac{y^2}{b^4}}$$

From equation (9-17.1), we conclude that the total torsional shear is constant on the contour

$$\frac{x^2}{a^4} + \frac{y^2}{b^4} = C \qquad (9\text{-}17.2)$$

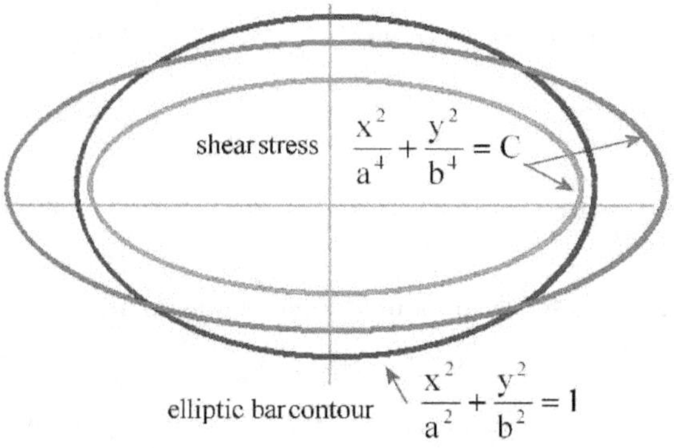

Figure 9-3. Distribution of net shear stress on the cross-section of en elliptic cylinder exposed to terminal couple.

Deformed coordinates

$$x' = x + u(x,y) = x - \left(\frac{M_t}{\mu J_p}\right)yz \qquad (9\text{-}18.1)$$

$$y' = y + v(x,y) = y + \left(\frac{M_t}{\mu J_p}\right)xz \qquad (9\text{-}18.2)$$

$$z' = z + w(x,y) = z + \left(\frac{b^2 - a^2}{a^2 + b^2}\right)\left(\frac{M_t}{\mu J_p}\right)xy \qquad (9\text{-}18.3)$$

At constant z, the cross sectional contour is obtained by substituting from (9-18) into (9-10.6) as follows

$$\frac{\left(x - \left(\dfrac{M_t}{\mu J_p}\right)yz\right)^2}{a^2} + \frac{\left(y + \left(\dfrac{M_t}{\mu J_p}\right)xz\right)^2}{b^2} = 1$$

$$\frac{x^2}{a^2} - \frac{2\left(\dfrac{M_t}{\mu J_p}\right)xyz}{a^2} + \frac{\left(\dfrac{M_t}{\mu J_p}\right)^2 y^2 z^2}{a^2} + \frac{y^2}{b^2} + \frac{2\left(\dfrac{M_t}{\mu J_p}\right)xyz}{b^2} + \frac{\left(\dfrac{M_t}{\mu J_p}\right)^2 x^2 z^2}{b^2} = 1 \qquad (9\text{-}19.1)$$

$$\frac{x^2}{a^2} + \frac{y^2}{b^2} + 2\left(\frac{1}{b^2} - \frac{1}{a^2}\right)\left(\frac{M_t}{\mu J_p}\right)xyz + \left(\frac{M_t}{\mu J_p}\right)^2 z^2\left(\frac{y^2}{a^2} + \frac{x^2}{b^2}\right) = 1$$

Putting $z = c$ and using equation (9-10.6) again, we can reduce (9-19.1) farther by replacing $\frac{x^2}{a^2} + \frac{y^2}{b^2}$ by a constant, thus

$$2\left(\frac{1}{b^2} - \frac{1}{a^2}\right)\left(\frac{M_t}{\mu J_p}\right)xyc + \left(\frac{M_t}{\mu J_p}\right)^2 c^2\left(\frac{y^2}{a^2} + \frac{x^2}{b^2}\right) = C \qquad (9\text{-}19.2)$$

Therefore, the equation of the **deformed cross-sectional contour of the elliptic cylindrical bar** takes the final form:

$$\frac{xy}{\dfrac{a^2 b^2}{2(a^2 - b^2)}\left(\dfrac{cM_t}{\mu J_p}\right)} + \frac{y^2}{\left(\dfrac{cM_t}{\mu J_p}\right)a^2} + \frac{x^2}{\left(\dfrac{cM_t}{\mu J_p}\right)b^2} = C' \qquad (9\text{-}20.1)$$

Equation (9-20.1) can also be obtained from equation (9-18.3) by putting $z = 0$ and noticing that the two terms that include x^2 and y^2 comprise an ellipse constant that could be set within C' as follows.

284

$$xy = z_n \left(\frac{a^2 + b^2}{b^2 - a^2} \right) \left(\frac{\mu J_p}{M_t} \right)$$ (9-20.2)

Equation (9-20.2) represents the hyperbolas of the cross-sectional counters after being deformed by the twisting couple M_t.

We also conclude that, by putting $a = b = r$, in equation (9-20.1), we could prove that the cross sections of a circular cylinder remains circular, as follows, Figure 9-4:

$$\frac{2(a^2 - b^2)xy}{a^2 b^2 \left(\frac{cM_t}{\mu J_p} \right)} + \frac{y^2}{\left(\frac{cM_t}{\mu J_p} \right) a^2} + \frac{x^2}{\left(\frac{cM_t}{\mu J_p} \right) b^2} = C'$$

$$\frac{0xy}{r^4 \left(\frac{cM_t}{\mu J_p} \right)} + \frac{y^2}{\left(\frac{cM_t}{\mu J_p} \right) r^2} + \frac{x^2}{\left(\frac{cM_t}{\mu J_p} \right) r^2} = C$$ (9-20.3)

Which is simplified to the equation of circles, as follows

$$y^2 + x^2 = C' \left(\frac{cM_t}{\mu J_p} \right) r^2$$ (9-20.4)

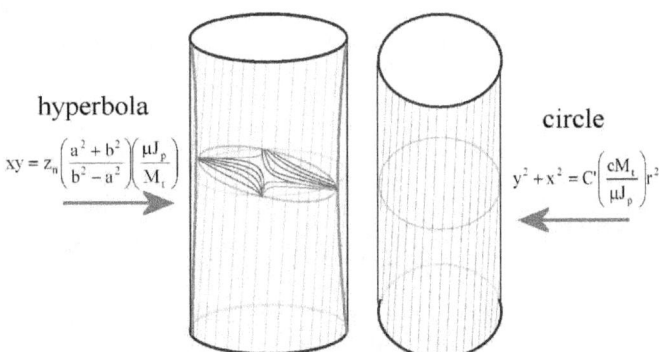

hyperbola

$$xy = z_n \left(\frac{a^2 + b^2}{b^2 - a^2} \right) \left(\frac{\mu J_p}{M_t} \right)$$

circle

$$y^2 + x^2 = C' \left(\frac{cM_t}{\mu J_p} \right) r^2$$

elliptic cylidner circular cylinder

Figure 9-4. Deformed cross-sections in elliptic and circular cylinders.

9-3. Triangular Prismatical Bar

Complex function representation of triangular geometry

As we did in equation (9-10.1), we will start our analysis of a prismatical bar with triangular cross-section by the complex function

$$-\psi(x,y) + i\varphi(x,y) = \lambda z^3(x,y)$$
$$= \lambda(x + iy)^3 \tag{9-21.1}$$
$$= \lambda(x^3 - 3xy^2) + \lambda i(3yx^2 - y^3)$$

Thus, the **stress function** is defined as
$$-\psi(x,y) = \lambda(x^3 - 3xy^2) \tag{9-21.2}$$

And the **torsion function** as

$$\varphi(x,y) = \lambda(3yx^2 - y^3) \tag{9-21.3}$$

Thus, equating the expression for ψ from equation (9-21.2) with that from equation (9-7.10), we get the equation of the contour of the cross section of the prismatic rod

$$-\lambda(x^3 - 3xy^2) = \left(\frac{x^2 + y^2}{2}\right) + C \tag{9-22}$$

i.e., $\qquad 2\lambda(x^3 - 3xy^2) + (x^2 + y^2) = -2C$

Which is a set of complete **hyperbola**, with three **asymptotes**, due to the terms of third power.

Let us determine the boundaries of our triangular cross-section bar on the asymptotes of equation (9-22). We note that the three asymptotes are obtained by equating the term of third power by zero as follows:

$$(x^3 - 3xy^2) = 0$$

Or, $\qquad x(x^2 - 3y^2) = 0 \tag{9-23}$

i.e., $\qquad x(x - \sqrt{3}y)(x + \sqrt{3}y) = 0$

Thus, the three asymptotes of equation (9-22) are

$$x = 0$$
$$x - \sqrt{3}y = 0 \qquad\qquad (9\text{-}24)$$
$$x + \sqrt{3}y = 0$$

We can define three lines parallel to the three asymptotes by adding constants to interceptions of the above three lines with the y- or x- axes, as follows:

$$x = a$$
$$x - \sqrt{3}y = b \qquad\qquad (9\text{-}25)$$
$$x + \sqrt{3}y = b$$

We could then reconstruct equation (9-22) from the three lines of equation (9-25) by the proper choice of the constants λ and C as follows.

First, from equation (9-25), we rewrite a third order hyperbolic equations entailing its three lines

$$\lambda(x - a)(x - \sqrt{3}y - b)(x + \sqrt{3}y - b) = 0 \qquad\qquad (9\text{-}26.1)$$

Or

$$\lambda(x - a)((x - b)^2 - 3y^2) = 0$$
$$\lambda(x^3 - 2x^2 b + xb^2 - 3xy^2 - ax^2 + 2axb - ab^2 + 3ay^2) = 0$$

i.e.,

$$\lambda[x^3 - (2b + a)x^2 + xb(b + 2a) - 3xy^2 - ab^2 + 3ay^2] = 0 \qquad\qquad (9\text{-}26.2)$$

Now, having devised equation (9-26.2) from the three parallel lines to the asymptotes of equation (9-22), we can determine the coefficients of terms of equal powers as follows

$$-\lambda(2b + a) = \frac{1}{2}$$
$$b + 2a = 0$$
$$a\lambda b^2 = -C \qquad\qquad (9\text{-}26.3)$$
$$\lambda 3a == \frac{1}{2}$$

i.e.,

$$\lambda = \frac{1}{6a}$$
$$b = -2a \qquad\qquad (9\text{-}26.4)$$
$$C = -\frac{2}{3}a^2$$

Thus, the three parallel lines, in equation (9-25), the tangents of our curve, are defined in terms of "a" alone, from (9-26.4), by putting $b = -2a$, as follows

$$x = a$$
$$x - \sqrt{3}y + 2a = 0 \qquad\qquad (9\text{-}27)$$
$$x + \sqrt{3}y + 2a = 0$$

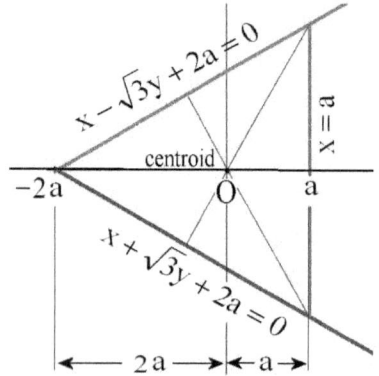

Figure 9-5. Equations of lines enclosing the area of triangle, the origin of the xOy is its centroid.

Displacement

From equations (9-10.4) and (9-21.3), the z-displacement is given by

$$w(x, y) = \tau\varphi(x, y)$$
$$= \tau\lambda\left(3x^2y - y^3\right) \qquad\qquad (9\text{-}28.1)$$

Substituting by λ from equation (9-26.4) we get

$$w(x, y) = \frac{\tau}{6a}\left(3x^2y - y^3\right) \qquad\qquad (9\text{-}28.2)$$

Thus, after deformation, the z-dependent coordinate of the bar becomes

288

$$z' = z + w(x,y)$$

$$= z + \frac{\tau}{6a}\left(3x^2y - y^3\right) \tag{9-29}$$

$$= z + \frac{\tau}{6a}y\left(\sqrt{3}x + y\right)\left(\sqrt{3}x - y\right)$$

Thus, at $z = $ constant, the cross-section of the bar retains its planar form with the three lines y, $\left(\sqrt{3}x + y\right)$, $\left(\sqrt{3}x - y\right)$ forming the sides of the triangle.

Shear stresses

From equations (9-21.2) and (9-12.1), we get

$$\tau_{zx} = \mu\tau\left(\frac{\partial\psi}{\partial y} - y\right) = \mu\tau\left[-\lambda\frac{\partial}{\partial y}\left(x^3 - 3xy^2\right) - y\right]$$

$$= \frac{\mu\tau y}{a}(x - a) \tag{9-30.1}$$

And

$$\tau_{zy} = \mu\tau\left(-\frac{\partial\psi}{\partial x} + x\right) = \mu\tau\left(\lambda\frac{\partial}{\partial x}\left(x^3 - 3xy^2\right) + x\right)$$

$$= \mu\tau\left[\lambda\left(3x^2 - 3y^2\right) + x\right] \tag{9-30.2}$$

$$= \frac{\mu\tau}{2a}\left[\left(x^2 - y^2\right) + 2ax\right]$$

Moment of couple

As in the case of **circular cylinder**, the couple tangential to wall of the triangular cross-section is the sum of the torques of all stresses on the outer wall (lateral wall) of the triangle. With all normal stresses vanishing, the only remaining shear stresses give the following sum (integral) couple.

$$M_t = -\iint_{cs}\left(\tau_{zy}x - \tau_{zx}y\right)dxdy$$

$$= -\iint_{cs}\left(\frac{\mu\tau}{2a}\left(x^2 - y^2 + 2ax\right)x - \frac{\mu\tau y^2}{a}(x - a)\right)dxdy \tag{9-31.1}$$

$$= -\frac{\mu\tau}{2a}\iint_{cs}\left(x^3 - 3xy^2 + 2ax^2 + 2ay^2\right)dxdy$$

First integrating with respect to y and setting the upper and lower limits according to equation (9-27), such that

$$y_{+ve} = \frac{x+2a}{\sqrt{3}}$$

$$y_{1ve} = -\frac{x+2a}{\sqrt{3}}$$

(9-31.2)

Therefore,

$$M_t = -\frac{\mu\tau}{2a}\int \left(x^3 y - xy^3 + 2ax^2 y + \frac{2}{3}ay^3\right)_{-ve}^{+ve} dx$$

$$= -\frac{\mu\tau}{a}\int \left[x^3\left(\frac{x+2a}{\sqrt{3}}\right) - x\left(\frac{x+2a}{\sqrt{3}}\right)^3 + 2ax^2\left(\frac{x+2a}{\sqrt{3}}\right) + \frac{2}{3}a\left(\frac{x+2a}{\sqrt{3}}\right)^3\right]dx$$

$$= -\frac{\mu\tau}{a\sqrt{3}}\int \left[x^3(x+2a) - \frac{1}{3}x(x+2a)^3 + 2ax^2(x+2a) + \frac{2a}{9}(x+2a)^3\right]dx$$

$$= -\frac{\mu\tau}{a\sqrt{3}}\int (x+2a)\left[x^3 - \frac{1}{3}x(x+2a)^2 + 2ax^2 + \frac{2a}{9}(x+2a)^2\right]dx$$

$$= -\frac{\mu\tau}{9a\sqrt{3}}\int (x+2a)\left(9x^3 - 3x(x^2+4ax+4a^2) + 18ax^2 + (2ax^2+8a^2x+8a^3)\right)dx$$

$$= -\frac{\mu\tau}{9a\sqrt{3}}\int (x+2a)\left[6x^3 + 8ax^2 - 4a^2x + 8a^3\right]dx$$

$$= -\frac{\mu\tau}{9a\sqrt{3}}\int \left[\begin{array}{c}6x^4 + 8ax^3 - 4a^2x^2 + 8xa^3 \\ +12ax^3 + 16a^2x^2 - 8a^3x + 16a^4\end{array}\right]dx$$

$$= -\frac{\mu\tau}{9a\sqrt{3}}\int \left[6x^4 + 20ax^3 + 12a^2x^2 + 16a^4\right]dx$$

(9-31.3)

Performing the second integration with the limits shown in Figure 9-5, i.e., x = -2a to x = a, we get:

$$M_t = -\frac{\mu\tau}{9a\sqrt{3}}\left[\frac{6}{5}x^5 + 5ax^4 + 4a^2x^3 + 16a^4x\right]_{-2a}^{+a}$$

$$= \frac{\mu\tau}{9a\sqrt{3}}\left[\begin{array}{c}\frac{6}{5}a^5 + 5a^5 + 4a^5 + 16a^5 \\ +\frac{6.32}{5}a^5 - 5.16a^5 + 4.8a^5 + 16.2a^5\end{array}\right]$$

290

$$= \frac{9\sqrt{3}\mu\tau a^4}{5} \qquad (9\text{-}31.4)$$

Angle of twist

$$\tau = \frac{M_t}{\mu\dfrac{9\sqrt{3}a^4}{5}} \qquad (9\text{-}31.5)$$

The area of the triangle, in Figure 9-5, is given by $3\sqrt{3}a^2$. The **torsional rigidity** is thus given by

$$\mu K = \mu\frac{9\sqrt{3}a^4}{5} \qquad (9\text{-}31.6)$$

9-4. Prismatical bar with rectangular cross-section

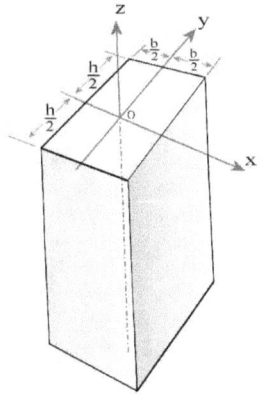

Figure 9-6. Coordinates of prismatical bar with rectangular cross-section.

(i) Separation of coordinate functions

The approach of choosing algebraic polynomial for the stress function $\varphi(x,y)$ starts from the Airy-Lévy's equation

291

$$\frac{\partial^4 \varphi(x,y)}{\partial x^4} + 2\frac{\partial^4 \varphi(x,y)}{\partial y^2 \partial x^2} + \frac{\partial^4 \varphi(x,y)}{\partial y^4} = 0 \tag{9-32.1}$$

And the solution

$$\varphi(x,y) = X(x)Y(y) \tag{9-32.2}$$

Therefore,

$$Y(y)\frac{d^4 X(x)}{dx^4} + 2\frac{d^2 X(x)}{dx^2} \cdot \frac{d^2 Y(y)}{dy^2} + X(x)\frac{d^4 Y(y)}{dy^4} = 0 \tag{9-32.3}$$

(ii) Choice of periodic polynomial functions

As we did before, we choose the periodic functions

$$\frac{d^4 X(x)}{dx^4} = k^4 X(x) \tag{9-32.4}$$

$$\frac{d^2 X(x)}{dx^2} = -k^2 X(x) \tag{9-32.5}$$

$$\frac{d^4 X(x)}{dx^4} = -k^2 \frac{d^2 X(x)}{dx^2} \tag{9-32.6}$$

And

$$\left(Y(y)k^4 - 2k^2 \cdot \frac{d^2 Y(y)}{dy^2} + \frac{d^4 Y(y)}{dy^4}\right) X(x) = 0 \tag{9-32.7}$$

And the two separate differential equations for X(x) and Y(y) as follows:

$$\frac{d^2 X(x)}{dx^2} + k^2 X(x) = 0 \tag{9-32.8}$$

$$Y(y)k^4 - 2k^2 \cdot \frac{d^2 Y(y)}{dy^2} + \frac{d^4 Y(y)}{dy^4} = 0 \tag{9-32.9}$$

Since k and C are arbitrary constants, we could express X(x) and Y(y) in terms of sum of sines, cosines, and hyperbolic functions, as follows:

$$X(x) = C_1 \sin kx + C_2 \cos kx \tag{9-32.10}$$
$$Y(y) = C_3 \cosh ky + C_4 \sinh ky \tag{9-32.11}$$

We have discussed the various alternatives of changing the cyclical variable k in such a way that allow us to write the desired solution of Airy-Lévy's equation as follows:

$$\varphi(x, y) = xy + \sum_{n=1}^{\infty} C_{1n} \sin \frac{n\pi x}{b} \left(C_{2n} \cosh \frac{n\pi y}{b} + C_{3n} \sinh \frac{n\pi y}{b} \right) \tag{9-32.12}$$

Where the xy term corresponds to the n = 0, as we have shown before.

(iii) Determination of coefficients of Fourier's series

The boundary condition at $x = \pm b/2$ and $y = \pm h/2$ are satisfied by using equations (9-8), by vanishing shearing stresses on the lateral walls of the rectangle, as follows.

At $x = \pm b/2$,

$$\tau_{zx} = \mu t \left(\frac{\partial \psi}{\partial y} - y \right) = 0$$

i.e., $\quad \dfrac{\partial \varphi}{\partial x} = \dfrac{\partial \psi}{\partial y} = y$ \hfill (9-33.1)

Differentiating equations (9-32.12) with respect to x, putting $x = \pm b/2$, and equating with (9-33.1), we get

$$\frac{\partial \varphi}{\partial x} = y + \sum_{n=1}^{\infty} \frac{n\pi}{b} C_{1n} \cos \frac{n\pi b}{2b} \left(C_{2n} \cosh \frac{n\pi y}{b} + C_{3n} \sinh \frac{n\pi y}{b} \right) = y$$

i.e., $\quad \displaystyle\sum_{n=1}^{\infty} \frac{n\pi}{b} C_{1n} \cos \frac{n\pi}{2} \left(C_{2n} \cosh \frac{n\pi y}{b} + C_{3n} \sinh \frac{n\pi y}{b} \right) = 0$ \hfill (9-33.2)

In order for $\cos(n\pi/2) = 0$, then

$$n = 1, 3, 5, . \tag{9-33.3}$$

At $y = \pm h/2$,

$$\tau_{zy} = \mu t \left(-\frac{\partial \psi}{\partial x} + x \right) = 0$$

$$\frac{\partial \varphi}{\partial y} = -\frac{\partial \psi}{\partial x} = -x \tag{9-33.4}$$

Differentiating equations (9-32.12) with respect to y, putting $x = \pm h/2$, and equating with (9-33.4), we get

293

$$\frac{\partial \varphi}{\partial y} = x + \sum_{n=1,3,5..}^{\infty} \frac{n\pi}{b} C_{1n} \sin \frac{n\pi x}{b} \left(C_{2n} \sinh \frac{n\pi h}{2b} + C_{3n} \cosh \frac{n\pi h}{2b} \right) = -x$$

i.e.,

$$\sum_{n=1,3,5..}^{\infty} \frac{n\pi}{b} C_{1n} \sin \frac{n\pi x}{b} \left(C_{2n} \sinh \frac{n\pi h}{2b} + C_{3n} \cosh \frac{n\pi h}{2b} \right) = -2x$$

(9-33.5)

Noting that

$$\sinh(n\pi h/2b) = - \sinh(-n\pi h/2b)$$
$$\cosh(n\pi h/2b) = \cosh(-n\pi h/2b)$$

We should eliminate the sinh term from equation (9-33.5) by setting $C_{2n} = 0$. We can also fuse C_{3n} with C_{1n}, since both are arbitrary constants.

$$\sum_{n=1,3,5..}^{\infty} \left(\frac{n\pi}{b} C_{1n} C_{3n} \cosh \frac{n\pi h}{2b} \right) \sin \frac{n\pi x}{b} = -2x$$

(9-33.5)

As usual, we will use **Euler's formula** in determining the constants C_{1n} through the following general formula

$$C_n = \frac{1}{l} \int_{-l}^{+l} f(x) \sin \frac{n\pi x}{2l} dx$$

(9-33.6)

Comparing the terms between equations (9-33.5) and (9-33.6), we have $f(x) = -2x$, $b = 2l$, and

$$C_n = \frac{n\pi}{b} C_{1n} C_{3n} \cosh \frac{n\pi h}{2b}$$

(9-33.7)

Therefore,

$$\frac{n\pi}{b} C_{1n} C_{3n} \cosh \frac{n\pi h}{2b} = \frac{2}{b} \int_{-\frac{b}{2}}^{+\frac{b}{2}} (-2x) \sin \frac{n\pi x}{b} dx$$

$$= \frac{4}{n\pi} \int_{-\frac{b}{2}}^{+\frac{b}{2}} x d\cos \frac{n\pi x}{b}$$

$$= \frac{4}{n\pi} \left(\left[x \cos \frac{n\pi x}{b} \right]_{-\frac{b}{2}}^{+\frac{b}{2}} - \frac{b}{n\pi} \left[\sin \frac{n\pi x}{b} \right]_{-\frac{b}{2}}^{+\frac{b}{2}} \right)$$

$$= \frac{4}{n\pi} \left(\left[x \cos \frac{n\pi x}{b} - \frac{b}{n\pi} \sin \frac{n\pi x}{b} \right]_{-\frac{b}{2}}^{+\frac{b}{2}} \right)$$

$$= -\frac{8b}{(n\pi)^2}(-1)^{\frac{n-1}{2}}$$

$$= \frac{8b}{(n\pi)^2}(-1)^{\frac{n+1}{2}}$$

(9-33.8)

Arranging, we get

$$\frac{n\pi}{b}C_{1n}C_{3n}\cosh\frac{n\pi h}{2b} = \frac{8b}{(n\pi)^2}(-1)^{\frac{n+1}{2}}$$

$$C_{1n}C_{3n} = \frac{8b^2(-1)^{\frac{n+1}{2}}}{(n\pi)^3\cosh\dfrac{n\pi h}{2b}}$$

(9-33.9)

(iv) Torsion function $\varphi(x,y)$

From equations (9-32.12), (9-33.3), and (9-33.9), we finally get the expression for the stress function

$$\varphi(x,y) = xy + \left(\frac{8b^2}{\pi^3}\right)\sum_{n=1,3,5...}^{\infty}\left(\frac{(-1)^{\frac{n+1}{2}}}{n^3\cosh\dfrac{n\pi h}{2b}}\right)\left(\sin\frac{n\pi x}{b}\sinh\frac{n\pi y}{b}\right)$$

(9-34)

(v) Stress function $\psi(x,y)$

Cauchy-Riemann condition, equation (9-7.7), allow us to conjugate equation (9-34) in order to obtain the stress function as follows

$$\frac{\partial\psi}{\partial x} = -\frac{\partial}{\partial y}\left(xy + \left(\frac{8b^2}{\pi^3}\right)\sum_{n=1,3,5...}^{\infty}\left(\frac{(-1)^{\frac{n+1}{2}}}{n^3\cosh\dfrac{n\pi h}{2b}}\right)\left(\sin\frac{n\pi x}{b}\sinh\frac{n\pi y}{b}\right)\right)$$

$$= -x - \left(\frac{8b}{\pi^2}\right)\sum_{n=1,3,5...}^{\infty}\left(\frac{(-1)^{\frac{n+1}{2}}}{n^2\cosh\dfrac{n\pi h}{2b}}\right)\left(\sin\frac{n\pi x}{b}\cosh\frac{n\pi y}{b}\right)$$

(9-35.1)

And

$$\frac{\partial \psi}{\partial y} = \frac{\partial}{\partial x}\left[xy + \left(\frac{8b^2}{\pi^3}\right)\sum_{n=1,3,5,\ldots}^{\infty}\left(\frac{(-1)^{\frac{n+1}{2}}}{n^3 \cosh\frac{n\pi h}{2b}}\right)\left(\sin\frac{n\pi x}{b}\sinh\frac{n\pi y}{b}\right)\right]$$

$$= y + \left(\frac{8b}{\pi^2}\right)\sum_{n=1,3,5,\ldots}^{\infty}\left(\frac{(-1)^{\frac{n+1}{2}}}{n^2 \cosh\frac{n\pi h}{2b}}\right)\left(\cos\frac{n\pi x}{b}\sinh\frac{n\pi y}{b}\right) \qquad (9\text{-}35.2)$$

Integrating equation (9-35.1) with respect to x, and equation (9-35.2) with respect to y, we get

$$\psi = -\frac{x^2}{2} + \left(\frac{8b^2}{\pi^3}\right)\sum_{n=1,3,5,\ldots}^{\infty}\left(\frac{(-1)^{\frac{n+1}{2}}}{n^3 \cosh\frac{n\pi h}{2b}}\right)\left(\cos\frac{n\pi x}{b}\cosh\frac{n\pi y}{b}\right) + f(y) \qquad (9\text{-}35.3)$$

And

$$\psi = \frac{y^2}{2} + \left(\frac{8b^2}{\pi^3}\right)\sum_{n=1,3,5,\ldots}^{\infty}\left(\frac{(-1)^{\frac{n+1}{2}}}{n^3 \cosh\frac{n\pi h}{2b}}\right)\left(\cos\frac{n\pi x}{b}\cosh\frac{n\pi y}{b}\right) + g(x) \qquad (9\text{-}35.4)$$

From (9-35.3) and (9-35.4) share the same long term in cos and cosh, are equal, but vary in x^2 and y^2. Therefore, we conclude that

$$g(x) = -\frac{x^2}{2} + C \qquad (9\text{-}35.5)$$

And

$$f(y) = \frac{y^2}{2} + C \qquad (9\text{-}35.6)$$

Where C is a constant determined by boundary conditions on the contour of the rectangle. Thus, either of (9-35.3) or (9-35.4) give

$$\psi = \frac{y^2 - x^2}{2} + \left(\frac{8b^2}{\pi^3}\right)\sum_{n=1,3,5,\ldots}^{\infty}\left(\frac{(-1)^{\frac{n+1}{2}}}{n^3 \cosh\frac{n\pi h}{2b}}\right)\left(\cos\frac{n\pi x}{b}\cosh\frac{n\pi y}{b}\right) + C) \qquad (9\text{-}35.7)$$

Stresses

From equations (9-9.1) and (9-35.7), the **Ludwig Prandtl's shear stress function** becomes

296

$$\Phi(x,y) = \mu\tau\left(\left(\left(\frac{y^2-x^2}{2}+\frac{8b^2}{\pi^3}\sum_{\substack{n=1,3\\.5...}}^{\infty}\left(\frac{(-1)^{\frac{n+1}{2}}}{n^3\cosh\frac{n\pi h}{2b}}\right)\cos\frac{n\pi x}{b}\cosh\frac{n\pi y}{b}+C\right)-\frac{x^2+y^2}{2}\right)\right) \qquad (9\text{-}36.1)$$

Choosing $C = (b/2)^2$ cancels the $(x/2)^2$ on the contour of the rectangle. Thus, Ludwig Prandtl's shear stress function becomes

$$\Phi(x,y) = \mu\tau\left(\frac{b^2}{4}-x^2+\frac{8b^2}{\pi^3}\sum_{\substack{n=1,3\\.5...}}^{\infty}\left(\frac{(-1)^{\frac{n+1}{2}}}{n^3\cosh\frac{n\pi h}{2b}}\right)\cos\frac{n\pi x}{b}\cosh\frac{n\pi y}{b}\right) \qquad (9\text{-}36.2)$$

The **shear stresses** are obtained by differentiating the shear stresses in equation (9-36.2), as follows

$$\tau_{zx} = \frac{\partial\Phi(x,y)}{\partial y} = \mu\tau\frac{8b}{\pi^2}\sum_{\substack{n=1,3\\.5...}}^{\infty}\left(\frac{(-1)^{\frac{n+1}{2}}}{n^2\cosh\frac{n\pi h}{2b}}\right)\cos\frac{n\pi x}{b}\sinh\frac{n\pi y}{b} \qquad (9\text{-}36.3)$$

$$\tau_{zy} = -\frac{\partial\Phi(x,y)}{\partial x} = 2\mu\tau\left(x+\frac{4b}{\pi^2}\sum_{\substack{n=1,3\\.5...}}^{\infty}\left(\frac{(-1)^{\frac{n+1}{2}}}{n^2\cosh\frac{n\pi h}{2b}}\right)\sin\frac{n\pi x}{b}\cosh\frac{n\pi y}{b}\right) \qquad (9\text{-}36.4)$$

Let us examine the maximum shear stress at the midsection, $y = 0$, the hyperbolic sinh vanishes. At $(x = \pm b/2)$, with $n = 1, 3, 5, ..$, the sine term becomes unity.

$$\tau_{zx} = \mu\tau\frac{8b}{\pi^2}\sum_{\substack{n=1,3\\.5...}}^{\infty}\left(\frac{(-1)^{\frac{n+1}{2}}}{n^2\cosh\frac{n\pi h}{2b}}\right)\cos\frac{n\pi x}{b}(0) = 0 \qquad (9\text{-}36.5)$$

As an approximation, we will ignore terms with $n > 1$, since the hyperbolic cosh is greater than 1. Thus, we have

$$\tau_{zy} = \mu\tau b\left(\pm 1 + \frac{8}{\pi^2}\sum_{\substack{n=1,3 \\ ,5,...}}^{\infty}\left(\frac{(-1)^{\frac{n+1}{2}}}{n^2\cosh\dfrac{n\pi h}{2b}}\right)\sin\frac{\pm n\pi}{2}(1)\right)$$

$$= \mu\tau b\left(\pm 1 - \frac{8}{\pi^2}\sum_{\substack{n=1,3 \\ ,5,...}}^{\infty}\left(\frac{\pm 1}{n^2\cosh\dfrac{n\pi h}{2b}}\right)\right)$$

(9-36.6)

Therefore, at $y = 0$ and $x = \pm b/2$, $\tau_{zx} = 0$ and τ_{zy} is maximum.

Moment couple

$$M_t = -\iint_{cs}\left(\tau_{zy}x - \tau_{zx}y\right)dxdy$$

$$= -\iint_{cs}\left[\begin{array}{l} 2\mu\tau x\left(x + \dfrac{4b}{\pi^2}\sum_{\substack{n=1,3 \\ ,5,...}}^{\infty}\dfrac{(-1)^{\frac{n+1}{2}}}{n^2\cosh\dfrac{n\pi h}{2b}}\sin\dfrac{n\pi x}{b}\cosh\dfrac{n\pi y}{b}\right) \\[4ex] -y\left(\mu\tau\dfrac{8b}{\pi^2}\sum_{\substack{n=1,3 \\ ,5,...}}^{\infty}\dfrac{(-1)^{\frac{n+1}{2}}}{n^2\cosh\dfrac{n\pi h}{2b}}\cos\dfrac{n\pi x}{b}\sinh\dfrac{n\pi y}{b}\right)\end{array}\right]dxdy$$

(9-37.1)

Performing the double integration, we get

$$M_t = \mu\tau\left(\frac{b^3h}{3} + \frac{64b^4}{\pi^5}\sum_{\substack{n=1,3 \\ ,5,...}}^{\infty}\left(\frac{(-1)^n}{n^5}\right)\tanh\frac{n\pi h}{2b}\right)$$

(9-37.2)

Angle of twist

$$\tau = \frac{M_t}{\mu b^3 h\left(\dfrac{1}{3} + \dfrac{64}{\pi^5}\left(\dfrac{b}{h}\right)\sum_{\substack{n=1,3 \\ ,5,...}}^{\infty}(-1)^n n^{-5}\tanh\dfrac{n\pi h}{2b}\right)}$$

(9-37.3)

298

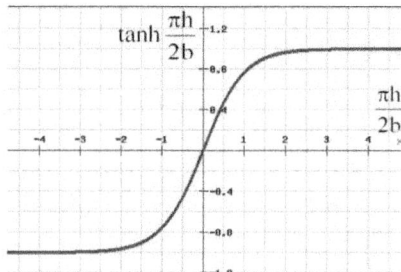

Figure 9-7. Hyperbolic tangent function variation with argument

From Figure 9-7, we notice that the hyperbolic tangent function varies between -1 and 1 for all values of h/b. Therefore, the bracketed term in the denominator of (9-37.3) is calculated approximately as follows:

$$\frac{1}{3} - \frac{64}{306}\left(\frac{b}{h}\right)\left(\tanh\frac{\pi h}{2b} + \frac{1}{243}\tanh\frac{3\pi h}{2b} - ..\right) \approx 0.33 \qquad (9\text{-}37.4)$$

i.e.,

$$\tau = \frac{M_t}{0.33\mu b^3 h} \qquad (9\text{-}37.5)$$

Displacements

$$u(x,y,z) = -\frac{M_t}{0.33\mu b^3 h}yz \qquad (9\text{-}38.1)$$

$$v(x,y,z) = +\frac{M_t}{0.33\mu b^3 h}xz \qquad (9\text{-}38.2)$$

$$w(x,y,z) = \frac{M_t}{0.33\mu b^3 h}\left(xy + \left(\frac{8b^2}{\pi^3}\right)\sum_{n=1,3,5..}^{\infty}\left(\frac{(-1)^{\frac{n+1}{2}}}{n^3\cosh\frac{n\pi h}{2b}}\right)\left(\sin\frac{n\pi x}{b}\sinh\frac{n\pi y}{b}\right)\right) \qquad (9\text{-}38.3)$$

9-5. Membrane surface tension with Ludwig Prandtl's stress function

From equations (9-8.1), (9-8.2), (9-9.2), and (9-9.3), we can prove that $\Phi(x,y)$ satisfies **Poisson's equation** as follows

$$\frac{\partial \Phi(x,y)}{\partial y} = \mu\tau\left(\frac{\partial \psi}{\partial y} - y\right) \tag{9-39.1}$$

$$-\frac{\partial \Phi(x,y)}{\partial x} = \mu\tau\left(-\frac{\partial \psi}{\partial x} + x\right) \tag{9-39.2}$$

Differentiating each side of the above two equations with respect to their respective derivatives we get

$$\frac{\partial^2 \Phi(x,y)}{\partial y^2} = \mu\tau\left(\frac{\partial^2 \psi}{\partial y^2} - 1\right) \tag{9-39.1a}$$

$$\frac{\partial^2 \Phi(x,y)}{\partial x^2} = \mu\tau\left(\frac{\partial^2 \psi}{\partial x^2} - 1\right) \tag{9-39.2a}$$

Adding the two equations by members and noting the vanishing **Poisson's term** of ψ, we get

$$\nabla^2 \Phi(x,y) = -2\mu\tau \tag{9-39.3}$$

In contrast to the constancy of the Poisson's function of **Prandtl's function** Φ on the contour, the **Poisson function** of the two conjugate stress functions φ and ψ vanish as follows

i.e.,
$$\nabla^2 \varphi(x,y) = 0 \tag{9-39.4}$$
$$\nabla^2 \psi(x,y) = 0 \tag{9-39.5}$$

From Hooke's law for shear stresses, equations (9-39.1a) and (9-39.2a) are also written as

$$\tau_{zx} = \frac{\partial \Phi(x,y)}{\partial y} = \mu\tau\left(\frac{\partial \psi}{\partial y} - y\right) = \mu\alpha_{zx} \tag{9-39.6}$$

$$\tau_{zy} = -\frac{\partial \Phi(x,y)}{\partial x} = \mu\tau\left(-\frac{\partial \psi}{\partial x} + x\right) = \mu\alpha_{zy} \tag{9-39.7}$$

Therefore, differentiating (9-39.6) and (9-39.7), we get

$$\frac{\partial \alpha_{zy}}{\partial x} - \frac{\partial \alpha_{zx}}{\partial y} = 2\tau \tag{9-39.8}$$

Summary of the implications of Prandtl's function

Poisson's equation of **torsion stress** function $\quad \nabla^2 \varphi(x,y) = 0$
(9-39.4)
Poisson's equation of **stress function** $\qquad\qquad \nabla^2 \psi(x,y) = 0$
(9-39.5)

Stress function   (9-7.10a)

$$\psi = \left(\frac{x^2 + y^2}{2}\right) + C = \frac{r^2}{2} + C$$

Prandtl's stress function   (9-9.1)

$$\Phi(x,y) = \mu\tau\left(\psi - \frac{x^2 + y^2}{2}\right)$$

Poisson's equation of **Prandtl's function**

$$\nabla^2\Phi(x,y) = -2\mu\tau$$

(9-39.3)

Shear strain zx   (9-39.6)

$$\alpha_{zx} = \frac{1}{\mu}\cdot\frac{\partial\Phi(x,y)}{\partial y}$$

Shear strain zy   (9-39.7)

$$\alpha_{zy} = -\frac{1}{\mu}\cdot\frac{\partial\Phi(x,y)}{\partial x}$$

Shear stress zx   (9-39.6)

$$\tau_{zx} = \frac{\partial\Phi(x,y)}{\partial y}$$

Shear stress zy   (9-39.7)

$$\tau_{zy} = -\frac{\partial\Phi(x,y)}{\partial x}$$

Rotational angle   (9-39.8)

$$\frac{\partial\alpha_{zy}}{\partial x} - \frac{\partial\alpha_{zx}}{\partial y} = 2\tau$$

From equation (9-9.1), the Prandtl's stress function takes the constant value on the **contour** of body under tension such that

$$\Phi(x,y) = \mu\tau C \qquad\qquad\qquad (9\text{-}39.9)$$

In the case of **closed contour**, without **holes**, the constant C could be set to zero as follows

$$\Phi(x,y) = 0 \qquad\qquad\qquad (9\text{-}39.10)$$

In the case of elliptical contour, Prandtl's modified equation (9-9.1) as follows

$$\Phi(x,y) = C'\mu\tau\left(\frac{x^2}{a^2} + \frac{y^2}{b^2} - 1\right) \qquad\qquad\qquad (9\text{-}39.11)$$

Where C' is determined from the boundary conditions.

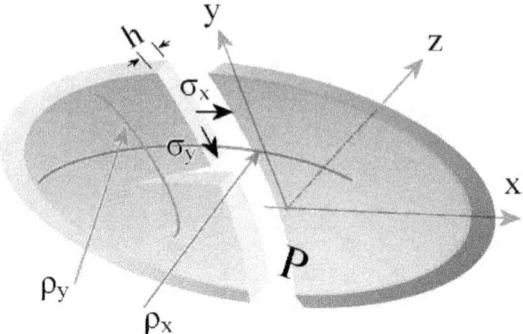

Figure 9-8. Laplace's membrane problem of equilibrium of surface tension with pressure imparted on the membrane. σ_x and σ_y are surface tensions in the x and y directions, ρ_x and ρ_y surface curvatures in the zx- and zy-planes, h thickness of membrane, P pressure.

Example 28

A membrane with thickness h, two curvatures ρ_x and ρ_y, is exposed to pressure P. Given the normal stresses σ_x and σ_y in the x and y directions. Pierre-Simon de Laplace gave the following equation of equilibrium of static stresses

$$\frac{\sigma_x}{\rho_x} + \frac{\sigma_y}{\rho_y} = \frac{P}{h} \qquad (9\text{-}40.1)$$

Assume that the surface tension is homogeneous such that

$$h\,\sigma_x = h\,\sigma_y = T \qquad (9\text{-}40.2)$$

Where, T is the surface tension of the membrane. Also, the two curvatures of the membranes are

$$\rho_x = \frac{1}{\sqrt{\left(\dfrac{\partial^2 z}{\partial x^2}\right)^2 + \left(\dfrac{\partial^2 z}{\partial y^2}\right)^2}} == \frac{1}{\dfrac{\partial^2 z}{\partial x^2}} \qquad (9\text{-}40.3)$$

$$\rho_y = \frac{1}{\sqrt{\left(\dfrac{\partial^2 z}{\partial x^2}\right)^2 + \left(\dfrac{\partial^2 z}{\partial y^2}\right)^2}} == \frac{1}{\dfrac{\partial^2 z}{\partial y^2}} \qquad (9\text{-}40.4)$$

Substituting from equations (9-40.2 through 40.4) in equation (9-40.1) we get

$$\frac{\partial^2 z}{\partial x^2} + \frac{\partial^2 z}{\partial y^2} = \frac{P}{T}$$

(9-40.5)

Comparing equations (9-40.5) with the **Poisson's equation** of **Prandtl's function** (9-39.3), we get

$$\Phi(x,y) = z(x,y)$$

(9-40.6)

$$P = -2\mu\tau T$$

(9-40.7)

Thus, equation (9-40.6) gives the Prandtl's stress function represented by the surface function of the membrane and equation (9-40.7) is Hooke's shear stress law, where the **strain** is represented by $2\tau T$. The twisting moment is equal to double volume bounded by the membrane given by (9-9.8a)

$$M_t = 2\mu\tau \iint \left(\psi - \frac{x^2 + y^2}{2} \right) dy dx$$

$$= 2\mu\tau \iint z(x,y) dy dx$$

(9-40.7)

$$= 2\mu\tau V$$

9-5. Membrane surface with holes with Ludwig Prandtl's and Bredt's stress function

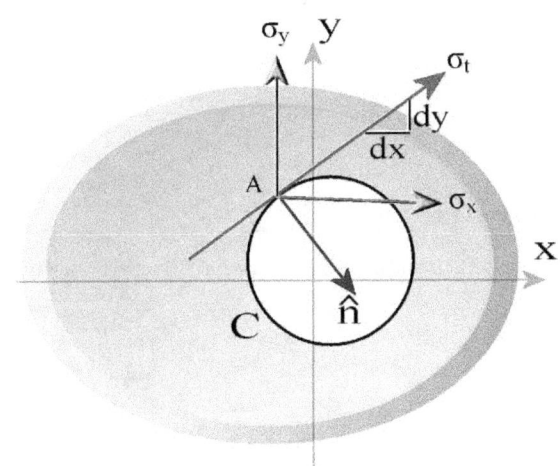

Figure 9-9. Laplace's membrane with holes treated Prandtl's stress function.

In Figure 9-9, a hole in Laplace's membrane is studies by finding the resultant stress on the contour of the hole by using the surface condition as follows

$$\sigma_t = \tau_{zx} l_t + \tau_{zy} m_t \tag{9-41.1}$$

Where l_t and m_t are the directional cosines of the tangent of the resultant stress.

From equations (9-39.6 and 7), we get

$$\sigma_t = \frac{\partial \Phi(x,y)}{\partial y} l_t - \frac{\partial \Phi(x,y)}{\partial x} m_t \tag{9-41.2}$$

From Figure 9-9, we conclude that

$$\begin{aligned}
\sigma_n &= \frac{\partial \Phi(x,y)}{\partial y} m_n - \frac{\partial \Phi(x,y)}{\partial x}(-l_n) \\
&= \frac{\partial \Phi(x,y)}{\partial x} l_n + \frac{\partial \Phi(x,y)}{\partial y} m_n \\
&= \frac{\partial \Phi(x,y)}{\partial n} \\
&= \nabla \Phi(x,y)
\end{aligned} \tag{9-41.3}$$

The equilibrium of stresses on the contour C, in Figure 9-9, is given by

$$\begin{aligned}
F &= \int_C \sigma_n ds \\
&= \int_C \nabla \Phi(x,y) ds = -2\mu\tau T_n
\end{aligned} \tag{9-41.4}$$

Where, T_n is the surface tension on the contour of nth contour.

The moment of the twisting couple on the n-holes in the main membrane is obtained by equation (9-9.6) as follows

$$\mathbf{M}_t = -\iint\limits_{cs} \left(-\frac{\partial \Phi(x,y)}{\partial x} x - \frac{\partial \Phi(x,y)}{\partial y} y \right) dxdy$$

$$= \iint\limits_{cs} \left(\frac{\partial \Phi(x,y)}{\partial x} x + \frac{\partial \Phi(x,y)}{\partial y} y \right) dxdy$$

$$= -\int \left([x_1 \Phi(x_1,y) - x_0 \Phi(x_0,y)] - \int \Phi dx \right) dy$$

$$\quad - \int \left([y_1 \Phi(x,y_1) - y_0 \Phi(x,y_0)] - \int \Phi dy \right) dx \qquad (9\text{-}41.5)$$

$$= 2C_o T_o - 2\sum_{n=1}^{m} C_n T_n - 2\iint \Phi ds$$

Where,

$C_o T_o$ denotes moment of couple due to outer contour.
$C_n T_n$ denote moment of couple due to inner contours.
Double integration is carried over the difference = main membrane – holes.

9-6. Membrane solution of plate elasticity: Marcus's solution

We have treated the membrane surface tension with Ludwig Prandtl's stress function using the Laplacian operator in equation (9-40.5). Since that equation entails **two bending curvatures** expressed in the second derivatives of z with respect to each of x and y, we could re-write the same governing equation in terms of sum of moments as follows

$$\nabla^2 z = -\frac{M_x + M_y}{(1+v)K} \qquad (9\text{-}42.1)$$

Where, the moments M_x and M_y, Poisson's ratio v, and flexural rigidity K (defined below) constantly appear in equations of deflections, and

$$K = \frac{Eh^3}{12(1-v^2)} \approx \frac{Eh^3}{12} \qquad (9\text{-}42.2)$$

K is called **flexural rigidity of plate.**

Thus, the **shearing forces**, equations (9-9.5 through 9.7), can be written as

$$F_{xz} = -K\frac{\partial}{\partial x}\left(\nabla^2 z\right) = \frac{1}{(1+v)}\frac{\partial}{\partial x}\left(M_x + M_y\right) \qquad (9\text{-}42.3)$$

$$F_{yz} = -K\frac{\partial}{\partial y}\left(\nabla^2 z\right) = \frac{1}{(1+v)}\frac{\partial}{\partial y}\left(M_x + M_y\right) \qquad (9\text{-}42.4)$$

Those forms simplify the concept that the **shearing forces** are the gradients of moments of bending. Put simply

305

$$F = \frac{dM}{dx} \tag{9-42.5}$$

As we will see later, the **Sophie Germain**'s equation, which generalizes the solution of bending of a rod to a moderately thick plate or thin slab, can be expressed in terms of the sums of momenta, equation (9-42.2), as follows

$$\nabla^2\nabla^2 w = \frac{q(x,y)}{K}$$
$$= -\frac{1}{(1+v)K}\nabla^2\left(M_x + M_y\right) \tag{9-42.6}$$

The term $q(x,y)$ is the intensity of force, or load, on the coordinates x and y.

Therefore,

$$q(x,y) = -\frac{1}{(1+v)}\nabla^2\left(M_x + M_y\right) \tag{9-42.7}$$

Equation (9-42.7) expresses the two dimensional **load** in terms of sum of Laplacians of moments, which is simplified in the case of a rod to

$$q = \frac{d^2 M}{dx^2} \tag{9-42.5}$$

Marcus's method relies on funicular polygon, or mesh of wires, to solve two-dimensional differential equations in terms of orthogonal curves connecting nodes of the mesh such that equations could be solved in one dimension and the results added according to sum of moments in (9-42.1)

9-7. Transverse bending of bar of arbitrary cross-section

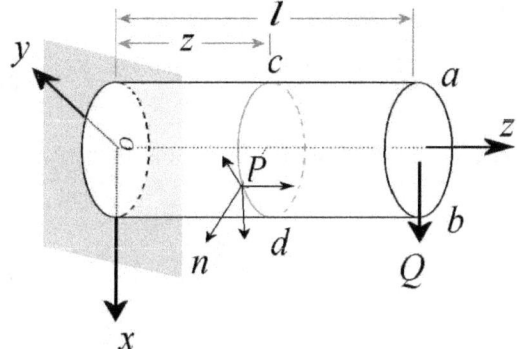

Figure 9-10. bar with arbitrary cross-section bent by shearing stress Q at one end. The length of the bar is l, its moment of inertia is I.

Proposed stresses

Saint Venant's inverse solution by plain stress or plain strain assumes the following expressions for stresses:

$$\sigma_z = \frac{Q(l-z)}{I_c} x \tag{9-43.1}$$

$$\sigma_x = \sigma_y = 0 \tag{9-43.2}$$

$$\tau_{xy} = \tau_{yx} = 0 \tag{9-43.3}$$

$$\tau_{xz}..? \qquad \tau_{yz}..? \tag{9-43.4}$$

$$\Theta = \frac{Q(l-z)}{I_c} x \tag{9-43.5}$$

Navier's equilibrium equations

Therefore, equation (1-3.6) yield the three **Navier's partial differential equations of equilibrium of stresses** of the three forces acting on the solid body as follows

$$\frac{\partial \tau_{xz}(x,y,z)}{\partial z} + \rho X_i = \rho \frac{\partial^2 u(x,y,z)}{\partial t^2} \tag{9-43.6}$$

307

$$\frac{\partial \tau_{yz}(x,y,z)}{\partial z} + \rho Y_i = \rho \frac{\partial^2 v(x,y,z)}{\partial t^2} \qquad (9\text{-}43.7)$$

$$\frac{\partial \tau_{zx}(x,y,z)}{\partial x} + \frac{\partial \tau_{zy}(x,y,z)}{\partial y} - \frac{Q}{I_c}x + \rho Z_i = \rho \frac{\partial^2 w(x,y,z)}{\partial t^2} \qquad (9\text{-}43.8)$$

The assumption that internal forces X_i, Y_i, and Z_i are ignored and the system in equilibrium, the above three equations give

$$\frac{\partial \tau_{xz}(x,y,z)}{\partial z} = 0 \qquad (9\text{-}43.9)$$

$$\frac{\partial \tau_{yz}(x,y,z)}{\partial z} = 0 \qquad (9\text{-}43.10)$$

$$\frac{\partial \tau_{zx}(x,y,z)}{\partial x} + \frac{\partial \tau_{zy}(x,y,z)}{\partial y} - \frac{Q}{I_c}x = 0 \qquad (9\text{-}43.11)$$

Continuity Saint Venant's equations

Farther, the assumption of **planar strain** implies that our stresses conform to **Maurice Lévy's equation**

$$(1+v)\nabla^2 \tau_{yz} = 0 \qquad (9\text{-}43.12)$$

$$(1+v)\nabla^2 \tau_{xz} + \frac{Q}{I_c} = 0 \qquad (9\text{-}43.13)$$

Or,

$$\nabla^2 \tau_{yz} = 0 \qquad (9\text{-}43.12a)$$

$$\nabla^2 \tau_{xz} = -\frac{Q}{I_c(1+v)} \qquad (9\text{-}43.13b)$$

Surface equations

On the surface wall, the shear stress vanish as follows

(i) Lateral wall:

$$\tau_{zx}(x,y,z)\big|_{wall} l + \tau_{zy}(x,y,z)\big|_{wall} m = 0 \qquad (9\text{-}43.14)$$

Where

$$l = \frac{dy}{ds}$$

$$m = -\frac{dx}{ds} \qquad (9\text{-}43.14a)$$

(ii) End of rod:

$$\tau_{zx}(x,y,z)\big|_{end} l + \tau_{zy}(x,y,z)\big|_{end} m = Q \qquad (9\text{-}43.15)$$

Summary of governing equations

Table 9-1. Governing equations of bending a bar with arbitrary cross section

$$\frac{\partial \tau_{zx}(x,y,z)}{\partial x} + \frac{\partial \tau_{zy}(x,y,z)}{\partial y} - \frac{Q}{I_c} x = 0 \qquad (9\text{-}43.11)$$

$$\nabla^2 \tau_{yz} = 0 \qquad (9\text{-}43.12a)$$

$$\nabla^2 \tau_{xz} = -\frac{Q}{I_c(1+v)} \qquad (9\text{-}43.13b)$$

Lateral wall: $\quad \tau_{zx}(x,y,z)\big|_{wall} l + \tau_{zy}(x,y,z)\big|_{wall} m = 0 \qquad (9\text{-}43.14)$

End of rod: $\quad \displaystyle\iint \tau_{zy}(x,y,z)\bigg|_{end} dxdy = 0 \qquad (9\text{-}43.15a)$

$$\iint \tau_{zx}(x,y,z)\bigg|_{end} dxdy = Q \qquad (9\text{-}43.15b)$$

Solution by proposing stress functions

(i) As we integrate equation (9-43.11) with respect to x, we will propose two stress functions that will be determined by solving the remaining governing equations as follows

$$\tau_{zx} = \frac{\partial \varphi}{\partial y} - \frac{Q}{2I_c} x^2 + f(y) \qquad (9\text{-}43.16)$$

$$\tau_{yz} = -\frac{\partial \varphi}{\partial x} \qquad (9\text{-}43.17)$$

(ii) Equations (9-43.16 and 17) could be differentiated to produce the governing equation (9-43.11). Substituting by the shear stresses from those equations into (9-43.12a and 13b), we get

$$\frac{\partial}{\partial x} \nabla^2 \varphi = 0 \qquad (9\text{-}43.18)$$

$$\frac{\partial}{\partial y} \nabla^2 \varphi - \frac{Q}{I_c} + \nabla^2 f''(y) = -\frac{Q}{I_c(1+v)}$$

Or,

$$\frac{\partial}{\partial y} \nabla^2 \varphi = \frac{Qv}{I_c(1+v)} - f''(y) \qquad (9\text{-}43.19)$$

Integrating equations (9-43.18 and 19) and adding by members, we get

$$\nabla^2 \varphi = \frac{Qv}{I_c(1+v)} y - f'(y) + C \tag{9-43.20}$$

Lateral wall surface conditions

Substituting by the shear stresses from (9-43.16 and 17), in the surface equation on the lateral wall, (9-43.14) and using (9-43.14a), we get

$$\left(\frac{\partial \varphi}{\partial y} - \frac{Q}{2I_c} x^2 + f(y) \right) \frac{dy}{ds} + \frac{\partial \varphi}{\partial x} \frac{dx}{ds} = 0 \tag{9-43.21}$$

Noting that

$$\frac{\partial \varphi}{\partial y} \frac{dy}{ds} + \frac{\partial \varphi}{\partial x} \frac{dx}{ds} = \frac{d\varphi}{ds} \tag{9-43.22}$$

Therefore, equation (9-43.21) becomes

$$\left. \frac{d\varphi}{ds} \right|_{wall} = \left(\frac{Q}{2I_c} x^2 - f(y) \right) \frac{dy}{ds} \tag{9-43.21}$$

Since, f(y) is an arbitrary function, we opt to set f(y) such that the bracketed term in the above equation vanishes. Therefore,

$$f(y)\big|_{wall} = \frac{Q}{2I_c} x^2 \tag{9-43.22}$$

Thus, integrating equation (9-43.21) over the surface contour, φ = constant, on the lateral wall.

End of bar surface conditions

Surface conditions at the **end of the bar**, equations (9-43.15a and 15b), are determined by the use of equations (9-43.16 and 17) as follows

End of rod:

$$\iint \tau_{zy}(x,y,z)\bigg|_{end} dxdy = -\iint \frac{\partial \varphi}{\partial x} dxdy$$

$$= -\int dy \int \frac{\partial \varphi}{\partial x} dx = 0$$

(9-43.23)

Thus, on the end of the bar, forces in the y-direction vanish.

In the x-direction we get

$$\iint \tau_{zx}(x,y,z)\bigg|_{end} dxdy = \iint \left(\frac{\partial \varphi}{\partial y} - \frac{Q}{2I_c} x^2 + f(y) \right) dxdy$$

$$= \iint \frac{\partial \varphi}{\partial y} dxdy - \frac{Q}{2I_c} \iint x^2 dxdy + \iint f(y) dxdy$$

(9-43.24)

The three terms in equation (9-43.24) are evaluated as follows:

(1) $\iint \frac{\partial \varphi}{\partial y} dxdy$ vanishes by equation (9-43.23)

(2) $-\frac{Q}{2I_c} \iint x^2 dxdy$, the integral is the moment of inertia about an axis parallel to the y-axis. Thus, it is equal to $-Q/2$.

(3) $\iint f(y) dxdy$, we note f(y) is defined by (9-43.22) on the lateral wall. But, on the end of the bar, we should first integrate f(y) on the x coordinate before substitute by the y value, as follows

$$\int \left(\int f(y) dx \right) dy = \int \left[xf(y) \right]_1^2 dy$$

$$= \int \left[x_2 f(y_2) - x_1 f(y_1) \right] dy$$

$$= \int \left[x_2 \frac{Qx_2^2}{2I_c} - x_1 \frac{Qx_1^2}{2I_c} \right] dy$$

(9-43.25)

$$= \frac{Q}{2I_c} \int \left[x_2^3 - x_1^3 \right] dy = \frac{3Q}{2I_c} \int \left[\frac{x_2^3 - x_1^3}{3} \right] dy$$

$$= \frac{3Q}{2I_c} [I_c] = \frac{3Q}{2}$$

Therefore, equation (9-43.24) becomes

$$\iint \tau_{zx}(x,y,z)\bigg|_{end} dxdy = -\frac{Q}{2} + \frac{3Q}{2} = Q \tag{9-43.26}$$

Thus, the shear stresses proposed in equations (9-43.16) and (9-43.17) satisfy the initially given boundary conditions with vanishing shear stresses on the lateral walls and downward resultant forces on one end.

Conclusion

The final solution of bending of a bar by terminal force is to determine **φ(x,y)** and **f(y)**, equation (9-43.20), which determine (9-43.16 and 17). The **constant C**, in (9-43.20) is determined by the symmetry around xOz.

9-8. Transverse bending of bar of circular cross-section

The equation of the **circular cross-section** in the xy-plane is

$$x^2 + y^2 - r^2 = 0 \tag{9-44.1}$$

Substituting by $x^2 = r^2 - y^2$ in equation (9-43.22), we get

$$f(y)\big|_{wall} = \frac{Q}{2I_c}\left(r^2 - y^2\right) \tag{9-44.2}$$

Differentiating equation (9-44.2) with respect to y and substituting in equation (9-43.20) we get

$$\begin{aligned} \nabla^2 \varphi &= \frac{Qv}{I_c(1+v)}y + \frac{Q}{I_c}y + C \\ &= \frac{(1+2v)Q}{(1+v)I_c}y + C \end{aligned} \tag{9-44.3}$$

We propose the function

$$\varphi(x,y) = m\left(r^2 - x^2 - y^2\right)y \tag{9-44.4}$$

Therefore,

$$\frac{\partial^2 \varphi}{\partial x^2} = -2my \tag{9-44.5a}$$

$$\frac{\partial^2 \varphi}{\partial y^2} = -6my \tag{9-44.5b}$$

Substituting from (9-44.4) and (9-44.5) in (9-44.3), and putting C = 0 for symmetrical circular rod, we get

$$m = -\frac{(1+2v)Q}{8(1+v)I_c} \tag{9-44.6}$$

The **stress function** (9-44.4) becomes

$$\varphi(x,y) = -\frac{(1+2v)Q}{8(1+v)I_c}\left(r^2 - x^2 - y^2\right)y \tag{9-44.7}$$

Substituting by the stress function φ, from equation (9-44.7) and f(y), equation (9-44.2), into equations (9-43.16 and 17), the **shearing stresses** are

$$\tau_{zx} = \frac{(3+2v)Q}{8(1+v)I_c}\left(r^2 - x^2 - \frac{(1-2v)}{(3+2v)}y^2\right) \tag{9-44.8}$$

$$\tau_{zy} = -\frac{(1+2v)Q}{4(1+v)I_c}xy \tag{9-44.9}$$

Thus, the shearing stresses depend on **Poisson's ratio**.
τ_{xy} vanished on the axis (x = y = 0).
τ_{zy} is constant on the axis.

The resultant shear stress is

$$\tau_z = \sqrt{\tau_{zx}^2 + \tau_{zx}^2}$$
$$= \frac{Q}{4(1+v)I_c}\sqrt{\frac{(3+2v)^2}{4}\left(r^2 - x^2 - \frac{(1-2v)}{(3+2v)}y^2\right)^2 + (1+2v)^2x^2y^2} \tag{9-44.10}$$

The principal strains are

$$\varepsilon_{xx} = -v\left(\frac{Q(l-z)}{I_cE}x\right)$$

$$\varepsilon_{yy} = -v\left(\frac{Q(l-z)}{I_cE}x\right)$$

$$\varepsilon_{zz} = \left(\frac{Q(l-z)}{I_cE}x\right) \tag{9-44.11}$$

$$\varepsilon = (1-2v)\left(\frac{Q(l-z)}{I_cE}x\right)$$

313

$$\sigma_x = 2\mu\frac{\partial u}{\partial x} + \lambda(1 - 2v)\left(\frac{Q(l - z)}{I_c E}x\right) = 0$$

$$\frac{\partial u}{\partial x} = -\lambda(1 - 2v)\left(\frac{Q(l - z)}{2\mu I_c E}x\right)$$

(9-44.12a)

Integrating we get

$$\mathbf{u} = -\lambda(1 - 2v)\left(\frac{Q(l - z)}{4\mu I_c E}x^2\right) + u_o(y, z)$$

(9-44.12b)

Substituting from (3-8.2) by

$$\lambda = \frac{vE}{(1 - 2v)(1 + v)}$$

$$\mathbf{u} = -\left(\frac{vQ(l - z)}{4\mu I_c(1 + v)}x^2\right) + u_o(y, z)$$

(9-44.12c)

And,

$$\sigma_y = 2\mu\frac{\partial v}{\partial y} + \lambda(1 - 2v)\left(\frac{Q(l - z)}{I_c E}x\right) = 0$$

$$\frac{\partial v}{\partial y} = -\lambda(1 - 2v)\left(\frac{Q(l - z)}{2\mu I_c E}x\right) + v_o(x, y)$$

(9-44.13a)

Integrating, we get

$$\mathbf{v} = -\lambda(1 - 2v)\left(\frac{Q(l - z)}{2\mu I_c E}xy\right) + v_o(x, z)$$

(9-44.13b)

Similarly,

$$\mathbf{v} = -\left(\frac{vQ(l - z)}{2\mu I_c(1 + v)}xy\right) + v_o(y, z)$$

(9-44.13c)

$$\sigma_z = 2\mu\frac{\partial w}{\partial z} + \lambda(1 - 2v)\left(\frac{Q(l - z)}{I_c E}x\right) = \left(\frac{Q(l - z)}{I_c}x\right)$$

$$\frac{\partial w}{\partial z} = \frac{Q(l - z)}{2\mu(1 + v)I_c}x$$

(9-44.14a)

Integrating, we get

314

$$W = \frac{Q(2l - z)}{4\mu I_c(1 + v)} zx + w_o(x, y)$$ (9-44.14a)

Summary of solution

9-7. Transverse bending of bar of circular cross-section

$$x^2 + y^2 - r^2 = 0$$ (9-44.1)

$$\varphi(x, y) = -\frac{(1 + 2v)Q}{8(1 + v)I_c}(r^2 - x^2 - y^2)y$$ (9-44.7)

$$\tau_{zx} = \frac{(3 + 2v)Q}{8(1 + v)I_c}\left(r^2 - x^2 - \frac{(1 - 2v)}{(3 + 2v)}y^2\right)$$ (9-44.8)

$$\tau_{zy} = -\frac{(1 + 2v)Q}{4(1 + v)I_c}xy$$ (9-44.9)

$$\tau_z = \sqrt{\tau_{zx}^2 + \tau_{zx}^2}$$

$$= \frac{Q}{4(1 + v)I_c}\sqrt{\frac{(3 + 2v)^2}{4}\left(r^2 - x^2 - \frac{(1 - 2v)}{(3 + 2v)}y^2\right)^2 + (1 + 2v)^2 x^2 y^2}$$ (9-44.10)

$$u = -\left(\frac{vQ(l - z)}{4\mu I_c(1 + v)}x^2\right) + u_o(y, z)$$ (9-44.12c)

$$v = -\left(\frac{vQ(l - z)}{2\mu I_c(1 + v)}xy\right) + v_o(y, z)$$ (9-44.13c)

$$W = \frac{Q(2l - z)}{4\mu I_c(1 + v)}zx + w_o(x, y)$$ (9-44.14a)

Chapter 10

GENERAL SOLUTION OF ELASTICITY PROBLEMS

General solution goes beyond the simplification of **planar stress** and **planar strain**. Here, we will ignore internal forces and external acceleration and solve the two sets of equations for **equilibrium of stresses** and **continuity of strains**, as follows.

10-1. Navier's partial differential equations of equilibrium of stresses

Navier's partial differential equations of equilibrium of stresses, equations (1-4), are written in the form

$$\tau_{ij,j} + \rho F_i = \rho \frac{\partial^2 u_i}{\partial t^2}$$

Where the dummy indices ij, i, and j denote the following coordinate designation and operational differentiation

ij : combinations of coordinates x, y, and z, and designate those coordinates. The double designation is attributed to the planar nature of stresses, perpendicular or parallel.

i : single indices of x, y, or z, denote variables without orientation on each of the three surfaces of the infinitesimal elements such as F and u.

j : designates the threes differential operators $\partial/\partial x$, $\partial/\partial y$, and $\partial/\partial z$.

τ : τ_{ij} designates normal and shear stresses such that
$\tau_{11} = \sigma_x$, $\tau_{22} = \sigma_y$, $\tau_{33} = \sigma_z$,
$\tau_{12} = \tau_{21} = \tau_{xy} = \tau_{yx}$, $\tau_{13} = \tau_{31} = \tau_{xz} = \tau_{zx}$, $\tau_{32} = \tau_{zy} = \tau_{yz}$.

In the special case of absent internal forces F_i and absent external accelerated motion, the above equilibrium equations are written in details as follows:

$$\frac{\partial \sigma_x(x,y,z)}{\partial x} + \frac{\partial \tau_{xy}(x,y,z)}{\partial y} + \frac{\partial \tau_{xz}(x,y,z)}{\partial z} = 0 \qquad (10\text{-}1.1)$$

$$\frac{\partial \tau_{yx}(x,y,z)}{\partial x} + \frac{\partial \sigma_y(x,y,z)}{\partial y} + \frac{\partial \tau_{yz}(x,y,z)}{\partial z} = 0 \qquad (10\text{-}1.2)$$

$$\frac{\partial \tau_{zx}(x,y,z)}{\partial x} + \frac{\partial \tau_{zy}(x,y,z)}{\partial y} + \frac{\partial \sigma_z(x,y,z)}{\partial z} = 0 \qquad (10\text{-}1.3)$$

Equations (10-1) contain the **nine** unknowns σ_i and τ_{ij} where j and i = x, y, and z, where i ≠ j. Solving for the nine unknowns requires the knowledge of the properties of matter in order to reduce the nine variables into three that can be solved by the three Navier's partial differential equations (10-1).

10-2. Equations of continuity of strain components (Saint –Venant's equation)

The equations of continuity of strains are written in the general index notation as follows:

$$\varepsilon_{i,j} + \varepsilon_{j,i} = 2\alpha_{ij}$$

Where the dummy indices i and j denote operational differentiation with respect to the three coordinates as follows

$$\frac{\partial^2 \varepsilon_{yy}}{\partial x^2} + \frac{\partial^2 \varepsilon_{xx}}{\partial y^2} = 2\frac{\partial^2 \alpha_{xy}}{\partial x \partial y} \tag{10-2.1}$$

$$\frac{\partial^2 \varepsilon_{zz}}{\partial y^2} + \frac{\partial^2 \varepsilon_{yy}}{\partial z^2} = 2\frac{\partial^2 \alpha_{yz}}{\partial y \partial z} \tag{10-2.2}$$

$$\frac{\partial^2 \varepsilon_{xx}}{\partial z^2} + \frac{\partial^2 \varepsilon_{zz}}{\partial x^2} = 2\frac{\partial^2 \alpha_{zx}}{\partial z \partial x} \tag{10-2.3}$$

$$\frac{\partial}{\partial x}\left(\frac{\partial \alpha_{xy}}{\partial z} + \frac{\partial \alpha_{xz}}{\partial y} - \frac{\partial \alpha_{yz}}{\partial x} \right) = \frac{\partial^2 \varepsilon_{xx}}{\partial y \partial z} \tag{10-2.4}$$

$$\frac{\partial}{\partial y}\left(\frac{\partial \alpha_{xy}}{\partial z} + \frac{\partial \alpha_{yz}}{\partial x} - \frac{\partial \alpha_{zx}}{\partial y} \right) = \frac{\partial^2 \varepsilon_{yy}}{\partial x \partial z} \tag{10-2.5}$$

$$\frac{\partial}{\partial z}\left(\frac{\partial \alpha_{yz}}{\partial x} + \frac{\partial \alpha_{xz}}{\partial y} - \frac{\partial \alpha_{xy}}{\partial z} \right) = \frac{\partial^2 \varepsilon_{zz}}{\partial x \partial y} \tag{10-2.6}$$

10-3. Beltrami-Michell Equations

Differentiating equation (10-1.1) with respect to x, we get

$$\frac{\partial^2 \sigma_x(x,y,z)}{\partial x^2} = -\frac{\partial^2 \tau_{xy}(x,y,z)}{\partial x \partial y} - \frac{\partial^2 \tau_{xz}(x,y,z)}{\partial x \partial z} \tag{10-3.1}$$

Substituting from (10-2) by the mixed derivatives of the shearing stresses and using **Hooke's law** for shear stresses we get

317

$$\frac{\partial^2 \sigma_x}{\partial x^2} = -\frac{\mu}{2}\left(\frac{\partial^2 \varepsilon_{yy}}{\partial x^2} + \frac{\partial^2 \varepsilon_{xx}}{\partial y^2} + \frac{\partial^2 \varepsilon_{xx}}{\partial z^2} + \frac{\partial^2 \varepsilon_{zz}}{\partial x^2} \right) \qquad (10\text{-}3.2)$$

Next, using the **volumetric Hooke's law** for normal stresses, equations (3-3.1 and 3.2) we get

$$\frac{\partial^2 \sigma_x}{\partial x^2} = -\frac{\mu}{2E}\left[\begin{array}{l} \dfrac{\partial^2 \sigma_y}{\partial x^2} - v\left(\dfrac{\partial^2 \sigma_x}{\partial x^2} + \dfrac{\partial^2 \sigma_z}{\partial x^2} \right) + \dfrac{\partial^2 \sigma_x}{\partial y^2} - v\left(\dfrac{\partial^2 \sigma_y}{\partial y^2} + \dfrac{\partial^2 \sigma_z}{\partial y^2} \right) \\[2mm] + \dfrac{\partial^2 \sigma_x}{\partial z^2} - v\left(\dfrac{\partial^2 \sigma_y}{\partial z^2} + \dfrac{\partial^2 \sigma_z}{\partial z^2} \right) + \dfrac{\partial^2 \sigma_z}{\partial x^2} - v\left(\dfrac{\partial^2 \sigma_y}{\partial x^2} + \dfrac{\partial^2 \sigma_x}{\partial x^2} \right) \end{array} \right] \qquad (10\text{-}3.3)$$

Arranging and making the substitutions, $\Xi = \sigma_x + \sigma_y + \sigma_z$ and $\mu = \dfrac{E}{2(1+v)}$, we get

$$-4(1+v)\frac{\partial^2 \sigma_x}{\partial x^2} + \frac{\partial^2 \sigma_x}{\partial x^2} = \left[\begin{array}{l} \left(\dfrac{\partial^2 \sigma_x}{\partial x^2} + \right)\dfrac{\partial^2 \sigma_y}{\partial x^2} + \dfrac{\partial^2 \sigma_z}{\partial x^2} + \dfrac{\partial^2 \sigma_x}{\partial y^2} + \dfrac{\partial^2 \sigma_x}{\partial z^2} \\[2mm] -v\left(\dfrac{\partial^2 \sigma_x}{\partial x^2} + \dfrac{\partial^2 \sigma_z}{\partial x^2} \right) - v\left(\dfrac{\partial^2 \sigma_y}{\partial y^2} + \dfrac{\partial^2 \sigma_z}{\partial y^2} \right) \\[2mm] -v\left(\dfrac{\partial^2 \sigma_y}{\partial z^2} + \dfrac{\partial^2 \sigma_z}{\partial z^2} \right) - v\left(\dfrac{\partial^2 \sigma_y}{\partial x^2} + \dfrac{\partial^2 \sigma_x}{\partial x^2} \right) \end{array} \right] \qquad (10\text{-}3.4)$$

Farther arranging, we get

$$-2(1+v)\frac{\partial^2 \sigma_x}{\partial x^2} = \frac{\partial^2 \Xi}{\partial x^2} + \nabla^2 \sigma_x - v\left(\frac{\partial^2}{\partial x^2} + \frac{\partial^2}{\partial y^2} + \frac{\partial^2}{\partial z^2} \right)(\Xi - \sigma_x) \qquad (10\text{-}3.5)$$

i.e.,

$$(1+v)\nabla^2 \sigma_x + \frac{\partial^2 \Xi}{\partial x^2} = v\nabla^2 \Xi - 2(1+v)\frac{\partial^2 \sigma_x}{\partial x^2} \qquad (10\text{-}3.6)$$

This can be written in the general form of dummy indices as follows

$$(1+v)\tau_{ij,kk} + \tau_{kk,ij} = -\frac{(1+v)v}{(1-v)}\delta_{ij}F_{k,k} - (1+v)\left(F_{j,i} + F_{i,j} \right) \qquad (10\text{-}3.7)$$

Equation (10-3.6) or (10-3.7) is one of the six equations of **Beltrami-Mitchell** for continuity expressed in terms of stresses.

In equation (10-3.7) the term $\tau_{ij,kk}$ represents the stresses τ_{ij} differentiated with respect to kk, of second derivatives of x, y, and z. Thus, is the $\tau_{ij,kk}$ represents the Laplacian term of σ_x in equation (10-3.6).

318

Also, the term $\tau_{kk,ij}$ represents the sum of the three stresses $\sigma_x + \sigma_y + \sigma_z$, differentiated with respect to ij. Thus, is the $\tau_{kk,ij}$ represents the second derivatives of the principal stress Ξ term in equation (10-3.6).

Similarly, we could prove the remaining five equations of **Beltrami-Michell's** equations, which are an alternative form of **Saint Venant's strain continuity** equations.

We now need to address the significance of the terms on the R.H.S. of equation (10-3.6). We could assume that in planar stress or planar strain problems, the R. H. S. vanishes on the assumption that equations (10-2) and (10-1) imply that only first derivatives of normal stresses exist along the directions of those stresses and only second derivatives exist in directions different from the directions of normal stresses.

Therefore,

$$(1+v)\nabla^2\sigma_x + \frac{\partial^2\Xi}{\partial x^2} = 0 \tag{10-4.1}$$

$$(1+v)\nabla^2\sigma_y + \frac{\partial^2\Xi}{\partial y^2} = 0 \tag{10-4.2}$$

$$(1+v)\nabla^2\sigma_z + \frac{\partial^2\Xi}{\partial z^2} = 0 \tag{10-4.3}$$

$$(1+v)\nabla^2\tau_{xy} + \frac{\partial^2\Xi}{\partial x\partial y} = 0 \tag{10-4.4}$$

$$(1+v)\nabla^2\tau_{yz} + \frac{\partial^2\Xi}{\partial y\partial z} = 0 \tag{10-4.5}$$

$$(1+v)\nabla^2\tau_{xz} + \frac{\partial^2\Xi}{\partial x\partial z} = 0 \tag{10-4.6}$$

10-4. Maxwell's stress functions

The general solution for stresses can be represented in terms of three arbitrary functions as follows

$$\varphi_1(x,y,z), \quad \varphi_2(x,y,z), \quad \varphi_3(x,y,z), \tag{10-5.1}$$

Such that the shearing stresses are defined as

$$\tau_{yz} = -\frac{\partial^2\varphi_1}{\partial y\partial z}, \quad \tau_{zx} = -\frac{\partial^2\varphi_2}{\partial x\partial z}, \quad \tau_{yx} = -\frac{\partial^2\varphi_3}{\partial x\partial y} \tag{10-5.2}$$

Thus, the three stress functions; φ_1, φ_2, and φ_3, correspond to the three coordinates in cyclical manner. We also acknowledge the condition of conjugation of shear stresses $\tau_{xy} = \tau_{yx}$, $\tau_{xz} = \tau_{zx}$, and $\tau_{zy} = \tau_{yz}$. It remains to determine the three proposed functions such that two sets of equations of **equilibrium**, (10-1), and **continuity**, (10-2) or (10-3), are satisfied.

Substituting from equation (10-5.2) into Navier's equations of equilibrium (10-1), we get

$$\frac{\partial \sigma_x}{\partial x} = \frac{\partial^3 \varphi_2}{\partial x \partial z^2} + \frac{\partial^3 \varphi_3}{\partial x \partial y^2} \tag{10-5.3}$$

Integrating with respect to x, we get

$$\sigma_x = \frac{\partial^2 \varphi_2}{\partial z^2} + \frac{\partial^2 \varphi_3}{\partial y^2} \tag{10-5.4}$$

Similarly, we get

$$\sigma_y = \frac{\partial^2 \varphi_1}{\partial z^2} + \frac{\partial^2 \varphi_3}{\partial x^2} \tag{10-5.5}$$

$$\sigma_z = \frac{\partial^2 \varphi_1}{\partial y^2} + \frac{\partial^2 \varphi_2}{\partial x^2} \tag{10-5.6}$$

The three stress functions are determined by double integration of equations (10-5.2) as follows

$$\varphi_1 = -\iint \tau_{yz} dy dz$$
$$\varphi_2 = -\iint \tau_{xz} dx dz \tag{10-5.7}$$
$$\varphi_3 = -\iint \tau_{yx} dy dx$$

We can easily reverse the process of differentiation to arrive to equations of stresses as follows.

$$\frac{\partial^2 \varphi_1}{\partial z^2} = -\int \frac{\partial \tau_{yz}}{\partial z} dy, \qquad \frac{\partial^2 \varphi_1}{\partial y^2} = -\int \frac{\partial \tau_{yz}}{\partial y} dz$$

$$\frac{\partial^2 \varphi_2}{\partial x^2} = -\int \frac{\partial \tau_{xz}}{\partial x} dz, \qquad \frac{\partial^2 \varphi_2}{\partial z^2} = -\int \frac{\partial \tau_{xz}}{\partial z} dx \tag{10-5.8}$$

$$\frac{\partial^2 \varphi_3}{\partial y^2} = -\int \frac{\partial \tau_{yx}}{\partial y} dx, \qquad \frac{\partial^2 \varphi_3}{\partial x^2} = -\int \frac{\partial \tau_{yx}}{\partial x} dy$$

Which could be substituted in equations (10-5.4 through 6) to get the normal stresses as follows

$$\sigma_x = \int \left(-\frac{\partial \tau_{xz}}{\partial z} - \frac{\partial \tau_{yx}}{\partial y} \right) dx$$

$$\frac{\partial \sigma_x}{\partial x} = -\frac{\partial \tau_{xz}}{\partial z} - \frac{\partial \tau_{yx}}{\partial y} \tag{10-5.9}$$

Which conform to Navier's equations (10-1). Similarly, we could derive the remaining two of Navier's equation.

Summary

Table 10-1. Summary of solution of general elasticity problems by Maxwell's stress functions

$\varphi_1(x,y,z)$, $\quad \varphi_2(x,y,z)$, $\quad \varphi_3(x,y,z)$,	(10-5.1)
$\tau_{yz} = -\dfrac{\partial^2 \varphi_1}{\partial y \partial z}$, $\quad \tau_{zx} = -\dfrac{\partial^2 \varphi_2}{\partial x \partial z}$, $\quad \tau_{yx} = -\dfrac{\partial^2 \varphi_3}{\partial x \partial y}$	(10-5.2)
$\sigma_x = \dfrac{\partial^2 \varphi_2}{\partial z^2} + \dfrac{\partial^2 \varphi_3}{\partial y^2}$	(10-5.4)
$\sigma_y = \dfrac{\partial^2 \varphi_1}{\partial z^2} + \dfrac{\partial^2 \varphi_3}{\partial x^2}$	(10-5.5)
$\sigma_z = \dfrac{\partial^2 \varphi_1}{\partial y^2} + \dfrac{\partial^2 \varphi_2}{\partial x^2}$	(10-5.6)
$\varphi_1 = -\iint \tau_{yz} dy dz$, $\quad \varphi_2 = -\iint \tau_{xz} dx dz$, $\quad \varphi_3 = -\iint \tau_{yx} dy dx$	(10-5.7)

10-5. Morera's stress functions

The general solution for stresses can be represented in terms of three arbitrary functions as follows

$$\psi_1(x,y,z), \quad \psi_2(x,y,z), \quad \psi_3(x,y,z), \tag{10-6.1}$$

Such that

$$\sigma_x = \frac{\partial^2 \psi_1}{\partial y \partial z}, \quad \sigma_y = \frac{\partial^2 \psi_2}{\partial x \partial z}, \quad \sigma_z = \frac{\partial^2 \psi_3}{\partial x \partial y} \tag{10-6.2}$$

Also, the three stress functions; ψ_1, ψ_2, and ψ_3, correspond to the three coordinates in cyclical manner. Substituting from equation (10-6.2) into Navier's equations of equilibrium (10-1), we get

$$
\begin{aligned}
\frac{\partial^3 \psi_1}{\partial x \partial y \partial z} &= -\frac{\partial \tau_{xy}}{\partial y} - \frac{\partial \tau_{xz}}{\partial z} \\
\frac{\partial^3 \psi_2}{\partial x \partial y \partial z} &= -\frac{\partial \tau_{xy}}{\partial x} - \frac{\partial \tau_{yz}}{\partial z} \\
\frac{\partial^3 \psi_3}{\partial x \partial y \partial z} &= -\frac{\partial \tau_{xz}}{\partial x} - \frac{\partial \tau_{yz}}{\partial y}
\end{aligned}
\tag{10-6.3}
$$

Differentiating the above three equations with respect to the third coordinate, absent on the right denominators, and adding by members according to the following scheme, we get

$$\frac{\partial}{\partial x}\left(\frac{\partial^3 \psi_1}{\partial x \partial y \partial z}+\frac{\partial \tau_{xy}}{\partial y}+\frac{\partial \tau_{xz}}{\partial z}\right)$$

$$+\frac{\partial}{\partial y}\left(\frac{\partial^3 \psi_2}{\partial x \partial y \partial z}+\frac{\partial \tau_{xy}}{\partial x}+\frac{\partial \tau_{yz}}{\partial z}\right)$$

$$-\frac{\partial}{\partial z}\left(\frac{\partial^3 \psi_3}{\partial x \partial y \partial z}+\frac{\partial \tau_{xz}}{\partial x}+\frac{\partial \tau_{yz}}{\partial y}\right)=0$$

Arranging, we get

$$2\frac{\partial^2 \tau_{xy}}{\partial x \partial y}=-\frac{\partial^3}{\partial x \partial y \partial z}\left(\frac{\partial \psi_1}{\partial x}+\frac{\partial \psi_2}{\partial y}-\frac{\partial \psi_3}{\partial z}\right) \qquad (10\text{-}6.4)$$

Integrating twice, we get

$$\tau_{xy}=-\frac{1}{2}\cdot\frac{\partial}{\partial z}\left(\frac{\partial \psi_1}{\partial x}+\frac{\partial \psi_2}{\partial y}-\frac{\partial \psi_3}{\partial z}\right) \qquad (10\text{-}6.5)$$

Similarly,

$$\tau_{yz}=-\frac{1}{2}\cdot\frac{\partial}{\partial x}\left(\frac{\partial \psi_2}{\partial y}+\frac{\partial \psi_3}{\partial z}-\frac{\partial \psi_1}{\partial x}\right) \qquad (10\text{-}6.6)$$

$$\tau_{xz}=-\frac{1}{2}\cdot\frac{\partial}{\partial y}\left(\frac{\partial \psi_1}{\partial x}+\frac{\partial \psi_3}{\partial z}-\frac{\partial \psi_2}{\partial y}\right) \qquad (10\text{-}6.7)$$

Summary

Table 10-2. Summary of solution of general elasticity problems by Morera's stress functions

$\psi_1(x,y,z),\quad \psi_2(x,y,z),\quad \psi_3(x,y,z),$	(10-6.1)
$\sigma_x=\dfrac{\partial^2 \psi_1}{\partial y \partial z},\quad \sigma_y=\dfrac{\partial^2 \psi_2}{\partial x \partial z},\quad \sigma_z=\dfrac{\partial^2 \psi_3}{\partial x \partial y}$	(10-6.2)
$\tau_{xy}=-\dfrac{1}{2}\cdot\dfrac{\partial}{\partial z}\left(\dfrac{\partial \psi_1}{\partial x}+\dfrac{\partial \psi_2}{\partial y}-\dfrac{\partial \psi_3}{\partial z}\right)$	(10-6.5)
$\tau_{yz}=-\dfrac{1}{2}\cdot\dfrac{\partial}{\partial x}\left(\dfrac{\partial \psi_2}{\partial y}+\dfrac{\partial \psi_3}{\partial z}-\dfrac{\partial \psi_1}{\partial x}\right)$	(10-6.6)

$$\tau_{xz} = -\frac{1}{2} \cdot \frac{\partial}{\partial y} \left(\frac{\partial \psi_1}{\partial x} + \frac{\partial \psi_3}{\partial z} - \frac{\partial \psi_2}{\partial y} \right) \qquad (10\text{-}6.7)$$

10-6. Generalized stress functions

Assuming an arbitrary function $f(x,y,z)$ and three constants a, b, and c, we could easily reconstruct the two set of functions of Maxwell and Morera, as follows

$$\sigma_x = (b+c)\frac{\partial^4 f}{\partial y^2 \partial z^2} \qquad (10\text{-}7.1)$$

$$\sigma_y = (a+c)\frac{\partial^4 f}{\partial x^2 \partial z^2} \qquad (10\text{-}7.2)$$

$$\sigma_z = (b+a)\frac{\partial^4 f}{\partial y^2 \partial x^2} \qquad (10\text{-}7.3)$$

$$\tau_{xy} = -c\frac{\partial^4 f}{\partial x \partial y \partial z^2} \qquad (10\text{-}7.4)$$

$$\tau_{yz} = -a\frac{\partial^4 f}{\partial z \partial y \partial x^2} \qquad (10\text{-}7.5)$$

$$\tau_{xz} = -b\frac{\partial^4 f}{\partial x \partial z \partial y^2} \qquad (10\text{-}7.6)$$

Equations (10-7) equally satisfy Navier's equations of equilibrium.

Comparing equations (10-7) with equations (10-6) and (10-5) we noticed that

Table 10-3. Comparison between Morera's and Maxwell's stress functions

Maxwell's stress functions	Morera's stress functions
$\displaystyle\sum_{n=1}^{\infty} a_n \frac{\partial^2 f_n}{\partial x^2} = \varphi_1(x,y,z)$	$\displaystyle\sum_{n=1}^{\infty} (c_n + b_n)\frac{\partial^2 f_n}{\partial z \partial y} = \psi_1(x,y,z)$
$\displaystyle\sum_{n=1}^{\infty} b_n \frac{\partial^2 f_n}{\partial y^2} = \varphi_2(x,y,z)$	$\displaystyle\sum_{n=1}^{\infty} (a_n + c_n)\frac{\partial^2 f_n}{\partial x \partial z} = \psi_2(x,y,z)$
$\displaystyle\sum_{n=1}^{\infty} c_n \frac{\partial^2 f_n}{\partial z^2} = \varphi_3(x,y,z)$	$\displaystyle\sum_{n=1}^{\infty} (a_n + b_n)\frac{\partial^2 f_n}{\partial x \partial y} = \psi_3(x,y,z)$

10-7. Navier's equilibrium equations in cylindrical coordinates

The three equations of static equilibrium with vanishing internal body forces were driven before as follows

$$\frac{\partial \sigma_r}{\partial r} + \frac{\partial \tau_{r\theta}}{r\partial \theta} + \frac{\sigma_r - \sigma_\theta}{r} + \frac{\partial \tau_{rz}}{\partial z} = 0 \tag{10-8.1}$$

$$\frac{\partial \sigma_\theta}{r\partial \theta} + \frac{\partial \tau_{r\theta}}{\partial r} + \frac{\partial \tau_{\theta z}}{\partial z} + \frac{2\tau_{r\theta}}{r} = 0 \tag{10-8.2}$$

$$\frac{\partial \sigma_z}{\partial z} + \frac{\partial \tau_{z\theta}}{r\partial \theta} + \frac{\partial \tau_{zr}}{\partial r} + \frac{\tau_{zr}}{r} = 0 \tag{10-8.3}$$

The general solutions of equations (10-8) are proposed as before in terms of three arbitrary functions as follows;

$$f_1(r,\theta,z), \qquad f_2(r,\theta,z), \qquad f_3(r,\theta,z) \tag{10-8.4}$$

$$\sigma_r = \frac{1}{r} \cdot \frac{\partial^2 f_1}{\partial z^2}$$

$$\sigma_\theta = \frac{\partial^2 f_2}{\partial z^2} \tag{10-8.5}$$

$$\tau_{r\theta} = \frac{1}{r} \cdot \frac{\partial^2 f_3}{\partial z^2}$$

Equations (10-8.5) determine the structure of the three functions (10-8.4).

(1) Substituting from (10-8.5a) into Navier's equation (10-8.1), we get

$$\frac{\partial}{\partial r}\left(\frac{1}{r} \cdot \frac{\partial^2 f_1}{\partial z^2}\right) + \frac{\partial}{r\partial \theta}\left(\frac{1}{r} \cdot \frac{\partial^2 f_3}{\partial z^2}\right) + \frac{1}{r}\left(\frac{1}{r} \cdot \frac{\partial^2 f_1}{\partial z^2} - \frac{\partial^2 f_2}{\partial z^2}\right) + \frac{\partial \tau_{rz}}{\partial z} = 0$$

Or,

$$\left(\frac{1}{r} \cdot \frac{\partial^3 f_1}{\partial r\partial z^2} - \frac{1}{r^2} \cdot \frac{\partial^2 f_1}{\partial z^2}\right) + \left(\frac{1}{r^2} \cdot \frac{\partial^3 f_3}{\partial \theta\partial z^2}\right) + \left(\frac{1}{r^2} \cdot \frac{\partial^2 f_1}{\partial z^2} - \frac{1}{r} \cdot \frac{\partial^2 f_2}{\partial z^2}\right) + \frac{\partial \tau_{rz}}{\partial z} = 0$$

i.e.,

$$\frac{1}{r} \cdot \frac{\partial^3 f_1}{\partial r\partial z^2} + \frac{1}{r^2} \cdot \frac{\partial^3 f_3}{\partial \theta\partial z^2} - \frac{1}{r} \frac{\partial^2 f_2}{\partial z^2} + \frac{\partial \tau_{rz}}{\partial z} = 0$$

$$\frac{\partial \tau_{rz}}{\partial z} = -\frac{1}{r} \cdot \frac{\partial^3 f_1}{\partial r\partial z^2} - \frac{1}{r^2} \cdot \frac{\partial^3 f_3}{\partial \theta\partial z^2} + \frac{1}{r} \frac{\partial^2 f_2}{\partial z^2}$$

Integrating with respect to z, we get

324

$$\tau_{rz} = -\frac{1}{r}\cdot\frac{\partial^2 f_1}{\partial r\partial z} - \frac{1}{r^2}\cdot\frac{\partial^2 f_3}{\partial\theta\partial z} + \frac{1}{r}\frac{\partial f_2}{\partial z}$$ (10-8.6)

(2) Substituting from (10-8.5a) into Navier's equation (10-8.2), we get

$$\frac{\partial}{r\partial\theta}\left(\frac{\partial^2 f_2}{\partial z^2}\right) + \frac{\partial}{\partial r}\left(\frac{1}{r}\frac{\partial^2 f_3}{\partial z^2}\right) + \frac{\partial\tau_{\theta z}}{\partial z} + \frac{2}{r}\left(\frac{1}{r}\frac{\partial^2 f_3}{\partial z^2}\right) = 0$$

Or,

$$\frac{1}{r}\frac{\partial^3 f_2}{\partial\theta\partial z^2} + \left(\frac{1}{r}\frac{\partial^3 f_3}{\partial r\partial z^2} - \frac{1}{r^2}\frac{\partial^2 f_3}{\partial z^2}\right) + \frac{\partial\tau_{\theta z}}{\partial z} + \frac{2}{r^2}\frac{\partial^2 f_3}{\partial z^2} = 0$$

Arranging, we get

$$\frac{\partial\tau_{\theta z}}{\partial z} = -\frac{1}{r}\frac{\partial^3 f_2}{\partial\theta\partial z^2} - \frac{1}{r}\frac{\partial^3 f_3}{\partial r\partial z^2} - \frac{1}{r^2}\frac{\partial^2 f_3}{\partial z^2}$$

Integrating, we get

$$\tau_{\theta z} = -\frac{1}{r}\frac{\partial^2 f_2}{\partial\theta\partial z} - \frac{1}{r}\frac{\partial^2 f_3}{\partial r\partial z} - \frac{1}{r^2}\frac{\partial f_3}{\partial z}$$ (10-8.7)

(3) Substituting from (10-8.6 and 8.7) into Navier's equation (10-8.3), we get

$$\frac{\partial\sigma_z}{\partial z} + \frac{\partial}{r\partial\theta}\left(-\frac{1}{r}\frac{\partial^2 f_2}{\partial\theta\partial z} - \frac{1}{r}\frac{\partial^2 f_3}{\partial r\partial z} - \frac{1}{r^2}\frac{\partial f_3}{\partial z}\right)$$
$$+ \left(\frac{\partial}{\partial r} + \frac{1}{r}\right)\left(-\frac{1}{r}\cdot\frac{\partial^2 f_1}{\partial r\partial z} - \frac{1}{r^2}\cdot\frac{\partial^2 f_3}{\partial\theta\partial z} + \frac{1}{r}\frac{\partial f_2}{\partial z}\right) = 0$$

Arranging, we get

$$\frac{\partial\sigma_z}{\partial z} - \frac{1}{r^2}\frac{\partial^3 f_2}{\partial\theta^2\partial z} - \frac{2}{r^2}\frac{\partial^3 f_3}{\partial\theta\partial r\partial z} + \frac{1}{r}\cdot\frac{\partial^3 f_1}{\partial r^2\partial z} + \frac{1}{r}\frac{\partial^2 f_2}{\partial r\partial z} = 0$$

Integrating with respect to z, we get

$$\sigma_z = \frac{1}{r^2}\frac{\partial^2 f_2}{\partial\theta^2} + \frac{2}{r^2}\frac{\partial^2 f_3}{\partial\theta\partial r} + \frac{1}{r}\cdot\frac{\partial^2 f_1}{\partial r^2} - \frac{1}{r}\frac{\partial f_2}{\partial r}$$ (10-8.8)

Summary of proposed solution

Table 10-4. Navier's equilibrium equations in cylindrical coordinates

$f_1(r,\theta,z),\qquad f_2(r,\theta,z),\qquad f_3(r,\theta,z)$	(10-8.4)
$\sigma_r = \dfrac{1}{r}\cdot\dfrac{\partial^2 f_1}{\partial z^2},\qquad \sigma_\theta = \dfrac{\partial^2 f_2}{\partial z^2},\qquad \tau_{r\theta} = \dfrac{1}{r}\cdot\dfrac{\partial^2 f_3}{\partial z^2}$	(10-8.5)
$\tau_{rz} = -\dfrac{1}{r}\cdot\dfrac{\partial^2 f_1}{\partial r\partial z} - \dfrac{1}{r^2}\cdot\dfrac{\partial^2 f_3}{\partial\theta\partial z} + \dfrac{1}{r}\dfrac{\partial f_2}{\partial z}$	(10-8.6)
$\tau_{\theta z} = -\dfrac{1}{r}\dfrac{\partial^2 f_2}{\partial\theta\partial z} - \dfrac{1}{r}\dfrac{\partial^2 f_3}{\partial r\partial z} - \dfrac{1}{r^2}\dfrac{\partial f_3}{\partial z}$	(10-8.7)
$\sigma_z = \dfrac{1}{r^2}\dfrac{\partial^2 f_2}{\partial\theta^2} + \dfrac{2}{r^2}\dfrac{\partial^2 f_3}{\partial\theta\partial r} + \dfrac{1}{r}\cdot\dfrac{\partial^2 f_1}{\partial r^2} - \dfrac{1}{r}\dfrac{\partial f_2}{\partial r}$	(10-8.8)

Plane stress in cylindrical coordinates

(1) Ignoring the angle dependent terms

The solution of problems of plain stress in cylindrical coordinates is obtained by ignoring the θ-dependent terms in the above equations as follows

$$\sigma_r = \frac{1}{r}\cdot\frac{\partial^2 f_1}{\partial z^2},\qquad \sigma_\theta = \frac{\partial^2 f_2}{\partial z^2},\qquad \tau_{r\theta} = \frac{1}{r}\cdot\frac{\partial^2 f_3}{\partial z^2} \tag{10-9.1}$$

$$\tau_{rz} = \frac{1}{r}\cdot\frac{\partial}{\partial z}\left(f_2 - \frac{\partial f_1}{\partial r}\right) \tag{10-9.2}$$

$$\tau_{\theta z} = -\frac{1}{r}\frac{\partial}{\partial z}\left(\frac{\partial f_3}{\partial r} + \frac{f_3}{r}\right) = -\frac{1}{r^2}\frac{\partial^2(rf_3)}{\partial z\partial r} \tag{10-9.3}$$

$$\sigma_z = -\frac{1}{r}\cdot\frac{\partial}{\partial r}\left(f_2 - \frac{\partial f_1}{\partial r}\right) \tag{10-9.4}$$

(2) Substitution of by arbitrary function

$$\psi = \frac{\partial^2 f_1}{\partial z^2} \tag{10-10.1}$$

$$\varphi = \left(f_2 - \frac{\partial f_1}{\partial r}\right) \tag{10-10.2}$$

$$\chi = \frac{\partial(rf_3)}{\partial z} \tag{10-10.3}$$

Equation (10-10.2) is re-written as

$$f_2 = \varphi + \frac{\partial f_1}{\partial r} \tag{10-10.4a}$$

Differentiating twice and substituting by (10-10.1), we get

$$\frac{\partial^2 f_2}{\partial z^2} = \frac{\partial^2 \varphi}{\partial z^2} + \frac{\partial \psi}{\partial r} \qquad\qquad (10\text{-}10.4b)$$

(3) Plane stresses in cylindrical coordinates

Substituting by the functions defined in equations (10-10) in the stress equations (10-9) we get

$$\sigma_r = \frac{\psi}{r} \qquad\qquad (10\text{-}11.1)$$

$$\sigma_\theta = \frac{\partial^2 \varphi}{\partial z^2} + \frac{\partial \psi}{\partial r} \qquad\qquad (10\text{-}11.2)$$

$$\tau_{r\theta} = \frac{1}{r^2} \cdot \frac{\partial \chi}{\partial z} \qquad\qquad (10\text{-}11.3)$$

$$\tau_{rz} = \frac{1}{r} \cdot \frac{\partial \varphi}{\partial z} \qquad\qquad (10\text{-}11.4)$$

$$\tau_{\theta z} = -\frac{1}{r^2} \cdot \frac{\partial \chi}{\partial r} \qquad\qquad (10\text{-}11.5)$$

$$\sigma_z = -\frac{1}{r} \cdot \frac{\partial \varphi}{\partial r} \qquad\qquad (10\text{-}11.6)$$

(4) Axi-symmetrical three-dimensional stresses

$$\chi = 0 \qquad \rightarrow \quad \tau_{r\theta} = \tau_{z\theta} = 0 \rightarrow \quad \tau_{rz} = \tau_{zr} \quad \text{remain}$$

$$\rightarrow \quad \alpha_{r\theta} = \frac{1}{\mu}\tau_{r\theta} = 0 \qquad\qquad (10\text{-}12)$$

$$\rightarrow \quad \alpha_{z\theta} = \frac{1}{\mu}\tau_{z\theta} = 0$$

Thus, when χ vanished, angular shear stresses vanish, only radial and axial shear stresses exist.

(5) Bars with body of revolution

$$\psi = \varphi = 0 \qquad \rightarrow \quad \sigma_r = \sigma_\theta = \tau_{rz} = 0 \rightarrow \quad \tau_{r\theta} = \tau_{z\theta} \quad \text{remain}$$

$$\rightarrow \quad \alpha_{rz} = \frac{1}{\mu}\tau_{rz} = 0 \qquad\qquad (10\text{-}13)$$

Thus, when $\psi = \varphi = 0$ vanished, angular shear stresses do not vanish.

327

10-8. Harmonic equation

(i) Derivative of a harmonic function with respect to its argument

A **harmonic function** or **Laplacian potential**, $\varphi(x,y,z,a,b,.t)$, satisfies the Laplacian equation

$$\nabla^2\varphi = 0 \tag{10-14.1}$$

Such that partial derivatives of φ with respect to any of the arguments of x, y, z, . t, will satisfy equation (10-14.1) as follows

$$\nabla^2\left(\frac{\partial\varphi}{\partial t}\right) = \frac{\partial}{\partial t}\left(\nabla^2\varphi\right) = 0 \tag{10-14.2}$$

The harmonic function has the following features:

1. Single-valued and finite
2. First derivative exists within the field
3. At least second derivative must exist
4. There exits infinite numbers of harmonic functions that satisfy Laplace equation

Example 29

$$\varphi = \frac{1}{\sqrt{x^2 + y^2 + z^2}} = \frac{1}{r} \tag{10-14.3}$$

i.e.,

$$r^2 = x^2 + y^2 + z^2$$

$$\frac{\partial}{\partial x}\left(r^2\right) = 2r\frac{\partial r}{\partial x} = 2x$$

Thus,

$$\frac{\partial r}{\partial x} = \frac{x}{r}, \quad \frac{\partial r}{\partial y} = \frac{y}{r}, \quad \frac{\partial r}{\partial z} = \frac{z}{r} \tag{10-14.4}$$

Also

$$\frac{\partial}{\partial x}\left(\frac{1}{r}\right) = -\frac{1}{r^2}\frac{\partial r}{\partial x} = -\frac{x}{r^3}$$

Thus, the three gradients are

$$\frac{\partial}{\partial x}\left(\frac{1}{r}\right) = -\frac{x}{r^3} \tag{10-14.5a}$$

$$\frac{\partial}{\partial y}\left(\frac{1}{r}\right) = -\frac{y}{r^3}$$ (10-14.5b)

$$\frac{\partial}{\partial z}\left(\frac{1}{r}\right) = -\frac{z}{r^3}$$ (10-14.5c)

The three gradients are written in terms of the three projections as follows

$$\frac{\partial}{\partial x}\left(\frac{1}{r}\right) = -\frac{1}{r^2}\left(\frac{x}{r}\right) = -\frac{1}{r^2}(\hat{r}.\hat{x})$$ (10-14.5d)

$$\frac{\partial}{\partial y}\left(\frac{1}{r}\right) = -\frac{1}{r^2}\left(\frac{y}{r}\right) = -\frac{1}{r^2}(\hat{r}.\hat{y})$$ (10-14.5e)

$$\frac{\partial}{\partial z}\left(\frac{1}{r}\right) = -\frac{1}{r^2}\left(\frac{z}{r}\right) = -\frac{1}{r^2}(\hat{r}.\hat{z})$$ (10-14.5f)

Therefore, $\frac{1}{r^2}$ is the magnitude of the gradient of the harmonic function $\varphi = \frac{1}{r}$, is directed along the vector $\hat{r}$.

Also, the three gradients, equations (10-14.5), are also harmonic functions, as we proved in equation (10-14.2).

The second derivatives are obtained in the same way as follows

$$\frac{\partial^2}{\partial x^2}\left(\frac{1}{r}\right) = -\frac{1}{r^3} + \frac{3x^2}{r^5}$$ (10-14.6a)

$$\frac{\partial^2}{\partial y^2}\left(\frac{1}{r}\right) = -\frac{1}{r^3} + \frac{3y^2}{r^5}$$ (10-14.6b)

$$\frac{\partial^2}{\partial z^2}\left(\frac{1}{r}\right) = -\frac{1}{r^3} + \frac{3z^2}{r^5}$$ (10-14.6c)

Adding equations (10-14.6), we get

$$\nabla^2\left(\frac{1}{r}\right) = -\frac{3}{r^3} + \frac{3(x^2 + y^2 + z^2)}{r^5} = 0$$ (10-14.7)

Similarly, the three gradients, equations (10-14.6), are also harmonic functions because

$$\nabla^2\left(\nabla^2\left(\frac{1}{r}\right)\right) = \nabla^2\left(-\frac{3}{r^3} + \frac{3(x^2 + y^2 + z^2)}{r^5}\right) = 0$$ (10-14.8)

Keep in mind that

$$\nabla^4 = \left(\frac{\partial^2}{\partial x^2} + \frac{\partial^2}{\partial y^2} + \frac{\partial^2}{\partial z^2}\right)\left(\frac{\partial^2}{\partial x^2} + \frac{\partial^2}{\partial y^2} + \frac{\partial^2}{\partial z^2}\right)$$
$$= \left(\frac{\partial^4}{\partial x^4} + \frac{\partial^4}{\partial y^4} + \frac{\partial^4}{\partial z^4} + \frac{2\partial^4}{\partial y^2 \partial z^2} + \frac{2\partial^2}{\partial x^2 \partial y^2} + \frac{2\partial^4}{\partial x^2 \partial z^2}\right)$$

(10-14.9)

.

(ii) Laplacian of a product of harmonic function by its argument

Prove that

$$\nabla^2[x\varphi(x,y,z)] = 2\frac{\partial\varphi}{\partial x}$$

(10-15.1)

Solution

First differentiation of $x\varphi$ gives

$$\frac{\partial}{\partial x}[x\varphi(x,y,z)] = \varphi + x\frac{\partial\varphi}{\partial x}$$
$$\frac{\partial}{\partial y}[x\varphi(x,y,z)] = x\frac{\partial\varphi}{\partial y}$$
$$\frac{\partial}{\partial z}[x\varphi(x,y,z)] = x\frac{\partial\varphi}{\partial z}$$

(10-15.2)

Second differentiation of $x\varphi$ gives

$$\frac{\partial}{\partial x}\left[\varphi + x\frac{\partial\varphi}{\partial x}\right] = 2\frac{\partial\varphi}{\partial x} + x\frac{\partial^2\varphi}{\partial x^2}$$
$$\frac{\partial}{\partial y}\left[x\frac{\partial\varphi}{\partial y}\right] = x\frac{\partial^2\varphi}{\partial y^2}$$
$$\frac{\partial}{\partial z}\left[x\frac{\partial\varphi}{\partial z}\right] = x\frac{\partial^2\varphi}{\partial z^2}$$

(10-15.3)

Adding the three equations by members we get

$$\nabla^2(x\varphi) = 2\frac{\partial\varphi}{\partial x} + x\left(\frac{\partial^2\varphi}{\partial x^2} + \frac{\partial^2\varphi}{\partial y^2} + \frac{\partial^2\varphi}{\partial z^2}\right)$$

(10-15.3)

Since, for an harmonic function φ, $\nabla^2\varphi = 0$, therefore, we have

330

$$\nabla^2(x\varphi) = 2\frac{\partial\varphi}{\partial x} \tag{10-15.4}$$

(iii) Harmonic function from a product of derivative of harmonic function by its argument

Since, from equation (10-14.2), the derivative of harmonic function, $(\partial\varphi/\partial t)$, is also harmonic, and also, since from equation (10-15.4), the Laplacian of the product $x\varphi$ is equal to the derivative $(2\partial\varphi/\partial x)$ of the harmonic function, therefore, we will prove that the following function is also harmonic

$$\psi = x\frac{\partial\varphi}{\partial x} + y\frac{\partial\varphi}{\partial y} + z\frac{\partial\varphi}{\partial z} \tag{10-16.1}$$

Start with equation (10-15.4), we could write the same equation for three arbitrary harmonic functions as follows

$$\nabla^2(x\varphi_1) = 2\frac{\partial\varphi_1}{\partial x}$$
$$\nabla^2(y\varphi_2) = 2\frac{\partial\varphi_2}{\partial y} \tag{10-16.2}$$
$$\nabla^2(z\varphi_3) = 2\frac{\partial\varphi_3}{\partial z}$$

Substitute in (10-16.2) by

$$\frac{\partial\varphi_1}{\partial x} = \frac{\partial}{\partial x}\left(\frac{\partial\varphi}{\partial x}\right)$$
$$\frac{\partial\varphi_2}{\partial y} = \frac{\partial}{\partial y}\left(\frac{\partial\varphi}{\partial y}\right) \tag{10-16.3}$$
$$\frac{\partial\varphi_3}{\partial z} = \frac{\partial}{\partial z}\left(\frac{\partial\varphi}{\partial z}\right)$$

Therefore,

$$\nabla^2\left(x\frac{\partial\varphi}{\partial x} + y\frac{\partial\varphi}{\partial y} + z\frac{\partial\varphi}{\partial z}\right) = 2\nabla^2\varphi = 0 \tag{10-16.4}$$

Therefore, if φ is harmonic function, then $\psi = x\frac{\partial\varphi}{\partial x} + y\frac{\partial\varphi}{\partial y} + z\frac{\partial\varphi}{\partial z}$ is also harmonic function.

(iv) Relation between Cartesian and Polar Harmonic functions

331

The harmonic function $\psi = x\dfrac{\partial\varphi}{\partial x} + y\dfrac{\partial\varphi}{\partial y} + z\dfrac{\partial\varphi}{\partial z}$ can be written in terms of its Cartesian projections by writing the Cartesian derivatives in terms of the directional cosines and derivatives of the normal as follows

$$x = |r|\hat{x}.\hat{r}$$
$$y = |r|\hat{y}.\hat{r}$$
$$z = |r|\hat{z}.\hat{r}$$

(10-17.1)

i.e.,

$$\frac{\partial\varphi}{\partial x} = \frac{\partial\varphi}{\partial n}\hat{x}.\hat{r}$$
$$\frac{\partial\varphi}{\partial y} = \frac{\partial\varphi}{\partial n}\hat{y}.\hat{r}$$
$$\frac{\partial\varphi}{\partial z} = \frac{\partial\varphi}{\partial n}\hat{z}.\hat{r}$$

(10-17.2)

Substituting from (10-17.1) and (10-17.2) in (10-16.1) we get

$$\begin{aligned}
\psi &= x\frac{\partial\varphi}{\partial x} + y\frac{\partial\varphi}{\partial y} + z\frac{\partial\varphi}{\partial z} \\
&= \left(|r|\hat{x}.\hat{r}\right)\frac{\partial\varphi}{\partial n}\hat{x}.\hat{r} + \left(|r|\hat{y}.\hat{r}\right)\frac{\partial\varphi}{\partial n}\hat{y}.\hat{r} + \left(|r|\hat{z}.\hat{r}\right)\frac{\partial\varphi}{\partial n}\hat{z}.\hat{r} \\
&= |r|\frac{\partial\varphi}{\partial n}\left((\hat{x}.\hat{r})^2 + (\hat{y}.\hat{r})^2 + (\hat{z}.\hat{r})^2\right) \\
&= r\frac{\partial\varphi}{\partial r}
\end{aligned}$$

(10-17.3)

i.e.,

$$\psi = x\frac{\partial\varphi}{\partial x} + y\frac{\partial\varphi}{\partial y} + z\frac{\partial\varphi}{\partial z} = r\frac{\partial\varphi}{\partial r}$$

(10-17.4)

Therefore, we have constructed the harmonic function ψ, which can be easily represented in terms of polar coordinates in terms of the harmonic function φ, represented in the Cartesian coordinates.

10-9. Bi-Harmonic equation

The bi-harmonic equation

$$\nabla^4\varphi(x,y) = \nabla^2\left(\nabla^2\varphi(x,y)\right) = 0$$

(10-18.1)

332

was shown to be satisfied by the stress function φ in order to fulfill the continuity condition. We will discuss six forms of harmonic functions as solution of the **bi-harmonic equation** (10-18.1) as follows.

(i) Harmonic functions satisfy both harmonic and bi-harmonic equations

A harmonic function φ that satisfied Laplace's equation $\nabla^2\varphi = 0$ will also satisfy the bi-harmonic equation $\nabla^2(\nabla^2\varphi) = 0$. For example

$$\nabla^2\left(\frac{1}{r}\right) = -\frac{3}{r^3} + \frac{3(x^2 + y^2 + z^2)}{r^5} = 0 \qquad (10\text{-}19.1)$$

$$\nabla^4\left(\frac{1}{r}\right) = \nabla^2\left(-\frac{3}{r^3} + \frac{3(x^2 + y^2 + z^2)}{r^5}\right) = 0 \qquad (10\text{-}19.2)$$

(ii) Bi-harmonic solution of a product of harmonic function by its argument

We proved in equation (10-15.4) that if φ was a harmonic function, its derivative with any of its arguments is also harmonic function, and that $\nabla^2(x\varphi) = 2\dfrac{\partial\varphi}{\partial x}$, therefore

$$\nabla^2\left[\nabla^2(x\varphi)\right] = 2\nabla^2\left(\frac{\partial\varphi}{\partial x}\right) = 0 \qquad (10\text{-}20)$$

Similarly, the same applies to y and z. i.e., $x\varphi$, $y\varphi$, and $z\varphi$ are solutions of the bi-harmonic equation.

(iii) Bi-harmonic solution by sum of two harmonic functions

If φ and ψ are two harmonic functions and solutions of the bi-harmonic equation then their sum is also a solution for bi-harmonic equation

$$\nabla^2(\varphi + \psi) = \nabla^2\varphi + \nabla^2\psi \qquad (10\text{-}21.1)$$

Also

$$\nabla^2\left[\nabla^2(\varphi + \psi)\right] = \nabla^2\left(\nabla^2\varphi + \nabla^2\psi\right)$$

$$= \nabla^2\nabla^2\varphi + \nabla^2\nabla^2\psi = 0 \qquad (10\text{-}21.2)$$

(iv) Bi-harmonic solution by sum of harmonic function and its product by its arguments

If φ is harmonic function, then a sum of the same function and its derivatives or of the same and a product of its argument and its derivatives are also solutions of the bi-harmonic equation. Thus

$$\nabla^2\left[\nabla^2(\varphi + x\varphi)\right] = \nabla^2\left(\nabla^2\varphi + \nabla^2(x\varphi)\right) = 0$$
$$\nabla^2\left[\nabla^2(\varphi + y\varphi)\right] = \nabla^2\left(\nabla^2\varphi + \nabla^2(y\varphi)\right) = 0 \qquad (10\text{-}22.1)$$
$$\nabla^2\left[\nabla^2(\varphi + z\varphi)\right] = \nabla^2\left(\nabla^2\varphi + \nabla^2(z\varphi)\right) = 0$$

Similarly,

$$\nabla^2\left[\nabla^2\left(\varphi + x\frac{\partial\varphi}{\partial x}\right)\right] = \nabla^2\left(\nabla^2\varphi + \nabla^2\left(x\frac{\partial\varphi}{\partial x}\right)\right) = 0$$

$$\nabla^2\left[\nabla^2\left(\varphi + y\frac{\partial\varphi}{\partial x}\right)\right] = \nabla^2\left(\nabla^2\varphi + \nabla^2\left(y\frac{\partial\varphi}{\partial x}\right)\right) = 0 \qquad (10\text{-}22.2)$$

$$\nabla^2\left[\nabla^2\left(\varphi + z\frac{\partial\varphi}{\partial x}\right)\right] = \nabla^2\left(\nabla^2\varphi + \nabla^2\left(z\frac{\partial\varphi}{\partial x}\right)\right) = 0$$

Similarly, we could obtain similar equations for derivatives for y and z.

(v) Bi-harmonic solution by product of harmonic function by r^2

If φ is a harmonic function, then $r^2\varphi$ is a solution of the bi-harmonic equation.

$$\frac{\partial}{\partial x}(r^2\varphi) = 2x\varphi + r^2\frac{\partial\varphi}{\partial x}$$

$$\frac{\partial}{\partial y}(r^2\varphi) = 2y\varphi + r^2\frac{\partial\varphi}{\partial y} \qquad (10\text{-}23.1)$$

$$\frac{\partial}{\partial z}(r^2\varphi) = 2z\varphi + r^2\frac{\partial\varphi}{\partial z}$$

Differentiating again, we get

$$\frac{\partial}{\partial x}\left(2x\varphi + r^2\frac{\partial\varphi}{\partial x}\right) = 2\varphi + 4x\frac{\partial\varphi}{\partial x} + r^2\frac{\partial^2\varphi}{\partial x^2}$$

$$\frac{\partial}{\partial y}\left(2y\varphi + r^2\frac{\partial\varphi}{\partial y}\right) = 2\varphi + 4y\frac{\partial\varphi}{\partial y} + r^2\frac{\partial^2\varphi}{\partial y^2} \qquad (10\text{-}23.2)$$

$$\frac{\partial}{\partial z}\left(2z\varphi + r^2\frac{\partial\varphi}{\partial z}\right) = 2\varphi + 4z\frac{\partial\varphi}{\partial z} + r^2\frac{\partial^2\varphi}{\partial z^2}$$

Adding, we get

$$\nabla^2(r^2\varphi) = 6\varphi + 4\left(x\frac{\partial\varphi}{\partial x} + y\frac{\partial\varphi}{\partial y} + z\frac{\partial\varphi}{\partial z}\right) + r^2\nabla^2\varphi$$

$$= 6\varphi + 4\left(x\frac{\partial\varphi}{\partial x} + y\frac{\partial\varphi}{\partial y} + z\frac{\partial\varphi}{\partial z}\right)$$

(10-23.3)

Where the term $\nabla^2\varphi = 0$ and the remaining two terms on the RHS of the above equation is also harmonic, because it consists of sums of harmonic function and products of argument with derivatives of the same harmonic function.

Therefore,

$$\nabla^4(r^2\varphi) = 0$$

(10-23.4)

(vi) Bi-harmonic solution by sums of product of derivatives of harmonic function by r^2

Similarly, the sums of the harmonic function φ and the product of its derivative with r^2 are harmonic function and solutions of the bi-harmonic equation. Thus

$$\nabla^2\left[\nabla^2\left(\varphi + r^2\frac{\partial\varphi}{\partial x}\right)\right] = \nabla^2\left(\nabla^2\varphi + \nabla^2\left(r^2\frac{\partial\varphi}{\partial x}\right)\right) = 0$$

(10-24.1)

$$\nabla^2\left[\nabla^2\left(\varphi + (r^2 - a^2)\frac{\partial\varphi}{\partial x}\right)\right] = \nabla^2\left(\nabla^2\varphi + \nabla^2\left((r^2 - a^2)\frac{\partial\varphi}{\partial x}\right)\right) = 0$$

(10-24.2)

Where, a is arbitrary constant. The harmonic function in equation (10-24.2) is used for solution of problems of **elastic sphere.**

Summary of solutions of bi-harmonic equation

Table 10-5. Properties of harmonic and bi-harmonic equations.

$F_1 = x\varphi$, $F_1 = \dfrac{\partial\varphi}{\partial x}$, $F_1 = \varphi + y\dfrac{\partial\varphi}{\partial x}$
$F_2 = \varphi + \psi$
$F_3 = \varphi + x\dfrac{\partial\varphi}{\partial x}$, $F_3 = \varphi + y\dfrac{\partial\varphi}{\partial x}$, $F_3 = \varphi + z\dfrac{\partial\varphi}{\partial x}$
$F_4 = r^2\varphi$
$F_5 = \varphi + r^2\dfrac{\partial\varphi}{\partial x}$
$F_6 = \varphi + (r^2 - a^2)\dfrac{\partial\varphi}{\partial x}$

10-10. Bi-Harmonic equation in elasticity

From equations (4-4), we recall the following **Lamé's equations**

$$\mu\nabla^2 u + (\mu+\lambda)\frac{\partial\varepsilon}{\partial x} + \rho X_i = \rho\frac{\partial^2 u}{\partial t^2} \qquad (10\text{-}25.1)$$

$$\mu\nabla^2 v + (\mu+\lambda)\frac{\partial\varepsilon}{\partial y} + \rho Y_i = \rho\frac{\partial^2 v}{\partial t^2} \qquad (10\text{-}25.2)$$

$$\mu\nabla^2 w + (\mu+\lambda)\frac{\partial\varepsilon}{\partial z} + \rho Z_i = \rho\frac{\partial^2 w}{\partial t^2} \qquad (10\text{-}25.3)$$

In the case of vanishing internal forces ($X_i = Y_i = Z_i = 0$) and external accelerations ($\frac{\partial^2 u}{\partial t^2} = \frac{\partial^2 v}{\partial t^2} = \frac{\partial^2 w}{\partial t^2} = 0$), **Lamé's equations** become

$$\nabla^2 u + (k+1)\frac{\partial\varepsilon}{\partial x} = 0 \qquad (10\text{-}26.1)$$

$$\nabla^2 v + (k+1)\frac{\partial\varepsilon}{\partial y} = 0 \qquad (10\text{-}26.2)$$

$$\nabla^2 w + (k+1)\frac{\partial\varepsilon}{\partial z} = 0 \qquad (10\text{-}26.3)$$

Where,

$$k = \frac{\lambda}{\mu} = \frac{2v}{(1-2v)} \qquad (10\text{-}26.4)$$

Thus, equations (10-26) are expressed solely in terms of Poisson's constant v, and represent homogeneous medium, free of internal forces and in equilibrium static equilibrium. Solutions of **Lamé's** equations (10-26) must satisfy the boundary conditions of cases under consideration in addition to the differential constraints of (10-26).

Thus, equations (3-7), are simplified as follows:

$$\sigma_x = \mu(2\varepsilon_{xx} + k\varepsilon) \qquad (10\text{-}27.1)$$
$$\sigma_y = \mu(2\varepsilon_{yy} + k\varepsilon) \qquad (10\text{-}27.2)$$
$$\sigma_z = \mu(2\varepsilon_{zz} + k\varepsilon) \qquad (10\text{-}27.3)$$
$$\tau_{xy} = \mu\alpha_{xy} \qquad (10\text{-}27.4)$$

$$\tau_{yz} = \mu\alpha_{yz} \tag{10-27.5}$$
$$\tau_{zx} = \mu\alpha_{zx} \tag{10-27.6}$$

From **Lamé's equations** (10-26) and equations (3-5.2), $\varepsilon = \dfrac{1-2v}{E}\Xi$, we get

$$\nabla^2\varepsilon = 0$$
$$\nabla^2\Xi = 0 \tag{10-28}$$

From Table 10-5, the properties of harmonic functions, equations (10-28) imply the following

(i)
$$\nabla^2\frac{\partial\varepsilon}{\partial x} = 0, \quad \nabla^2\frac{\partial\varepsilon}{\partial y} = 0, \quad \nabla^2\frac{\partial\varepsilon}{\partial z} = 0$$
$$\nabla^2\frac{\partial\Xi}{\partial x} = 0, \quad \nabla^2\frac{\partial\Xi}{\partial y} = 0, \quad \nabla^2\frac{\partial\Xi}{\partial z} = 0 \tag{10-28}$$

Farther, we the derivatives of the harmonic functions ε and Ξ are also harmonic functions as follows

(ii)
$$\frac{\partial\varepsilon}{\partial x}, \quad \frac{\partial\varepsilon}{\partial y}, \quad \frac{\partial\varepsilon}{\partial z}$$
$$\frac{\partial\Xi}{\partial x}, \quad \frac{\partial\Xi}{\partial y}, \quad \frac{\partial\Xi}{\partial z} \tag{10-29}$$

Thus, the properties of harmonic functions enable us to prove that the **Lamé's** and **Beltrami's** equations which relate geometrical stresses to material strains in the absence of body forces and under static equilibrium satisfy the bi-harmonic equations as follows

(iii)
$$\nabla^2\nabla^2\varepsilon = 0$$
$$\nabla^2\nabla^2 u = 0$$
$$\nabla^2\nabla^2 v = 0$$
$$\nabla^2\nabla^2 w = 0$$
$$\nabla^2\nabla^2\sigma_x = 0$$
$$\nabla^2\nabla^2\sigma_y = 0$$
$$\nabla^2\nabla^2\sigma_z = 0$$
$$\nabla^2\nabla^2\tau_{xy} = 0$$
$$\nabla^2\nabla^2\tau_{xz} = 0$$
$$\nabla^2\nabla^2\tau_{zy} = 0 \tag{10-30}$$

Because, equations (3-4) and (3-5) give

$$\Xi = \frac{\varepsilon E}{(1 - 2v)}$$

$$\Xi = \sigma_x + \sigma_y + \sigma_z$$
$$\varepsilon = \varepsilon_{xx} + \varepsilon_{yy} + \varepsilon_{zz}$$

However, even though **Lamé's equations** (10-26.3) satisfy the **bi-harmonic equations** (10-30), which have higher orders, does not imply that the solution of the bi-harmonic equations must satisfy **Lamé's** equations unless those solutions also satisfy the lesser order equations (10-26.3).

10-11. Solving bi-harmonic form of Lamé's equations

Boussinesq's method

We will study the two solutions of the forms (10-22) and (10-24) as follows

(i)
$$u = \varphi_1 + z\frac{\partial \varphi}{\partial x}$$
$$v = \varphi_2 + z\frac{\partial \varphi}{\partial y}$$
$$w = \varphi_3 + z\frac{\partial \varphi}{\partial z}$$
(10-31)

and

(ii)
$$u = \varphi_1 + r^2\frac{\partial \varphi}{\partial x}$$
$$v = \varphi_2 + r^2\frac{\partial \varphi}{\partial y}$$
$$w = \varphi_3 + r^2\frac{\partial \varphi}{\partial z}$$
(10-32)

Reverse substitution from (10-31) into Lamé's equations

First differentiating equations (10-31), we get

$$\varepsilon_{xx} = \frac{\partial u}{\partial x} = \frac{\partial \varphi_1}{\partial x} + z\frac{\partial^2 \varphi}{\partial x^2}$$

$$\varepsilon_{yy} = \frac{\partial v}{\partial y} = \frac{\partial \varphi_2}{\partial y} + z\frac{\partial^2 \varphi}{\partial y^2}$$

$$\varepsilon_{zz} = \frac{\partial w}{\partial z} = \frac{\partial \varphi_3}{\partial z} + z\frac{\partial^2 \varphi}{\partial z^2} + \frac{\partial \varphi}{\partial z}$$

(10-33.1)

Adding the three strains, we get

$$\varepsilon = \frac{\partial u}{\partial x} + \frac{\partial v}{\partial y} + \frac{\partial w}{\partial z}$$

$$= \frac{\partial \varphi_1}{\partial x} + \frac{\partial \varphi_2}{\partial y} + \frac{\partial \varphi_3}{\partial z} + \frac{\partial \varphi}{\partial z} + z\nabla^2 \varphi$$

$$= \frac{\partial \varphi_1}{\partial x} + \frac{\partial \varphi_2}{\partial y} + \frac{\partial \varphi_3}{\partial z} + \frac{\partial \varphi}{\partial z} + 0$$

(10-33.2)

Second, taking the Laplacian of (10-31), we get

$$\nabla^2 u = \nabla^2 \varphi_1 + \nabla^2\left(z\frac{\partial \varphi}{\partial x}\right) = \nabla^2\left(z\frac{\partial \varphi}{\partial x}\right)$$

$$\nabla^2 v = \nabla^2 \varphi_2 + \nabla^2\left(z\frac{\partial \varphi}{\partial y}\right) = \nabla^2\left(z\frac{\partial \varphi}{\partial y}\right)$$

$$\nabla^2 w = \nabla^2 \varphi_3 + \nabla^2\left(z\frac{\partial \varphi}{\partial z}\right) = \nabla^2\left(z\frac{\partial \varphi}{\partial z}\right)$$

(10-33.3)

Applying the property of harmonic functions in equation (10-16.2), equations (10-33.3) become

$$\nabla^2 u = 2\frac{\partial}{\partial x}\left(\frac{\partial \varphi}{\partial z}\right)$$

$$\nabla^2 v = 2\frac{\partial}{\partial y}\left(\frac{\partial \varphi}{\partial z}\right)$$

$$\nabla^2 w = 2\frac{\partial}{\partial z}\left(\frac{\partial \varphi}{\partial z}\right)$$

(10-33.4)

Substituting from equations (10-33) into **Lamé's equations** (10-26.1), we get

$$\nabla^2 u + (k+1)\frac{\partial \varepsilon}{\partial x} = \frac{\partial}{\partial x}\left((k+3)\frac{\partial \varphi}{\partial z} + (k+1)\left(\frac{\partial \varphi_1}{\partial x} + \frac{\partial \varphi_2}{\partial y} + \frac{\partial \varphi_3}{\partial z}\right)\right) = 0$$

(10-34.1)

Substituting from equations (10-33) into **Lamé's equations** (10-26.2), we get

$$\nabla^2 \mathbf{v} + (k+1)\frac{\partial \varepsilon}{\partial y} = \frac{\partial}{\partial y}\left((k+3)\frac{\partial \varphi}{\partial z} + (k+1)\left(\frac{\partial \varphi_1}{\partial x} + \frac{\partial \varphi_2}{\partial y} + \frac{\partial \varphi_3}{\partial z}\right)\right) = 0 \qquad (10\text{-}34.2)$$

Substituting from equations (10-33) into **Lamé's equations** (10-26.3), we get

$$\nabla^2 \mathbf{w} + (k+1)\frac{\partial \varepsilon}{\partial z} = \frac{\partial}{\partial z}\left((k+3)\frac{\partial \varphi}{\partial z} + (k+1)\left(\frac{\partial \varphi_1}{\partial x} + \frac{\partial \varphi_2}{\partial y} + \frac{\partial \varphi_3}{\partial z}\right)\right) = 0 \qquad (10\text{-}34.3)$$

From equations (10-34), we conclude that

$$(k+3)\frac{\partial \varphi}{\partial z} + (k+1)\left(\frac{\partial \varphi_1}{\partial x} + \frac{\partial \varphi_2}{\partial y} + \frac{\partial \varphi_3}{\partial z}\right) = C \qquad (10\text{-}35.1)$$

Thus, **equation (10-35.1) defines the condition** under which the proposed solution in equations (10-31) of the bi-harmonic equation could satisfy **Lamé's equation.** Since the constant C is arbitrary, we could define the derivative φ with respect to the three arbitrary harmonic functions φ_1, φ_2, and φ_3 as follows

$$\frac{\partial \varphi}{\partial z} = -\frac{k+1}{k+3}\left(\frac{\partial \varphi_1}{\partial x} + \frac{\partial \varphi_2}{\partial y} + \frac{\partial \varphi_3}{\partial z}\right) \qquad (10\text{-}35.2)$$

Reverse substitution from (10-32) into Lamé's equations

First differentiating equations (10-32), we get

$$\varepsilon_{xx} = \frac{\partial u}{\partial x} = \frac{\partial \varphi_1}{\partial x} + \frac{\partial}{\partial x}\left(r^2\frac{\partial \varphi}{\partial x}\right)$$

$$\varepsilon_{yy} = \frac{\partial v}{\partial y} = \frac{\partial \varphi_2}{\partial y} + \frac{\partial}{\partial y}\left(r^2\frac{\partial \varphi}{\partial y}\right) \qquad (10\text{-}36.1)$$

$$\varepsilon_{zz} = \frac{\partial w}{\partial z} = \frac{\partial \varphi_3}{\partial z} + \frac{\partial}{\partial z}\left(r^2\frac{\partial \varphi}{\partial z}\right)$$

Farther manipulation, we get

$$\varepsilon_{xx} = \frac{\partial\varphi_1}{\partial x} + r^2\frac{\partial^2\varphi}{\partial x^2} + 2x\frac{\partial\varphi}{\partial x}$$

$$\varepsilon_{yy} = \frac{\partial\varphi_2}{\partial y} + r^2\frac{\partial^2\varphi}{\partial y^2} + 2y\frac{\partial\varphi}{\partial y} \qquad\qquad (10\text{-}36.2)$$

$$\varepsilon_{zz} = \frac{\partial\varphi_3}{\partial z} + r^2\frac{\partial^2\varphi}{\partial z^2} + 2z\frac{\partial\varphi}{\partial z}$$

Adding the three strains, we get

$$\varepsilon = \frac{\partial\varphi_1}{\partial x} + \frac{\partial\varphi_2}{\partial y} + \frac{\partial\varphi_3}{\partial z} + r^2\left(\frac{\partial^2\varphi}{\partial x^2} + \frac{\partial^2\varphi}{\partial y^2} + \frac{\partial^2\varphi}{\partial z^2}\right)$$

$$+ 2\left(x\frac{\partial\varphi}{\partial x} + y\frac{\partial\varphi}{\partial y} + z\frac{\partial\varphi}{\partial z}\right)$$

i.e.,

$$\varepsilon = \varepsilon_0 + 2\Phi \qquad\qquad (10\text{-}36.3)$$

Where,

$$\nabla^2\varphi = 0 \qquad\qquad (10\text{-}36.3a)$$

$$\Phi = x\frac{\partial\varphi}{\partial x} + y\frac{\partial\varphi}{\partial y} + z\frac{\partial\varphi}{\partial z} \qquad\qquad (10\text{-}36.3b)$$

$$\varepsilon_0 = \frac{\partial\varphi_1}{\partial x} + \frac{\partial\varphi_2}{\partial y} + \frac{\partial\varphi_3}{\partial z} \qquad\qquad (10\text{-}36.3c)$$

Second, taking the Laplacian of (10-32), we get

$$\nabla^2 u = \nabla^2\varphi_1 + \nabla^2\left(r^2\frac{\partial\varphi}{\partial x}\right)$$

$$\nabla^2 v = \nabla^2\varphi_2 + \nabla^2\left(r^2\frac{\partial\varphi}{\partial y}\right) \qquad\qquad (10\text{-}37.1)$$

$$\nabla^2 w = \nabla^2\varphi_3 + \nabla^2\left(r^2\frac{\partial\varphi}{\partial z}\right)$$

Applying the property of harmonic functions in equation (10-23.3), (yet substituting the harmonic function φ by the derivatives of φ) and noticing the vanishing Laplacians of the three arbitrary functions φ_1, φ_2, and φ_3, as follows, equations (10-37) become

$$\nabla^2 u = \nabla^2\left(r^2\frac{\partial\varphi}{\partial x}\right) = 6\frac{\partial\varphi}{\partial x} + 4\left(x\frac{\partial^2\varphi}{\partial x^2} + y\frac{\partial^2\varphi}{\partial y\partial x} + z\frac{\partial^2\varphi}{\partial z\partial x}\right)$$

$$\nabla^2 v = \nabla^2\left(r^2\frac{\partial\varphi}{\partial y}\right) = 6\frac{\partial\varphi}{\partial y} + 4\left(y\frac{\partial^2\varphi}{\partial y^2} + x\frac{\partial^2\varphi}{\partial y\partial x} + z\frac{\partial^2\varphi}{\partial z\partial y}\right) \qquad (10\text{-}37.2)$$

$$\nabla^2 w = \nabla^2\left(r^2\frac{\partial\varphi}{\partial z}\right) = 6\frac{\partial\varphi}{\partial z} + 4\left(z\frac{\partial^2\varphi}{\partial z^2} + x\frac{\partial^2\varphi}{\partial z\partial x} + y\frac{\partial^2\varphi}{\partial z\partial y}\right)$$

Let us rearrange the terms in equations (10-37.2) such that we could use Φ, equation (10-36.3b), as follows

$$\nabla^2 u = 6\frac{\partial\varphi}{\partial x} + 4\left(\frac{\partial}{\partial x}\left(x\frac{\partial\varphi}{\partial x} + y\frac{\partial\varphi}{\partial y} + z\frac{\partial\varphi}{\partial z}\right) - \frac{\partial\varphi}{\partial x}\right)$$

$$\nabla^2 v = 6\frac{\partial\varphi}{\partial y} + 4\left(\frac{\partial}{\partial y}\left(x\frac{\partial\varphi}{\partial x} + y\frac{\partial\varphi}{\partial y} + z\frac{\partial\varphi}{\partial z}\right) - \frac{\partial\varphi}{\partial y}\right) \qquad (10\text{-}37.3)$$

$$\nabla^2 w = 6\frac{\partial\varphi}{\partial z} + 4\left(\frac{\partial}{\partial z}\left(x\frac{\partial\varphi}{\partial x} + y\frac{\partial\varphi}{\partial y} + z\frac{\partial\varphi}{\partial z}\right) - \frac{\partial\varphi}{\partial z}\right)$$

Then, we substitute by Φ from equation (10-36.3b) into (10-37.3), we get

$$\nabla^2 u = 2\frac{\partial}{\partial x}(\varphi + 2\Phi)$$

$$V^2 v = 2\frac{\partial}{\partial y}(\varphi + 2\Phi) \qquad (10\text{-}37.4)$$

$$V^2 w = 2\frac{\partial}{\partial z}(\varphi + 2\Phi)$$

Substituting from equations (10-36 and 37) into **Lamé's equations** (10-26.1), we get

$$\nabla^2 u + (k+1)\frac{\partial\varepsilon}{\partial x} = \frac{\partial}{\partial x}\left(2\varphi + 2\Phi(3+k) + (k+1)\varepsilon_0\right) = 0 \qquad (10\text{-}38.1)$$

Substituting from equations (10-36 and 37) into **Lamé's equations** (10-26.2), we get

$$\nabla^2 v + (k+1)\frac{\partial\varepsilon}{\partial y} = \frac{\partial}{\partial y}\left(2\varphi + 2\Phi(3+k) + (k+1)\varepsilon_0\right) = 0 \qquad (10\text{-}38.2)$$

Substituting from equations (10-36 and 37) into **Lamé's equations** (10-26.3), we get

$$\nabla^2 \mathbf{w} + (k+1)\frac{\partial \varepsilon}{\partial z} = \frac{\partial}{\partial z}\left(2\varphi + 2\Phi(3+k) + (k+1)\varepsilon_0\right) = 0 \qquad (10\text{-}38.3)$$

From equations (10-38), we conclude that

$$2\varphi + 2\Phi(3+k) + (k+1)\varepsilon_0 = C \qquad (10\text{-}38.4)$$

Thus, **equation (10-38.4) defines the condition** under which the proposed solution in equations (10-32) of the bi-harmonic equation could satisfy **Lamé's equation.** Since the constant C is arbitrary, it could be assumed zero. Therefore,

$$\Phi + \varphi \frac{1}{(3+k)} = -\frac{(k+1)}{2(3+k)}\varepsilon_0 \qquad (10\text{-}38.5)$$

Using the property of harmonic function, equation (10-17.3), we could replace Φ by (r $\partial\varphi/\partial r$) as follows

$$r\frac{\partial \varphi}{\partial r} + \varphi \frac{1}{(3+k)} = -\frac{(k+1)}{2(3+k)}\varepsilon_0 \qquad (10\text{-}38.6)$$

The integration constant of equation (10-38.6) depends on the remaining coordinates, θ, or z.

Solution of equation (10-38.6)

Assume that the harmonic function φ is given by the product of two arbitrary functions of r as follows

$$\varphi(r) = \varphi_1(r)\varphi_2(r) \qquad (10\text{-}39.1)$$

Substituting from (10-39.1) in (10-38.6), we get

$$r\varphi_1'\varphi_2 + r\varphi_1\varphi_2' + \varphi_1\varphi_2 \frac{1}{(3+k)} = -\frac{(k+1)}{2(3+k)}\varepsilon_0 \qquad (10\text{-}39.2)$$

Arrange the terms such that we could set the value of the either φ_1 or φ_2 such that it could farther simplify the above equation, we get

$$r\varphi_1'\varphi_2 + \varphi_1\left(r\varphi_2' + \varphi_2\frac{1}{(3+k)}\right) = -\frac{(k+1)}{2(3+k)}\varepsilon_0 \qquad (10\text{-}39.3)$$

If we set the bracketed term to zero, we get

$$r\varphi_2' + \varphi_2 \frac{1}{(3+k)} = 0$$

$$\int \frac{d\varphi_2}{\varphi_2} = -\frac{1}{(3+k)} \int \frac{dr}{r}$$

$$\ln \varphi_2 = -\frac{1}{(3+k)} \ln r \qquad\qquad (10\text{-}39.4)$$

$$\varphi_2 = r^{-\frac{1}{(3+k)}}$$

And

$$r\varphi_1' \varphi_2 = -\frac{(k+1)}{2(3+k)} \varepsilon_0$$

$$\int d\varphi_1 = -\frac{(k+1)}{2(3+k)} \int \frac{\varepsilon_0}{r\varphi_2} dr \qquad\qquad (10\text{-}39.5)$$

Substituting by φ_2, from (10-39.4), into (10-39.5), we get

$$\varphi_1 = -\frac{(k+1)}{2(3+k)} \int \varepsilon_0 r^{-\frac{2+k}{(3+k)}} dr + C \qquad\qquad (10\text{-}39.6)$$

Thus, the final solution is obtained by substituting from (10-39.4) and (10-39.6) in (10-39.1)

$$\varphi(r) = \varphi_1(r)\varphi_2(r)$$

$$= -\frac{(k+1)}{2(3+k)} r^{-\frac{1}{(3+k)}} \int \varepsilon_0 r^{-\frac{2+k}{(3+k)}} dr + C \qquad\qquad (10\text{-}39.7)$$

10-12. Stress distribution in loaded half-space medium

Boussinesq's problem

Example 30

In Figure 10-1, an elastic medium filling half-space, bounded by the xOy surface and loaded in the negative direction o z-axis. Find the general solution for an arbitrary distribution of load on the surface given either of the following **boundary conditions**:

(i) Boundary conditions

At z = 0

(1) Displacements: u(x,y,0), v(x,y,0) and w(x,y,0)   (10-40.1)

Those boundary condition are introduced in equations (10-31), where the three harmonic functions φ_1, φ_2 and φ_3 are to be determined, as follows.

$$u(x,y,z) = \varphi_1(x,y,0) + z\frac{\partial\varphi}{\partial x} = f_1(x,y) + z\frac{\partial\varphi}{\partial x}$$

$$v(x,y,z) = \varphi_2(x,y,0) + z\frac{\partial\varphi}{\partial y} = f_2(x,y) + z\frac{\partial\varphi}{\partial y} \qquad (10\text{-}40.2)$$

$$w(x,y,z) = \varphi_3(x,y,0) + z\frac{\partial\varphi}{\partial z} = f_3(x,y) + z\frac{\partial\varphi}{\partial z}$$

Or,

(2) Stresses: $\qquad\qquad\qquad \sigma_z(x,y,0),\ \tau_{zx}(x,y,0),\ \tau_{zy}(x,y,0), \qquad\qquad\qquad (10\text{-}41.1)$

Which are introduced in (10-27. 3, 27.5 and 27.6) to give

$$\sigma_z = \mu(2\varepsilon_{zz} + k\varepsilon) = \mu\left(2\frac{\partial w}{\partial z} + k\varepsilon\right) = \Phi_1(x,y)$$

$$\tau_{zx} = \mu\alpha_{zx} \qquad = \mu\left(\frac{\partial w}{\partial x} + \frac{\partial u}{\partial z}\right) = \Phi_2(x,y) \qquad (10\text{-}41.2)$$

$$\tau_{zy} = \mu\alpha_{zy} \qquad = \mu\left(\frac{\partial w}{\partial y} + \frac{\partial v}{\partial z}\right) = \Phi_3(x,y)$$

Where three harmonic functions ψ_1, ψ_2 and ψ_3 are sought as solution as follows

$$\mu\psi_1(x,y,0) = \Phi_1(x,y)$$
$$\mu\psi_2(x,y,0) = \Phi_2(x,y) \qquad (10\text{-}41.3)$$
$$\mu\psi_3(x,y,0) = \Phi_3(x,y)$$

(ii) Solution by harmonic functions equations (10-31)

Substituting by the derivatives of displacements u, v, and w, from equations (10-40.2) and principal elongation strain ε, equation (10-33.2), into equation (10-41.2), we get

$$\psi_1 = 2\left(\frac{\partial\varphi_3}{\partial z} + z\frac{\partial^2\varphi}{\partial z^2} + \frac{\partial\varphi}{\partial z}\right) + k\left(\frac{\partial\varphi_1}{\partial x} + \frac{\partial\varphi_2}{\partial y} + \frac{\partial\varphi_3}{\partial z} + \frac{\partial\varphi}{\partial z}\right)$$

$$\psi_2 = \frac{\partial\varphi_3}{\partial x} + 2z\frac{\partial^2\varphi}{\partial x\partial z} + \frac{\partial\varphi_1}{\partial z} + \frac{\partial\varphi}{\partial x} \qquad (10\text{-}42)$$

$$\psi_3 = \frac{\partial\varphi_3}{\partial y} + 2z\frac{\partial^2\varphi}{\partial z\partial y} + \frac{\partial\varphi_2}{\partial z} + \frac{\partial\varphi}{\partial y}$$

Substituting from equation (10-35.2) and putting z = 0, at the boundary, equation (10-42) become

$$\psi_1 = 2\frac{\partial \varphi_3}{\partial z} + \frac{2}{k+1}\frac{\partial \varphi}{\partial z}$$

$$\psi_2 = \frac{\partial \varphi_3}{\partial x} + \frac{\partial \varphi_1}{\partial z} + \frac{\partial \varphi}{\partial x} \tag{10-43}$$

$$\psi_3 = \frac{\partial \varphi_3}{\partial y} + \frac{\partial \varphi_2}{\partial z} + \frac{\partial \varphi}{\partial y}$$

Interpretation of equation (10-43)

1. The R.H.S. of equation (10-43) represents the harmonic functions of the boundary stresses on the plane xOy, which are needed to express the sought unknown φ, which is the ultimate goal of solution as the **displacement function** within the half-space elastic medium.

2. The three harmonic functions φ1, φ1, and φ1, are the boundary conditions of displacements proposed in equation (10-40-2) and which must be eliminated during the determination of φ.

3. We will assume, without proof, that **when two harmonic functions coincide on the boundary of a region, they are valid throughout the region**. Thus, equations (10-43) are valid throughout the half-space elastic region.

Differentiating equations (10-43) (ψ_1 with respect to z, ψ_2 with respect to x, ψ_3 with respect to y) and adding the three equation by members (noting the vanishing of Laplacian of harmonic functions), we get

$$\frac{\partial \psi_1}{\partial z} + \frac{\partial \psi_2}{\partial x} + \frac{\partial \psi_3}{\partial y} = \begin{pmatrix} +\dfrac{\partial^2 \varphi_3}{\partial x^2} + \dfrac{\partial^2 \varphi_3}{\partial y^2} + 2\dfrac{\partial^2 \varphi_3}{\partial z^2} \\[2mm] +\dfrac{\partial^2 \varphi}{\partial x^2} + \dfrac{\partial^2 \varphi}{\partial y^2} + \dfrac{2}{k+1}\dfrac{\partial^2 \varphi}{\partial z^2} \\[2mm] +\dfrac{\partial^2 \varphi_2}{\partial y \partial z} + \dfrac{\partial^2 \varphi_1}{\partial x \partial z} \end{pmatrix}$$

Substituting by vanishing Laplacians of harmonic functions, we get

$$\frac{\partial \psi_1}{\partial z} + \frac{\partial \psi_2}{\partial x} + \frac{\partial \psi_3}{\partial y} = \left(\frac{\partial^2 \varphi_3}{\partial z^2} + \frac{\partial^2 \varphi_2}{\partial y \partial z} + \frac{\partial^2 \varphi_1}{\partial x \partial z} - \frac{k-1}{k+1}\frac{\partial^2 \varphi}{\partial z^2} \right)$$

$$= \frac{\partial}{\partial z}\left(\frac{\partial \varphi_3}{\partial z} + \frac{\partial \varphi_2}{\partial y} + \frac{\partial \varphi_1}{\partial x} \right) - \frac{k-1}{k+1}\frac{\partial^2 \varphi}{\partial z^2} \tag{10-44.1}$$

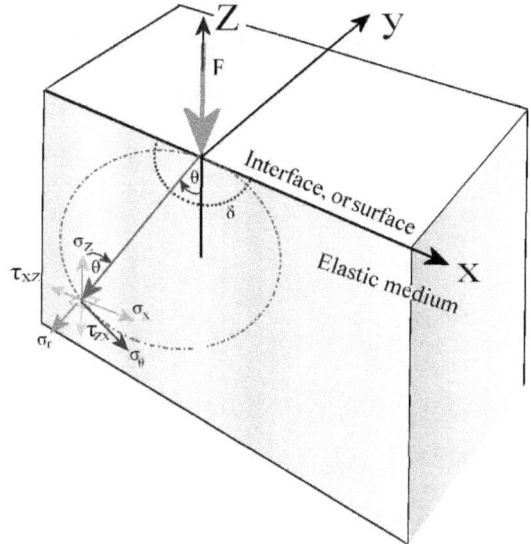

Figure 10-1. General solution of half-space bounded medium loaded on the surface xOy in the negative direction of z-direction.

Once again, substitute from equation (10-35.2) into equation (10-44.1), we get

$$\frac{\partial\psi_1}{\partial z}+\frac{\partial\psi_2}{\partial x}+\frac{\partial\psi_3}{\partial y}=-2\frac{\partial^2\varphi}{\partial z^2} \qquad (10\text{-}44.2)$$

Equation (10-44) describes the relation of the **displacement function** φ, equation (10-40.2), in the half-space medium in terms of the gradients of the boundary stresses (10-41).

Integrating equation (10-44.2), we get

$$\frac{\partial\varphi}{\partial z}=-\frac{1}{2}\int_{-\infty}^{z}\left(\frac{\partial\psi_1}{\partial z}+\frac{\partial\psi_2}{\partial x}+\frac{\partial\psi_3}{\partial y}\right)dz+f_1(x,y) \qquad (10\text{-}45.1)$$

Integrating again, we get

$$\varphi=-\frac{1}{2}\int_{-\infty}^{z}\left(\int_{-\infty}^{z}\left(\frac{\partial\psi_1}{\partial z}+\frac{\partial\psi_2}{\partial x}+\frac{\partial\psi_3}{\partial y}\right)dz\right)dz+zf_1(x,y)+f_2(x,y) \qquad (10\text{-}45.2)$$

The integration of the displacement function φ are subjected finite displacements and strains (derivatives of displacements) as z becomes infinitely large, i.e., in remote locations from the boundary. Thus, $f_1(x,y)$ and $f_2(x,y)$ are taken as zeros.

10-13. Concentrated load on half-space medium

Boundary conditions

$$\sigma_z = \Phi_1(x,y) = \text{infinite} \quad \text{at} \quad x = y = 0 \rightarrow \psi_2(x,y,0) = \text{infinite}$$

$$= 0 \qquad \text{at} \quad x, \ y \ \text{not} \ \text{zero} \rightarrow \psi_2(x,y,0) = 0$$

$$\tau_{zx} = \Phi_2(x,y) = 0 \quad \text{at} \quad xOy \ \text{plane} \rightarrow \psi_2(x,y,0) = 0$$

$$\tau_{zy} = \Phi_3(x,y) = 0 \quad \text{at} \quad xOy \ \text{plane} \rightarrow \psi_3(x,y,0) = 0$$

(10-46.1)

Quantifying the finite stress σ_z on the boundary

As we have done before in equation (7-78.1) in dealing with finite load, and from Figure 10-1, the external pressure (load) applied at $x = y = 0$ can be measured on a finite circle (sphere) or radius δ, as follows

$$\int \sigma_{-z} dA = 2\pi\delta^2 \overline{\sigma}_{-z} = F$$

i.e., $$\overline{\sigma}_{-z} = \frac{F}{2\pi\delta^2}$$

(10-46.2)

Where, A is the area of the half sphere of radius δ and $\overline{\sigma}_{-z}$ is an average value of the stress on the area of the hemisphere. Therefore, since the finite stress $\overline{\sigma}_{-z}$ is proportional to the **inverse square of the radius** δ, thus the displacement functions ψ_1, in equation (10-41.3), is equally proportional to the inverse square of radius and to the directional cosine made with the direction of load. Thus, we could propose the displacement function of the form

$$\psi_1(x,y,z) = C\frac{1}{r^2}\cdot\frac{z}{r}$$

$$\psi_2(x,y,z) = 0$$

$$\psi_3(x,y,z) = 0$$

(10-46.3)

Equation (10-45.2) is written in terms of the boundary conditions defined in equation (10-46.3) as follows

$$\varphi = -\frac{1}{2}\int_{-\infty}^{z}\left(\int_{-\infty}^{z}\frac{\partial\psi_1}{\partial z}dz\right)dz$$

$$= -\frac{1}{2}\int_{-\infty}^{z}\psi_1 dz \qquad\qquad\qquad (10\text{-}47.1)$$

$$= -\frac{1}{2}\int_{-\infty}^{z}\frac{Cz}{r^3}dz$$

Substituting by the derivative of $(-1/r)$ given by

$$\frac{\partial}{\partial z}\left(-\frac{1}{r}\right) = \frac{\partial}{\partial z}\left(-\left(x^2+y^2+z^2\right)^{-\frac{1}{2}}\right)$$

$$= z\left(x^2+y^2+z^2\right)^{-\frac{3}{2}} \qquad\qquad (10\text{-}47.2)$$

$$= \frac{z}{r^3}$$

Equation (10-47.1) becomes

$$\varphi = \frac{C}{2}\int_{-\infty}^{z}\frac{\partial}{\partial z}\left(\frac{1}{r}\right)dz$$

$$= \frac{C}{2}\int_{-\infty}^{z}d\left(\frac{1}{r}\right) \qquad\qquad\qquad (10\text{-}47.3)$$

$$= \frac{C}{2}\frac{1}{r}$$

Displacement harmonic functions φ_1, φ_2 and φ_3, equations (10-43), by substituting by the derivatives of φ from (10-47.1), as follows

$$\psi_1 = 2\frac{\partial\varphi_3}{\partial z} + \frac{2}{k+1}\frac{\partial}{\partial z}\left(\frac{C}{2r}\right) = \frac{zC}{r^3}$$

$$\psi_2 = \frac{\partial\varphi_3}{\partial x} + \frac{\partial\varphi_1}{\partial z} + \frac{\partial}{\partial x}\left(\frac{C}{2r}\right) = 0 \qquad (10\text{-}47.4)$$

$$\psi_3 = \frac{\partial\varphi_3}{\partial y} + \frac{\partial\varphi_2}{\partial z} + \frac{\partial}{\partial y}\left(\frac{C}{2r}\right) = 0$$

Performing the differentiation and arranging, we get

$$2\frac{\partial \varphi_3}{\partial z} = \frac{k+2}{k+1}\left(\frac{Cz}{r^3}\right)$$

$$\frac{\partial \varphi_3}{\partial x} + \frac{\partial \varphi_1}{\partial z} = \left(\frac{Cx}{2r^3}\right) \qquad (10\text{-}47.5)$$

$$\frac{\partial \varphi_3}{\partial y} + \frac{\partial \varphi_2}{\partial z} = \left(\frac{Cy}{2r^3}\right)$$

Before performing successive integrations to determine the three displacement harmonic functions, we will use equation (10-26.4) to express k by **Poisson's ratio**, as follows

$$\frac{k+2}{k+1} = 2(1-v) \qquad (10\text{-}47.6)$$

$$k = \frac{2v}{(1-2v)} \qquad (10\text{-}47.6a)$$

Harmonic displacement function φ_3

Thus, integrating the **derivative of φ_3**, in equation (10-47.5), we get

$$\frac{\partial \varphi_3}{\partial z} = (1-v)\left(\frac{Cz}{r^3}\right)$$

$$\varphi_3 = C(1-v) \int \left(\frac{z}{r^3}\right)dz$$

$$= -C(1-v) \int \frac{\partial}{\partial z}\left(\frac{1}{r}\right)dz \qquad (10\text{-}47.7)$$

$$= -\frac{C(1-v)}{r}$$

Harmonic displacement function φ_1

Substituting from equation (10-47.7), by φ_3, in equation (10-47.5), we get

$$\frac{\partial \varphi_1}{\partial z} = \left(\frac{Cx}{2r^3}\right) - \frac{\partial \varphi_3}{\partial x}$$

$$= \left(\frac{Cx}{2r^3}\right) + \frac{\partial}{\partial x}\left(\frac{C(1-v)}{r}\right) \qquad (10\text{-}47.8)$$

$$= -\frac{(1-2v)}{2r^3}xC$$

Integrating, we get

$$\varphi_1 = -\frac{(1-2v)xC}{2} \int \frac{1}{r^3}dz \qquad (10\text{-}47.9)$$

In order to perform the above integration, we first need to manipulate the integrand as follows

(1) Euler's substitution $r = z + t$ gives

$$\begin{aligned} r^2 &= x^2 + y^2 + z^2 \\ &= (z+t)^2 \\ &= z^2 + 2zt + t^2 \end{aligned} \qquad (10\text{-}48.1)$$

Organizing, we get

$$\begin{aligned} x^2 + y^2 &= 2zt + t^2 \\ z &= \frac{x^2 + y^2 - t^2}{2t} \end{aligned} \qquad (10\text{-}48.2)$$

Substituting by this expression in Euler's expression $r = t + z$, get

$$\begin{aligned} r &= t + z = t + \frac{x^2 + y^2 - t^2}{2t} \\ r &= \frac{t^2 + x^2 + y^2}{2t} \end{aligned} \qquad (10\text{-}48.3)$$

Differentiating z with respect to t, we get

$$dz = \frac{-\left(x^2 + y^2 + t^2\right)}{2t^2}dt \qquad (10\text{-}48.4)$$

(2) Using the above expression in equation (10-47.9), we get

$$\begin{aligned} \varphi_1 &= -\frac{(1-2v)xC}{2} \int \frac{-4t}{\left(x^2 + y^2 + t^2\right)^2}dt \\ &= -\frac{(1-2v)xC}{2} \int \frac{-2}{\left(x^2 + y^2 + t^2\right)^2}dt^2 \\ &= -\frac{(1-2v)xC}{\left(x^2 + y^2 + t^2\right)} \\ &= -\frac{(1-2v)xC}{\left(x^2 + y^2 + (r-z)^2\right)} \end{aligned} \qquad (10\text{-}49.1)$$

i.e.,

$$\varphi_1 = -\frac{(1-2v)xC}{(2r^2 - 2rz)}$$
$$= -\frac{(1-2v)xC}{2r(r-z)}$$

(10-49.2)

Harmonic displacement function φ_2

The last equation if (10-47.5) gives the last displacement function φ_2 after substituting by φ_3, as follows

$$\frac{\partial}{\partial y}\left(-\frac{C(1-v)}{r}\right) + \frac{\partial \varphi_2}{\partial z} = \left(\frac{Cy}{2r^3}\right)$$
$$\frac{\partial \varphi_2}{\partial z} = -C(1-2v)\left(\frac{y}{2r^3}\right)$$

(10-50.1)

This can be integrated in the same fashion as we did in equation (10-47.9). Thus, we get

$$\varphi_2 = -\left(\frac{C(1-2v)y}{2r(r-z)}\right)$$

(10-50.1)

Summary of the displacement harmonic functions

Table 6. Harmonic functions of displacements φ_1, φ_2, and φ_3, obtained from the harmonic functions of stresses on the boundary of the medium ψ_1, ψ_2, and ψ_3, together with the harmonic function φ that define the condition imposed on biharmonic equations in order to conform to the equation of elasticity; Lamé's and Beltrami's equations

$\varphi_1 = -\dfrac{(1-2v)xC}{2r(r-z)}$	(10-49.2)
$\varphi_2 = -\left(\dfrac{C(1-2v)y}{2r(r-z)}\right)$	(10-50.1)
$\varphi_3 = -\dfrac{C(1-v)}{r}$	(10-47.7)
$\psi_1 = \dfrac{zC}{r^3}, \quad \psi_2 = 0, \quad \psi_3 = 0$	(10-47.4)
$\varphi = \dfrac{C}{2}\dfrac{1}{r}$	(10-47.3)

Displacements

352

The postulated equations of displacements (10-40.2) entail the three harmonic functions φ_1, φ_2, and φ_3, and the derivatives of the conditional harmonic functions φ. Table 10-6 summarizes the four functions to be used in determining **Boussinesq's solution** for three displacements as follows

$$u(x,y,z) = \varphi_1(x,y,z) + z\frac{\partial\varphi}{\partial x} = -\frac{C(1-2v)x}{2r(r-z)} - \frac{C}{2}\frac{xz}{r^3}$$

$$v(x,y,z) = \varphi_2(x,y,z) + z\frac{\partial\varphi}{\partial y} = -\frac{C(1-2v)y}{2r(r-z)} - \frac{C}{2}\frac{yz}{r^3}$$

$$w(x,y,z) = \varphi_3(x,y,z) + z\frac{\partial\varphi}{\partial z} = -\frac{C(1-v)}{r} - \frac{C}{2}\frac{z^2}{r^3}$$

(10-51)

Where, the constant C is determined from the boundary conditions of stresses under finite load F as follows.

Strain expressed in terms of the four harmonic functions

Equation (10-33.2) expressed the principal dilatational strain as follows

$$\varepsilon = \frac{\partial\varphi_1}{\partial x} + \frac{\partial\varphi_2}{\partial y} + \frac{\partial\varphi_3}{\partial z} + \frac{\partial\varphi}{\partial z}$$

(10-52.1)

Which, upon substitution from equation (10-35.21), gives

$$\varepsilon = -\frac{2}{k+1}\frac{\partial\varphi}{\partial z}$$

(10-52.2)

Substituting by the derivative of φ from equation (10-47.3), and k, from equation (10-47.6a), the dilatational strain becomes

$$\varepsilon = C(1-2v)\frac{z}{r^3}$$

(10-52.3)

Distribution of stresses in the half-space medium

Substituting by the expressions of displacements equations (10-51) and strain equation (10-52.2), and k, from equation (10-47.6a), in the equations of Hooke's law, equations (10-27), we get

$$\sigma_x = \mu\left(2\varepsilon_{xx} + k\varepsilon\right)$$

$$= \mu\left(-C\frac{\partial}{\partial x}\left(\frac{(1-2v)x}{r(r-z)} + \frac{xz}{r^3}\right) + \frac{2v}{(1-2v)}C(1-2v)\frac{z}{r^3}\right) \tag{10-53.1}$$

$$= \mu C\left(\frac{3x^2z}{r^5} - (1-2v)\left(\frac{r^2+rz-z^2}{(r-z)r^3} - \frac{x^2(2r-z)}{r^3(r-z)^2}\right)\right)$$

$$\sigma_y = \mu\left(2\varepsilon_{yy} + k\varepsilon\right)$$

$$= \mu\left(-C\frac{\partial}{\partial y}\left(\frac{(1-2v)y}{r(r-z)} + \frac{yz}{r^3}\right) + \frac{2v}{(1-2v)}C(1-2v)\frac{z}{r^3}\right) \tag{10-53.2}$$

$$= \mu C\left(\frac{3y^2z}{r^5} - (1-2v)\left(\frac{r^2+rz-z^2}{(r-z)r^3} - \frac{y^2(2r-z)}{r^3(r-z)^2}\right)\right)$$

$$\sigma_z = \mu\left(2\varepsilon_{zz} + k\varepsilon\right)$$

$$= \mu\left(-2C\frac{\partial}{\partial z}\left(\frac{(1-v)}{r} + \frac{z^2}{2r^3}\right) + \frac{2v}{(1-2v)}C(1-2v)\frac{z}{r^3}\right) \tag{10-53.3}$$

$$= \mu C\left(\frac{3z^3}{r^5}\right)$$

$$\boldsymbol{\tau}_{xy} = \mu\alpha_{xy}$$

$$= \frac{\mu}{2}\left(\frac{\partial u}{\partial y} + \frac{\partial v}{\partial x}\right) \tag{10-53.4}$$

$$= C\mu\left(\frac{3xyz}{r^5} + (1-2v)\frac{xy}{r^3}\frac{(2r-z)}{(r-z)^2}\right)$$

$$\boldsymbol{\tau}_{yz} = \mu\alpha_{yz}$$

$$= \frac{\mu}{2}\left(\frac{\partial w}{\partial y} + \frac{\partial v}{\partial z}\right) \tag{10-53.5}$$

$$= \mu C\left(\frac{3yz^2}{r^5}\right)$$

$$\boldsymbol{\tau}_{zx} = \mu\alpha_{zx}$$

$$= \frac{\mu}{2}\left(\frac{\partial w}{\partial x} + \frac{\partial u}{\partial z}\right) \tag{10-53.6}$$

$$= \mu C\left(\frac{3xz^2}{r^5}\right)$$

Table 7. **Boussinesq's solution** for stresses. Distribution of normal and shear stresses in the half-space bounded medium loaded by concentrated force at x= y = 0.

$\sigma_x = \mu C\left(\dfrac{3x^2z}{r^5} - (1-2v)\left(\dfrac{r^2 + rz - z^2}{(r-z)r^3} - \dfrac{x^2(2r-z)}{r^3(r-z)^2}\right)\right)$	(10-53.1)
$\sigma_y = \mu C\left(\dfrac{3y^2z}{r^5} - (1-2v)\left(\dfrac{r^2 + rz - z^2}{(r-z)r^3} - \dfrac{y^2(2r-z)}{r^3(r-z)^2}\right)\right)$	(10-53.2)
$\sigma_z = \mu C\left(\dfrac{3z^3}{r^5}\right)$ 53.3)	(10-
$\tau_{xy} = C\mu\left(\dfrac{3xyz}{r^5} + (1-2v)\dfrac{xy}{r^3}\dfrac{(2r-z)}{(r-z)^2}\right)$	(10-53.4)
$\tau_{yz} = \mu C\left(\dfrac{3yz^2}{r^5}\right)$	(10-53.5)
$\tau_{zx} = \mu C\left(\dfrac{3xz^2}{r^5}\right)$	(10-53.6)

Determining the constant C from the boundary conditions

Equation (10-46.2) quantify the finite average stress in the neighborhood of the loading spot x = y = 0. Substituting from (10-53.3) into (10-46.2) we get

$$\int \sigma_{-z}dA = \sigma_z = -3\mu C\int\left(\dfrac{z^3}{r^5}\right)dA = F \qquad (10\text{-}54.1)$$

The following assumptions are made in integrating equation (10-54.1):

1. The negative sign in equation (10-54.1) accounts for the equal but opposite reaction to the stress

$$+\sigma_{+z} = -\sigma_{-z}$$

2. Assume that z = -h, which is a constant depth inside the medium.

3. The element of area dA is determined as follows

$$r^2 = x^2 + y^2 + z^2$$
$$= \rho^2 + h^2 \qquad (10\text{-}54.2)$$

$$dA = \rho d\rho d\theta \qquad (10\text{-}54.3)$$

355

4. **Boussinesq's solution** does not hold in the near vicinity of application of load due to the excess load beyond the **cut-off** limit below which Hooke's law holds.

5. Equation (10-35.1) limits the solution via the three harmonic functions φ_1, φ_2, φ_3, in terms of φ, which is imposes conditions on the application of bi-harmonic equation to the **Lamé's** and **Beltrami's equations** of continuity.

Thus, equation (10-54.1) becomes

$$\int \sigma_{-z}dA = -3\mu C(-h)^3 \int\limits_{\rho=0}^{\infty} \int\limits_{\theta=0}^{2\pi} \int \left(\rho^2 + h^2\right)^{-\frac{5}{2}}\rho\,d\rho\,d\theta$$

$$= \frac{3}{2}\mu Ch^3(2\pi)\int\limits_{\rho=0}^{\infty} \left(\rho^2 + h^2\right)^{-\frac{5}{2}}d\rho^2$$

$$= (3\pi)\mu Ch^3\left[-\frac{2}{3}\left(\rho^2 + h^2\right)^{-\frac{3}{2}}\right]_0^{\infty} \qquad (10\text{-}54.4)$$

$$= (2\pi)\mu Ch^3\left[(h)^{-3}\right]$$

$$= (2\pi)\mu C = F$$

i.e.,

$$C = \frac{F}{2\pi\mu} \qquad (10\text{-}54.5)$$

10-14. Distributed load on half-space medium

Boundary conditions

The load on the surface xOy can be described by $q(\xi, \eta)$ and the effects of the differential elements

$$q(\xi, \eta)dA \qquad (10\text{-}55.1)$$

are added to comprise the net displacements and stresses.
The differential load at the surface coordinates (ξ, η) displaces an arbitrary point $P(x,y,0)$ on the surface according to equation (10-51) by taking the differential of w as follows

$$\delta w(x, y, 0) = -\left(\frac{q(\xi, \eta)dA}{2\pi\mu}\right)\left(\frac{(1-v)}{\sqrt{(x-\xi)^2 + (y-\eta)^2}} + \frac{0}{2\left(\sqrt{(x-\xi)^2 + (y-\eta)^2}\right)^3}\right) \qquad (10\text{-}55.2)$$

356

Where the load F, in equation (10-54.5), is replaced by the function $q(\xi, \eta)dA$.

Integrating equation (10-51), we get the vertical displacement of the boundary xOy under the arbitrary load $q(\xi, \eta)$.

$$w(x,y,0) = -\frac{(1-v)}{2\pi\mu} \iint \frac{q(\xi,\eta)}{\sqrt{(x-\xi,)^2 + (y-\eta)^2}} d\xi d\eta \qquad (10\text{-}55.3)$$

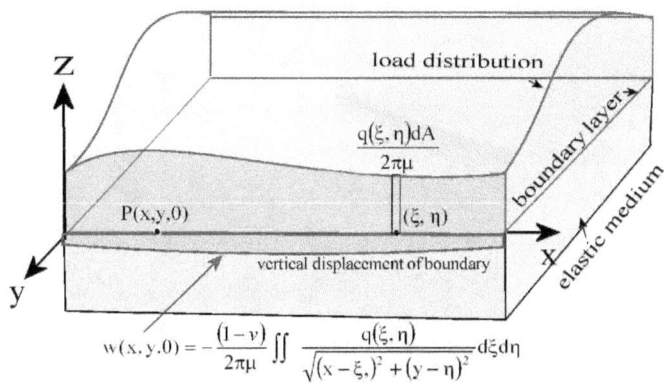

Figure 10-2. Vertical displacement of boundary layer under arbitrary load distribution using Boussinesq's solution.

10-15. Filon's solution of plain stress problem by complex variables

(i) Complex variable representation of orthogonal variables

Let us define the two conjugate complex variables

$$\begin{aligned} z &= x + iy \\ \bar{z} &= x - iy \end{aligned} \qquad (10\text{-}56.1)$$

The derivatives of the two conjugate variables have the following characteristic structures

$$\frac{\partial z}{\partial x} = 1, \qquad \frac{\partial z}{\partial y} = i, \qquad \frac{\partial \bar{z}}{\partial x} = 1, \qquad \frac{\partial \bar{z}}{\partial y} = -i \qquad (10\text{-}56.2)$$

Therefore, the independent variables x and y of a function f(x,y) can be represented in terms of the two conjugate variables in equation (10-56.1) by the following replacements

$$x = \frac{z + \bar{z}}{2}$$
$$y = \frac{z - \bar{z}}{2i}$$

(10-56.3a)

From equations (10-56.1 through 56.3) we conclude the following

$$\frac{\partial z}{\partial \bar{z}} = \frac{\partial x}{\partial \bar{z}} + \frac{\partial x}{\partial z}\frac{\partial z}{\partial \bar{z}} + i\frac{\partial y}{\partial \bar{z}} + i\frac{\partial y}{\partial z}\frac{\partial z}{\partial \bar{z}}$$
$$= 1 + \frac{\partial z}{\partial \bar{z}} - 1 + \frac{\partial z}{\partial \bar{z}} = 0$$
$$\frac{\partial \bar{z}}{\partial z} = \frac{\partial x}{\partial z} + \frac{\partial x}{\partial \bar{z}}\frac{\partial \bar{z}}{\partial z} - i\frac{\partial y}{\partial z} - i\frac{\partial y}{\partial \bar{z}}\frac{\partial \bar{z}}{\partial z}$$
$$= 1 + \frac{\partial z}{\partial \bar{z}} - 1 + \frac{\partial z}{\partial \bar{z}} = 0$$

(10-56.3b)

Therefore, we can prove the following complex variable representations of the f(x,y), its derivatives, and Laplacian:

1. Composite function

$$f(x, y) = f\left(\frac{z + \bar{z}}{2}, \frac{z - \bar{z}}{2i}\right)$$

(10-56.4a)

2. Composite derivatives

The consequences of such replacements on the derivatives of f(x,y) are shown by differentiating the composite function, as follows

$$\frac{\partial f}{\partial x} = \frac{\partial f}{\partial z}\frac{\partial z}{\partial x} + \frac{\partial f}{\partial \bar{z}}\frac{\partial \bar{z}}{\partial x} = \frac{\partial f}{\partial z} + \frac{\partial f}{\partial \bar{z}}$$

(10-56.4b)

And

$$\frac{\partial f}{\partial y} = \frac{\partial f}{\partial z}\frac{\partial z}{\partial y} + \frac{\partial f}{\partial \bar{z}}\frac{\partial \bar{z}}{\partial y} = i\frac{\partial f}{\partial z} - i\frac{\partial f}{\partial \bar{z}}$$

(10-56.4c)

3. Conjugate functions

Multiplying equations (10-56.4b) by i and adding the two equations by members, we get

$$\frac{\partial f}{\partial x} + i\frac{\partial f}{\partial y} = 2\frac{\partial f}{\partial \overline{z}}$$

(10-56.5a)

Multiplying equations (10-56.4b) by i and subtracting the two equations by members, we get

$$\frac{\partial f}{\partial x} - i\frac{\partial f}{\partial y} = 2\frac{\partial f}{\partial z}$$

(10-56.5b)

4. Laplace 's equation

Differentiating equations (10-56.4) we get

$$\frac{\partial^2 f}{\partial x^2} = \frac{\partial^2 f}{\partial z^2}\frac{\partial z}{\partial x} + \frac{\partial^2 f}{\partial z \partial \overline{z}}\frac{\partial \overline{z}}{\partial x} + \frac{\partial^2 f}{\partial \overline{z}^2}\frac{\partial \overline{z}}{\partial x} + \frac{\partial^2 f}{\partial \overline{z} \partial z}\frac{\partial z}{\partial x}$$
$$= \frac{\partial^2 f}{\partial z^2} + 2\frac{\partial^2 f}{\partial z \partial \overline{z}} + \frac{\partial^2 f}{\partial \overline{z}^2}$$

(10-56.6a)

And

$$\frac{\partial^2 f}{\partial y^2} = i\left(\frac{\partial^2 f}{\partial z^2}\frac{\partial z}{\partial y} + \frac{\partial^2 f}{\partial z \partial \overline{z}}\frac{\partial \overline{z}}{\partial y}\right) - i\left(\frac{\partial^2 f}{\partial \overline{z}^2}\frac{\partial \overline{z}}{\partial y} + \frac{\partial^2 f}{\partial \overline{z} \partial z}\frac{\partial z}{\partial y}\right)$$
$$= -\frac{\partial^2 f}{\partial z^2} + 2\frac{\partial^2 f}{\partial z \partial \overline{z}} - \frac{\partial^2 f}{\partial \overline{z}^2}$$

(10-56.6b)

Adding equations (10-56) gives the Laplacian of the composite function as follows

$$\nabla^2 f = \frac{\partial^2 f}{\partial x^2} + \frac{\partial^2 f}{\partial y^2}$$
$$= 4\frac{\partial^2 f}{\partial z \partial \overline{z}}$$

(10-56.7)

If f(x,y) is harmonic, then

$$\frac{\partial^2 f}{\partial z \partial \overline{z}} = 0$$

(10-56.8)

5. Representation of harmonic function by sum of two functions of conjugate variables

Let $f(x,y)$ be represented by the sum of the two arbitrary functions as follows

$$f(z,\bar{z}) = f_1(z) + f_2(\bar{z}) \tag{10-56.9}$$

Differentiate with respect to the two variables, and use the property proved in (10-56.3b), we get

$$\frac{\partial f}{\partial z} = \frac{\partial f_1}{\partial z} + \frac{\partial f_1}{\partial \bar{z}}\frac{\partial \bar{z}}{\partial z} = \frac{\partial f_1}{\partial z}$$

$$\frac{\partial^2 f}{\partial z \partial \bar{z}} = \frac{\partial^2 f_1}{\partial z^2}\frac{\partial z}{\partial \bar{z}} = 0 \tag{10-56.10a}$$

Similarly,

$$\frac{\partial f}{\partial \bar{z}} = \frac{\partial f_2}{\partial \bar{z}}$$

$$\frac{\partial^2 f}{\partial z \partial \bar{z}} = \frac{\partial^2 f_2}{\partial \bar{z}^2}\frac{\partial \bar{z}}{\partial z} = 0 \tag{10-56.10b}$$

Table 8. Complex variable representation of harmonic function in terms of two conjugate variables

$z = x + iy, \quad \bar{z} = x - iy$	(10-56.1)
$\dfrac{\partial z}{\partial x} = 1, \quad \dfrac{\partial z}{\partial y} = i, \quad \dfrac{\partial \bar{z}}{\partial x} = 1, \quad \dfrac{\partial \bar{z}}{\partial y} = -i$	(10-56.2)
$x = \dfrac{z + \bar{z}}{2}, \quad y = \dfrac{z - \bar{z}}{2i}$	(10-56.3a)
$\dfrac{\partial z}{\partial \bar{z}} = 0, \quad \dfrac{\partial \bar{z}}{\partial z} = 0$	(10-56.3b)
$f(x,y) = f\left(\dfrac{z+\bar{z}}{2}, \dfrac{z-\bar{z}}{2i}\right)$	(10-56.4a)
$\dfrac{\partial f}{\partial x} = \dfrac{\partial f}{\partial z} + \dfrac{\partial f}{\partial \bar{z}}$	(10-56.4b)
$\dfrac{\partial f}{\partial y} = i\dfrac{\partial f}{\partial z} - i\dfrac{\partial f}{\partial \bar{z}}$	(10-56.4c)
$\dfrac{\partial f}{\partial x} + i\dfrac{\partial f}{\partial y} = 2\dfrac{\partial f}{\partial \bar{z}}$	(10-56.5a)
$\dfrac{\partial f}{\partial x} - i\dfrac{\partial f}{\partial y} = 2\dfrac{\partial f}{\partial z}$	(10-56.5b)

$\nabla^2 f = 4\dfrac{\partial^2 f}{\partial z \partial \overline{z}}$	(10-56.7)
$f(z,\overline{z}) = f_1(z) + f_2(\overline{z})$	(10-56.9)

(ii) Filon's solution of plane strain problems using conjugate complex variables

In the absence of internal body forces and external acceleration, **Lamé's equations** for plane strain are obtained from equations (10-26) by ignoring the z-derivative and z-displacement as follows

$$\nabla^2 u + (k+1)\frac{\partial \varepsilon}{\partial x} = 0 \qquad (10\text{-}57.1)$$

$$\nabla^2 v + (k+1)\frac{\partial \varepsilon}{\partial y} = 0 \qquad (10\text{-}57.2)$$

In order to use **conjugate complex variables** is solving those equations, we will multiply equation (10-57.2) by the complex number "i", then add, by member, the multiplied equation to equation (10-57.1), to get

$$\nabla^2 (u + iv) + (k+1)\left(\frac{\partial \varepsilon}{\partial x} + i\frac{\partial \varepsilon}{\partial y}\right) = 0 \qquad (10\text{-}58.1)$$

Then subtract, by member, the multiplied equation from equation (10-57.1), to get

$$\nabla^2 (u - iv) + (k+1)\left(\frac{\partial \varepsilon}{\partial x} - i\frac{\partial \varepsilon}{\partial y}\right) = 0 \qquad (10\text{-}58.2)$$

We will simplify the four variables in complex variable form of **Lamé's equations** (10-58) into complex conjugate variables through <u>four steps</u> as follows:

(i) First, denote the **complex conjugate displacement functions** as follows

$$\begin{aligned} \overline{\chi} &= u - iv \\ \chi &= u + iv \end{aligned} \qquad (10\text{-}59.1)$$

(ii) Representing the **displacement functions**, u and v, in terms of the complex conjugates, equations (10-56.3a), as follows

$$u(x,y) = u\left(\frac{z+\bar{z}}{2}, \frac{z-\bar{z}}{2i}\right)$$
$$v(x,y) = v\left(\frac{z+\bar{z}}{2}, \frac{z-\bar{z}}{2i}\right)$$

(10-59.2)

Therefore, the complex conjugates of strains are obtained by differentiating equations (10-59.2), and substituting from equation (10-56.2), as follows

$$\frac{\partial u}{\partial x} = \frac{\partial u}{\partial z}\frac{\partial z}{\partial x} + \frac{\partial u}{\partial \bar{z}}\frac{\partial \bar{z}}{\partial x} = \frac{\partial u}{\partial z} + \frac{\partial u}{\partial \bar{z}}$$
$$\frac{\partial v}{\partial y} = \frac{\partial v}{\partial z}\frac{\partial z}{\partial y} + \frac{\partial v}{\partial \bar{z}}\frac{\partial \bar{z}}{\partial y} = \frac{\partial v}{\partial z}i - \frac{\partial v}{\partial \bar{z}}i$$

(10-59.3)

(iii) Thus, the complex conjugate function for the **dilatational strain** is obtained by adding the two equations (10-58.5) by members, arranging, and substituting from equation (10-58.3), to get

$$\varepsilon = \frac{\partial u}{\partial x} + \frac{\partial v}{\partial y}$$
$$= \frac{\partial u}{\partial z} + \frac{\partial u}{\partial \bar{z}} + \frac{\partial v}{\partial z}i - \frac{\partial v}{\partial \bar{z}}i$$
$$= \frac{\partial}{\partial z}(u+iv) + \frac{\partial}{\partial \bar{z}}(u-iv)$$
$$= \frac{\partial \chi}{\partial z} + \frac{\partial \bar{\chi}}{\partial \bar{z}}$$

(10-59.4)

(iv) Express the derivatives in x and y in terms of the two complex conjugates as follows

$$\frac{\partial}{\partial x} + i\frac{\partial}{\partial y} = \frac{\partial}{\partial z}\frac{\partial z}{\partial x} + \frac{\partial}{\partial \bar{z}}\frac{\partial \bar{z}}{\partial x} + i\left(\frac{\partial}{\partial z}\frac{\partial z}{\partial y} + \frac{\partial}{\partial \bar{z}}\frac{\partial \bar{z}}{\partial y}\right)$$
$$= 2\frac{\partial}{\partial \bar{z}}$$
$$\frac{\partial}{\partial x} - i\frac{\partial}{\partial y} = \frac{\partial}{\partial z}\frac{\partial z}{\partial x} + \frac{\partial}{\partial \bar{z}}\frac{\partial \bar{z}}{\partial x} - i\left(\frac{\partial}{\partial z}\frac{\partial z}{\partial y} + \frac{\partial}{\partial \bar{z}}\frac{\partial \bar{z}}{\partial y}\right)$$
$$= 2\frac{\partial}{\partial z}$$

(10-59.5)

Table 9. Summary of steps of representing functions in terms of complex conjugate variables

$\bar{\chi} = u - iv$, $\chi = u + iv$		(10-59.1)
$u(x,y) = u\left(\dfrac{z+\bar{z}}{2}, \dfrac{z-\bar{z}}{2i}\right)$, $v(x,y) = v\left(\dfrac{z+\bar{z}}{2}, \dfrac{z-\bar{z}}{2i}\right)$		(10-59.2)

$\dfrac{\partial u}{\partial x} = \dfrac{\partial u}{\partial z} + \dfrac{\partial u}{\partial \bar{z}}$, $\qquad \dfrac{\partial v}{\partial y} = \dfrac{\partial v}{\partial z}i - \dfrac{\partial v}{\partial \bar{z}}i$	(10-59.3)
$\varepsilon = \dfrac{\partial \chi}{\partial z} + \dfrac{\partial \bar{\chi}}{\partial \bar{z}}$	(10-59.4)
$\dfrac{\partial}{\partial x} + i\dfrac{\partial}{\partial y} = 2\dfrac{\partial}{\partial \bar{z}}$, $\qquad \dfrac{\partial}{\partial x} - i\dfrac{\partial}{\partial y} = 2\dfrac{\partial}{\partial z}$	(10-59.5)

Thus, substituting from equations (10-59), Table 8, into the complex variable form of **Lamé's equations** (10-58.1), we get

$$\nabla^2\chi + 2(k+1)\frac{\partial}{\partial \bar{z}}\left(\frac{\partial \chi}{\partial z} + \frac{\partial \bar{\chi}}{\partial \bar{z}}\right) = 0 \qquad (10\text{-}60.1)$$

Substituting from equation (10-56.7) by $\nabla^2 f = 4\dfrac{\partial^2 f}{\partial z \partial \bar{z}}$ and performing the differentiation

we get

$$4\frac{\partial^2\chi}{\partial z\partial \bar{z}} + 2(k+1)\left(\frac{\partial^2\chi}{\partial z\partial \bar{z}} + \frac{\partial^2\bar{\chi}}{\partial \bar{z}^2}\right) = 0$$

$$2(k+3)\frac{\partial^2\chi}{\partial z\partial \bar{z}} + 2(k+1)\frac{\partial^2\bar{\chi}}{\partial \bar{z}^2} = 0 \qquad (10\text{-}60.2)$$

$$\frac{\partial}{\partial \bar{z}}\left(2(k+3)\frac{\partial\chi}{\partial z} + 2(k+1)\frac{\partial\bar{\chi}}{\partial \bar{z}}\right) = 0$$

Integrating, we get

$$2(k+3)\frac{\partial\chi}{\partial z} + 2(k+1)\frac{\partial\bar{\chi}}{\partial \bar{z}} = f_1(z) \qquad (10\text{-}60.3)$$

Similarly, substituting from equations (10-59), Table 8, into the complex variable form of **Lamé's equations** (10-58.2), we get

$$\nabla^2 \overline{\chi} + 2(k+1)\frac{\partial}{\partial z}\left(\frac{\partial \chi}{\partial z} + \frac{\partial \overline{\chi}}{\partial \overline{z}}\right) = 0$$

$$4\frac{\partial^2 \overline{\chi}}{\partial z \partial \overline{z}} + 2(k+1)\left(\frac{\partial^2 \chi}{\partial z^2} + \frac{\partial^2 \overline{\chi}}{\partial \overline{z} \partial z}\right) = 0 \qquad (10\text{-}60.4)$$

$$\frac{\partial}{\partial z}\left(2(k+1)\frac{\partial \chi}{\partial z} + 2(k+3)\frac{\partial \overline{\chi}}{\partial \overline{z}}\right) = 0$$

Integrating, we get

$$2(k+1)\frac{\partial \chi}{\partial z} + 2(k+3)\frac{\partial \overline{\chi}}{\partial \overline{z}} = f_2(\overline{z}) \qquad (10\text{-}60.5)$$

We can organize equations (10-60.3) and (10-60.5) in such manner that facilitate their solution as follows

$$(k+3)\frac{\partial \chi}{\partial z} + (k+1)\frac{\partial \overline{\chi}}{\partial \overline{z}} = \frac{\partial g_1(z)}{\partial z} \qquad (10\text{-}61.1)$$

$$(k+1)\frac{\partial \chi}{\partial z} + (k+3)\frac{\partial \overline{\chi}}{\partial \overline{z}} = \frac{\partial g_2(\overline{z})}{\partial \overline{z}} \qquad (10\text{-}61.2)$$

Where, f_1, f_2, g_1, and g_2 are arbitrary analytical functions. Equations (10-61) are complex conjugate form of **Lamé's equations** for plain strain.

Solution of complex conjugate Lamé's equations for plain strain

Equations (10-61) are solved for the derivatives of the complex displacements fin two steps:

(i) Solving the simultaneous linear equations for the derivatives

$$\frac{\partial \overline{\chi}}{\partial \overline{z}} = \frac{1}{4(k+2)}\left(-(k+1)\frac{\partial g_1(z)}{\partial z} + (k+3)\frac{\partial g_2(\overline{z})}{\partial \overline{z}}\right) \qquad (10\text{-}62.1)$$

$$\frac{\partial \chi}{\partial z} = \frac{1}{4(k+2)}\left((k+3)\frac{\partial g_1(z)}{\partial z} - (k+1)\frac{\partial g_2(\overline{z})}{\partial \overline{z}}\right) \qquad (10\text{-}62.2)$$

(ii) Integrating the two equations, we get

$$\overline{\chi} = \frac{1}{4(k+2)}\left(-(k+1)\overline{z}\frac{\partial g_1(z)}{\partial z} + (k+3)g_2(\overline{z}) + h_1(z)\right) \qquad (10\text{-}63.1)$$

$$\chi = \frac{1}{4(k+2)}\left((k+3)g_1(z) - (k+1)z\frac{\partial g_2(\overline{z})}{\partial \overline{z}} + h_2(\overline{z})\right) \qquad (10\text{-}63.2)$$

Equations (10-63) give the complex conjugate displacement function and solution of **Lamé's equations** (10-60.3 and 60.5) in terms of the four arbitrary functions g's and h's.

Solution for displacements

Substituting from equations (10-63) in equations (10-59), we obtain the solution for u and v in terms of the four arbitrary functions g's and h's as follows

$$u = \frac{\overline{\chi} + \chi}{2}$$

$$= \frac{1}{8(k+2)}\left(-(k+1)\left(\overline{z}\frac{\partial g_1(z)}{\partial z} + z\frac{\partial g_2(\overline{z})}{\partial \overline{z}}\right) + (k+3)[g_2(\overline{z}) + g_1(z)] + h_1(z) + h_2(\overline{z})\right)$$

(10-64.1)

$$v = \frac{-\overline{\chi} + \chi}{2i}$$

$$= \frac{i}{8(k+2)}\left(-(k+1)\left(\overline{z}\frac{\partial g_1(z)}{\partial z} - z\frac{\partial g_2(\overline{z})}{\partial \overline{z}}\right) - (k+3)[g_1(z) - g_2(\overline{z})] + h_1(z) - h_2(\overline{z})\right)$$

(10-64.2)

The principal dilatational strain is obtained by substitution from equations (10-62) in equation (10-59.4) as follows

$$\varepsilon = \frac{\partial \chi}{\partial z} + \frac{\partial \overline{\chi}}{\partial \overline{z}}$$

$$= \frac{1}{2(k+2)}\left(\frac{\partial g_1(z)}{\partial z} + \frac{\partial g_2(\overline{z})}{\partial \overline{z}}\right)$$

(10-64.3)

Solution for stresses from Hooke's law equations (3-7)

$$\sigma_x = \mu\left(2\frac{\partial u}{\partial x} + k\varepsilon\right)$$

$$= \mu\left(2\frac{\partial u}{\partial z} + 2\frac{\partial u}{\partial \overline{z}} + k\varepsilon\right)$$

$$= \frac{\mu}{4(k+2)}\begin{pmatrix}(-(k+1)(\overline{z}g_1''(z) + g_2'(\overline{z})) + (k+3)[g_1'(z)] + h_1'(z)) \\ (-(k+1)(g_1'(z) + zg_2''(\overline{z})) + (k+3)[g_2'(\overline{z})] + h_2'(\overline{z})) \\ 2k(g_1'(z) + g_2'(\overline{z}))\end{pmatrix}$$

$$= \frac{\mu}{4(k+2)}\left[2(k+1)\big(g_1{'}(z)+g{'}_2(\overline{z})\big)-(k+1)\big(zg_2{''}(\overline{z})+\overline{z}g_1{''}(z)\big)+h_1{'}(z)+h_2{'}(\overline{z})\right] \qquad (10\text{-}65.1)$$

$$\sigma_y = \mu\left(2\frac{\partial v}{\partial y}+k\varepsilon\right)$$

$$= \mu\left(2\frac{\partial v}{\partial z}i-2i\frac{\partial v}{\partial \overline{z}}+k\varepsilon\right)$$

$$= \frac{\mu}{4(k+2)}\begin{pmatrix} +(k+1)\big(\overline{z}g_1{''}(z)-g_2{'}(\overline{z})\big)+(k+3)\big[g_1{'}(z)\big]-h_1{'}(z) \\ -(k+1)\big(g_1{'}(z)-zg_2{''}(\overline{z})\big)+(k+3)\big[g_2{'}(\overline{z})\big]-h_2{'}(\overline{z}) \\ +2k\big(g_1{'}(z)+g_2{'}(\overline{z})\big) \end{pmatrix}$$

$$= \frac{\mu}{4(k+2)}\left[(k+1)\big(\overline{z}g_1{''}(z)+zg_2{''}(\overline{z})\big)+2(k+1)\big[g_1{'}(z)+g_2{'}(\overline{z})\big]-h_1{'}(z)-h_2{'}(\overline{z})\right] \qquad (10\text{-}65.2)$$

$$\sigma_z = \mu\left(2\frac{\partial w}{\partial z}+k\varepsilon\right)$$

$$= \mu k\varepsilon \qquad (10\text{-}65.3)$$

$$= \frac{k\mu}{2(k+2)}\big(g_1{'}(z)+g_2{'}(\overline{z})\big)$$

$$\tau_{xy} = \mu\left(\frac{\partial u}{\partial y}+\frac{\partial v}{\partial x}\right)$$

$$= \mu\left(\frac{\partial u}{\partial z}\frac{\partial z}{\partial y}+\frac{\partial u}{\partial \overline{z}}\frac{\partial \overline{z}}{\partial y}+\frac{\partial v}{\partial z}\frac{\partial z}{\partial x}+\frac{\partial v}{\partial \overline{z}}\frac{\partial \overline{z}}{\partial x}\right)$$

$$= \mu\left(\frac{\partial u}{\partial z}i-i\frac{\partial u}{\partial \overline{z}}+\frac{\partial v}{\partial z}+\frac{\partial v}{\partial \overline{z}}\right)$$

$$= \mu\left(\frac{i}{8(k+2)}\begin{pmatrix} -(k+1)\big(\overline{z}g_1{''}(z)+g_2{'}(\overline{z})\big)+(k+3)\big[g_1{'}(z)\big]+h_1{'}(z) \\ +(k+1)\big(g_1{'}(z)+zg_2{''}(\overline{z})\big)-(k+3)\big[g_2{'}(\overline{z})\big]-h_2{'}(\overline{z}) \\ -(k+1)\big(\overline{z}g_1{''}(z)-g_2{'}(\overline{z})\big)-(k+3)\big[g_1{'}(z)\big]+h_1{'}(z) \\ -(k+1)\big(g_1{'}(z)-zg_2{''}(\overline{z})\big)-(k+3)\big[-g_2{'}(\overline{z})\big]-h_2{'}(\overline{z}) \end{pmatrix}\right)$$

$$= \frac{i\mu}{4(k+2)}\big((k+1)\big(zg_2{''}(\overline{z})-\overline{z}g_1{''}(z)\big)+h_1{'}(z)-h_2{'}(\overline{z})\big) \qquad (10\text{-}65.4)$$

366

$$\tau_{yz} = \mu\left(\frac{\partial w}{\partial y} + \frac{\partial v}{\partial z}\right) = \mu(0+0) = 0 \tag{10-65.5}$$

$$\tau_{zx} = \mu\left(\frac{\partial w}{\partial x} + \frac{\partial u}{\partial z}\right) = \mu(0+0) = 0 \tag{10-65.6}$$

Table 10. Summary of solutions for plain strain problem using Filon's method

$\sigma_x = \dfrac{\mu}{4(k+2)}\left[2(k+1)\left(g_1{'}(z)+g{'}_2(\overline{z})\right) - (k+1)\left(zg_2{''}(\overline{z})+\overline{z}g_1{''}(z)\right) + h_1{'}(z) + h_2{'}(\overline{z})\right]$	(10-65.1)
$\sigma_y = \dfrac{\mu}{4(k+2)}\left[(k+1)\left(\overline{z}g_1{''}(z)+zg_2{''}(\overline{z})\right) + 2(k+1)\left[g_1{'}(z)+g_2{'}(\overline{z})\right] - h_1{'}(z) - h_2{'}(\overline{z})\right]$	(10-65.2)
$\sigma_z = \dfrac{k\mu}{2(k+2)}\left(g_1{'}(z)+g_2{'}(\overline{z})\right)$	(10-65.3)
$\tau_{xy} = \dfrac{i\mu}{4(k+2)}\left((k+1)\left(zg_2{''}(\overline{z}) - \overline{z}g_1{''}(z)\right) + h_1{'}(z) - h_2{'}(\overline{z})\right)$	(10-65.4)
$\tau_{yz} = 0$	(10-65.5)
$\tau_{zx} = 0$	(10-65.6)
$u = \dfrac{1}{8(k+2)}\left(-(k+1)\left(\overline{z}g_1{'}(z)+zg_2{'}(\overline{z})\right) + (k+3)\left[g_1(z)+g_2(\overline{z})\right] + h_1(z) + h_2(\overline{z})\right)$	(10-64.1)
$v = \dfrac{i}{8(k+2)}\left(-(k+1)\left(\overline{z}g_1{'}(z)-zg_2{'}(\overline{z})\right) - (k+3)\left[g_1(z)-g_2(\overline{z})\right] + h_1(z) - h_2(\overline{z})\right)$	(10-64.2)
$\varepsilon = \dfrac{1}{2(k+2)}\left(g_1{'}(z)+g_2{'}(\overline{z})\right)$	(10-64.3)

Determination of the four arbitrary functions g's and h's

The four arbitrary functions f_1, f_2, g_1, and g_2 must be set such that imaginary terms in equations (10-64.2) and (10-65.4) result into net real values.

$$\overline{h}_1(\overline{z}) = h_2(\overline{z})$$
$$\overline{g}_1(\overline{z}) = g_2(\overline{z}) \tag{10-66.1}$$

Where,

$$\overline{h}_1(\overline{z}) = \text{conjugate}\left|h_1(\overline{z})\right|$$
$$\overline{g}_1(\overline{z}) = \text{conjugate}\left|g_1(\overline{z})\right| \tag{10-66.2}$$

Therefore, four functions can be written in the complex form as follows

$$h_1(z) = \varphi_1(x,y) + i\psi_1(x,y)$$
$$g_1(z) = \varphi_2(x,y) + i\psi_2(x,y)$$

(10-66.3)

And the conjugates, in equations (10-66.1), become

$$h_2(\overline{z}) = \overline{h}_1(\overline{z}) = \varphi_1(x,y) - i\psi_1(x,y)$$
$$g_2(\overline{z}) = \overline{g}_1(\overline{z}) = \varphi_2(x,y) - i\psi_2(x,y)$$

(10-66.4)

In order to prove that the constraints imposed of the arbitrary functions g's and h's, equations (10-66.1) will achieve the purpose of rendering our solution real quantities, we will first obtain the derivatives of h_1 and h_2, equations (10-66.3) and (10-66.4), and substitute those in the equations for displacements (10-64.1 through 64.3), as follows:

Differentiate the first equation of (10-66.3) with respect to x, we get

$$\frac{\partial h_1(z)}{\partial z}\frac{\partial z}{\partial x} = \frac{\partial \varphi_1}{\partial x} + i\frac{\partial \psi_1}{\partial x}$$

(10-67.1)

Differentiate first equation of (10-66.3) with respect to y, we get

$$\frac{\partial h_1(z)}{\partial z}\frac{\partial z}{\partial y} = \frac{\partial \varphi_1}{\partial y} + i\frac{\partial \psi_1}{\partial y}$$

(10-67.2)

Similarly, differentiate second equation of (10-66.3) with respect to x, we get

$$\frac{\partial g_1(z)}{\partial z}\frac{\partial z}{\partial x} = \frac{\partial \varphi_2}{\partial x} + i\frac{\partial \psi_2}{\partial x}$$

(10-67.3)

Differentiate second equation of (10-66.3) with respect to y, we get

$$\frac{\partial g_1(z)}{\partial z}\frac{\partial z}{\partial y} = \frac{\partial \varphi_z}{\partial y} + i\frac{\partial \psi_2}{\partial y}$$

(10-67.4)

Substituting from equation (10-56.2) in equations (10-67), and separating real and imaginary terms, we get

$$\frac{\partial \varphi_1}{\partial x} = \frac{\partial \psi_1}{\partial y}, \qquad \frac{\partial \varphi_1}{\partial y} = -\frac{\partial \psi_1}{\partial x}$$

(10-68.1)

$$\frac{\partial \varphi_2}{\partial x} = \frac{\partial \psi_2}{\partial y}, \qquad \frac{\partial \varphi_z}{\partial y} = -\frac{\partial \psi_2}{\partial x}$$

(10-68.2)

Differentiate the first equation of (10-66.4) with respect to x, we get

$$\frac{\partial h_2(\bar{z})}{\partial \bar{z}} = \frac{\partial \varphi_1}{\partial x} - i\frac{\partial \psi_1}{\partial x} \qquad (10\text{-}69.1)$$

Differentiate second equation of (10-66.4) with respect to x, we get

$$\frac{\partial g_2(\bar{z})}{\partial \bar{z}} = \frac{\partial \varphi_2}{\partial x} - i\frac{\partial \psi_2}{\partial x} \qquad (10\text{-}69.2)$$

Table 11. Summary of derivatives of conjugate functions

$h_1(z) = \varphi_1(x,y) + i\psi_1(x,y)$ $g_1(z) = \varphi_2(x,y) + i\psi_2(x,y)$	(10-66.3)
$h_2(\bar{z}) = \bar{h}_1(\bar{z}) = \varphi_1(x,y) - i\psi_1(x,y)$ $g_2(\bar{z}) = \bar{g}_1(\bar{z}) = \varphi_2(x,y) - i\psi_2(x,y)$	(10-66.4)
$h_1{}'(z) = \dfrac{\partial \varphi_1}{\partial x} + i\dfrac{\partial \psi_1}{\partial x}$	(10-67.1)
$g_1{}'(z) = \dfrac{\partial \varphi_2}{\partial x} + i\dfrac{\partial \psi_2}{\partial x}$	(10-67.3)
$h_2{}'(\bar{z}) = \dfrac{\partial \varphi_1}{\partial x} - i\dfrac{\partial \psi_1}{\partial x}$	(10-69.1)
$g_2{}'(\bar{z}) = \dfrac{\partial \varphi_2}{\partial x} - i\dfrac{\partial \psi_2}{\partial x}$	(10-69.2)
$\dfrac{\partial \varphi_1}{\partial x} = \dfrac{\partial \psi_1}{\partial y}, \qquad \dfrac{\partial \varphi_1}{\partial y} = -\dfrac{\partial \psi_1}{\partial x}$	(10-68.1)
$\dfrac{\partial \varphi_2}{\partial x} = \dfrac{\partial \psi_2}{\partial y}, \qquad \dfrac{\partial \varphi_z}{\partial y} = -\dfrac{\partial \psi_2}{\partial x}$	(10-68.2)

Reconstruction of the arbitrary terms in equations (10-64) and (10-65)

Using Table 11, we reconstruct the arbitrary terms in g's and h's as follows:

$$zg_2{}''(\bar{z}) - \bar{z}g_1{}''(z) = (x+iy)\left(\frac{\partial^2 \varphi_2}{\partial x^2} - i\frac{\partial^2 \psi_2}{\partial x^2}\right) - (x-iy)\left(\frac{\partial^2 \varphi_2}{\partial x^2} + i\frac{\partial^2 \psi_2}{\partial x^2}\right)$$

$$= 2i\left(y\frac{\partial^2 \varphi_2}{\partial x^2} - x\frac{\partial^2 \psi_2}{\partial x^2}\right) \qquad (10\text{-}70.1)$$

369

$$zg_2''(\overline{z}) + \overline{z}g_1''(z) = (x + iy)\left(\frac{\partial^2\varphi_2}{\partial x^2} - i\frac{\partial^2\psi_2}{\partial x^2}\right) + (x - iy)\left(\frac{\partial^2\varphi_2}{\partial x^2} + i\frac{\partial^2\psi_2}{\partial x^2}\right)$$

$$= 2\left(x\frac{\partial^2\varphi_2}{\partial x^2} + y\frac{\partial^2\psi_2}{\partial x^2}\right)$$

(10-70.2)

$$h_1'(z) - h_2'(\overline{z}) = 2i\frac{\partial\psi_1}{\partial x}$$

(10-70.3)

$$h_1'(z) + h_2'(\overline{z}) = 2\frac{\partial\varphi_1}{\partial x}$$

(10-70.4)

$$\overline{z}g_1'(z) + zg_2'(\overline{z}) = (x - iy)\left(\frac{\partial\varphi_2}{\partial x} + i\frac{\partial\psi_2}{\partial x}\right) + (x + iy)\left(\frac{\partial\varphi_2}{\partial x} - i\frac{\partial\psi_2}{\partial x}\right)$$

$$= 2\left(x\frac{\partial\varphi_2}{\partial x} + y\frac{\partial\psi_2}{\partial x}\right)$$

(10-70.5)

$$g_1'(z) + g_2'(\overline{z}) = \left(\frac{\partial\varphi_2}{\partial x} + i\frac{\partial\psi_2}{\partial x}\right) + \left(\frac{\partial\varphi_2}{\partial x} - i\frac{\partial\psi_2}{\partial x}\right)$$

$$= 2\frac{\partial\varphi_2}{\partial x}$$

$$g_1'(z) - g_2'(\overline{z}) = \left(\frac{\partial\varphi_2}{\partial x} + i\frac{\partial\psi_2}{\partial x}\right) + \left(\frac{\partial\varphi_2}{\partial x} - i\frac{\partial\psi_2}{\partial x}\right)$$

$$= 2i\frac{\partial\psi_2}{\partial x}$$

$$g_1(z) + g_2(\overline{z}) = 2\varphi_2(x, y)$$
$$g_1(z) - g_2(\overline{z}) = 2i\psi_2(x, y)$$
$$h_1(z) + h_2(\overline{z}) = 2\varphi_1(x, y)$$
$$h_1(z) - h_2(\overline{z}) = 2i\psi_1(x, y)$$

(10-70.6)

$$zg_2'(\overline{z}) - \overline{z}g_1'(z) = (x + iy)\left(\frac{\partial\varphi_2}{\partial x} - i\frac{\partial\psi_2}{\partial x}\right) - (x - iy)\left(\frac{\partial\varphi_2}{\partial x} + i\frac{\partial\psi_2}{\partial x}\right)$$

$$= 2i\left(y\frac{\partial\varphi_2}{\partial x} - x\frac{\partial\psi_2}{\partial x}\right)$$

(10-70.7)

$$zg_2'(\overline{z}) + \overline{z}g_1'(z) = (x + iy)\left(\frac{\partial\varphi_2}{\partial x} - i\frac{\partial\psi_2}{\partial x}\right) + (x - iy)\left(\frac{\partial\varphi_2}{\partial x} + i\frac{\partial\psi_2}{\partial x}\right)$$

$$= 2\left(x\frac{\partial\varphi_2}{\partial x} + y\frac{\partial\psi_2}{\partial x}\right)$$

(10-70.8)

Stresses and displacements in terms of real quantities

370

$$\sigma_x = \frac{\mu}{4(k+2)}\left[4(k+1)\frac{\partial\varphi_2}{\partial x} - 2(k+1)\left(x\frac{\partial^2\varphi_2}{\partial x^2} + y\frac{\partial^2\psi_2}{\partial x^2}\right) + 2\frac{\partial\varphi_1}{\partial x}\right] \tag{10-71.1}$$

$$\sigma_y = \frac{\mu}{4(k+2)}\left[2(k+1)\left(x\frac{\partial^2\varphi_2}{\partial x^2} + y\frac{\partial^2\psi_2}{\partial x^2}\right) + 4(k+1)\frac{\partial\varphi_2}{\partial x} - 2\frac{\partial\varphi_1}{\partial x}\right] \tag{10-71.2}$$

$$\sigma_z = \frac{k\mu}{(k+2)}\frac{\partial\varphi_2}{\partial x} \tag{10-71.3}$$

$$\tau_{xy} = \frac{-\mu}{4(k+2)}\left((k+1)\left(y\frac{\partial^2\varphi_2}{\partial x^2} - x\frac{\partial^2\psi_2}{\partial x^2}\right) + \frac{\partial\psi_1}{\partial x}\right) \tag{10-71.4}$$

$$\tau_{yz} = 0 \tag{10-71.5}$$

$$\tau_{zx} = 0 \tag{10-71.6}$$

$$u = \frac{1}{4(k+2)}\left(-(k+1)\left(x\frac{\partial\varphi_2}{\partial x} + y\frac{\partial\psi_2}{\partial x}\right) + (k+3)\varphi_2 + \varphi_1\right) \tag{10-71.7}$$

$$v = \frac{1}{4(k+2)}\left(-(k+1)\left(y\frac{\partial\varphi_2}{\partial x} - x\frac{\partial\psi_2}{\partial x}\right) + (k+3)\psi_2 - \psi_1\right) \tag{10-71.8}$$

$$\varepsilon = \frac{1}{(k+2)}\frac{\partial\varphi_2}{\partial x} \tag{10-71.9}$$

In equations (10-71), we have thus eliminated all imaginary terms from the final expressions of displacements and stresses.

Rotation strain

The derivatives of the displacements u and v, equations (10-71.7) and (10-71.8), yield the rotational strain ω_z as follows

$$2\omega_z = \frac{\partial v}{\partial x} - \frac{\partial u}{\partial y} \tag{10-72.1}$$

First, we use the Cauchy-Riemann equations, (10-68.2), to represent u in terms of y-derivatives as follows

$$u = \frac{1}{4(k+2)}\left(-(k+1)\left(x\frac{\partial\psi_2}{\partial y} - y\frac{\partial\varphi_2}{\partial y}\right) + (k+3)\varphi_2 + \varphi_1\right) \tag{10-72.2}$$

Second, differentiate (10-72.2) to get

$$\frac{\partial u}{\partial y} = \frac{1}{4(k+2)}\left(-(k+1)\left(x\frac{\partial^2\psi_2}{\partial y^2} - y\frac{\partial^2\varphi_2}{\partial y^2} - \frac{\partial\varphi_2}{\partial y}\right) + (k+3)\frac{\partial\varphi_2}{\partial y} + \frac{\partial\varphi_1}{\partial y}\right) \qquad (10\text{-}72.3)$$

Third, differentiate (10-71.8) to get

$$\frac{\partial v}{\partial x} = \frac{1}{4(k+2)}\left(-(k+1)\left(y\frac{\partial^2\varphi_2}{\partial x^2} - x\frac{\partial^2\psi_2}{\partial x^2} - \frac{\partial\psi_2}{\partial x}\right) + (k+3)\frac{\partial\psi_2}{\partial x} - \frac{\partial\psi_1}{\partial x}\right) \qquad (10\text{-}72.4)$$

Substituting from (10-72.3) and (10-72.4) into (10-72.1), we get

$$2\omega_z = \frac{\partial v}{\partial x} - \frac{\partial u}{\partial y}$$

$$= \frac{1}{4(k+2)}\left(\begin{array}{l} -(k+1)\left(y\left(\frac{\partial^2\varphi_2}{\partial x^2} + \frac{\partial^2\varphi_2}{\partial y^2}\right) - x\left(\frac{\partial^2\psi_2}{\partial x^2} + \frac{\partial^2\psi_2}{\partial y^2}\right) - \frac{\partial\psi_2}{\partial x} + \frac{\partial\varphi_2}{\partial y}\right) \\ +(k+3)\left(\frac{\partial\psi_2}{\partial x} - \frac{\partial\varphi_2}{\partial y}\right) - \left(\frac{\partial\psi_1}{\partial x} + \frac{\partial\varphi_1}{\partial y}\right) \end{array}\right) \qquad (10\text{-}72.5)$$

$$2\omega_z = \frac{\partial v}{\partial x} - \frac{\partial u}{\partial y}$$

$$= \frac{1}{4(k+2)}\left(\begin{array}{l} -(k+1)\left(\begin{array}{l} +y\left(\frac{\partial^2\varphi_2}{\partial x^2} + \frac{\partial^2\varphi_2}{\partial y^2}\right) - x\left(\frac{\partial^2\psi_2}{\partial x^2} + \frac{\partial^2\psi_2}{\partial y^2}\right) \\ -\frac{\partial\psi_2}{\partial x} + \frac{\partial\varphi_2}{\partial y} \end{array}\right) \\ +(k+3)\left(\frac{\partial\psi_2}{\partial x} - \frac{\partial\varphi_2}{\partial y}\right) - \frac{\partial\psi_1}{\partial x} - \frac{\partial\varphi_1}{\partial y} \end{array}\right) \qquad (10\text{-}72.5)$$

Substituting from **Cauchy-Riemann's equations** (10-68.1) and (10-68.2) and noting the vanishing Laplacian of the two harmonic functions (all four functions φ_1, φ_2, ψ_1, ψ_2, are harmonic), we get

$$2\omega_z = \frac{1}{4(k+2)}\left(-(k+1)\left(\frac{\partial\varphi_2}{\partial y} - \frac{\partial\psi_2}{\partial x}\right) + (k+3)\left(\frac{\partial\psi_2}{\partial x} - \frac{\partial\varphi_2}{\partial y}\right) - \frac{\partial\psi_1}{\partial x} - \frac{\partial\varphi_1}{\partial y}\right)$$

$$= \frac{\partial\psi_2}{\partial x} \qquad (10\text{-}72.6)$$

Equation (10-72.6)

$$\omega_z = \frac{1}{(k+2)}\left(-\frac{\partial \psi_2}{\partial x}\right)$$ (10-72.5)

, confers a meaning on the derivative ψ_2 in terms of the rotational twit angle $2\omega_z$.

Relationship between rotational strain ω_z and dilatational strain ε

The simple appearance of the derivatives of ψ_2 and ϕ_2 in the expressions of rotational strain equation (10-72.6) and dilatational strain equation (10-71.9) makes it possible to construct the harmonic function $g_1(z)$, equation (10-66.4), by adding equations (10-72.6) and (10-71.9) by members as follows

$$i \quad x \qquad 2\omega_z = \frac{\partial \psi_2}{\partial x}$$

$$+ \qquad (k+2)\varepsilon = \frac{\partial \phi_2}{\partial x}$$

Add _____

$$(k+2)\varepsilon + i2\omega_z = \frac{\partial \phi_2}{\partial x} + i\frac{\partial \psi_2}{\partial x}$$

$$= g_1{}'(z)$$

(10-73.1)

Generalization of complex presentation of solution of plain strain problems

(i) The **complex displacement variable**, $u + iv$, is obtained from equations (10-64.1) and (10-64.2) by multiplying v by i and adding the two equations by members to get:

$$u + iv = \frac{1}{4(k+2)}\left(-(k+1)\bar{z}g_1{}'(z) + (k+3)g_2(\bar{z}) + h_1(z)\right)$$ (10-74.1)

(ii) Sum of surface stresses are obtained by adding equations (10-65.1) and (10-65.2) by members as follows

$$\sigma_x + \sigma_y = \mu \frac{(k+1)}{(k+2)}\left[(g_1{}'(z) + g'_2(\bar{z}))\right]$$ (10-74.2)

(iii) Surfaces stresses, equations (10-65.1) and (10-65.2), and shear stress, equation (10-65.4), yield the following complex form

$$\sigma_y - \sigma_x + 2i\tau_{xy} = \frac{\mu}{4(k+2)}\begin{bmatrix}(k+1)(\overline{z}g_1''(z)+zg_2''(\overline{z}))+2(k+1)[g_1'(z)+g_2'(\overline{z})]-h_1'(z)-h_2'(\overline{z})\\ -2(k+1)(g_1'(z)+g_2'(\overline{z}))+(k+1)(zg_2''(\overline{z})+\overline{z}g_1''(z))-h_1'(z)-h_2'(\overline{z})\\ -2(k+1)(zg_2''(\overline{z})-\overline{z}g_1''(z))-2h_1'(z)+2h_2'(\overline{z})\end{bmatrix}$$

$$= \frac{\mu}{(k+2)}[(k+1)(zg_2''(\overline{z}))-h_1'(z)] \tag{10-74.3}$$

Airy stress function with complex harmonic function

Equations (7-17), (10-56.7) and (10-74.2) give

$$\sigma_x + \sigma_y = \mu\frac{(k+1)}{(k+2)}[(g_1'(z)+g_2'(\overline{z}))]$$

$$= \frac{\partial^2\varphi(x,y)}{\partial x^2}+\frac{\partial^2\varphi(x,y)}{\partial y^2} \tag{10-75.1}$$

$$= 4\frac{\partial^2\varphi(z,\overline{z})}{\partial z\partial\overline{z}}$$

Equation (10-75) offers Airy's stress function in terms of complex harmonic functions. We shall write it in more transparent integrable form as follows

$$\frac{\partial^2\varphi(z,\overline{z})}{\partial z\partial\overline{z}} = \mu\frac{(k+1)}{4(k+2)}[(g_1'(z)+g_2'(\overline{z}))] \tag{10-75.2}$$

Integrating twice, we get

$$\varphi(z,\overline{z}) = \mu\frac{(k+1)}{4(k+2)}[(\overline{z}g_1(z)+zg_2(\overline{z})+k_1(z)+k_2(\overline{z}))] \tag{10-75.3}$$

In order to ensure that all terms in the equation (10-75.3) are real, equation (10-75.3) is used to replace g_2 by its conjugate g_1, and the two arbitrary functions k_1 and k_2 are chosen similarly as follows

$$k_1(\overline{z}) = \overline{k}_2(\overline{z}) \tag{10-75.4}$$

Thus, equation (10-75.3) becomes

$$\varphi(z,\overline{z}) = \mu\frac{(k+1)}{4(k+2)}[(\overline{z}g_1(z)+z\overline{g}_1(\overline{z})+\overline{k}_2(\overline{z})+k_2(\overline{z}))] \tag{10-75.5}$$

374

Which is the complex form of **Airy's stress function** for plain strain obtained in terms of harmonic functions

Using equations (7-17) and (10-75.5), we obtain the plain stresses as follows

$$\sigma_x = \frac{\partial^2 \varphi(x,y)}{\partial y^2}$$
$$= \mu \frac{(k+1)}{4(k+2)}\left[\left(\begin{array}{c} 2\left(g_1{}'(z)+\overline{g}_1{}'(\overline{z})\right)-\left(\overline{z}g_1{}''(z)+z\overline{g}_1{}''(\overline{z})\right) \\ -\left(\overline{k}_2{}''(\overline{z})+k_2{}''(\overline{z})\right) \end{array}\right)\right]$$

(10-76.1)

$$\sigma_y = \frac{\partial^2 \varphi(x,y)}{\partial x^2}$$
$$= \mu \frac{(k+1)}{4(k+2)}\left[\left(\begin{array}{c} 2\left(g_1{}'(z)+\overline{g}_1{}'(\overline{z})\right)+\left(\overline{z}g_1{}''(z)+z\overline{g}_1{}''(\overline{z})\right) \\ +\overline{k}_2{}''(\overline{z})+k_2{}''(\overline{z}) \end{array}\right)\right]$$

(10-76.2)

$$\tau_{xy} = -\frac{\partial^2 \varphi(x,y)}{\partial x \partial y}$$
$$= \mu \frac{i(k+1)}{4(k+2)}\left(\overline{z}g_1{}''(z)-z\overline{g}_1{}''(\overline{z})-\overline{k}_2{}''(\overline{z})-k_2{}''(\overline{z})\right)$$

(10-76.3)

10-16. Elastic vibrational waves

(i) General wave equation

Vibrational motions in elastic medium were treated by equations (5-1), which are **the Lamé's equations**, that involve the spatial derivatives of displacements and dilatational strain weighed with the Lamé's coefficients, internal body forces, and external inertial forces as follows

$$\mu \nabla^2 \begin{bmatrix} u \\ v \\ w \end{bmatrix} + (\mu+\lambda)\begin{bmatrix} \partial\varepsilon/\partial x \\ \partial\varepsilon/\partial y \\ \partial\varepsilon/\partial z \end{bmatrix} + \rho \begin{bmatrix} X_i \\ Y_i \\ Z_i \end{bmatrix} = \rho \frac{\partial^2}{\partial t^2}\begin{bmatrix} u \\ v \\ w \end{bmatrix}$$

(10-77.1)

(ii) Dilatational wave equation

In order to express equation (10-77.1) in terms of **dilatational strain**, we will differentiate each of the three equations with respect to each respective coordinate and then add the three equations to produce a single partial differential equation as follows

$$\mu\nabla^2\begin{bmatrix}\partial u/\partial x\\ \partial v/\partial y\\ \partial w/\partial z\end{bmatrix}+(\mu+\lambda)\begin{bmatrix}\partial^2\varepsilon/\partial x^2\\ \partial^2\varepsilon/\partial y^2\\ \partial\varepsilon^2/\partial z^2\end{bmatrix}=\rho\frac{\partial^2}{\partial t^2}\begin{bmatrix}\partial u/\partial x\\ \partial v/\partial y\\ \partial w/\partial z\end{bmatrix}$$ (10-77.2)

Adding the three equations

$$\mu\nabla^2\left[\frac{\partial u}{\partial x}+\frac{\partial v}{\partial y}+\frac{\partial w}{\partial z}\right]+(\mu+\lambda)\nabla^2\varepsilon=\rho\frac{\partial^2}{\partial t^2}\left[\frac{\partial u}{\partial x}+\frac{\partial v}{\partial y}+\frac{\partial w}{\partial z}\right]$$ (10-77.3)

Where the **divergence of displacements** (the bracketed sums of derivatives) comprises the dilatational strain thus simplifying the wave equation to

$$(2\mu+\lambda)\nabla^2\varepsilon=\rho\frac{\partial^2\varepsilon}{\partial t^2}$$ (10-77.4)

The velocity of the displacement wave is defined

$$c_L=\pm\sqrt{\frac{2\mu+\lambda}{\rho}}$$ (10-77.5)

Equation (10-77.4a) can be written in more representative form in terms of a **displacement vector u**(u,v,w) as follows

$$\mu\nabla^2[\nabla.\mathbf{u}]+(\mu+\lambda)\nabla^2\varepsilon=\rho\frac{\partial^2}{\partial t^2}[\nabla.\mathbf{u}]$$ (10-77.6)

Equivoluminal S-waves

If the dilatational strain ε vanishes, then equation (10-77.6) becomes

$$\mu\nabla^2[\nabla.\mathbf{u}]\varepsilon=\rho\frac{\partial^2}{\partial t^2}[\nabla.\mathbf{u}]$$ (10-77.7)

Which represents vibrations in displacements with net vanishing dilatation. Thus, the speed of propagation is mainly of distortion or rotation or shear strains (S-waves) given by

$$c_S = \pm\sqrt{\frac{\mu}{\rho}} \qquad (10\text{-}77.8)$$

(iii) Distortional or solenoidal wave equation

In order to derive the **rotational or solenoidal wave equation**, we differentiate the second equation with respect to z, the third with respect to y, we get

$$\mu V^2\begin{bmatrix}\partial v/\partial z \\ \partial w/\partial y\end{bmatrix} + (\mu+\lambda)\begin{bmatrix}\partial^2\varepsilon/\partial y\partial z \\ \partial^2\varepsilon/\partial y\partial z\end{bmatrix} = \rho\frac{\partial^2}{\partial t^2}\begin{bmatrix}\partial v/\partial z \\ \partial w/\partial y\end{bmatrix} \qquad (10\text{-}78.1)$$

Subtracting the two equations by members we get

$$\mu V^2\left(\frac{\partial w}{\partial y} - \frac{\partial v}{\partial z}\right) + (\mu+\lambda)\left[\frac{\partial^2\varepsilon}{\partial y\partial z} - \frac{\partial^2\varepsilon}{\partial y\partial z}\right] = \rho\frac{\partial^2}{\partial t^2}\left(\frac{\partial w}{\partial y} - \frac{\partial v}{\partial z}\right) \qquad (10\text{-}78.2)$$

Replacing the bracketed differences of derivatives by rotational strains, we get

$$\mu\nabla^2\omega_x = \rho\frac{\partial^2\omega_x}{\partial t^2} \qquad (10\text{-}78.3)$$

Similarly, we can obtain the remaining two equations for y- and z- dependent rotational strains

$$\mu\nabla^2\omega_x = \rho\frac{\partial^2\omega_x}{\partial t^2}$$
$$\mu\nabla^2\omega_y = \rho\frac{\partial^2\omega_y}{\partial t^2} \qquad (10\text{-}78.4)$$
$$\mu\nabla^2\omega_z = \rho\frac{\partial^2\omega_z}{\partial t^2}$$

The velocity of the **rotational waves** is defined as follows

$$c_S = \pm\sqrt{\frac{\mu}{\rho}} \qquad (10\text{-}78.5)$$

Equations (10-78.4) is also put into more representative form in terms of a **solenoid vector ω** (ω_x, ω_y, ω_z)

$$\mu\nabla^2(\nabla\times\mathbf{u}) = \rho\frac{\partial^2}{\partial t^2}(\nabla\times\mathbf{u}) \qquad (10\text{-}78.6)$$

377

Where

$$\boldsymbol{\omega}\left(\omega_x, \omega_y, \omega_z\right) = \nabla \times \mathbf{u} \qquad\qquad (10\text{-}78.7)$$

Equation (10-78.6) is called the wave equation for:

Rotational waves
Equivoluminal waves
Distortional waves
Shear waves
Transverse waves
S-waves

(iv) Irrotational or P-wave equation

$$\frac{\partial w}{\partial y} - \frac{\partial v}{\partial z} = 0$$

$$\frac{\partial w}{\partial x} - \frac{\partial u}{\partial z} = 0 \qquad\qquad (10\text{-}79.1)$$

$$\frac{\partial u}{\partial y} - \frac{\partial v}{\partial x} = 0$$

Therefore, we could assume that the three displacements are derivatives of the same arbitrary potential Φ, as follows

$$u = \frac{\partial \Phi}{\partial x}, \qquad v = \frac{\partial \Phi}{\partial y}, \qquad w = \frac{\partial \Phi}{\partial z} \qquad\qquad (10\text{-}79.2)$$

Therefore, the potential dilatational strain becomes

$$\begin{aligned} \varepsilon &= \frac{\partial u}{\partial x} + \frac{\partial v}{\partial y} + \frac{\partial w}{\partial z} \\ &= \frac{\partial^2 \Phi}{\partial x^2} + \frac{\partial^2 \Phi}{\partial y^2} + \frac{\partial^2 \Phi}{\partial z^2} \\ &= \nabla^2 \Phi \end{aligned} \qquad\qquad (10\text{-}79.3)$$

From, equations (10-79.2) and (10-79.3), the derivatives of the dilatational strain can be represented as

$$\frac{\partial \varepsilon}{\partial x} = \frac{\partial}{\partial x}\left(\nabla^2 \Phi\right)$$

$$= \nabla^2 \left(\frac{\partial \Phi}{\partial x}\right) \tag{10-79.4}$$

$$= \nabla^2 u$$

Or,

$$\frac{\partial \varepsilon}{\partial x} = \nabla^2 u, \qquad \frac{\partial \varepsilon}{\partial y} = \nabla^2 v, \qquad \frac{\partial \varepsilon}{\partial z} = \nabla^2 w \tag{10-79.5}$$

Substituting from equations (10-79.5) into (10-77.1), the potential wave equation becomes

$$\mu \nabla^2 \begin{bmatrix} u \\ v \\ w \end{bmatrix} + (\mu + \lambda)\nabla^2 \begin{bmatrix} u \\ v \\ w \end{bmatrix} = \rho \frac{\partial^2}{\partial t^2} \begin{bmatrix} u \\ v \\ w \end{bmatrix} \tag{10-79.6}$$

Or

$$(2\mu + \lambda)\nabla^2 \begin{bmatrix} u \\ v \\ w \end{bmatrix} = \rho \frac{\partial^2}{\partial t^2} \begin{bmatrix} u \\ v \\ w \end{bmatrix} \tag{10-79.7}$$

Which is a wave equation for displacements, given any of the following names:

1. Longitudinal waves
2. P-waves
3. Bulk waves
4. Dilatational waves
5. Irrotational waves

(v) Spherical wave equation

Similarities between Laplace equation and wave equation

Consider the general form of wave equation

$$\frac{\partial^2 f(x, y, z, t)}{\partial t^2} = c^2 \nabla^2 f(x, y, z, t) \tag{10-80.1}$$

The solution of wave equation entails its integration, which can be facilitated by studying the similarities of the harmonic functions of Laplace's equation with wave equations, which have non-vanishing second derivative for time.

Consider the general solution for equation (10-80.1) in the form

$$f = f_o(x, y, z, t, \alpha, \beta, \xi, \dots) \qquad (10\text{-}80.2)$$

Differentiate equation (10-80.1) with respect to any of the arguments of f_0, as follows

$$\frac{\partial}{\partial \xi}\left(\frac{\partial^2 f_0}{\partial t^2}\right) = c^2 \frac{\partial}{\partial \xi}\left(\nabla^2 f_0\right)$$

$$\frac{\partial^2}{\partial t^2}\left(\frac{\partial f_0}{\partial \xi}\right) = c^2 \nabla^2\left(\frac{\partial f_0}{\partial \xi}\right) \qquad (10\text{-}80.3)$$

Therefore, we conclude that the **derivatives of the wave function** with respect to its derivatives are also solutions of the wave equation.

The simple case of spherical wave can be represented by the radial and temporal coordinates of the form

$$f = f_o(r, t)$$

where,

$$r = \sqrt{x^2 + y^2 + z^2} \qquad (10\text{-}80.4)$$

The derivatives of f_0 are obtained as follows

$$\frac{\partial r}{\partial x} = \frac{x}{r}$$

$$\frac{\partial f_0}{\partial x} = \frac{\partial f_0}{\partial r}\frac{\partial r}{\partial x} = \frac{x}{r}\frac{\partial f_0}{\partial r} \qquad (10\text{-}80.5)$$

Thus, the second derivatives with respect to x, and similarly, for y- and z- derivatives, are

$$\frac{\partial^2 f_0}{\partial x^2} = \frac{x^2}{r^2}\frac{\partial^2 f_0}{\partial r^2} + \frac{r^2 - x^2}{r^3}\frac{\partial f_0}{\partial r}$$

$$\frac{\partial^2 f_0}{\partial y^2} = \frac{y^2}{r^2}\frac{\partial^2 f_0}{\partial r^2} + \frac{r^2 - y^2}{r^3}\frac{\partial f_0}{\partial r} \qquad (10\text{-}80.6)$$

$$\frac{\partial^2 f_0}{\partial z^2} = \frac{z^2}{r^2}\frac{\partial^2 f_0}{\partial r^2} + \frac{r^2 - z^2}{r^3}\frac{\partial f_0}{\partial r}$$

Adding equations (10-80.6) by members we get

$$\nabla^2 f_0 = \frac{\partial^2 f_0}{\partial r^2} + \frac{2}{r}\frac{\partial f_0}{\partial r}$$

$$= \frac{1}{r}\frac{\partial^2}{\partial r^2}(rf_0)$$

(10-80.7)

Thus, from equations (10-80.1) and (10-80.7), the spherical wave equation becomes

$$\frac{\partial^2 f_0}{\partial t^2} = \frac{c^2}{r}\frac{\partial^2}{\partial r^2}(rf_0)$$

(10-80.8)

Since r is an independent variable, equation (10-80.8) can be written as

$$\frac{\partial^2}{\partial t^2}(rf_0) = c^2\frac{\partial^2}{\partial r^2}(rf_0)$$

(10-80.9)

Equation (10-80.9) can be solved by separation of variables as follows

$$f_0 = R(r)T(t)$$

(10-80.10)

Such that

$$RT'' = c^2 TR''$$

(10-80.11)

Or $\qquad \dfrac{T''}{T} = c^2\dfrac{R''}{R}$

The two separate equation in R and T, are written as usual in the form

$$T'' + \lambda^2 T = 0$$

$$R'' + \frac{\lambda^2}{c^2}R = 0$$

(10-80.12)

CHAPTER 11

THIN SLAB

SOLUTION BY PLANE APPROXIMATION

11-1. Bending of rod versus bending of thin slab

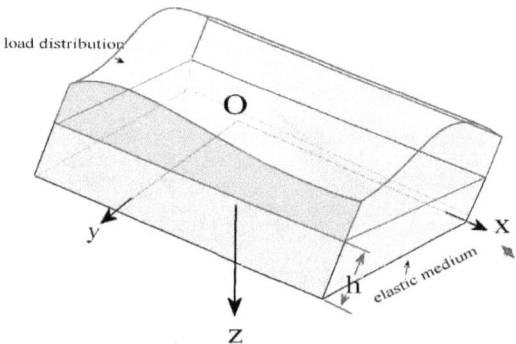

Figure 11-1. Prismatical plate with center of coordinates taken at its half thickness, z-axis perpendicular downward and xOy plane on the widest plane of the plate.

The following approximations are made in formulating the solution of elasticity equations for plate loading:

1. The plate is considered a rod with **thin height** compared to the other two dimensions. Figure 11-1. (Thin slab, moderately thick plate)

2. If the load is below certain limit, the center of the plate **does not deflect**.

3. If the load is normal and reaction normal to the middle plane, the center of the plate **deflects**.

4. Bending is accompanied by **torsion**.

5. Both stress and stain in the direction of the thickness are **neglected[MFE1]** $\varepsilon_{zz} = 0$ and $\sigma_z = 0$. Figure 11-2.

6. Cylindrical bending of the plate does not alter the **plane cross-sections,** which remain normal to middle plane of the plate. (Cross sections does not deform from planar forms. Hypothesis of plane sections.)

7. Non-cylindrical bending does not alter the **linear vertical** of cross-section perpendicular to the middle plane. (Hypothesis of linear element)

8. **Middle plane's displacement** is assumed purely vertical, i.e., $u = v = 0$, and $w = f(x,y)$.

9. **Extension and shear** are absent in the middle plane.

10. The planar normal strains, off-middle plane are given by

$$\varepsilon_{xx} = E^{-1}\left[\sigma_x - \nu\sigma_y\right]$$
$$\varepsilon_{yy} = E^{-1}\left[\sigma_y - \nu\sigma_x\right]$$

(11-1)

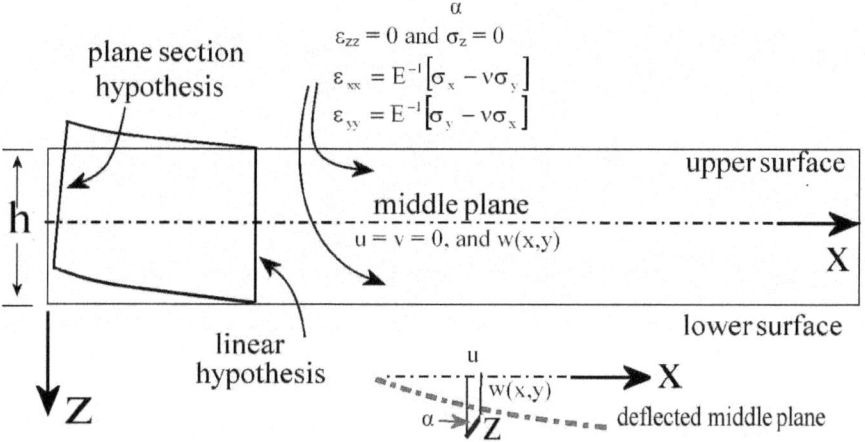

Figure 11-2. Cross section in the prismatical plate showing the assumption made on the middle plane and the vertical stresses and strains.

383

11-2. Sophie Germain's equation for bending and torsion of thin slab

The assumptions made above enable us to derive expressions for the displacements above and below the middle line, in terms of the vertical displacement w(x,y) and the vertical deflection z as follows, Figure 11-2,

$$u = -z\frac{\partial w}{\partial x}, \qquad v = -z\frac{\partial w}{\partial y} \qquad (11\text{-}2)$$

We will use **Cauchy's formula** to obtain the three members of the **strain tensor** from equation (11-2) as follows

$$\varepsilon_{xx} = \frac{\partial u}{\partial x} = -z\frac{\partial^2 w}{\partial x^2}$$

$$\varepsilon_{yy} = \frac{\partial v}{\partial y} = -z\frac{\partial^2 w}{\partial y^2} \qquad (11\text{-}3)$$

$$\varepsilon_{xy} = \frac{\partial u}{\partial y} + \frac{\partial v}{\partial x} = -2z\frac{\partial^2 w}{\partial x \partial y}$$

Equations (11-3) are used with **Hooke's law** to give the three members of the **stress tensor** as follows

$$\sigma_x = \frac{E}{1-v^2}\left(\varepsilon_{xx} + v\varepsilon_{yy}\right) = -\frac{zE}{1-v^2}\left(\frac{\partial^2 w}{\partial x^2} + v\frac{\partial^2 w}{\partial y^2}\right)$$

$$\sigma_y = \frac{E}{1-v^2}\left(\varepsilon_{yy} + v\varepsilon_{xx}\right) = -\frac{zE}{1-v^2}\left(\frac{\partial^2 w}{\partial y^2} + v\frac{\partial^2 w}{\partial x^2}\right) \qquad (11\text{-}4)$$

$$\tau_{xy} = \frac{(1-v)E}{2(1-v^2)}\varepsilon_{xy} = -\frac{zE}{1-v^2}(1-v)\frac{\partial^2 w}{\partial y \partial x}$$

Therefore, the two normal stresses have the following features:

1. Normal stresses vary linearly with z, which is the coordinate in the direction of the thickness of the plate, change sign from negative above the middle plane to positive sign below the middle plane.

2. Normal stresses vary proportionally to the curvatures (the second derivatives with respect to each stresses direction and the curvature in the perpendicular direction weighed by Poisson's ratio).

3. Shear stress also varies linearly with z, and changes signs from above to below the middle plane.

4. Shear stresses are greater with smaller Poisson's ratio (more rigid material). Thus, Concrete (v = 0.1 - 0.2) is subjected to greater shear than Aluminum (v = 0.334).

384

5. Shear stresses vary with relative curvatures of the two perpendicular planes zOx and zOy. The mixed derivative $\dfrac{\partial^2 w}{\partial y \partial x}$ represent the gradients of the angular orientations of the two tangents $\dfrac{\partial w}{\partial y}$ and $\dfrac{\partial w}{\partial x}$. Thus, $\dfrac{\partial}{\partial x}\left(\dfrac{\partial w}{\partial y}\right)$ and $\dfrac{\partial}{\partial y}\left(\dfrac{\partial w}{\partial x}\right)$ represent the torsion in the xOy plane around the z-axis.

The three remaining stresses σ_z, τ_{zx}, and τ_{zy}, are obtained by substituting from equation (11-4) into **Navier's equations** of equilibrium, (1-4), as follows

$$\frac{\partial \tau_{xz}}{\partial z} = \frac{zE}{1-v^2}\frac{\partial}{\partial x}\left(\frac{\partial^2 w}{\partial x^2} + v\frac{\partial^2 w}{\partial y^2}\right) + \frac{zE}{1-v^2}(1-v)\frac{\partial}{\partial y}\frac{\partial^2 w}{\partial y \partial x} \qquad (11\text{-}5.1)$$

$$\frac{\partial \tau_{yz}}{\partial z} = \frac{zE}{1-v^2}(1-v)\frac{\partial}{\partial x}\frac{\partial^2 w}{\partial y \partial x} + \frac{zE}{1-v^2}\frac{\partial}{\partial y}\left(\frac{\partial^2 w}{\partial y^2} + v\frac{\partial^2 w}{\partial x^2}\right) \qquad (11\text{-}5.2)$$

$$\frac{\partial \sigma_z}{\partial z} = -\frac{\partial \tau_{zx}}{\partial x} - \frac{\partial \tau_{zy}}{\partial y} \qquad (11\text{-}5.3)$$

Even though, we have assumed that the normal stress $\sigma_z = 0$, we will realistically use the negligibly stress for **other purposes**, as shown below.

In order to determine σ_z, in equation (11-5.3), we must first arrange and integrate equations (11-5.1 and 11-5.2) to get

$$\frac{\partial \tau_{xz}}{\partial z} = \frac{zE}{1-v^2}\frac{\partial}{\partial x}\left(\nabla^2 w\right) \qquad (11\text{-}5.4)$$

$$\frac{\partial \tau_{yz}}{\partial z} = \frac{zE}{1-v^2}\frac{\partial}{\partial y}\left(\nabla^2 w\right) \qquad (11\text{-}5.5)$$

Which upon integrating (noting that the RHS are independent of z), we get

$$\tau_{xz} = \left(\frac{z^2}{2} + C\right)\frac{z^2 E}{2(1-v^2)}\frac{\partial}{\partial x}\left(\nabla^2 w\right) \qquad (11\text{-}6.1)$$

$$\tau_{yz} = \left(\frac{z^2}{2} + C\right)\frac{E}{2(1-v^2)}\frac{\partial}{\partial y}\left(\nabla^2 w\right) \qquad (11\text{-}6.2)$$

The constant C is determined on the boundary conditions z[MFE2] $= \pm h/2$, given that

$$\tau_{xz} = \tau_{xz} = 0 \qquad (11\text{-}7.1)$$

Thus,

$$0 = \frac{1}{2}\left(\frac{h}{2}\right)^2 + C \tag{11-7.2}$$

Or

$$C = -\frac{h^2}{8} \tag{11-7.3}$$

Therefore, equations (11-6.1) and (11-6.2) becomes

$$\tau_{xz} = \left(\frac{z^2}{2} - \frac{h^2}{8}\right)\frac{E}{(1-v^2)}\frac{\partial}{\partial x}(\nabla^2 w) \tag{11-8.1}$$

$$\tau_{yz} = \left(\frac{z^2}{2} - \frac{h^2}{8}\right)\frac{E}{(1-v^2)}\frac{\partial}{\partial y}(\nabla^2 w) \tag{11-8.2}$$

The third remaining stress, in equation (11-5.3), in obtained by substituting from (11-8.1 and 8.2), as follows

$$\frac{\partial \sigma_z}{\partial z} = -\frac{\partial \tau_{zx}}{\partial x} - \frac{\partial \tau_{zy}}{\partial y}$$

$$= -\frac{E}{2(1-v^2)}\left(z^2 - \frac{h^2}{4}\right)\left(\frac{\partial^2}{\partial x^2}(\nabla^2 w) + \frac{\partial^2}{\partial y^2}(\nabla^2 w)\right) \tag{11-9.1}$$

$$= -\frac{E}{2(1-v^2)}\left(z^2 - \frac{h^2}{4}\right)\nabla^2\nabla^2 w$$

Integrating, we get

$$\sigma_z\left(\frac{h}{2}\right) - \sigma_z\left(-\frac{h}{2}\right) = -\frac{E}{2(1-v^2)}\left(\frac{z^3}{3} - \frac{zh^2}{4}\right)\Big|_{-h/2}^{+h/2}\nabla^2\nabla^2 w$$

$$= \frac{Eh^3}{12(1-v^2)}\nabla^2\nabla^2 w \tag{11-9.1}$$

From Figure 11-2, the stress on upper surface $-\sigma_z(-h/2)$ is equal to the external load, say, q(x,y). The lower surface is assumed free from stress on the z-direction. Thus, $-\sigma_z(h/2) = 0$. Therefore, equation (11-9.1) becomes

$$q(x, y) = \frac{Eh^3}{12(1-v^2)}\nabla^2\nabla^2 w$$

Or,

$$\nabla^2\nabla^2 w = \frac{12(1-v^2)q(x,y)}{Eh^3} \tag{11-9.2}$$

Equation (11-9.2) is known as **Sophie Germain's equation** for approximate solution of bending a thin slab. and is written in details as

$$\left(\frac{\partial^4}{\partial x^4} + 2\frac{\partial^4}{\partial x^2 \partial y^2} + \frac{\partial^4}{\partial y^4}\right) w(x,y) = \frac{q(x,y)}{K} \tag{11-9.3}$$

Where,

$$K = \frac{Eh^3}{12(1-v^2)} \approx \frac{Eh^3}{12} \tag{11-9.4}$$

K is called **flexural rigidity of plate.**

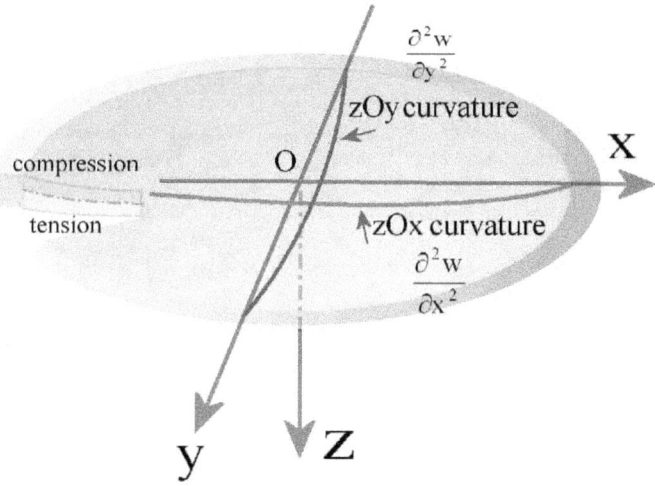

Figure 11-3. Explanation of the terms in as Sophie Germain's equation of approximate bending of a plate.

Orthogonal bending moments

The two curvatures, Figure 11-3, in the zOx and zOy planes, are accompanied by **moments of inertia**, which are determined by integrating the moments of **bending couples** in each of the two directions as follows

$$M_x = \int_{-h/2}^{+h/2} \sigma_x z dz, \qquad M_y = \int_{-h/2}^{+h/2} \sigma_y z dz \qquad (11\text{-}10.1)$$

Substituting by the expressions for stresses from equations (11-4) into (11-10), we get

$$
\begin{aligned}
M_x &= \frac{E}{1-v^2}\left(\frac{\partial^2 w}{\partial x^2} + v\frac{\partial^2 w}{\partial y^2}\right)\int_{-h/2}^{+h/2} -z^2 dz = -\frac{h^3 E}{12(1-v^2)}\left(\frac{\partial^2 w}{\partial x^2} + v\frac{\partial^2 w}{\partial y^2}\right) \\
M_y &= \frac{E}{1-v^2}\left(\frac{\partial^2 w}{\partial y^2} + v\frac{\partial^2 w}{\partial x^2}\right)\int_{-h/2}^{+h/2} -z^2 dz = -\frac{h^3 E}{12(1-v^2)}\left(\frac{\partial^2 w}{\partial y^2} + v\frac{\partial^2 w}{\partial x^2}\right)
\end{aligned}
\qquad (11\text{-}10.2)
$$

M_x and M_y are the **bending moment per unit length of cross section** of the slab.

Shearing forces

The shearing forces impose no bending couple but cause torsion and are determined by integrating the shear stresses on the surface they act upon

(i) Shear force across the thickness of the plate in the xOz plane:

$$
\begin{aligned}
F_{xz} &= \int_{-h/2}^{+h/2} \tau_{xz} dz \\
&= \frac{E}{(1-v^2)}\frac{\partial}{\partial x}\left(\nabla^2 w\right)\int_{-h/2}^{+h/2}\left(\frac{z^2}{2} - \frac{h^2}{8}\right)dz \\
&= -\frac{Eh^3}{12(1-v^2)}\frac{\partial}{\partial x}\left(\nabla^2 w\right)
\end{aligned}
\qquad (11\text{-}11.1)
$$

(ii) Shear force across the thickness of the plate in the yOz plane:

$$
\begin{aligned}
F_{yz} &= \int_{-h/2}^{+h/2} \tau_{yz} dz \\
&= \frac{E}{(1-v^2)}\frac{\partial}{\partial y}\left(\nabla^2 w\right)\int_{-h/2}^{+h/2}\left(\frac{z^2}{2} - \frac{h^2}{8}\right)dz \\
&= -\frac{Eh^3}{12(1-v^2)}\frac{\partial}{\partial y}\left(\nabla^2 w\right)
\end{aligned}
\qquad (11\text{-}11.2)
$$

(iii) Torsion couple or twisting moment around the z axis (normal to xOy-plane):

$$M_z = \int_{-h/2}^{+h/2} \tau_{xy} z\,dz$$

$$= -\frac{E}{(1-v^2)}(1-v)\frac{\partial^2 w}{\partial y \partial x}\int_{-h/2}^{+h/2} z^2 dz \qquad (11\text{-}11\text{-}3)$$

$$= -\frac{h^3 E}{12(1-v^2)}(1-v)\frac{\partial^2 w}{\partial y \partial x}$$

Summary of solution of thin slab

Table 11-1. Summary of expressions for stresses, shear forces, bending couples, and governing equations for a thin slab

$\sigma_x = \dfrac{E}{1-v^2}\left(\varepsilon_{xx} + v\varepsilon_{yy}\right) = -\dfrac{zE}{1-v^2}\left(\dfrac{\partial^2 w}{\partial x^2} + v\dfrac{\partial^2 w}{\partial y^2}\right)$	(11-4)
$\sigma_y = \dfrac{E}{1-v^2}\left(\varepsilon_{yy} + v\varepsilon_{xx}\right) = -\dfrac{zE}{1-v^2}\left(\dfrac{\partial^2 w}{\partial y^2} + v\dfrac{\partial^2 w}{\partial x^2}\right)$	(11-4)
$\tau_{xy} = \dfrac{(1-v)E}{2(1-v^2)}\varepsilon_{xy} = -\dfrac{zE}{1-v^2}(1-v)\dfrac{\partial^2 w}{\partial y \partial x}$	(11-4)
$\tau_{xz} = \left(\dfrac{z^2}{2}+C\right)\dfrac{z^2 E}{2(1-v^2)}\dfrac{\partial}{\partial x}\left(\nabla^2 w\right)$	(11-6.1)
$\tau_{yz} = \left(\dfrac{z^2}{2}+C\right)\dfrac{E}{2(1-v^2)}\dfrac{\partial}{\partial y}\left(\nabla^2 w\right)$	(11-6.2)
$\tau_{xz} = \left(\dfrac{z^2}{2}-\dfrac{h^2}{8}\right)\dfrac{E}{(1-v^2)}\dfrac{\partial}{\partial x}\left(\nabla^2 w\right)$	(11-8.1)
$\tau_{yz} = \left(\dfrac{z^2}{2}-\dfrac{h^2}{8}\right)\dfrac{E}{(1-v^2)}\dfrac{\partial}{\partial y}\left(\nabla^2 w\right)$	(11-8.2)
$\nabla^2\nabla^2 w = \dfrac{12(1-v^2)q(x,y)}{Eh^3}$	(11-9.2)
$M_x = \dfrac{E}{1-v^2}\left(\dfrac{\partial^2 w}{\partial x^2} + v\dfrac{\partial^2 w}{\partial y^2}\right)\int_{-h/2}^{+h/2} -z^2 dz = -\dfrac{h^3 E}{12(1-v^2)}\left(\dfrac{\partial^2 w}{\partial x^2} + v\dfrac{\partial^2 w}{\partial y^2}\right)$	(11-10.2)
$M_y = \dfrac{E}{1-v^2}\left(\dfrac{\partial^2 w}{\partial y^2} + v\dfrac{\partial^2 w}{\partial x^2}\right)\int_{-h/2}^{+h/2} -z^2 dz = -\dfrac{h^3 E}{12(1-v^2)}\left(\dfrac{\partial^2 w}{\partial y^2} + v\dfrac{\partial^2 w}{\partial x^2}\right)$	(11-10.2)
$F_{xz} = -\dfrac{Eh^3}{12(1-v^2)}\dfrac{\partial}{\partial x}\left(\nabla^2 w\right)$	(11-11.1)

$F_{yz} = -\dfrac{Eh^3}{12(1-v^2)}\dfrac{\partial}{\partial y}\left(\nabla^2 w\right)$	(11-11.2)
$M_z = -\dfrac{h^3 E}{12(1-v^2)}(1-v)\dfrac{\partial^2 w}{\partial y \partial x}$	(11-11.3)

Special cases and approximations

Case 1: Pure torsion of thin slab

Vanishing shear stresses and normal stresses in the plane of the slab by replacing the solid plate by set of rods able to twist and rotate.

Approximations:

$$M_{xy} = constant$$
$$M_x = M_y = F_{xz} = F_{yz} = 0 \qquad (11\text{-}12\text{-}1)$$

Equation (11-11.3), which represents the torsion due to the surface shear stress, is integrated as follows

$$w(x, y) == -\frac{12(1+v)M_z}{h^3 E}xy \qquad (11\text{-}12.2)$$

Thus, the plate takes the form of **oblique plate**, or **hyperbolic paraboloid**.

In this particular case of pure torsion, $w(x,y)$ represents the **angle of twist** τ per unit length times the distances x and y. Substituting by the **Lamé's coefficient** from equation (3-8.1), $\mu = \dfrac{E}{2(1+v)}$, in equation (11-12.2), we get

$$w(x, y) = \tau xy \qquad (11\text{-}12.3)$$

Where

$$\tau = -\frac{M_z}{\dfrac{\mu h^3}{6}} \qquad (11\text{-}12.4)$$

Case 2: Pure bending of thin slab

Vanishing shear stresses and torsion stresses in the plane of the slab.

Approximations:

$$M_x = M_y = M \ constant$$

$$M_{xy} = F_{xz} = F_{yz} = 0 \qquad (11\text{-}13\text{-}1)$$

Equations (11-10.2) are written and integrated as follows

$$\frac{\partial^2 w}{\partial x^2} + v\frac{\partial^2 w}{\partial y^2} = -\frac{12(1-v^2)M_x}{h^3 E} \qquad (11\text{-}13.2a)$$

$$\frac{\partial^2 w}{\partial y^2} + v\frac{\partial^2 w}{\partial x^2} = -\frac{12(1-v^2)M_y}{h^3 E} \qquad (11\text{-}13.2b)$$

Solving the two simultaneous equations, we get the following integrable forms

$$\frac{\partial^2 w}{\partial x^2} = -\frac{12(1-v^2)M_x}{(1+v)h^3 E}$$
$$\frac{\partial^2 w}{\partial y^2} = -\frac{12(1-v^2)M_x}{(1+v)h^3 E} \qquad (11\text{-}13.3)$$

Integrating and arranging the constants of integration, we get

$$w(x,y) = -\frac{12(1-v^2)M_x}{(1+v)h^3 E}\left(x^2 + y^2\right) \qquad (11\text{-}13.4)$$

Thus, the middle plane converts to **paraboloid of revolution** where the radius of curvature is given by

$$\rho(x,y) = -\frac{(1+v)h^3 E}{12(1-v^2)M_x} \qquad (11\text{-}13.5)$$

Such ideal condition of pure bending could simulate a round slab loaded uniformly at its periphery by bending moments of intensity M per unit length.

11-2. Elliptic plate

Assume the vertical deflection of the form

$$w(x,y) = cZ^2 \qquad (11\text{-}14.1)$$

Where,

$$Z = \frac{x^2}{a^2} + \frac{y^2}{b^2} - 1 \qquad (11\text{-}14.2)$$

Equation (11-14) implies that **deflection** vanishes on the contour, defined by

$$Z = \frac{x^2}{a^2} + \frac{y^2}{b^2} - 1 = 0 \qquad (11\text{-}14.3)$$

Farther, the derivatives of w(x,y), which constitute the **angles of deflections** also vanish on the contour as follows

$$\frac{\partial w}{\partial x} = 2cZ\frac{\partial Z}{\partial x}$$
$$\frac{\partial w}{\partial y} = 2cZ\frac{\partial Z}{\partial y} \qquad (11\text{-}14.4)$$

The vanishing of angles of deflection of the elliptic plate on the contour fulfills to the boundary conditions of **clamping the edges** of the slab.

We now determine the derivatives of the displacement function, equation (11-14.1), which are required to solve as **Sophie Germain's equation** (11-9.3), which will define the unknown constant c, in the displacement function. From equation (11-14.1), we get

$$\frac{\partial^4 w}{\partial x^4} = 24\frac{c}{a^4}$$
$$\frac{\partial^4 w}{\partial x^2 \partial y^2} = 8\frac{c}{a^2 b^2} \qquad (11\text{-}14.5)$$
$$\frac{\partial^4 w}{\partial y^4} = 24\frac{c}{b^4}$$

Substituting from (11-14.5) into (11-9.3), we get

$$\left(24\frac{c}{a^4} + 16\frac{c}{a^2 b^2} + 24\frac{c}{b^4}\right) = \frac{q(x,y)}{K} \qquad (11\text{-}14.6)$$

Therefore, the constant c, in equation (11-14.1), is defined by

$$c = \frac{q(x,y)}{K\left(\dfrac{24}{a^4} + \dfrac{16}{a^2 b^2} + \dfrac{24c}{b^4}\right)} \qquad (11\text{-}14.7)$$

Thus, the displacement equation (11-14.1) becomes

$$w(x,y) = \frac{q(x,y)}{K\left(\dfrac{24}{a^4} + \dfrac{16}{a^2 b^2} + \dfrac{24c}{b^4}\right)}\left(\frac{x^2}{a^2} + \frac{y^2}{b^2} - 1\right)$$ (11-14.8)

Where, K is defined by equation (11-9.4).

Substituting by the derivatives of w, from equation (11-14.8) into the expressions listed in Table 11-1, we can determined the stresses, shearing forces, and bending couples in the case of elliptic slab fixed at it contour.

11-3. Circular plate

The polar parametric equation of circle is

$$x = r\cos\theta, \qquad\qquad y = r\sin\theta$$ (11-15.1)

From equation (8-1), we write **Sophie Germain's equation** (11-9.3) in polar coordinates as follows

$$\nabla^2\left(\nabla^2 w(x,y)\right) = \frac{q(x,y)}{K}$$

$$\left(\frac{1}{r}\frac{\partial}{\partial r} + \frac{\partial^2}{\partial r^2} + \frac{1}{r^2}\frac{\partial^2}{\partial\theta^2}\right)\left(\frac{1}{r}\frac{\partial}{\partial r} + \frac{\partial^2}{\partial r^2} + \frac{1}{r^2}\frac{\partial^2}{\partial\theta^2}\right)w(r,\theta) = \frac{q(r,\theta)}{K}$$ (11-15.2)

In order to present a complete yet simple solution, we will assume that the load q is independent of the polar angle θ. Thus, we get

$$\frac{\partial^4 w}{\partial r^4} + \frac{2}{r}\frac{\partial^3 w}{\partial r^3} - \frac{1}{r^2}\frac{\partial^2 w}{\partial r^2} + \frac{1}{r^3}\frac{\partial w}{\partial r} = \frac{q}{K}$$ (11-15.3)

The **general solution** of the equation (11-15.3) under **vanishing load** was given in equation (8-2.6b) as follows

$$\text{G.S.} = C_1 r^2 \ln r + C_3 \ln r + C_2 r^2 + C_4$$ (11-15.4)

The **particular solution** under load is added to the expression in (11-15.4) to give the general solution with load

$$w(r,\theta) = \text{G.S.} + \text{P..S}$$
$$= \left(C_1 r^2 \ln r + C_3 \ln r + C_2 r^2 + C_4\right) + \text{P..S}$$ (11-15.4)

The fourth power of equation (11-15.3) implies that a particular solution, with **constant load**, takes the form

$$P.S. = C_5 r^4 \qquad\qquad (11-15.5)$$

Substituting P.S. in lieu of w(x,y) in equation (11-15.3), we get

$$\left(\frac{\partial^4 w}{\partial r^4} + \frac{2}{r}\frac{\partial^3 w}{\partial r^3} - \frac{1}{r^2}\frac{\partial^2 w}{\partial r^2} + \frac{1}{r^3}\frac{\partial w}{\partial r}\right)C_5 r^4 = \frac{q}{K}$$

$$C_5(24 + 48 - 12 + 4) = \frac{q}{K} \qquad\qquad (11-15.6)$$

$$C_5 = \frac{q}{64K}$$

Substituting from (11-15.5) and (11-15.6) into (11-15.4), we get

$$w(r,\theta) = \left(C_1 r^2 \ln r + C_3 \ln r + C_2 r^2 + C_4\right) + \frac{q}{64K}r^4 \qquad\qquad (11-15.7)$$

Moments of bending and twisting couples

Equations (11-10.2 and 10.3) are transformed to polar coordinates by aligning the polar direction with x-axis, the perpendicular to polar with y-axis such that

Bending moment normal to radius:

$$M_r = -\frac{h^3 E}{12(1-v^2)}\left(\frac{\partial^2 w}{\partial r^2} + \frac{v}{r}\frac{\partial w}{\partial r}\right) \qquad\qquad (11-16.1)$$

Bending moment over radial section:

$$M_\perp = -\frac{h^3 E}{12(1-v^2)}\left(\frac{1}{r}\frac{\partial w}{\partial r} + v\frac{\partial^2 w}{\partial r^2}\right) \qquad\qquad (11-16.2)$$

Twisting moment is absent due to assumed symmetry.

$$M_z = 0 \qquad\qquad (11-16.3)$$

Boundary conditions

(i) Singularities of deflection and displacements at r = 0

Equation (11-15.7) must be devoid of the logarithmic terms. Thus, $C_1 = C_3 = 0$.

i.e., $$w(r) = C_2 r^2 + C_4 + \frac{q}{64K} r^4 \qquad (11\text{-}17.1)$$

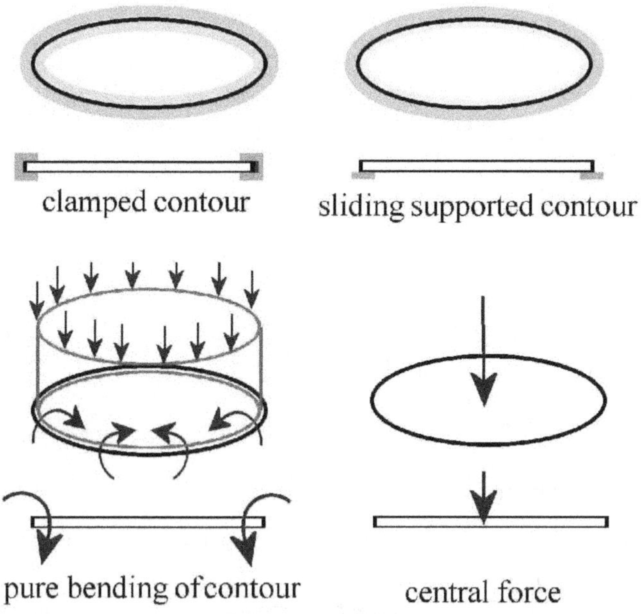

clamped contour   sliding supported contour

pure bending of contour   central force

Figure 11-4. Four configurations of loading and bending of circular slab

(ii) Fixed-in contour

The clamping of the contour of the slab prevents both deflection and displacements:

At r = a → Displacement: $w(a) = 0$

and → Deflection : $\dfrac{\partial w}{\partial r} = 0$ $\qquad (11\text{-}17.2)$

Substituting in the displacement w(r), equation (11-17.1), we get

$$0 = C_2 a^2 + C_4 + \frac{q}{64K} a^4$$

$$0 = 2C_2 a + \frac{q}{16K} a^3$$

(11-17.2a)

i.e.,

$$C_4 = \frac{q}{64K} a^4$$

$$C_2 = -\frac{q}{32K} a^2$$

(11-17.2b)

Thus, equation (11-17.1) becomes

$$w(r) = -\frac{q}{32K} a^2 r^2 + \frac{q}{64K} a^4 + \frac{q}{64K} r^4$$

$$= \frac{q}{64K}\left(a^2 - r^2\right)^2$$

(11-17.3)

Moments of bending couples, equations (11-16) become

$$M_r = \frac{q}{16}\left[a^2(v+1) - r^2(3+v)\right]$$

(11-18.1)

$$M_\perp = \frac{q}{16}\left[a^2(v+1) - r^2(1+3v)\right]$$

(11-18.2)

(iii) Supported contour (free motion)

The unclamped contour of the slab prevents both displacement and couple as the ends are free to slide:

At r = a   → Displacement: w(a) = 0

and   → Twisting couple: M_r = 0

(11-19.1)

Substituting in the displacement w(r), equation (11-17.1), we get

$$0 = C_2 a^2 + C_4 + \frac{q}{64K} a^4$$

(11-19.2)

Similarly, equation (11-16.1) is equated to zero to give

$$\left(\frac{\partial^2 w}{\partial r^2} + \frac{v}{r}\frac{\partial w}{\partial r}\right) = 2C_2(v+1) + \frac{qa^2}{16K}(v+3) = 0$$

(11-19.3)

396

Thus, equations (11-19.2) and (11-19.3) give

$$C_2 = -\frac{qa^2}{32K}\frac{(v+3)}{(v+1)}$$

(11-19.4)

$$C_4 = -\frac{q}{64K}\frac{(v+5)}{(v+1)}a^4$$

(11-19.5)

Displacement

Therefore, equation (11-17.1) becomes

$$w(r) = -\frac{2qa^2}{64K}\frac{(v+3)}{(v+1)}r^2 - \frac{qa^4}{64K}\frac{(v+5)}{(v+1)} + \frac{q}{64K}r^4$$

i.e.,
$$= \frac{q}{64K}\left(-2a^2r^2\left(1+\frac{2}{(v+1)}\right) - a^4\left(1+\frac{4}{(v+1)}\right) + r^4\right)$$

(11-19.6)

$$= \frac{q}{64K}\left((a^2-r^2)^2 + \frac{4a^2}{(v+1)}(a^2-r^2)\right)$$

Bending moments

The derivatives of w(r) are

$$\frac{1}{r}\frac{\partial w}{\partial r} = \frac{q}{64K}\left(4r^2 - \frac{4a^2(3+v)}{(v+1)}\right)$$

$$\frac{\partial^2 w}{\partial r^2} = \frac{q}{64K}\left(12r^2 - \frac{4a^2(3+v)}{(v+1)}\right)$$

(11-19.7)

Bending moment normal to radius, equation (11-16.1) becomes

$$M_r = \frac{q}{16}\left[a^2(3+v) - r^2(3+v)\right]$$

(11-19.8)

Bending moment over radial section, equation (11-16.2) becomes

$$M_\perp = \frac{q}{16}\left[a^2(3+v) - r^2(1+3v)\right]$$

(11-19.9)

(iv) Pure bending of a slab by torqued contour, unloaded slab

The contour is subjected to uniform moment while the load is assumed zero.

At r = a   → Displacement: w(a) = 0

		→ Bending couple	$M_r = M$	
and	At r =any	→ vanishing load	$q(r) = 0$	(11-20.1)

Substituting by q = 0 in equation (11-17.1), we get

$$w(r) = C_2 r^2 + C_4 \tag{11-20.2}$$

Substituting by the w(a) = 0, we get

$$0 = C_2 a^2 + C_4 \tag{11-20.3}$$

Substituting by the $M_r(a) = M$, in equation (11-16.1), we get

$$M_r = -\frac{h^3 E}{12(1-v^2)} 2C_2(v+1) = M$$

i.e.,

$$C_2 = -\frac{6M(1-v)}{h^3 E} \tag{11-20.4}$$

$$C_4 = \frac{6M(1-v)}{h^3 E} a^2 \tag{11-20.5}$$

Displacement

Therefore, equation (11-20.2) becomes

$$
\begin{aligned}
w(r) &= \frac{6M(1-v)}{h^3 E}(a^2 - r^2) \\
&= \frac{M}{2K(1+v)}(a^2 - r^2)
\end{aligned}
\tag{11-20.2}
$$

Again, we substituted by K from equation (11-9.4).

(v) Central concentrated load with fixed-in contour

The clamping of the contour of the slab prevents both deflection and displacements:

At r = a	→ Displacement:	$w(a) = 0$	
and	→ Deflection :	$\dfrac{\partial w}{\partial r} = 0$	(11-21.1)
At r = 0	→ Load :	$q(0) = F$	

Equation (11-15.7) is simplified to

$$w(r) = C_1 r^2 \ln r + C_2 r^2 + C_4 \tag{11-21.2}$$

398

Substituting, w(a) = 0 and q(a) = 0, in the displacement w(r), equation (11-21.2), we get

$$0 = C_1 a^2 \ln a + C_2 a^2 + C_4 \qquad (11\text{-}21.3)$$

Substituting by the clamping condition w'(a) = 0 in the displacement w(r), equation (11-21.2), we get

$$\frac{\partial w}{\partial r} = 2C_1 a \ln a + C_1 a + 2C_2 a = 0 \qquad (11\text{-}21.4)$$

From equations (11-21.3 and 21.4), we can represent the three constants in terms of one constant, which will be determined from the central loading condition as follows.

$$C_4 = \frac{a^2}{2} C_1$$
$$C_2 = -C_1 \frac{(2 \ln a + 1)}{2} \qquad (11\text{-}21.5)$$

$$w(r) = \frac{C_1}{4} \left(\frac{1}{2} \left(a^2 - r^2\right) + r^2 \ln \frac{a}{r} \right) \qquad (11\text{-}21.6)$$

From equation (10-54.4)

$$(2\pi)KC = F \qquad (11\text{-}21.7)$$

Therefore,

$$w(r) = \frac{F}{(8\pi)K} \left(\frac{1}{2} \left(a^2 - r^2\right) + r^2 \ln \frac{a}{r} \right) \qquad (11\text{-}21.8)$$

(vi) Central concentrated load with freely supported contour

The clamping of the contour of the slab prevents both deflection and displacements:

At r = a	→ Displacement:	w(a) = 0
and	→ Twisting couple:	M_r =0

$$(11\text{-}22.1)$$

At r = 0 → Load : q(0) = F

Equation (11-15.7) is simplified to

$$w(r) = C_1 r^2 \ln r + C_2 r^2 + C_4 \qquad (11\text{-}22.2)$$

Substituting, w(a) = 0 and q(a) = 0, in the displacement w(r), equation (11-22.2), we get

399

$$0 = C_1a^2 \ln a + C_2a^2 + C_4 \qquad (11\text{-}22.3)$$

Substituting by the vanishing moment $M_r(a) = 0$, in equation (11-16.1), we get

$$M_r = -\frac{h^3 E}{12(1-v^2)} \left(\frac{\partial^2 w}{\partial r^2} + \frac{v}{r} \frac{\partial w}{\partial r} \right) = 0 \qquad (11\text{-}22.4)$$

Since,

$$\frac{\partial w}{\partial r} = 2C_1 r \ln r + C_1 r + 2C_2 r \qquad (11\text{-}22.5)$$

$$\frac{\partial^2 w}{\partial r^2} = 3C_1 + 2C_1 \ln r + 2C_2 \qquad (11\text{-}22.6)$$

Therefore,

$$M_r = -\frac{h^3 E}{12(1-v^2)} \left(\frac{\partial^2 w}{\partial r^2} + \frac{v}{r} \frac{\partial w}{\partial r} \right) = 0$$

$$= -\frac{h^3 E}{12(1-v^2)} (3C_1 + 2C_1 \ln r + 2C_2 + v(2C_1 \ln r + C_1 + 2C_2))$$

i.e.,

$$C_2 = -C_1 \frac{((3+v) + 2 \ln a(1+v))}{2(1+v)} \qquad (11\text{-}22.7)$$

Substituting in equation (11-22.3), we get

$$C_4 = -C_1a^2 \ln a + a^2 C_1 \frac{((3+v) + 2 \ln a(1+v))}{2(1+v)}$$

Or, $\qquad C_4 = \frac{C_1 a^2 (3+v)}{2(1+v)} \qquad (11\text{-}22.8)$

Having determined C_2 and C_4 in terms of C_1, we can now write the displacement equation (11-22.2) as follows.

$$w(r) = C_1 r^2 \ln r - C_1 \frac{((3+v) + 2 \ln a(1+v))}{2(1+v)} r^2 + \frac{C_1 a^2 (3+v)}{2(1+v)}$$

i.e., $\qquad w(r) = C_1 \left[\frac{(3+v)}{2(1+v)} (a^2 - r^2) - r^2 \ln \frac{a}{r} \right] \qquad (11\text{-}22.9)$

From equation (10-54.4), and accounting for the constants of our present problem, the constant C_1 becomes

$$(2\pi)KC_1 = F \tag{11-22.10}$$

Therefore,

$$w(r) = \frac{F}{(8\pi)K}\left[\frac{(3+v)}{2(1+v)}(a^2 - r^2) - r^2 \ln\frac{a}{r}\right] \tag{11-22.11}$$

Table 11-2. Summary of expressions for solution of circular slab

(ii) Fixed-in contour	$w(r) = \frac{q}{64K}(a^2 - r^2)^2$	(11-17.3)
	$M_r = \frac{q}{16}\left[a^2(v+1) - r^2(3+v)\right]$	(11-18.1)
	$M_\perp = \frac{q}{16}\left[a^2(v+1) - r^2(1+3v)\right]$	(11-18.2)
(iii) Supported contour (free motion	$w(r) = \frac{q}{64K}\left((a^2 - r^2)^2 + \frac{4a^2}{(v+1)}(a^2 - r^2)\right)$	(11-19.6)
	$M_r = \frac{q}{16}\left[a^2(3+v) - r^2(3+v)\right]$	(11-19.8)
	$M_\perp = \frac{q}{16}\left[a^2(3+v) - r^2(1+3v)\right]$	(11-19.9)
(iv) Pure bending of a slab	$w(r) = \frac{M}{2K(1+v)}(a^2 - r^2)$	(11-20.2)
(v) concentrated load with fixed-in contour	$w(r) = \frac{F}{(8\pi)K}\left(\frac{1}{2}(a^2 - r^2) + r^2 \ln\frac{a}{r}\right)$	(11-21.8)
(vi) concentrated load freely supported contour	$w(r) = \frac{F}{(8\pi)K}\left[\frac{(3+v)}{2(1+v)}(a^2 - r^2) - r^2 \ln\frac{a}{r}\right]$	(11-22.11)

11-4. Rectangular plate

a. Navier's method

A rectangular plate **supported** at it four-sided contour and exposed to **arbitrary load** is treated by **Sophie Germain's equation** (11-9.3) in terms of sums of infinite series of trigonometric functions of the form or equation (7-44), as follows

$$w(x,y) = \sum_{m=1}^{\infty} \sum_{n=1}^{\infty} A_{mn} \sin\left(m\pi\frac{x}{a}\right) \sin\left(n\pi\frac{y}{b}\right) \tag{11-23.1}$$

Where, a and b are the width and breadth of the rectangle.

Substituting by the derivatives of w(x,y) from (11-23.1) into equations (11-9.3), we get

$$
\left(\frac{\partial^4}{\partial x^4} + 2\frac{\partial^4}{\partial x^2 \partial y^2} + \frac{\partial^4}{\partial y^4} \right) w(x,y) = \frac{q(x,y)}{K}
$$

$$
\pi^4 K \sum_{m=1}^{\infty} \sum_{n=1}^{\infty} A_{mn} \left(\frac{m^2}{a^2} + \frac{n^2}{b^2} \right)^2 \sin\left(m\pi\frac{x}{a} \right) \sin\left(n\pi\frac{y}{b} \right) = q(x,y)
$$

(11-23.2)

Since A_{mn} is a constant to be determined from the boundary conditions, wee could as well fuse the constants into a new symbol as follows

$$
\sum_{m=1}^{\infty} \sum_{n=1}^{\infty} B_{mn} \sin\left(m\pi\frac{x}{a} \right) \sin\left(n\pi\frac{y}{b} \right) = q(x,y)
$$

(11-23.3)

Determining B_{mn} by Euler's formulas:

(i) Eliminating the y-terms

We have previously discussed the use of Euler's formula in equation (5-12.17a). Similarly, multiply both sides of equation (11-23.3) by

$$
\sin\frac{j\pi y}{b} dy
$$

(11-23.4)

then integrate both sides with respect to y over the period 0 to b, we get

$$
\sum_{m=1}^{\infty} \sum_{n=1}^{\infty} B_{mn} \sin\frac{m\pi x}{a} \int_0^b \sin\frac{n\pi y}{b} \sin\frac{j\pi y}{b} dy = \int_0^b q(x,y)\sin\frac{j\pi y}{b} dy
$$

(11-23.5)

We have already established the property of Euler's formula in equation (5-12.18c) in the two cases: $n = j$ and $n \neq j$, as follows

$$
\sum_{m=1}^{\infty} B_{mn} \sin\frac{m\pi x}{a} = \frac{2}{b}\int_0^b q(x,y)\sin\frac{n\pi y}{b} dy
$$

(11-23.6)

(ii) Eliminating the x-terms

402

We will repeat the use of Euler's formula in equation (5-12.17a) by multiplying both sides of equation (11-23.6) by

$$\sin\frac{k\pi x}{a}dx \qquad (11\text{-}23.7)$$

Then integrate both sides with respect to x over the period 0 to a, we get

$$\sum_{m=1}^{\infty} B_{mn} \int_0^a \sin\frac{m\pi x}{a}\sin\frac{k\pi x}{a}dx = \frac{2}{b}\int_0^a\int_0^b q(x,y)\sin\frac{n\pi y}{b}\sin\frac{k\pi x}{a}dxdy \qquad (11\text{-}23.8)$$

Similarly, we used the property of Euler's formula in equation (5-12.18c) in the two cases: m = k and m ≠ k, as follows

$$B_{mn} = \frac{4}{ba}\int_0^a\int_0^b q(x,y)\sin\frac{n\pi y}{b}\sin\frac{m\pi x}{a}dxdy \qquad (11\text{-}23.9)$$

We could now revert to express the coefficient A_{mn}, in equation (11-23.2) in terms of B_{mn}, (11-23.9) as follows

$$A_{mn} = \frac{4}{\pi^4 baK\left(\dfrac{m^2}{a^2}+\dfrac{n^2}{b^2}\right)^2}\int_0^a\int_0^b q(x,y)\sin\frac{n\pi y}{b}\sin\frac{m\pi x}{a}dxdy \qquad (11\text{-}23.10)$$

Final expression for deflection of rectangular slab

Substituting from (11-23.10) into (11-23.1), we get

$$w(x,y) = \sum_{m=1}^{\infty}\sum_{n=1}^{\infty}\left[\frac{4}{\pi^4 baK\left(\dfrac{m^2}{a^2}+\dfrac{n^2}{b^2}\right)^2}\int_0^a\int_0^b q(x,y)\sin\frac{n\pi y}{b}\sin\frac{m\pi x}{a}dxdy\right]\sin\left(m\pi\frac{x}{a}\right)\sin\left(n\pi\frac{y}{b}\right) \qquad (11\text{-}24)$$

Equation (11-24) expresses the vertical deflection w(x,y) in terms of the load distribution q(x,y) and the breadth and width of the rectangle a and b.

Case 1: Uniformly loaded rectangular slab, q(x,y) = q

Executing the integrals in equation (11-23.10) with constant q, we get

403

$$A_{mn} = \cfrac{4}{\pi^4 baK\left(\cfrac{m^2}{a^2} + \cfrac{n^2}{b^2}\right)^2} \int_0^a \int_0^b q(x,y)\sin\frac{n\pi y}{b}\sin\frac{m\pi x}{a}\,dxdy$$

$$= \cfrac{4q}{\pi^4 baK\left(\cfrac{m^2}{a^2} + \cfrac{n^2}{b^2}\right)^2}\left(\int_0^b \sin\frac{n\pi y}{b}\,dy\right)\left(\int_0^a \sin\frac{m\pi x}{a}\,dx\right) \qquad (11\text{-}25.1)$$

$$= \cfrac{4q}{mn\pi^6 K\left(\cfrac{m^2}{a^2} + \cfrac{n^2}{b^2}\right)^2}\left(1 - \cos\frac{n\pi y}{b}\right)\left(1 - \cos\frac{m\pi x}{a}\right)$$

In equation (11-25.1), if n or m is even, A_{mn} vanishes. Therefore, **only when m and n are odd** that A_{mn} has the following non-vanishing value

$$A_{mn} = \cfrac{4q}{mn\pi^6 K\left(\cfrac{m^2}{a^2} + \cfrac{n^2}{b^2}\right)^2}(2)(2)$$

$$= \cfrac{16q}{mn\pi^6 K\left(\cfrac{m^2}{a^2} + \cfrac{n^2}{b^2}\right)^2} \qquad (11\text{-}25.2)$$

(i) Therefore, the **deflection equation** (11-24) becomes

$$w(x,y) = \frac{16q}{\pi^6 K}\sum_{m=1,3,5}^{\infty}\sum_{n=1,3,5}^{\infty}\left[\cfrac{\sin\left(m\pi\dfrac{x}{a}\right)\sin\left(n\pi\dfrac{y}{b}\right)}{mn\left(\dfrac{m^2}{a^2} + \dfrac{n^2}{b^2}\right)^2}\right] \qquad (11\text{-}25.3)$$

Equation (11-25.3) is sketched in Figure 11-5.

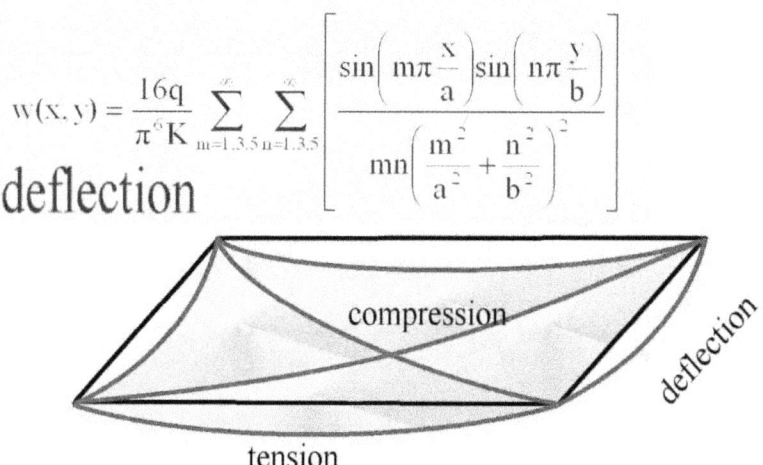

$$w(x,y) = \frac{16q}{\pi^6 K} \sum_{m=1,3,5}^{\infty} \sum_{n=1,3,5}^{\infty} \left[\frac{\sin\left(m\pi\frac{x}{a}\right)\sin\left(n\pi\frac{y}{b}\right)}{mn\left(\frac{m^2}{a^2}+\frac{n^2}{b^2}\right)^2} \right]$$

deflection

compression

deflection

tension

Figure 11-5. Deflection of uniformly loaded rectangular slab.

(ii) The **bending moments per unit length of cross section** of the slab, equations (11-10.2) are obtained substituting by derivatives from (11-25.3), as follows

$$M_x = -K\left(\frac{\partial^2 w}{\partial x^2} + v\frac{\partial^2 w}{\partial y^2}\right)$$

$$= \frac{16q}{\pi^4} \sum_{m=1,3,5}^{\infty} \sum_{n=1,3,5}^{\infty} \left[\frac{\frac{m^2}{a^2}+v\frac{n^2}{b^2}}{mn\left(\frac{m^2}{a^2}+\frac{n^2}{b^2}\right)^2} \right] \sin\left(m\pi\frac{x}{a}\right)\sin\left(n\pi\frac{y}{b}\right) \qquad (11\text{-}25.4)$$

$$M_y = -K\left(\frac{\partial^2 w}{\partial y^2} + v\frac{\partial^2 w}{\partial x^2}\right)$$

$$= \frac{16q}{\pi^4} \sum_{m=1,3,5}^{\infty} \sum_{n=1,3,5}^{\infty} \left[\frac{v\frac{m^2}{a^2}+\frac{n^2}{b^2}}{mn\left(\frac{m^2}{a^2}+\frac{n^2}{b^2}\right)^2} \right] \sin\left(m\pi\frac{x}{a}\right)\sin\left(n\pi\frac{y}{b}\right) \qquad (11\text{-}25.5)$$

(iii) The **shearing forces unit length of cross section** of the slab, equations (11-11) are obtained substituting by derivatives from (11-25.3), as follows

$$F_{xz} = -K\frac{\partial}{\partial x}\left(\nabla^2 w\right)$$

$$= -\frac{16q}{a\pi^3}\sum_{m=1,3,5}^{\infty}\sum_{n=1,3,5}^{\infty}\left[\frac{\cos\left(m\pi\frac{x}{a}\right)\sin\left(n\pi\frac{y}{b}\right)}{n\left(\frac{m^2}{a^2}+\frac{n^2}{b^2}\right)}\right] \qquad (11\text{-}25.6)$$

$$F_{yz} = -K\frac{\partial}{\partial y}\left(\nabla^2 w\right)$$

$$= -\frac{16q}{b\pi^3}\sum_{m=1,3,5}^{\infty}\sum_{n=1,3,5}^{\infty}\left[\frac{\sin\left(m\pi\frac{x}{a}\right)\cos\left(n\pi\frac{y}{b}\right)}{m\left(\frac{m^2}{a^2}+\frac{n^2}{b^2}\right)}\right] \qquad (11\text{-}25.7)$$

The two shearing forces F_{xz} and F_{yz} are maximal at the centers of the four sides, Figure 11-82

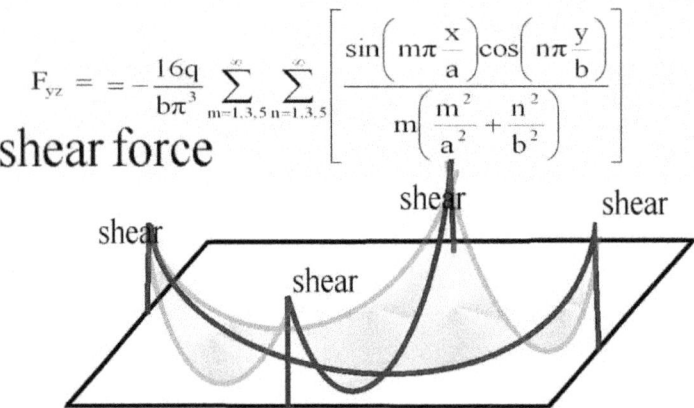

Figure 11-6. Maximum shear forces of uniformly loaded rectangular slab.

(iv) The **torsion moment per unit length of cross section** of the slab, equations (11-11) are obtained substituting by derivatives from (11-25.3), as follows

$$M_z = -K(1-v)\frac{\partial^2 w}{\partial y \partial x}$$

$$= -\frac{16q}{ab\pi^4}(1-v)\sum_{m=1,3,5}^{\infty}\sum_{n=1,3,5}^{\infty}\left[\frac{\cos\left(m\pi\frac{x}{a}\right)\cos\left(n\pi\frac{y}{b}\right)}{\left(\frac{m^2}{a^2}+\frac{n^2}{b^2}\right)^2}\right] \qquad (11-25.8)$$

The **twisting moment per unit length of cross section**, equation (11-25.8) is maximal at the corners of the contour, minimal on the centers of the plate and centers of sides. Figure 11-7.

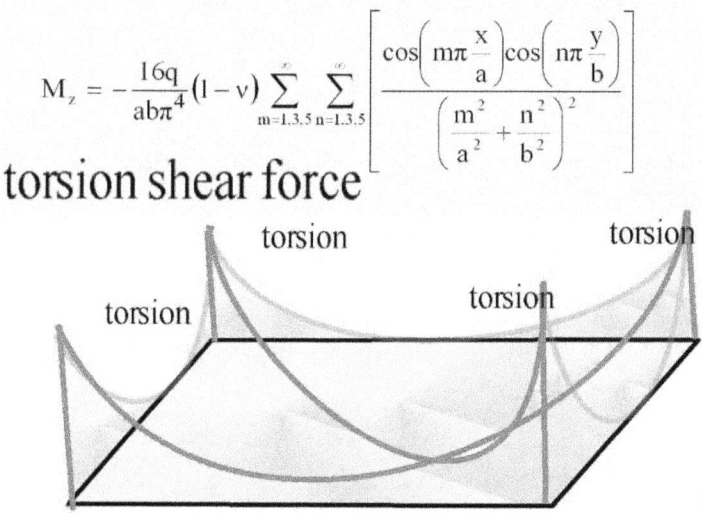

$$M_z = -\frac{16q}{ab\pi^4}(1-v)\sum_{m=1,3,5}^{\infty}\sum_{n=1,3,5}^{\infty}\left[\frac{\cos\left(m\pi\frac{x}{a}\right)\cos\left(n\pi\frac{y}{b}\right)}{\left(\frac{m^2}{a^2}+\frac{n^2}{b^2}\right)^2}\right]$$

torsion shear force

torsion

torsion

torsion

torsion

torsion

Figure 11-7. Maximum torsion forces of uniformly loaded rectangular slab.

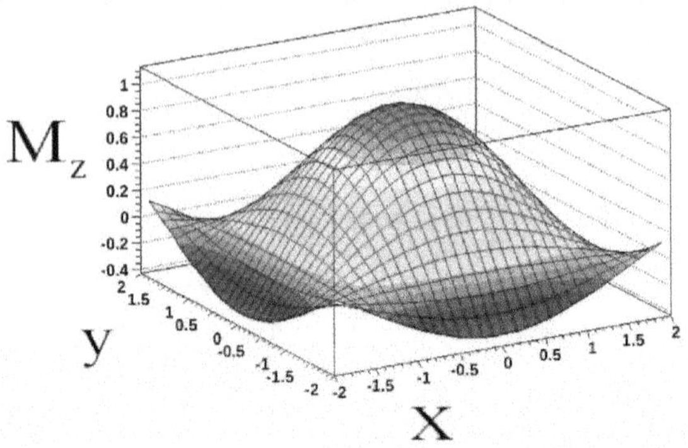

$$M_z = -\frac{16q}{ab\pi^4}(1-v)\sum_{m=1,3,5}^{\infty}\sum_{n=1,3,5}^{\infty}\left[\frac{\cos\left(m\pi\frac{x}{a}\right)\cos\left(n\pi\frac{y}{b}\right)}{\left(\frac{m^2}{a^2}+\frac{n^2}{b^2}\right)^2}\right]$$

Figure 11-8. Graphing rotational moment on the cross section of the rectangular plate

Maximum shearing forces, deflections, moments of bending

(i) Maximum deflection: $x = a/2$ and $y = b/2$

Equation (11-25.3) expresses the distribution of the vertical deflection (z-dependent) of the slab over the x and y coordinates.

$$
\begin{aligned}
w\left(\frac{a}{2},\frac{b}{2}\right) &= \frac{16q}{\pi^6 K}\sum_{m=1,3,5}^{\infty}\sum_{n=1,3,5}^{\infty}\left[\frac{\sin\left(\frac{m\pi}{2}\right)\sin\left(\frac{n\pi}{2}\right)}{mn\left(\frac{m^2}{a^2}+\frac{n^2}{b^2}\right)^2}\right] \\
&= \frac{16q}{\pi^6 K}\sum_{m=1,3,5}^{\infty}\sum_{n=1,3,5}^{\infty}\left[\frac{(-1)^{\frac{m+n}{2}-1}}{mn\left(\frac{m^2}{a^2}+\frac{n^2}{b^2}\right)^2}\right] \\
&= \frac{16q}{\pi^6 K}\Gamma
\end{aligned}
\qquad (11\text{-}26.1)
$$

Where,

$$\Gamma = \sum_{m=1,3,5}^{\infty} \sum_{n=1,3,5}^{\infty} \left[\frac{(-1)^{\frac{m+n}{2}-1}}{mn\left(\frac{m^2}{a^2} + \frac{n^2}{b^2}\right)^2} \right] \qquad (11\text{-}26.2)$$

(ii) Maximum bending moments

Equations (11-25.4 and 25.5) express the distribution of bending moments over the x and y coordinates. The maximum bending moments per unit length of the cross section occur at the maximum deflection, $x = a/2$ and $y = b/2$, as follows:

$$M_x\left(\frac{a}{2},\frac{b}{2}\right) = \frac{16q}{\pi^4} \sum_{m=1,3,5}^{\infty} \sum_{n=1,3,5}^{\infty} \left[\frac{\frac{m^2}{a^2} + v\frac{n^2}{b^2}}{mn\left(\frac{m^2}{a^2} + \frac{n^2}{b^2}\right)^2} \right] (-1)^{\frac{m+n}{2}-1} \qquad (11\text{-}26.3)$$

$$M_y\left(\frac{a}{2},\frac{b}{2}\right) = \frac{16q}{\pi^4} \sum_{m=1,3,5}^{\infty} \sum_{n=1,3,5}^{\infty} \left[\frac{v\frac{m^2}{a^2} + \frac{n^2}{b^2}}{mn\left(\frac{m^2}{a^2} + \frac{n^2}{b^2}\right)^2} \right] (-1)^{\frac{m+n}{2}-1} \qquad (11\text{-}26.4)$$

(iii) Minimum and maximum shearing forces

Equations (11-25.6 and 25.7) express the distribution of shearing forces over the x and y coordinates. The minimum shear forces occur at center of plate: $x = a/2$, $y = b/2$

$$F_{xz}\left(\frac{a}{2},\frac{b}{2}\right) = -\frac{16q}{a\pi^3} \sum_{m=1,3,5}^{\infty} \sum_{n=1,3,5}^{\infty} \left[\frac{\cos\left(\frac{m\pi}{2}\right)\sin\left(\frac{n\pi}{2}\right)}{n\left(\frac{m^2}{a^2} + \frac{n^2}{b^2}\right)} \right] = 0 \qquad (11\text{-}26.5)$$

Maximum shear forces: $x = a$, $y = b/2$

$$F_{xz}\left(\pm a, \frac{b}{2}\right) = \frac{16q}{a\pi^3} \sum_{m=1,3,5}^{\infty} \sum_{n=1,3,5}^{\infty} \left[\frac{\sin\left(\frac{n\pi}{2}\right)}{n\left(\frac{m^2}{a^2} + \frac{n^2}{b^2}\right)} \right]$$

(11-26.6)

Put simply,

$$F_{xz}\left(\pm a, \frac{b}{2}\right) = \frac{16q}{a\pi^3} \Gamma_s$$

(11-26.7)

Where,

$$\Gamma_s = \sum_{m=1,3,5}^{\infty} \sum_{n=1,3,5}^{\infty} \left[\frac{(-1)^{\frac{n-1}{2}}}{mn\left(\frac{m^2}{a^2} + \frac{n^2}{b^2}\right)^2} \right]$$

(11-26.8)

(iv) Minimum and maximum torsion couples and shearing forces

Equation (11-25.8) gives the **twisting moment** per unit length of cross section. M_z is maximal at the corners of the contour, minimal on the centers of the plate and centers of sides.

$$M_z = -K(1-v)\frac{\partial^2 w}{\partial y \partial x}$$

$$= -\frac{16q}{ab\pi^4}(1-v) \sum_{m=1,3,5}^{\infty} \sum_{n=1,3,5}^{\infty} \left[\frac{\cos\left(m\pi\frac{x}{a}\right)\cos\left(n\pi\frac{y}{b}\right)}{\left(\frac{m^2}{a^2} + \frac{n^2}{b^2}\right)^2} \right]$$

(11-27.1)

The **shearing forces** due to **twisting moment per unit length** of cross section

$$T_{zx} = \frac{\partial M_z}{\partial y} = -K(1-v)\frac{\partial^3 w}{\partial y^2 \partial x}$$

$$= \frac{16q}{ab^2\pi^3}(1-v) \sum_{m=1,3,5}^{\infty} \sum_{n=1,3,5}^{\infty} \left[\frac{n\cos\left(m\pi\frac{x}{a}\right)\sin\left(n\pi\frac{y}{b}\right)}{\left(\frac{m^2}{a^2} + \frac{n^2}{b^2}\right)^2} \right]$$

(11-27.2)

$$T_{zy} = \frac{\partial M_z}{\partial x} = -\frac{h^3 E}{12(1-v^2)}(1-v)\frac{\partial^3 w}{\partial y \partial x^2}$$

$$= \frac{16q}{a^2 b \pi^3}(1-v)\sum_{m=1,3,5}^{\infty}\sum_{n=1,3,5}^{\infty}\left[\frac{m\sin\left(m\pi\frac{x}{a}\right)\cos\left(n\pi\frac{y}{b}\right)}{\left(\frac{m^2}{a^2}+\frac{n^2}{b^2}\right)^2}\right] \tag{11-27.3}$$

(v) Resultant shear forces

The resultant shearing forces on the contour is the sum of shearing due to bending and shearing due to twisting. Thus,

$$F_{xz} + T_{zx} = -K\frac{\partial}{\partial x}\left(\nabla^2 w\right) - K(1-v)\frac{\partial^3 w}{\partial y^2 \partial x}$$

$$= \frac{16q}{a\pi^3}\sum_{m=1,3,5}^{\infty}\sum_{n=1,3,5}^{\infty}\left[\frac{\frac{m^2}{a^2}+(2-v)\frac{n^2}{b^2}}{n\left(\frac{m^2}{a^2}+\frac{n^2}{b^2}\right)^2}\right]\cos\left(m\pi\frac{x}{a}\right)\sin\left(n\pi\frac{y}{b}\right) \tag{11-28.1}$$

And

$$F_{yz} + T_{zy} = -K\frac{\partial}{\partial y}\left(\nabla^2 w\right) - K(1-v)\frac{\partial^3 w}{\partial y \partial x^2}$$

$$= \frac{16q}{b\pi^3}\sum_{m=1,3,5}^{\infty}\sum_{n=1,3,5}^{\infty}\left[\frac{(2-v)\frac{m^2}{a^2}+\frac{n^2}{b^2}}{m\left(\frac{m^2}{a^2}+\frac{n^2}{b^2}\right)^2}\right]\sin\left(m\pi\frac{x}{a}\right)\cos\left(n\pi\frac{y}{b}\right) \tag{11-28.2}$$

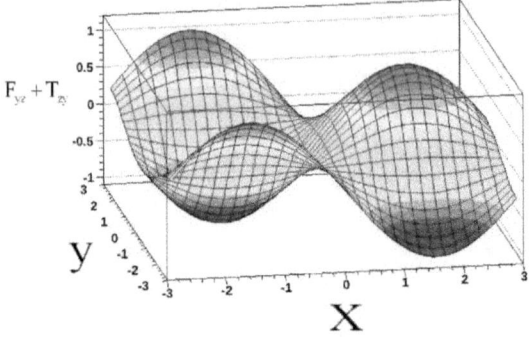

$$F_{yz} + T_{zy} = \frac{16q}{b\pi^3} \sum_{m=1,3,5}^{\infty} \sum_{n=1,3,5}^{\infty} \left[\frac{(2-v)\dfrac{m^2}{a^2} + \dfrac{n^2}{b^2}}{m\left(\dfrac{m^2}{a^2} + \dfrac{n^2}{b^2}\right)^2} \right] \sin\left(m\pi\frac{x}{a}\right)\cos\left(n\pi\frac{y}{b}\right)$$

Figure 11-9. Resultant shear forces on thin slab.

Case 2: Concentrated loaded rectangular slab, q(c,d) = F

A concentrated load given by

$$q(x,y) \qquad = F \qquad\qquad x = c \text{ and } y = d$$
$$= 0 \qquad\qquad x \neq c \text{ and } y \neq d \qquad\qquad (11\text{-}29.1)$$

Equation (11-25.1) becomes

$$A_{mn} = \frac{4}{\pi^4 baK\left(\dfrac{m^2}{a^2} + \dfrac{n^2}{b^2}\right)^2} \int_0^a \int_0^b P\delta(c,d)\sin\frac{n\pi y}{b}\sin\frac{m\pi x}{a}\,dxdy$$

$$= \frac{4P}{\pi^4 baK\left(\dfrac{m^2}{a^2} + \dfrac{n^2}{b^2}\right)^2}\sin\frac{m\pi c}{a}\sin\frac{n\pi d}{b} \qquad\qquad (11\text{-}29.2a)$$

Where, the **Kronecker Delta function** defined by

$$\delta(c,d) \qquad = 1, \qquad\qquad x = c \text{ and } y = d$$

$$\delta(c,d) \qquad = 0, \qquad x \neq c \text{ and } y \neq d \qquad (11\text{-}29.2\text{b})$$

Therefore, the **deflection equation** (11-24) becomes

$$w(x,y) = \frac{4P}{\pi^4 baK} \sum_{m=1}^{\infty} \sum_{n=1}^{\infty} \left[\frac{\sin\dfrac{m\pi c}{a}\sin\dfrac{n\pi d}{b}}{\left(\dfrac{m^2}{a^2}+\dfrac{n^2}{b^2}\right)^2} \right] \sin\!\left(m\pi\frac{x}{a}\right)\sin\!\left(n\pi\frac{y}{b}\right) \qquad (11\text{-}29.3)$$

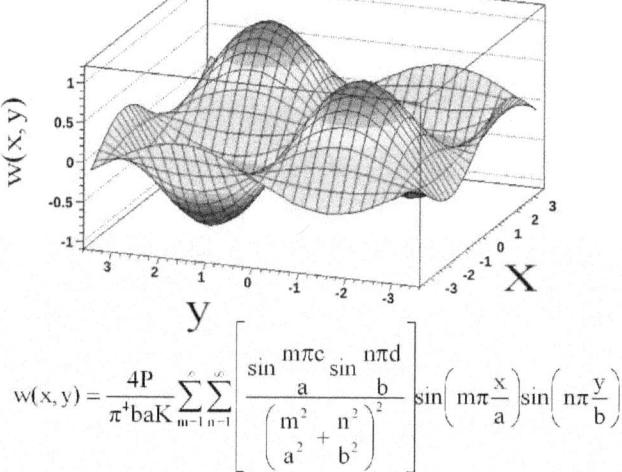

$$w(x,y) = \frac{4P}{\pi^4 baK} \sum_{m=1}^{\infty} \sum_{n=1}^{\infty} \left[\frac{\sin\dfrac{m\pi c}{a}\sin\dfrac{n\pi d}{b}}{\left(\dfrac{m^2}{a^2}+\dfrac{n^2}{b^2}\right)^2} \right] \sin\!\left(m\pi\frac{x}{a}\right)\sin\!\left(n\pi\frac{y}{b}\right)$$

Figure 11-10. Graphing a single Fourier's sine product on the cross-section of the rectangular plate.

b. Levy's method

Imposing different boundary conditions, than in **Navier's** problem (supported **four sides** of contour), requires different formulation of **Fourier's series** expansions. Figure 11-11.

413

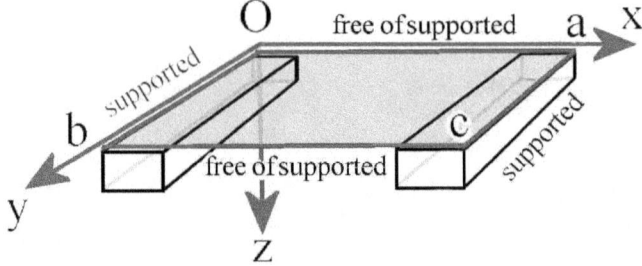

Figure 11-11. Boundary conditions in Levy's problem and solution.

Equations of elasticity

Deflection bi-harmonic equation:

$$\nabla^2\nabla^2 w(x,y) = \frac{q(x,y)}{K}$$
(11-30.1)

Stresses bi-harmonic function:

$$\nabla^2\nabla^2\varphi(x,y) = 0$$
(11-30.2)

Boundary conditions

Figure 11-11, sides $x = 0$ and $x = a$ are **supported**

$$w(0,y) = w(a,y) = 0$$
$$\frac{\partial^2 w(0,y)}{\partial x^2} = \frac{\partial^2 w(a,y)}{\partial x^2} = 0$$
(11-31.1)

Fourier's solution of infinite series

$$w(x,y) = \sum_{m=1}^{\infty} Y_m(y)\sin\left(m\pi\frac{x}{a}\right)$$
(11-31.2)

Equation (11-31.2) satisfies the boundary conditions, equation (11-31.1), as the sine function vanishes at the two sides, $x = 0$ and $x = a$.

Substituting by $w(x,y)$, from equation (11-31.2), in equation (11-30.1), we get

414

$$\sum_{m=1}^{\infty} \left(\left(\frac{m\pi}{a} \right)^4 Y_m + Y_m'''' - 2\left(\frac{m\pi}{a} \right)^2 Y_m'' \right) \sin\left(m\pi \frac{x}{a} \right) = \frac{q(x,y)}{K} \tag{11-31.3}$$

Using the property of integral of the sine product in (5-12.17a) and equation (5-12.18a), equation (11-31.3) is multiplied by $\sin\left(n\pi \frac{x}{a} \right)$, and integrating the two sides, we get the value $\left(\frac{a}{2} \right)$ only when m = n, zero when $m \neq n$[MFE3].

Therefore,

$$\left(\frac{m\pi}{a} \right)^4 Y_m + Y_m'''' - 2\left(\frac{m\pi}{a} \right)^2 Y_m'' = \frac{2}{aK} \int_0^a q(x,y)\sin\left(m\pi \frac{x}{a} \right) dx \tag{11-31.4}$$

General and particular solutions

We have already solved obtained **the general solution** (G.S.) of equation (7-38.2) in the form

$$Y(y)k^4 - 2k^2 \cdot \frac{d^2 Y(y)}{dy^2} + \frac{d^4 Y(y)}{dy^4} = 0 \tag{11-31.5}$$

As follows

$$\text{G.S.} = A_m \cosh y_1 + B_m \sinh y_1 + C_m y_1 \cosh y_1 + D_m y_1 \sinh y_1 \tag{11-31.6}$$

Where,

Therefore, the load term q(x,y) comprises the **particular solution** added to Y_m, in equation (11-31.6) as follows

$$Y_m = \text{G.S.} + \text{P.S.}$$
$$= A_m \cosh y_1 + B_m \sinh y_1 + C_m y \cosh y_1 + D_m y \sinh y_1 + \text{P.S.} \tag{11-31.7a}$$

Where,

$$y_1 = \frac{m\pi y}{a} \tag{11-31.7b}$$

Thus, the deflection equation (11-31.2) becomes

$$w(x,y) = \sum_{m=1}^{\infty} \left[A_m \cosh y_1 + B_m \sinh y_1 + C_m y \cosh y_1 + D_m y \sinh y_1 + \text{P.S.} \right] \sin\left(m\pi \frac{x}{a} \right) \tag{11-31.8}$$

Boundary conditions

Supported remaining two sides (OA and bc), Figure 11-11.

$$w(x,0) = w(x,b) = 0$$

$$\frac{\partial^2 w(x,0)}{\partial y^2} = \frac{\partial^2 w(x,b)}{\partial y^2} = 0 \qquad (11\text{-}32.1)$$

Hence, we have the four Fourier coefficients A_m, B_m, C_m, and D_m which are satisfied at the four sides of the rectangular slab.

Substituting by the deflection $w(x,y)$, from equation (11-31.8), into the boundary conditions, equations (11-32.1), we get the following four equations, required to determine the four constants, as follows:

Displacements:

(1)****** $w(x,0) = 0$ ******

$$w(x,0) = A_m + P.S.(0) = 0 \qquad (11\text{-}32.2)$$

(2) ****** $w(x,b) = 0$ ******

$$A_m \cosh\frac{m\pi b}{a} + B_m \sinh\frac{m\pi b}{a} + C_m b\cosh\frac{m\pi b}{a} + D_m b\sinh\frac{m\pi b}{a} + P.S.(b) = 0 \qquad (11\text{-}32.3)$$

Curvatures:

(3) ****** $\dfrac{\partial^2 w(x,0)}{\partial y^2} = 0$ ******

$$\frac{\partial^2 w(x,0)}{\partial y^2} = A_m\left(\frac{m\pi}{a}\right)^2 + 2D_m\left(\frac{m\pi}{a}\right) + P.S.''(0) = 0 \qquad (11\text{-}32.4)$$

(4) ****** $\dfrac{\partial^2 w(x,b)}{\partial y^2} = 0$ ******

$$A_m\left(\frac{m\pi}{a}\right)^2 \cosh\frac{m\pi b}{a} + B_m\left(\frac{m\pi}{a}\right)^2 \sinh\frac{m\pi b}{a} + 2C_m\left(\frac{m\pi}{a}\right)\sinh y$$

$$+ C_m\left(\frac{m\pi}{a}\right)^2 y\cosh y + 2D_m\left(\frac{m\pi}{a}\right)\cosh y + D_m\left(\frac{m\pi}{a}\right)^2 y\sinh y + P.S.''(b) = 0 \qquad (11\text{-}32.5)$$

Suggested simplification

We will assume that the particular solutions P.S.(0) and its second derivative P.S."[MFE4][MFE5](0), thus

$$P.S.'0) = P.S.''(0) = 0 \tag{11-33.1}$$

That greatly simplifies the determination of the four constants such that, equation (11-32.2) gives

$$A_m = 0 \tag{11-33.2}$$

Equation (11-32.4) gives

$$D_m = 0 \tag{11-33.3}$$

Thus, equations (11-32.3) and (11-32.5) become

$$B_m \sinh \frac{m\pi b}{a} + C_m b \cosh \frac{m\pi b}{a} + P.S.(b) = 0 \tag{11-33.4}$$

$$B_m \left(\frac{m\pi}{a}\right)^2 \sinh \frac{m\pi b}{a} + 2C_m \left(\frac{m\pi}{a}\right) \sinh \frac{m\pi}{a} + C_m \left(\frac{m\pi}{a}\right)^2 b \cosh \frac{m\pi}{a} + P.S.''(b) = 0 \tag{11-33.5}$$

Solving the two equations, we obtain expressions for B_m and C_m in terms of P.S. as follows:

$$C_m = \frac{\left(\frac{m\pi}{a}\right)^2 P.S.(b) - P.S.''(b)}{2\left(\frac{m\pi}{a}\right) \sinh \frac{m\pi}{a}} \tag{11-33.6}$$

$$B_m = \frac{P.S.''(b)\left(b\cosh \frac{m\pi b}{a}\right) - P.S.(b)\left(\left(\frac{m\pi}{a}\right)^2 b\cosh \frac{m\pi b}{a} + 2\left(\frac{m\pi}{a}\right)\sinh \frac{m\pi}{a}\right)}{2\left(\frac{m\pi}{a}\right)\left(\sinh \frac{m\pi}{a}\right)^2} \tag{11-33.7}$$

Particular solution by Cauchy's method

The particular solution P.S.(y), of equation (11-31.4) is assumed to fulfill the following boundary conditions

$$Y_{ps}(0) = Y_{ps}(0)' = Y_{ps}(0)'' = 0$$

$$Y_{ps}(0)''' = 1 \tag{11-34.1}$$

A proposed particular solution to equation (11-31.4) that satisfies the conditions in (11-34.1) takes a form similar to (11-31.6)

$$Y_{ps} = a_m y_1 \cosh y_1 + b_m \sinh y_1 \tag{11-34.2}$$

Our reasons for selecting those terms from the general solution of (11-31.4) will be evident from the following three derivatives of Y_{ps}, beside positioning of y in Y_{ps}, equation (11-34.2).

Derivatives of (11-34.2)

$$Y_{ps}' = a_m\left(\cosh y_1 + y\frac{m\pi}{a}\sinh\frac{m\pi y}{a}\right) + b_m\frac{m\pi}{a}\cosh\frac{m\pi y}{a} \tag{11-34.3a}$$

$$Y_{ps}'' = a_m\left(2\frac{m\pi}{a}\sinh\frac{m\pi y}{a} + y\left(\frac{m\pi}{a}\right)^2\cosh\frac{m\pi y}{a}\right) + b_m\left(\frac{m\pi}{a}\right)^2\sinh\frac{m\pi y}{a} \tag{11-34.3b}$$

$$Y_{ps}''' = a_m\left(3\left(\frac{m\pi}{a}\right)^2\cosh\frac{m\pi y}{a} + y\left(\frac{m\pi}{a}\right)^3\sinh\frac{m\pi y}{a}\right) + b_m\left(\frac{m\pi}{a}\right)^3\cosh\frac{m\pi y}{a} \tag{11-34.3c}$$

Now, substituting the boundary condition, (11-34.1), in the above derivatives, we get

(i) The first condition, $Y_{ps}(0)$, is satisfied by the vanishing sinh-term and the product of y by cosh.

(ii) The second condition, $Y'_{ps}(0) = 0$, in equation (11-34.3a), gives

$$Y'_{ps}. = a_m + b_m\frac{n\pi}{a} = 0 \tag{11-34.4}$$

(iii) The third condition, $Y''_{ps}(0) = 0$, is also satisfied as the first by looking at the structure of equation (11-34.3b).

(iv) The fourth condition, $Y'''_{ps}(0) = 1$, in equation (11-34.3c), gives.

$$\left(\frac{2n\pi}{a}+1\right)a_m\frac{n\pi}{a}\cosh\frac{n\pi y}{a} + b_m\left(\frac{n\pi}{a}\right)^3\cosh\frac{n\pi}{a} = 1 \tag{11-34.5}$$

Solving (11-34.4) and (11-34.5), we get

418

$$a_m = \frac{1}{2}\left(\frac{a}{m\pi}\right)^2$$

$$b_m = -\frac{1}{2}\left(\frac{a}{m\pi}\right)^3$$

(11-34.6)

Thus, equation (11-34.2) becomes

$$Y_{ps}(y) = \frac{1}{2}\left(\frac{a}{m\pi}\right)^2 \left(y\cosh\frac{m\pi y}{a} - \left(\frac{a}{m\pi}\right)\sinh\frac{m\pi y}{a} \right)$$

(11-34.7)

Cauchy's method of solution of the particular solution of (11-31.4)

First, denote the load term in (11-31.4) by

$$F(y) = \frac{2}{aK}\int_0^a q(x,y)\sin\left(m\pi\frac{x}{a}\right)dx$$

(11-35.1)

Cauchy's method implies that

$$P.S.(y) = \int_0^y Y_{ps}(y-\xi)F(\xi)d\xi$$

(11-35.2)

Summary of Levy's method

The **particular solution**, equation (11-35.2), is obtained by integration of the load function (11-35.1) and the **Cauchy-Euler's solution** (11-34.7).

The particular solution P.S.(y) is then used to determined the **Fourier's coefficients** B_m and C_m (11-33.6 and 33.7).

The Fourier coefficients ate then used in the **deflection equation** (11-31.8), which is the main key to determining **strains, stresses, moments of bending couples**, and **shear forces**.

Table 11-3. Summary of expressions used in Levy's method of solution of rectangular plate

$$w(x,y) = \sum_{m=1}^{\infty} \left[B_m\sinh y_1 + C_m y\cosh y_1 + P.S. \right]\sin\left(m\pi\frac{x}{a}\right)$$

(11-31.8)

419

$$B_m = \frac{P.S.''(b)\left(b\cosh\dfrac{m\pi b}{a}\right) - P.S.(b)\left(\left(\dfrac{m\pi}{a}\right)^2 b\cosh\dfrac{m\pi b}{a} + 2\left(\dfrac{m\pi}{a}\right)\sinh\dfrac{m\pi}{a}\right)}{2\left(\dfrac{m\pi}{a}\right)\left(\sinh\dfrac{m\pi}{a}\right)^2} \qquad (11\text{-}33.7)$$

$$C_m = \frac{\left(\dfrac{m\pi}{a}\right)^2 P.S.(b) - P.S.''(b)}{2\left(\dfrac{m\pi}{a}\right)\sinh\dfrac{m\pi}{a}} \qquad (11\text{-}33.6)$$

$$Y_{ps}(y) = \frac{1}{2}\left(\frac{a}{m\pi}\right)^2\left(y\cosh\frac{m\pi y}{a} - \left(\frac{a}{m\pi}\right)\sinh\frac{m\pi y}{a}\right) \qquad (11\text{-}34.7)$$

$$F(y) = \frac{2}{aK}\int_0^a q(x,y)\sin\left(m\pi\frac{x}{a}\right)dx \qquad (11\text{-}35.1)$$

$$P.S.(y) = \int_0^y Y_{ps}(y-\xi)F(\xi)d\xi \qquad (11\text{-}35.2)$$

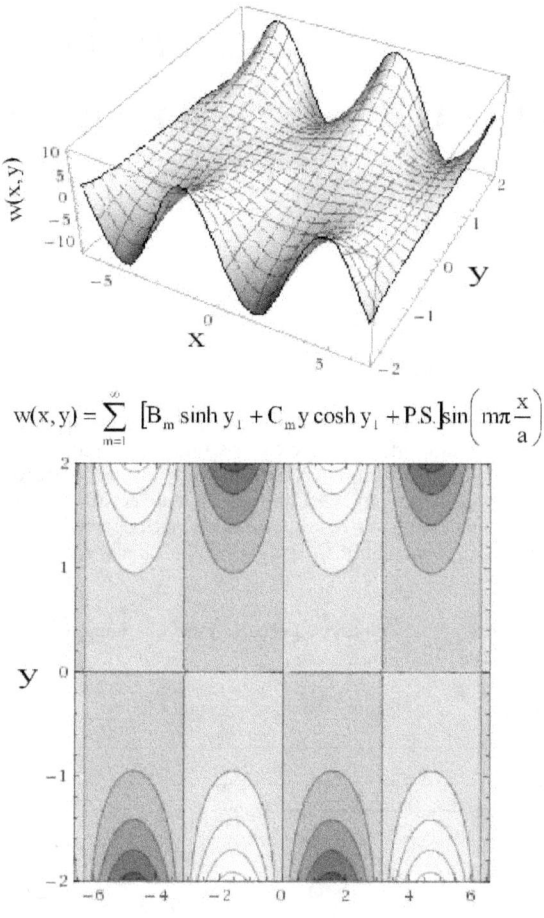

$$w(x,y) = \sum_{m=1}^{\infty} \left[B_m \sinh y_1 + C_m y \cosh y_1 + P.S. \right] \sin\left(m\pi \frac{x}{a} \right)$$

Figure 11-12. Graphing a single Fourier's product of trigonometric and hyperbolic functions on the cross-section of the rectangular plate.

CHAPTER 12

VARIATIONAL METHOD OF SOLUTION IN PLANAR ELASTICITY

12-1. Clapeyron's Theorem in Linear Elasticity

Elasticity of infinitesimal elements

To add another meaning to the **state of motion** within elastic medium under the **effect of force**, the variations of the relative coordinates of molecules or particles, or **strains** and **displacements**, are related to those **external forces**, resulted **internal stresses**, through the net or **integral energy** or **Lagrangian** of the system. This is accomplished through the **conservation of mass and energy** through surface- and volume-integration of displacements accompanied by their inducing stresses, i.e. net **work-done** by deformation of the elastic medium under the effect of force.

Let us start by the set of equations of elasticity which are required to form one integral quantity that characterizes a state elastic equilibrium.

Table 12-1. Equations of elasticity without Saint-Venant's continuity conditions

Stresses	$\sigma_x, \sigma_y, \sigma_z, \tau_{xy}, \tau_{zy}, \tau_{xz}$		(12-1.1)
Displacements	$u(x,y,z),\ v(x,y,z),\ w(x,y,z)$		(12-1.2)
Strains	$\varepsilon_{xx},\ \varepsilon_{yy},\ \varepsilon_{zz},\ \alpha_{xy},\ \alpha_{zy},\ \alpha_{xz}$		(12-1.3)
Navier's partial differential equations of equilibrium of stresses	$\dfrac{\partial \sigma_x}{\partial x} + \dfrac{\partial \tau_{xy}}{\partial y} + \dfrac{\partial \tau_{xz}}{\partial z} + \rho X_i = \rho \dfrac{\partial^2 u}{\partial t^2}$		(1-4.1)
	$\dfrac{\partial \tau_{yx}}{\partial x} + \dfrac{\partial \sigma_y}{\partial y} + \dfrac{\partial \tau_{yz}}{\partial z} + \rho Y_i = \rho \dfrac{\partial^2 v}{\partial t^2}$		(1-4.2)
	$\dfrac{\partial \tau_{zx}}{\partial x} + \dfrac{\partial \tau_{zy}}{\partial y} + \dfrac{\partial \sigma_z}{\partial z} + \rho Z_i = \rho \dfrac{\partial^2 w}{\partial t^2}$		(1-4.3)
Surface conditions linking **external and internal** stresses	$X_n = \sigma_x\, l + \tau_{xy}\, m + \tau_{xz}\, n$		(1-8.1)
	$Y_n = \tau_{yx}\, l + \sigma_y\, m + \tau_{yz}\, n$		(1-8.2)
	$Z_n = \tau_{zx}\, l + \tau_{zy}\, m + \sigma_z\, n$		(1-8.3)
Cauchy's equations for displacement, elongation, shear, and rotational	$\varepsilon_{xx} = \dfrac{\partial u}{\partial x}$ (2-2.4)	$2\alpha_{xy} = \dfrac{\partial u}{\partial y} + \dfrac{\partial v}{\partial x}$ (2-3.1)	

strains	$\varepsilon_{yy} = \dfrac{\partial v}{\partial y}$ (2-2.5)	$2\alpha_{yz} = \dfrac{\partial v}{\partial z} + \dfrac{\partial w}{\partial y}$ (2-3.2)
	$\varepsilon_{zz} = \dfrac{\partial w}{\partial z}$ (2-2.6)	$2\alpha_{zx} = \dfrac{\partial w}{\partial x} + \dfrac{\partial u}{\partial z}$ (2-3.3)

Table 12-1 represents **internal behavior** or state of internal motion of the elastic medium. The absence of **Saint Venant's equations of continuity** will **not be needed** in the study of variation, since differential variation is an exercise in the continuity of the state of motion of the medium.

Let us assume **external behavior** in the form volume displacements functions, which are continuous and have continuous first and second derivatives throughout the body volume as follows:

Table 12-2. Overall volume displacements, strains, and external forces

External stresses	X_n, Y_n, Z_n		(12-2.1)
Volume displacements	$\hat{u}(x, y, x, t), \hat{v}(x, y, x, t), \hat{w}(x, y, x, t)$		(12-2.2)
Volume Cauchy's equations for displacement, elongation, shear, and rotational strains	$\hat{\varepsilon}_{xx} = \dfrac{\partial \hat{u}}{\partial x}$ (12-2.3)	$2\hat{\alpha}_{xy} = \dfrac{\partial \hat{u}}{\partial y} + \dfrac{\partial \hat{v}}{\partial x}$ (12-2.6)	
	$\hat{\varepsilon}_{yy} = \dfrac{\partial \hat{v}}{\partial y}$ (12-2.4)	$2\hat{\alpha}_{yz} = \dfrac{\partial \hat{v}}{\partial z} + \dfrac{\partial \hat{w}}{\partial y}$ (12-2.7)	
	$\hat{\varepsilon}_{zz} = \dfrac{\partial \hat{w}}{\partial z}$ (12-2.5)	$2\hat{\alpha}_{zx} = \dfrac{\partial \hat{w}}{\partial x} + \dfrac{\partial \hat{u}}{\partial z}$ (12-2.8)	

Integrals of surface work-done

The products of **external stresses** on the surface by the **infinitesimal surface area** and **surface displacements** comprise the **infinitesimal work-done** by the deformation of the elastic medium under the effect of external stresses. The integration of that infinitesimal work-done is written as follows:

$$W.D. = \iint_S \left(X_n \hat{u} + Y_n \hat{v} + Z_n \hat{w} \right) dA \tag{12-3.1}$$

Substituting from the surface conditions (1-8) into (12-3.1), we get

$$W.D. = \iint_S \left(\begin{array}{c} (\sigma_x l + \tau_{xy} m + \tau_{xz} n)\hat{u} + (\tau_{xy} l + \sigma_y m + \tau_{yz} n)\hat{v} \\ + (\tau_{xz} l + \tau_{yz} m + \sigma_z n)\hat{w} \end{array} \right) dA$$

$$= \iint_S \left(\begin{array}{c} (\sigma_x \hat{u} + \tau_{xy} \hat{v} + \tau_{xz} \hat{w})l + (\tau_{xy} \hat{u} + \sigma_y \hat{v} + \tau_{yz} \hat{w})m \\ + (\tau_{xz} \hat{u} + +\tau_{yz} \hat{v} + \sigma_z \hat{w})n \end{array} \right) dA \tag{12-3.2a}$$

$$\text{W.D.} = \iint_S \left(Pl + Qm + Rn \right) dA \qquad (12\text{-}3.2b)$$

Where,

$$P = \sigma_x \hat{u} + \tau_{xy} \hat{v} + \tau_{xz} \hat{w}$$
$$Q = \tau_{xy} \hat{u} + \sigma_y \hat{v} + \tau_{yz} \hat{w} \qquad (12\text{-}3.3)$$
$$R = \tau_{xz} \hat{u} + \tau_{yz} \hat{v} + \sigma_z \hat{w}$$

Integrals of volume work-done

The surface-integral in equation (12-3.2b) is converted to volume-integral by the **divergence theorem** (Gauss's theorem)

$$\text{W.D.} = \iint_S \left(F.n \right) dA = \iiint_V \left(\nabla.F \right) dV \qquad (12\text{-}3.4a)$$

Therefore

$$\text{W.D.} = \iiint_V \left(\frac{\partial P}{\partial x} + \frac{\partial Q}{\partial y} + \frac{\partial R}{\partial z} \right) dV \qquad (12\text{-}3.4b)$$

Where, each of the three gradients has six components obtained by differentiating equations (12-3.3). Three of those six components are gradients of stresses and three are gradients of displacements (strains). Thus, in total, we have 18 components in the bracket of equation (12-3.4), nine are gradients of stresses, and nine strains, as follows:

(I)	(II)	(III)	(IV)	(V)	(VI)
Navier's Equations			Cauchy's Equations		(12-3.5)

$$\frac{\partial P}{\partial x} = \frac{\partial \sigma_x}{\partial x}\hat{u} + \frac{\partial \tau_{xy}}{\partial x}\hat{v} + \frac{\partial \tau_{xz}}{\partial x}\hat{w} + \frac{\partial \hat{u}}{\partial x}\sigma_x + \frac{\partial \hat{v}}{\partial x}\tau_{xy} + \frac{\partial \hat{w}}{\partial x}\tau_{xz}$$

$$\frac{\partial Q}{\partial y} = \frac{\partial \tau_{xy}}{\partial y}\hat{u} + \frac{\partial \sigma_y}{\partial y}\hat{v} + \frac{\partial \tau_{yz}}{\partial y}\hat{w} + \frac{\partial \hat{u}}{\partial y}\tau_{xy} + \frac{\partial \hat{v}}{\partial y}\sigma_y + \frac{\partial \hat{w}}{\partial y}\tau_{yz}$$

$$\frac{\partial R}{\partial z} = \frac{\partial \tau_{xz}}{\partial z}\hat{u} + \frac{\partial \tau_{yz}}{\partial z}\hat{v} + \frac{\partial \sigma_z}{\partial z}\hat{w} + \frac{\partial \hat{u}}{\partial z}\tau_{xz} + \frac{\partial \hat{v}}{\partial z}\tau_{yz} + \frac{\partial \hat{w}}{\partial z}\sigma_z$$

Adding the three equations, by members, then substituting by **Navier's equations** (1-4) and **Cauchy's equations** (2-2 and 2-3), we get

$$\frac{\partial P}{\partial x} + \frac{\partial Q}{\partial y} + \frac{\partial R}{\partial z} = \left(\frac{\partial^2 u}{\partial t^2} - X_i \right)\rho\hat{u} + \left(\frac{\partial^2 v}{\partial t^2} - Y_i \right)\rho\hat{v} + \left(\frac{\partial^2 w}{\partial t^2} - Z_i \right)\rho\hat{w}$$
$$+ \sigma_x \hat{\varepsilon}_{xx} + \sigma_y \hat{\varepsilon}_{yy} + \sigma_z \hat{\varepsilon}_{zz} + \tau_{xy} \hat{\alpha}_{xy} + \tau_{yz} \hat{\alpha}_{yz} + \tau_{xz} \hat{\alpha}_{xz} \qquad (12\text{-}3.6)$$

From equations (12-3.1) and (12-3.4b), we get

$$\iint_S \left(X_n\hat{u} + Y_n\hat{v} + Z_n\hat{w}\right)dA - \iiint_V \rho\left[\left(\frac{\partial^2 u}{\partial t^2} - X_i\right)\hat{u} + \left(\frac{\partial^2 v}{\partial t^2} - Y_i\right)\hat{v} + \left(\frac{\partial^2 w}{\partial t^2} - Z_i\right)\hat{w}\right]dV$$

$$= \iiint_V \left[\sigma_x\hat{\varepsilon}_{xx} + \sigma_y\hat{\varepsilon}_{yy} + \sigma_z\hat{\varepsilon}_{zz} + \tau_{xy}\hat{\alpha}_{xy} + \tau_{yz}\hat{\alpha}_{yz} + \tau_{xz}\hat{\alpha}_{xz}\right]dV \qquad (12\text{-}3.7)$$

Equation (12-3.7) represents the integral entity that satisfies **Navier's equation of equilibrium** and **Cauchy's equations of strains**.

In the case of **equilibrium**, the three accelerations (second derivatives with respect to time) vanish, rendering internal displacements equal to volume displacements, we get

$$\hat{u} = u, \qquad\qquad \hat{v} = v, \qquad\qquad \hat{w} = w$$

$$\frac{\partial^2 u}{\partial t^2} = 0, \qquad \frac{\partial^2 v}{\partial t^2} = 0, \qquad \frac{\partial^2 w}{\partial t^2} = 0 \qquad (12\text{-}3.8)$$

Thus, (12-3.7) yield **Clapeyron's theorem**, summed up in the form

$$\iint_S \left(X_n u + Y_n v + Z_n w\right)dA + \iiint_V \rho[(X_i u + Y_i v + Z_i w)]dV$$

$$= \iiint_V \left[\sigma_x\hat{\varepsilon}_{xx} + \sigma_y\hat{\varepsilon}_{yy} + \sigma_z\hat{\varepsilon}_{zz} + \tau_{xy}\hat{\alpha}_{xy} + \tau_{yz}\hat{\alpha}_{yz} + \tau_{xz}\hat{\alpha}_{xz}\right]dV \qquad (12\text{-}3.9)$$

Which has been obtained before in equation (3-14.1).

12-2. Lagrange's geometrical variation

Lagrange's equation of variation depends on infinitesimal vibration of coordinates of the medium particles (strains and displacements).

Vibrational perturbation of displacements and strains

If we assume that both external and internal **stresses** (σ_x, σ_y, σ_z, τ_{xy}, τ_{zy}, τ_{xz}) and **inertial forces** ($\rho\overset{**}{u}$, $\rho\overset{**}{v}$, $\rho\overset{**}{w}$) and **body forces** (ρX_i, ρY_i, ρZ_i) are all kept constants, we could describe **elastic vibration** by replacing, in equation (12-3.7), the whole body displacements by perturbation in infinitesimal displacements such that

$$\hat{u} = \delta u, \qquad \hat{v} = \delta v, \qquad \hat{w} = \delta w \qquad\qquad (12\text{-}4.1)$$

Where, δ is the infinitesimal **variation operator** signifying small changes in the operand.

Thus, **strains** also are perturbed according to **Cauchy's equations** (2-2) and (2-3).

Equation (12-3.7), becomes,

$$\iint_S \left(X_n \delta u + Y_n \delta v + Z_n \delta w \right) dA$$

$$- \iiint_V \rho\left[\left(\frac{\partial^2 u}{\partial t^2} - X_i \right) \delta u + \left(\frac{\partial^2 v}{\partial t^2} - Y_i \right) \delta v + \left(\frac{\partial^2 w}{\partial t^2} - Z_i \right) \delta w \right] dV \qquad (12\text{-}4.2)$$

$$= \iiint_V \left[\sigma_x \delta\varepsilon_{xx} + \sigma_y \delta\varepsilon_{yy} + \sigma_z \delta\varepsilon_{zz} + \tau_{xy} \delta\alpha_{xy} + \tau_{yz} \delta\alpha_{yz} + \tau_{xz} \delta\alpha_{xz} \right] dV$$

Elastic body energy

Replacing the products of stresses by vibrations in strains, R.H.S. in the above equation, by **change of elastic energy[MFE6]** δW (3-14.1), we get

$$\iint_S \left(X_n \delta u + Y_n \delta v + Z_n \delta w \right) dA$$

$$- \iiint_V \rho\left[\left(\frac{\partial^2 u}{\partial t^2} - X_i \right) \delta u + \left(\frac{\partial^2 v}{\partial t^2} - Y_i \right) \delta v + \left(\frac{\partial^2 w}{\partial t^2} - Z_i \right) \delta w \right] dV \qquad (12\text{-}5)$$

$$= \iiint_V [\delta W] dV$$

Equation (12-5) represents **D'Alembert's general principle** for elastic body. This means that the work-done by elastic deformation, in addition to the work-done by internal forces, minus inertial forces, is equal to the change in elastic energy of the body.

In case of equilibrium, we get **Lagrange's principle of virtual displacements** in the form

$$\iint_S \left(X_n \delta u + Y_n \delta v + Z_n \delta w \right) dA + \iiint_V \rho[(X_i \delta u + Y_i \delta v + Z_i \delta w)] dV = \iiint_V [\delta W] dV \qquad (12\text{-}6.1)$$

By virtue of the assumption that both external and internal forces are not perturbed, we could move the variation operator δ outside the integration operation, as follows

$$\delta \iint_S \left(X_n u + Y_n v + Z_n w \right) dA + \delta \iiint_V \rho[(X_i u + Y_i v + Z_i w)] dV = \delta \iiint_V W dV \qquad (12\text{-}6.2)$$

426

Virtual work done

Therefore, the **work-done,** U, by all external forces applied to the elastic body on **virtual displacements** (meaning no external acceleration), is given by

$$U = \iint_S (X_n u + Y_n v + Z_n w) dA + \iiint_V \rho[(X_i u + Y_i v + Z_i w)] dV$$

(12-6.3)

External forces work   Internal forces work

As we have stated, in equation (12-6.1), accelerations vanish in equilibrium such that forces increase infinitely slowly such that only **virtual displacements** in the form of elastic deformation dominates equilibrium.

Equation (12-6.2) can be written is a form that shows the **potential energy** as follows:

$$\delta\left(U - \iiint_V W dV\right) = 0$$

(12-6.4)

Where,

$$F = U - \iiint_V W dV$$

(12-6.5)

is the **force function** generated by the difference **virtual work done,** U, by external forces and internal energy, W, consumed in **elastic deformation.**

Stability of equilibrium signifies that the negative term in (12-6.5) is great enough to keep force function, F, minimum.

Instability of equilibrium occurs when the body is too thin to resist positive term in equation (12-6.5), which makes the force function, F, maximum.

Limited surface vibration

In the absence of internal variation in displacements, and with vanishing internal body forces, equation (12-6.2) gives

$$\delta U = \delta \iiint_V W dV = 0$$

(12-7.1)

Which comprises **minimum total of elastic energy** of the medium.

Minimum force function

427

The minimum force function, equation (12-6.5), is determined from equations (12-4.2) or (3-14.1), where

$$\delta W = \tfrac{1}{2}\left[\sigma_x \delta\varepsilon_{xx} + \sigma_y \delta\varepsilon_{yy} + \sigma_z \delta\varepsilon_{zz} + \tau_{xy}\delta\alpha_{xy} + \tau_{yz}\delta\alpha_{yz} + \tau_{xz}\delta\alpha_{xz}\right] \qquad (12\text{-}8.1)$$

$$\begin{aligned}
dW &= dW_n + dW_s \\
&= \tfrac{1}{2}\left(\left(\sigma_x \varepsilon_{xx} + \sigma_y \varepsilon_{yy} + \sigma_z \varepsilon_{zz}\right) + \tau_{xy}\varepsilon_{xy} + \tau_{yz}\varepsilon_{yx} + \tau_{zx}\varepsilon_{zx}\right) dx\,dy\,dz
\end{aligned} \qquad (12\text{-}8.2)$$

Comparing the two equations, the elastic energy of the medium is the average ($\tfrac{1}{2}$) of the energies of **bending** and **shear forces** and **twisting moments** of infinitesimal elements.

Plane cross-section approximations in thick media

Equations (12-8) are used to determine the **three energies** for **minimal force function**, equation (12-6.5), of infinitesimal elements.

However, thick plates, rods, and wedges are not infinitesimal and are approximated by the **hypothesis of planar cross section** during elastic work. i.e., cross sections of medium retain their plane form during bending, twisting, and shearing.

Such hypothesis isolates bending of a plane from shear by adjacent elements. The following examples show the utility of the planar cross sections.

(1) Energy of bending of straight bar by the planar approximation

An element of length, dx, bent by moment of couple, M, curves with a radius of curvature, ρ, having Young's modulus E and moment of Inertia, I_x, bends an angle $d\varphi = dx/\rho$, will have bending energy simplified as follows

$$\begin{aligned}
dW &= \tfrac{1}{2}M\,d\varphi \\
&= \tfrac{1}{2}M\frac{dx}{\rho} \\
&= \tfrac{1}{2}\left(\frac{EI_x}{\rho}\right)\frac{dx}{\rho}
\end{aligned} \qquad (12\text{-}9.1)$$

Where, relationship between moment and deflection, equation (6-2.16), gives $M = \dfrac{EI}{\rho}$.

The energy of bending per unit length of the bar is

428

$$\frac{dW}{dx} = \frac{EI_x}{2}\left(\frac{1}{\rho}\right)^2$$

$$= \frac{EI_x}{2}\left(\frac{d^2v(x)}{dx^2}\right)^2$$

(12-9.2)

Where, W is used interchangeably with dW/dx.

In order to determine the minimal force function, the **virtual work-done**, U, equation (12-6.3), is written in terms of the **load function**, q(x), multiplied by an **unknown optimization function**, v(x), as follows;

$$U = \int_0^l q(x)v(x)dx$$

(12-9.3)

Substituting from (12-9.2 and 9.3), in equations (12-6.4 and 6.5), we get the condition for determining the unknown function v(x), which should minimize the force function F to achieve maximum stability.

$$F = U - \int_0^l \frac{dW}{dx}dx$$

$$= \text{External /Internal forces} - \text{Elastic Energy}$$

(12-9.4)

$$= \int_0^l \left(q(x)v(x) - \frac{EI_x}{2}\left(\frac{1}{\rho}\right)^2\right)dx$$

Ritz-Timoshenko's method of determination of the optimization deflection function

The optimization deflection function, v(x), is expressed in terms of arbitrary constants, which are determined from boundary conditions of clamping of ends or static allowances of sliding or fixation. Thus,

$$v(x) = \sum_{m=1}^{\infty} c_m \sin\frac{m\pi x}{l}$$

(12-10.1)

Subjected to the boundary conditions

429

Fixed one − end conditions $\quad v_i(0) \quad = 0, \qquad \dfrac{\partial v_i}{\partial x}\bigg|_{x=0} = 0$

$$\text{(12-10.2)}$$

Fixed two − ends conditions $\quad v_i\big|_{x=0,l} \quad = 0, \qquad \dfrac{\partial v_i}{\partial x}\bigg|_{x=0,l} = 0$

(i) The **virtual work-done**, U, equation (12-9.3), is determined for a constant load, $q(x) = q$, as follows

$$U = q\int_0^l \sum_{m=1}^{\infty} c_m \sin\frac{m\pi x}{l}\,dx$$

$$= \frac{ql}{\pi}\sum_{m=1}^{\infty}\frac{c_m}{m}(1 - \cos m\pi)$$

$$= \frac{2ql}{\pi}\sum_{m=1,3,5}^{\infty}\frac{c_m}{m}, \qquad m = 1,3,5,..$$

$$\text{(12-10.3)}$$

(ii) The **elastic energy** W, equation (12-9.2), becomes

$$\int_0^l Wdx = \frac{EI_x}{2}\int_0^l\left(\frac{d^2v(x)}{dx^2}\right)^2 dx$$

$$= \frac{EI_x}{2}\frac{\pi^4}{l^4}\int_0^l\left[\sum_{m=1}^{\infty}\left(-c_m m^2 \sin\frac{m\pi x}{l}\right)\right]^2 dx$$

$$= \frac{EI_x\pi^4}{4l^3}\sum_{m=1}^{\infty}c_m^2 m^4$$

$$\text{(12-10.4)}$$

(iii) **Force function** F, equation (12-6.5)

Substituting from equations (12-10.3 and 10.4) in (12-6.5), we get

$$F = U - \iiint_V WdV$$

$$= \frac{2ql}{\pi}\sum_{m=1}^{\infty}\frac{c_m}{m} - \frac{EI_x\pi^4}{4l^3}\sum_{m=1}^{\infty}c_m^2 m^4, \qquad m = 1,3,5,..$$

$$\text{(12-10.5)}$$

(iv) **Minimizing the force function**

The minimum value of F is obtained from the vanishing derivative of F as follows

$$\frac{\partial F}{\partial c_m} = \frac{2ql}{\pi m} - \frac{EI_x\pi^4}{2l^3}c_m m^4 = 0$$

$$\text{(12-10.6)}$$

Thus,

$$c_m = \frac{4ql^4}{\pi^5 m^5 EI_x}$$ (12-10.7)

(v) Optimum deflection function for maximal stability

$$v(x) = \frac{4ql^4}{\pi^5 m^5 EI_x} \sum_{m=1,3,5}^{\infty} \frac{1}{m^5} \sin \frac{m\pi x}{l}$$ (12-10.8)

Maximum deflection occurs at $x = l/2$, as follows

$$v_{max}\left(\frac{l}{2}\right) = \frac{4ql^4}{\pi^5 EI_x}\left(1 - \frac{1}{3^5} + \frac{1}{5^5} - \right)$$

$$\approx \frac{4ql^4}{\pi^5 EI_x}$$ (12-10.9)

(2) Energy of bending of thin slab by the planar approximation

An element of area of length, dx, width dy, bent by moments of couple, M_x and M_y, curves with radii of curvatures, ρ_x and ρ_y, having flexural rigidity, K, and twisted with force F_{xy} and angle[MFE7], τ, will have bending energy simplified as follows

$$W = \tfrac{1}{2}\left(M_x d\varphi_x + M_y d\varphi_y + M_z \tau\right)$$

$$= \tfrac{1}{2}\left(M_x \frac{1}{\rho_x} + M_y \frac{1}{\rho_y} + M_z \tau\right)$$

$$= \tfrac{1}{2}\left(M_x \frac{\partial^2 w}{\partial x^2} + M_y \frac{\partial^2 w}{\partial y^2} + 2M_z \frac{\partial^2 w}{\partial x \partial y}\right)$$ (12-11.1)

$$\underbrace{\qquad\qquad\qquad}_{\text{Bending Energies}} \quad \underbrace{\qquad}_{\text{Twisting Energy}}$$

Where, from equations (11-10.2) and (11-11-3), give

$$M_x = -K\left(\frac{\partial^2 w}{\partial x^2} + v\frac{\partial^2 w}{\partial y^2}\right)$$

$$M_y = -K\left(\frac{\partial^2 w}{\partial y^2} + v\frac{\partial^2 w}{\partial x^2}\right)$$ (12-11.2)

$$M_z = -K(1-v)\frac{\partial^2 w}{\partial y \partial x}$$ (12-11.3)

431

Substituting by the moments of bending and twisting from equations (12-11.2 and 10.3) into (12-11.1), we get

$$W = \tfrac{1}{2}K\left(\left(\frac{\partial^2 w}{\partial x^2} + v\frac{\partial^2 w}{\partial y^2}\right)\frac{\partial^2 w}{\partial x^2} + \left(v\frac{\partial^2 w}{\partial x^2} + \frac{\partial^2 w}{\partial y^2}\right)\frac{\partial^2 w}{\partial y^2} + 2(1-v)\left(\frac{\partial^2 w}{\partial x \partial y}\right)^2\right) \qquad (12\text{-}11.4a)$$

Bending Energies   Bending Energies   Twisting Energy

i.e.,

$$W = \tfrac{1}{2}K\left(\left(\frac{\partial^2 w}{\partial x^2}\right)^2 + \left(\frac{\partial^2 w}{\partial y^2}\right)^2 + 2v\left(\frac{\partial^2 w}{\partial y^2}\frac{\partial^2 w}{\partial x^2}\right) + 2(1-v)\left(\frac{\partial^2 w}{\partial x \partial y}\right)^2\right) \qquad (12\text{-}11.4b)$$

Bending Energies   Bending Energies   Twisting Energy

Thus, the **force function**, equation (12-6.5), then becomes

$$F = U - \int_0^l \frac{dW}{dx}dx$$

$$= \int_0^l \left(q(x)v(x) - \tfrac{1}{2}K\left(\left(\frac{\partial^2 w}{\partial x^2}\right)^2 + \left(\frac{\partial^2 w}{\partial y^2}\right)^2 + 2v\left(\frac{\partial^2 w}{\partial y^2}\frac{\partial^2 w}{\partial x^2}\right) + 2(1-v)\left(\frac{\partial^2 w}{\partial x \partial y}\right)^2\right)\right)dx \quad (12\text{-}11.5)$$

External Energy   Bending Energies   Bending Energies   Twisting Energy

Equation (12-11.5) has the unknown $v(x)$, which is the **minimization function**, which achieves minimal **force function** required for maximal stability.

Ritz-Timoshenko's method of determination of the optimization deflection function

Similar to equation (12-10.1), the optimization deflection function $w(x,y)$ is expressed in terms of arbitrary constants, which are determined from boundary conditions of clamping of ends or static allowances of sliding or fixation. Thus,

$$w(x,y) = \sum_{m=1}^{\infty}\sum_{n=1}^{\infty} c_{mn}\sin\frac{m\pi x}{a}\sin\frac{n\pi x}{b} \qquad (12\text{-}12.1)$$

(i) The **virtual work-done**, U, equation (12-9.3), is determined for a concentrated load,

$$q(x,y[\text{MFE8}]) \qquad = q, \qquad\qquad\qquad x = \xi \text{ and } y = \eta$$
$$= 0, \qquad x \neq \xi \text{ and } y \neq \eta \qquad\qquad (12\text{-}12.2)$$

as follows

$$U = q \sum_{m=1,3,5,..}^{\infty} \sum_{n=1,3,5,..}^{\infty} c_{mn} \sin\frac{m\pi\xi}{a}\sin\frac{n\pi\eta}{b}$$ (12-12.3)

(ii) The **elastic energy** W, is obtained by substituting from equation (12-12.1) in equation (12-11.4b), and integrating, to get

$$W = \frac{K\pi^4}{8}ab \sum_{m=1,3,5,..}^{\infty} \sum_{n=1,3,5,..}^{\infty} c_{mn}^2\left(\frac{m^2}{a^2}+\frac{n^2}{b^2}\right)^2$$ (12-12.4)

(iii) **Force function** F, equation (12-6.5)

Substituting from equations (12-12.3 and 12.4) in (12-6.5), we get

$$F = U - \iiint_V W dV$$

$$= \sum_{m=1,3,5,..}^{\infty} \sum_{n=1,3,5,..}^{\infty} \left[q\left(c_{mn} \sin\frac{m\pi\xi}{a}\sin\frac{n\pi\eta}{b}\right) - \frac{K\pi^4 ab}{8}c_{mn}^2\left(\frac{m^2}{a^2}+\frac{n^2}{b^2}\right)^2 \right]$$ (12-12.5)

(iv) **Minimizing the force function**

The minimum value of F is obtained from the vanishing derivative of F as follows

$$\frac{\partial F}{\partial c_m} = \sum_{m=1,3,5,..}^{\infty} \sum_{n=1,3,5,..}^{\infty}\left[q\sin\frac{m\pi\xi}{a}\sin\frac{n\pi\eta}{b} - \frac{K\pi^4}{4}abc_{mn}\left(\frac{m^2}{a^2}+\frac{n^2}{b^2}\right)^2 \right] = 0$$ (12-12.6)

Thus,

$$c_{mn} = \frac{4qb^3c^3}{K\pi^4\left(m^2b^2+a^2n^2\right)^2}\sin\frac{m\pi\xi}{a}\sin\frac{n\pi\eta}{b}$$ (12-12.7)

(v) **Optimum deflection function for maximal stability**

$$w(x,y) = \frac{4qb^3c^3}{K\pi^4} \sum_{m=1,3,5,..}^{\infty} \sum_{n=1,3,5,..}^{\infty}\left[\frac{\sin\frac{m\pi\xi}{a}\sin\frac{n\pi\eta}{b}}{\left(m^2b^2+a^2n^2\right)^2} \right]\sin\frac{m\pi x}{a}\sin\frac{n\pi x}{b}$$ (12-12.8)

433

Maximum deflection. Consider $\xi = a/2$ and $\eta = b/2$. The first term in the expansion of equation (12-12.8) gives

$$w\left(\frac{a}{2},\frac{b}{2}\right) = \frac{4qb^3c^3}{K\pi^4\left(b^2+a^2\right)^2} \tag{12-12.9}$$

12-3. Lagrange's equation for three-dimensional arbitrary body

(i) Boundary conditions in three-dimensional body

The displacements **on the surface** or in specified **parts of the body** could be assumed of the form

$$u = \overline{u}(x,y,z), \qquad v = \overline{v}(x,y,z), \qquad w = \overline{w}(x,y,z) \tag{12-13.1}$$

Let the displacements in the 3D elastic body comprise of the following two sets of arbitrary functions:

1. Arbitrary displacements functions yield **constants** equal to those in (12-13.1) on the boundary conditions, but comprise the non-varying components of the displacements. Those are assumed in the forms:

$$u_0(x,y,z), \qquad v_0(x,y,z), \qquad w_0(x,y,z) \tag{12-13.2}$$

2. Arbitrary displacements functions **vanishing** on the boundary conditions, but comprise the factors of the varying components of the displacements. Those are assumed in the forms:

$$\varphi(x,y,z), \qquad \psi(x,y,z), \qquad \chi(x,y,z) \tag{12-13.3}$$

(ii) Series expansion of displacements

The two sets of arbitrary functions, in equations (12-13.2) and (12-13.3), comprise the general forms of displacements as follows

$$u = u_0(x,y,z) + \sum_{m=1}^{\infty} a_m \varphi_m(x,y,z)$$

$$v = v_0(x,y,z) + \sum_{m=1}^{\infty} b_m \psi_m(x,y,z) \tag{12-13.4}$$

$$w = w_0(x,y,z) + \sum_{m=1}^{\infty} c_m \chi_m(x,y,z)$$

434

The coefficients a_m, b_m, and c_m are varied to satisfy the boundary conditions (12-13.1), while the six members of the two sets of arbitrary functions are kept un-perturbed.

(iii) Perturbation of displacements

The states of vibration in the body are described by equations (12-13.4) as follows:

$$\delta u = \sum_{m=1}^{\infty} \varphi_m(x,y,z)\delta a_m$$

$$\delta v = \sum_{m=1}^{\infty} \psi_m(x,y,z)\delta b_m \qquad (12\text{-}13.5)$$

$$\delta w = \sum_{m=1}^{\infty} \chi_m(x,y,z)\delta c_m$$

(iv) Energy balance of equilibrium in terms of strains and displacements

Volumetric Hooke's law, equations (3-7), is used to express equation (12-8.2) in terms of strains and displacements as follows

$$W = \tfrac{1}{2}\left(2\mu\left(\varepsilon_{xx}^2 + \varepsilon_{yy}^2 + \varepsilon_{zz}^2\right) + \lambda\varepsilon^2 + \mu\left(\alpha_{xy}^2 + \alpha_{zy}^2 + \alpha_{xz}^2\right)\right) \qquad (12\text{-}14.1)$$

We will denote the volume integral of W as follows

$$\Psi\left(a_m, b_m, c_m\right) = \iiint_V W dV \qquad (12\text{-}14.2)$$

Where, $\Psi\left(a_m, b_m, c_m\right)$ has the following characteristics:

1. Is a **homogeneous** function.
2. Of the **second degree** in the constant coefficients a_m, b_m, and c_m.
3. Its **first derivatives** with respect to a_m, b_m, and c_m are of the **first degree** in those coefficients.

Substituting by the variations of displacements, equations (12-13.5), in the equation of energy balance of equilibrium, (12-6.2), we get

$$\sum_{m=1}^{\infty} \left[\begin{array}{l} \iint_S \left(X_n\varphi_m\delta a_m + Y_n\psi_m\delta b_m + Z_n\chi_m\delta c_m\right)dA \\ + \iiint_V \rho\left[\left(X_i\varphi_m\delta a_m + Y_i\psi_m\delta b_m + Z_i\chi_m\delta c_m\right)\right]dV \end{array} \right]$$

$$= \sum_{m=1}^{\infty} \frac{\partial\Psi}{\partial a_m}\delta a_m + \frac{\partial\Psi}{\partial b_m}\delta b_m + \frac{\partial\Psi}{\partial c_m}\delta c_m \qquad (12\text{-}14.3)$$

Since the variations in three arbitrary coefficients δa_m, δb_m, and δc_m are independent, we can equate the factors of each of the three variations as follows:

$$\delta a_m : \quad \frac{\partial \Psi}{\partial a_m} = \iint_S (X_n \varphi_m)\, dA + \iiint_V \rho(X_i \varphi_m)\, dV$$

$$\delta b_m : \quad \frac{\partial \Psi}{\partial b_m} = \iint_S (Y_n \psi_m)\, dA + \iiint_V \rho(Y_i \psi_m)\, dV \qquad (12\text{-}14.4)$$

$$\delta c_m : \quad \frac{\partial \Psi}{\partial c_m} = \iint_S (Z_n \chi_m)\, dA + \iiint_V \rho(Z_i \chi_m \delta c_m)\, dV$$

Equations (12-14.4) represent the sums of **external forces** on the surface of the body and sums of **internal forces** in the mass of the body, which determine the derivatives of the homogeneous function Ψ, needed to determine the **displacement coefficients** a_m, b_m, and c_m.

Having solved the (3 x m)-equations (12-14.4) and obtained the (3 x m) constant coefficients needed in equations (12-13.4) to determine the general displacements (u, v, w), we solve the distributions of the six stresses and six strains by **Cauchy's equations for strains**, equations (2-2 and 2-3), the **volumetric Hooke's law**, equations (3-7. 1), and the surface equations, equations (1-8).

12-4. Castigliano's static variation

Castiglianoe's equation of variation depends on infinitesimal vibration of **stresses and forces** from within and without the medium particles (stresses of tension, shear, twisting, internal body stresses, and external stresses).

(i) Variation of stresses

Here, we keep the geometrical variables unperturbed, vary the stresses and forces, in equation (12-1), as we did to displacements in equation (12-15.1). Thus,

Elastic stresses:	$\delta\sigma_x$,	$\delta\sigma_y$,	$\delta\sigma_z$,	$\delta\tau_{zy}$,	$\delta\tau_{zy}$,	$\delta\tau_{zz}$ (12-15.1)
External stresses:	δX_n,	δY_n,	δZ_n			(12-15.2)
Internal body stresses:	δX_i,	δY_i,	δZ_i			(12-15.3)

Substituting by variations in equation (12-15.1) in **Navier's equation** (1-4.1), with vanishing acceleration (equilibrium), we get

$$\frac{\partial(\sigma_x + \delta\sigma_x)}{\partial x} + \frac{\partial(\tau_{xy} + \delta\tau_{xy})}{\partial y} + \frac{\partial(\tau_{xz} + \delta\tau_{xz})}{\partial z} + \rho X_i = 0 \qquad (12\text{-}15.4a)$$

Or,

$$\frac{\partial\sigma_x}{\partial x} + \frac{\partial(\delta\sigma_x)}{\partial x} + \frac{\partial\tau_{xy}}{\partial y} + \frac{\partial(\delta\tau_{xy})}{\partial y} + \frac{\partial\tau_{xz}}{\partial z} + \frac{\partial(\delta\tau_{xz})}{\partial z} + \rho X_i = 0 \qquad (12\text{-}15.4b)$$

Substituting from (1-4.1) in (12-15.4b), we get

$$\frac{\partial(\delta\sigma_x)}{\partial x} + \frac{\partial(\delta\tau_{xy})}{\partial y} + \frac{\partial(\delta\tau_{xz})}{\partial z} = 0 \qquad (12\text{-}15.4c)$$

Similarly, we could prove the remaining two equations of Navier, for variations in the y- and z-oriented stresses

We now turn to the surface conditions.

Substituting by variations in equation (12-15.2) in **surface equation** (1-8.1), with vanishing acceleration (equilibrium), we get

$$X_n + \delta X_n = (\sigma_x + \delta\sigma_x)l + (\tau_{xy} + \delta\tau_{xy})m + (\tau_{xz} + \delta\tau_{xz})n \qquad (12\text{-}15.5a)$$

Or,

$$X_n + \delta X_n = [(\sigma_x)l + (\tau_{xy})m + (\tau_{xz})n] + [(\delta\sigma_x)l + (\delta\tau_{xy})m + (\delta\tau_{xz})n] \qquad (12\text{-}15.5b)$$

Substituting from (1-8.1) in (12-15.5b), we get

$$\delta X_n = (\delta\sigma_x)l + (\delta\tau_{xy})m + (\delta\tau_{xz})n \qquad (12\text{-}15.5c)$$

Similarly, can prove that δY_n and δZ_n satisfy Navier's equations.

From equations (12-15.4c and 15.5c), we conclude that the variations of stresses satisfy both Navier's equation of equilibrium and the surface conditions.

We then use **Volumetric Hooke's law** (3-3.1 and 3.2) and equation (3-3.1) to express **Clapeyron's theorem**, equation (12-3.9), in terms of stresses, by replacing the strains terms by their stress equivalents. Since we assume that strains are kept unvarying, the RHS of equations (12-3.9) becomes.

$$\delta\iiint_V W dV = \iiint_V \left[\delta\sigma_x \varepsilon_{xx} + \delta\sigma_y \varepsilon_{yy} + \delta\sigma_z \varepsilon_{zz} + \delta\tau_{xy}\alpha_{xy} + \delta\tau_{yz}\alpha_{yz} + \delta\tau_{xz}\alpha_{xz}\right] dV$$

$$= \iiint_V \left[\begin{array}{l} \delta\sigma_x E^{-1}[\sigma_x - \nu(\sigma_y + \sigma_z)] + \delta\sigma_y E^{-1}[\sigma_y - \nu(\sigma_x + \sigma_z)] \\ + \delta\sigma_z E^{-1}[\sigma_z - \nu(\sigma_x + \sigma_y)] + \mu^{-1}[\tau_{xy}\delta\tau_{xy} + \tau_{yz}\delta\tau_{yz} + \tau_{xz}\delta\tau_{xz}] \end{array} \right] dV \qquad (12\text{-}16.1)$$

Substituting from (3-3.1), by $\mu = \dfrac{E}{2(1+v)}$, and arranging we get

$$\delta\iiint_V WdV = \frac{1}{E}\iiint_V \left[\begin{array}{l} \delta\sigma_x\left[\sigma_x - v\left(\sigma_y + \sigma_z\right)\right] + \delta\sigma_y\left[\sigma_y - v\left(\sigma_x + \sigma_z\right)\right] \\ + \delta\sigma_z\left[\sigma_z - v\left(\sigma_x + \sigma_z\right)\right] + 2(1+v)\left[\tau_{xy}\delta\tau_{xy} + \tau_{yz}\delta\tau_{yz} + \tau_{xz}\delta\tau_{xz}\right] \end{array} \right] dV$$

$$= \frac{1}{E}\iiint_V \left[\begin{array}{l} \left(\sigma_x\delta\sigma_x + \sigma_y\delta\sigma_y + \sigma_z\delta\sigma_z\right) \\ - v\left[\delta\sigma_x\left(\sigma_y + \sigma_z\right) + \delta\sigma_y\left(\sigma_x + \sigma_z\right) + \delta\sigma_z\left(\sigma_x + \sigma_z\right)\right] \\ + 2(1+v)\left(\tau_{xy}\delta\tau_{xy} + \tau_{yz}\delta\tau_{yz} + \tau_{xz}\delta\tau_{xz}\right) \end{array} \right] dV \qquad (12\text{-}16.2)$$

Re-writing the variation operator δ such that we could lump the variants in a single term, we get

$$\delta\iiint_V WdV == \frac{1}{E}\iiint_V \left[\begin{array}{l} \frac{1}{2}\delta\left(\sigma_x^2 + \sigma_y^2 + \sigma_z^2\right) - v\delta\left(\sigma_x\sigma_y + \sigma_y\sigma_z + \sigma_z\sigma_x\right) \\ + (1+v)\delta\left(\tau_{xy}^2 + \tau_{yz}^2 + \tau_{xz}^2\right) \end{array} \right] dV \qquad (12\text{-}16.3)$$

With farther arrangements, we get

$$\delta\iiint_V WdV = \delta\left(\frac{1}{2E}\iiint_V \left[\begin{array}{l} \left(\sigma_x^2 + \sigma_y^2 + \sigma_z^2\right) - 2v\left(\sigma_x\sigma_y + \sigma_y\sigma_z + \sigma_z\sigma_x\right) \\ + 2(1+v)\left(\tau_{xy}^2 + \tau_{yz}^2 + \tau_{xz}^2\right) \end{array} \right] dV \right) \qquad (12\text{-}16.3)$$

Therefore, equation (12-3.9), **Clapeyron's theorem**, becomes

$$\iint_S \left(\delta X_n u + \delta Y_n v + \delta Z_n w\right)dA + \iiint_V \left(\delta(\rho X_i)u + \delta(\rho Y_i)v + \delta(\rho Z_i)w\right)dV$$

$$= \delta\left(\frac{1}{2E}\iiint_V \left[\begin{array}{l} \left(\sigma_x^2 + \sigma_y^2 + \sigma_z^2\right) - 2v\left(\sigma_x\sigma_y + \sigma_y\sigma_z + \sigma_z\sigma_x\right) \\ + 2(1+v)\left(\tau_{xy}^2 + \tau_{yz}^2 + \tau_{xz}^2\right) \end{array} \right] dV \right) \qquad (12\text{-}16.4)$$

Which is **Castigliano's variation equation**

Torsion of prismatical rod

438

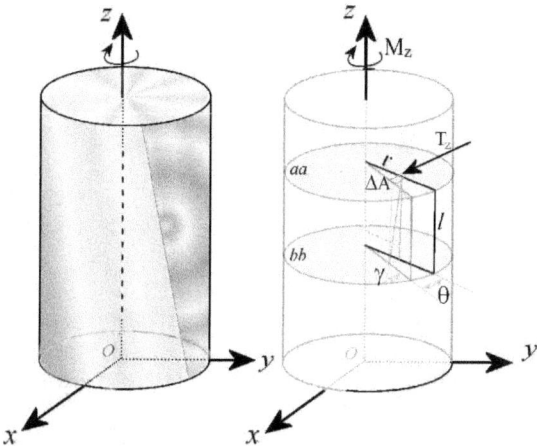

Figure 12-1. Torsion of prismatical rod solved by Castigliano's variation method.

Torsion of prismatical rod has been studied by the used of **Prandtl's stress function**, Φ, equation (9-9.1) and Figure 12-1.

Boundary conditions

1. The rod is considered simply connected medium.
2. Resultant forces on bases vanish.
3. Three normal stresses vanish on **the lateral walls**.
4. Shear stress in the xy-plane vanish.
5. Ignore internal body forces.

Elastic stresses : $\qquad \sigma_x = \sigma_y = \sigma_z = \tau_{xy} = 0$

External stresses : $\qquad X_n = Y_n = Z_n = 0 \qquad ---> \text{Lateral} \quad \text{walls}$ $\qquad$ (12-17.1)

Interanl body stresses : $\qquad X_i = Y_i = Z_i = 0 \qquad ---> \text{Medium} \quad \text{interior}$

$$\tau_{zx} = \frac{\partial \Phi(x,y)}{\partial y} \qquad\qquad \tau_{zy} = -\frac{\partial \Phi(x,y)}{\partial x} \qquad (12\text{-}17.2)$$

The twisting moment on the end, Figure 12-1,

$$\mathbf{M}_z = 2\int \left(\int \Phi\, dx\right) dy$$
$$= 2\int\int\left(\psi - \frac{x^2+y^2}{2}\right) dy\, dx \qquad (12\text{-}17.3)$$

The displacements were defined by equations (9-2) as follows

$$u(x,y,z) = -\tau\, yz \qquad\qquad (12\text{-}18.1)$$
$$v(x,y,z) = \tau\, xz \qquad\qquad (12\text{-}18.2)$$
$$w(x,y,z) = \tau\, \varphi(x,y) \qquad\qquad (12\text{-}18.3)$$

Variation of stresses, equations (12-17.1 and 17.2)

Applying the variation operator on the defining stresses of torsion, equations (12-17.1 and 17.2), we get

$$\delta\sigma_x = \delta\sigma_y = \delta\sigma_z = \delta\tau_{xy} = 0 \qquad\qquad (12\text{-}19.1)$$

$$\delta\tau_{zx} = \frac{\partial}{\partial y}(\delta\Phi) \qquad\qquad \delta\tau_{zy} = -\frac{\partial}{\partial x}(\delta\Phi) \qquad\qquad (12\text{-}19.2)$$

Castigliano's variation equation for torsion of rod

Substituting by the defining variations from (12-19) into **Castigliano's variation equation** (12-16.4), we get

$$\iint\limits_S \left(\delta X_n u + \delta Y_n v + \delta Z_n w\right) dA = \delta\left(\frac{(1+v)}{E}\iiint\limits_V \left[\left(\tau_{yz}^2 + \tau_{xz}^2\right)\right] dV\right) \qquad (12\text{-}20.1)$$

Substituting from (12-17.2), we get

$$\iint\limits_S \left(\delta X_n u + \delta Y_n v + \delta Z_n w\right) dA = \delta\left(\frac{(1+v)}{E}\iiint\limits_V \left[\left(\frac{\partial\Phi}{\partial x}\right)^2 + \left(\frac{\partial\Phi}{\partial y}\right)^2\right] dV\right) \qquad (12\text{-}20.2)$$

Boundary condition on lower base of rod

Consider a fixed base at $z = 0$.
Equations (12-18) gives $u = v = 0$.
Then, the integrals in equation (12-20.2) vanish.

Boundary condition on upper base of rod

Consider a freely twisted upper end at $z = l$.
Equations (12-18) give

$$u(x,y, l) = -\tau\, y\, l \tag{12-21.1}$$
$$v(x,y, l) = \tau\, x\, l \tag{12-21.2}$$
$$w(x,y, l) = \tau\, \varphi(x,y) \tag{12-21.3}$$

This is the only part of the surface where external forces exist in the form of shear stresses given by equations (12-17.2) in terms of the derivatives of the **stress function of Prandtl.**

Then, the LHS integral in equation (12-20.1), is determined from the torquing conditions in equation (12-19.2) and displacement conditions (12-21), as follows

$$\iint_S \left(\delta X_n u + \delta Y_n v + \delta Z_n w\right) dA = \iint_S \left[(-l\tau y)\frac{\partial}{\partial y}(\delta\Phi) + (l\tau x)\frac{\partial}{\partial x}(-\delta\Phi) + 0 \right] dA$$

$$= -l\tau\iint_S \left(y\frac{\partial}{\partial y}(\delta\Phi) + x\frac{\partial}{\partial x}(\delta\Phi) \right) dA \tag{12-22.1a}$$

We can manipulate the last integral by "**integration by parts**" and arrange keeping in mind that we only deal with "**first variation**" of first differential of function $\Phi(x,y)$ with respect to its independent variables x and y, such that

$$d\Phi = \frac{\partial\Phi}{\partial x}dx + \frac{\partial\Phi}{\partial y}dy \tag{12-22.1b}$$

Therefore, equation (12-22.1a) becomes

$$\iint_S \left(\delta X_n u + \delta Y_n v + \delta Z_n w\right) dA = -l\tau\iint_S \left(y\frac{\partial}{\partial y}(\delta\Phi) + x\frac{\partial}{\partial x}(\delta\Phi) \right) dxdy$$

$$= -l\tau\left(\left(\oint dx \int y\frac{\partial}{\partial y}(\delta\Phi)dy \right) + \left(\oint dy \int x\frac{\partial}{\partial x}(\delta\Phi)dx \right) \right) \tag{12-22.2}$$

$$= -l\tau\left[\oint dx \int yd(\delta\Phi) + \oint dy \int xd(\delta\Phi) \right]$$

$$= -l\tau\left[\oint \left[[y(\delta\Phi)]_{y_1}^{y_2} - \int (\delta\Phi)dy \right] dx + \oint \left[[x(\delta\Phi)]_{x_1}^{x_2} - \int (\delta\Phi)dx \right] dy \right]$$

We note that the two terms $y(\delta\Phi)$ and $x(\delta\Phi)$ vanish upon integration as $\delta\Phi$ vanish on the contour per equation (9-9.1), where

$$\Phi(x,y) = \mu\tau \left(\psi - \frac{x^2+y^2}{2} \right) \tag{12-22.3}$$

Thus, equation (12-22.2) becomes

$$\iint\limits_{S} (\delta X_n u + \delta Y_n v + \delta Z_n w)dA = 2l\tau \iint (\delta\Phi)dydx \tag{12-22.4}$$

Substituting by the variation of torsion energy, equation (12-22.4), into **Castigliano's variation equation** (12-20.2), we get

$$\delta\left(\frac{(1+v)}{E} \iiint\limits_{V} \left[\left(\frac{\partial\Phi}{\partial x}\right)^2 + \left(\frac{\partial\Phi}{\partial y}\right)^2 \right]dV \right) = 2l\tau\iint (\delta\Phi)dydx \tag{12-23.1}$$

Since the **Prandtl's function** Φ is independent of z, we get

$$\delta\left(\frac{(1+v)}{E} \iint \left[\left(\frac{\partial\Phi}{\partial x}\right)^2 + \left(\frac{\partial\Phi}{\partial y}\right)^2 \right]ds \right)\int\limits_{0}^{l} dz = 2l\tau\iint (\delta\Phi)dydx \tag{12-23.2}$$

Substituting from equation (3-8.1), by $\mu = \dfrac{E}{2(1+v)}$, and arranging the variation operator δ, we get

$$\delta\left(\iint \left[\left(\frac{\partial\Phi}{\partial x}\right)^2 + \left(\frac{\partial\Phi}{\partial y}\right)^2 - 4\tau\mu\Phi \right]dydx \right) = 0 \tag{12-23.3}$$

Laplace's form of Castigliano's variation equation for torsion of rod

Equation (12-23.3) is the **Castigliano's variation** equation, which determines the **Prandtl's function** that minimizes the surface integral and enables us to solve torsion problems for rods of **complex cross sections**.

Equation (12-23.3) can be transformed to the form of Laplace's equation as we perform the variation operation as follows:

$$\iint \left[\frac{\partial\Phi}{\partial x}\delta\left(\frac{\partial\Phi}{\partial x}\right) + \frac{\partial\Phi}{\partial y}\delta\left(\frac{\partial\Phi}{\partial y}\right) - 2\tau\mu(\delta\Phi) \right]dydx = 0 \tag{12-24.1}$$

We could arrange the integrands in equation (12-24.1) and perform the integration-by-parts as follows

$$\int dy \left(\int \frac{\partial \Phi}{\partial x}\left(\frac{\partial}{\partial x}\delta\Phi\right)dx \right) + \int dx \left(\int \frac{\partial \Phi}{\partial y}\left(\frac{\partial}{\partial y}\delta\Phi\right)dy \right) - 2\tau\mu\iint (\delta\Phi)dydx = 0$$

$$\int dy \left(\int \frac{\partial \Phi}{\partial x}\partial(\delta\Phi) \right) + \int dx \left(\int \frac{\partial \Phi}{\partial y}\partial(\delta\Phi) \right) - 2\tau\mu\iint (\delta\Phi)dydx = 0$$

(12-24.2)

Integrating by parts, we get

$$\int dy \left(\left[\frac{\partial \Phi}{\partial x}\delta\Phi \right]_{x_1}^{x_2} - \int \delta\Phi \frac{\partial^2 \Phi}{\partial x^2}dx \right)$$

$$+ \int dx \left(\left[\frac{\partial \Phi}{\partial y}\delta\Phi \right]_{y_1}^{y_2} - \int \delta\Phi \frac{\partial^2 \Phi}{\partial y^2}dy \right) - 2\tau\mu\iint (\delta\Phi)dydx = 0$$

(12-24.3)

As Prandtl's function vanishes on the contour, we get rid of two terms in the above equation, as follows

$$-\iint \delta\Phi\left(\frac{\partial^2 \Phi}{\partial x^2} + \frac{\partial^2 \Phi}{\partial y^2} + 2\tau\mu \right)dxdy = 0$$

(12-24.4)

Since the variation $\delta\Phi$ is arbitrary, therefore, the bracketed term must vanish.

Therefore,

$$\nabla^2\Phi = -2\tau\mu$$

(12-24.5)

The **Prandtl's stress function** Φ can take the form similar to (12-13.4) such that the arbitrary functions and arbitrary constants satisfy the boundary conditions of an arbitrary contour, as follows:

Rod cross section contour : $F(x, y) = 0$

Prandtl's stress function : $\Phi(x, y) = \sum_{m=1}^{\infty} C_m \psi_m(x, y, z)$

(12-25)

First problem of elasticity with first theory of minimum or first differential of stress function

The first differential derivative of variation of work-done by external stresses and elastic deformation, equation (12-22.1b), determines the minimum or maximum of the elastic energy.

443

$$d\Phi = \frac{\partial \Phi}{\partial x}dx + \frac{\partial \Phi}{\partial y}dy$$

In arriving at the equation (12-23.2) of balance of **volume sum of variations** of elastic energy ($W\delta\tau$) with the **sum of variations** of work-down by external forces ($\delta X_n u + \delta Y_n v + \delta Z_n w$) we concluded, from equation (12-24.5), that $\nabla^2 \Phi = -2\tau\mu$ would achieve minimum or maximum of elastic energy.

Therefore, equation (12-23.2), becomes

$$
\begin{aligned}
2/\tau \!\int\!\!\int (\delta\Phi)dydx &= \delta\!\left(\frac{(1+v)}{E}\!\int\!\!\int\left[\left(\frac{\partial\Phi}{\partial x}\right)^2 + \left(\frac{\partial\Phi}{\partial y}\right)^2\right]ds\right)_0^l \int dz \\
&= \delta\!\left(l\,\frac{(1+v)}{E}\!\int\!\!\int \left[\nabla^2\Phi\right]ds\right) \\
&= \delta\!\left(l\,\frac{(1+v)}{E}\!\int\!\!\int \left[-2\tau\mu\right]ds\right) \qquad\qquad (12\text{-}26)\\
&= -2\tau\mu\,\frac{(1+v)}{E}\,\delta(sl) \\
&= 0
\end{aligned}
$$

Therefore, we **conclude** that the **first problem of elasticity** comprises of retaining the volume of the medium, $\delta(sl)$, constant by ensuring that the next **work-done by variations of external forces** ($\delta X_n u + \delta Y_n v + \delta Z_n w$) **vanish** over the whole surface of the body. This ensures that the **elastic energy**, W, would have either **maximum** or **minimum** value. Figure 12-2.

First problem of elasticity with second theory of minimum or second differential of stress function

The second differential derivative of variation of work-done by external stresses and elastic deformation determines the **either** the minimum or the maximum of the elastic energy, but not **both**. Figure 12-2.

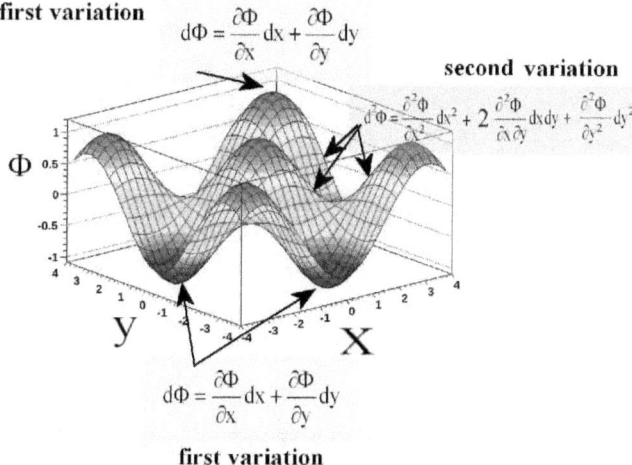

first variation

$$d\Phi = \frac{\partial\Phi}{\partial x}dx + \frac{\partial\Phi}{\partial y}dy$$

second variation

$$d^2\Phi = \frac{\partial^2\Phi}{\partial x^2}dx^2 + 2\frac{\partial^2\Phi}{\partial x\partial y}dxdy + \frac{\partial^2\Phi}{\partial y^2}dy^2$$

$$d\Phi = \frac{\partial\Phi}{\partial x}dx + \frac{\partial\Phi}{\partial y}dy$$

first variation

Figure 12-2. First and second theorems of minimum, one determines peaks and bottoms, the other determines reflection.

The second differential of a function is obtained by considering the coordinate differentials dx and dy constants. Therefore

$$d^2\Phi = \frac{\partial^2\Phi}{\partial x^2}dx^2 + 2\frac{\partial^2\Phi}{\partial x\partial y}dxdy + \frac{\partial^2\Phi}{\partial y^2}dy^2 \qquad (12\text{-}27.1)$$

The second variation of elastic energy $\delta^2 W$ is obtained from (12-16.2) by performing the variation operation while keeping the first variation of stresses ($\delta\sigma_x$, $\delta\sigma_y$, $\delta\sigma_z$, $\delta\tau_{xy}$, $\delta\tau_{zy}$, $\delta\tau_{xz}$) constants. Thus, equation (12-16.2) gives

$$\delta^2 \iiint_V WdV = \delta\left\{\frac{1}{E}\iiint_V \begin{bmatrix} \left(\sigma_x\delta\sigma_x + \sigma_y\delta\sigma_y + \sigma_z\delta\sigma_z\right) \\ -v\left[\delta\sigma_x\left(\sigma_y + \sigma_z\right) + \delta\sigma_y\left(\sigma_x + \sigma_z\right) + \delta\sigma_z\left(\sigma_x + \sigma_y\right)\right] \\ +2(1+v)\left(\tau_{xy}\delta\tau_{xy} + \tau_{yz}\delta\tau_{yz} + \tau_{xz}\delta\tau_{xz}\right) \end{bmatrix}dV\right\} \qquad (12\text{-}27.2)$$

Performing the variation of operation on the bracketed expression, we get

$$\delta^2 \iiint_V WdV = \delta\left\{\frac{1}{E}\iiint_V \begin{bmatrix} \left(\delta\sigma_x\right)^2 + \left(\delta\sigma_y\right)^2 + \left(\delta\sigma_z\right)^2 \\ -2v\left[\delta\sigma_x\delta\sigma_y + \delta\sigma_x\delta\sigma_z + \delta\sigma_y\delta\sigma_z\right] \\ +2(1+v)\left[\left(\delta\tau_{xy}\right)^2 + \left(\delta\tau_{yz}\right)^2 + \left(\delta\tau_{xz}\right)^2\right] \end{bmatrix}dV\right\} \qquad (12\text{-}27.3)$$

As the **first variation** of the total elastic energy

445

$$\delta\iiint_V WdV = 0 \qquad\qquad (12\text{-}27.4)$$

determines the maximum or minimum of the elastic energy of the medium, necessary for equilibrium.

The **second variation** of the total elastic energy

$$\delta^2\iiint_V WdV > 0 \qquad\qquad (12\text{-}27.5)$$

determines the minimum elastic energy (**principle of least work**) needed for equilibrium.

12-5. Practical approximate solution of elasticity by method of variation of elastic energy

We have seen in equations (12-16.3) and (12-27.3) that the elastic energy is expressed in terms of either internal elastic stresses or strains. But, in practice, we need discernable variables that could be easily ascertained and manipulated. Hence, the **hypothesis of planar sections** facilitates the sorting out of the portions of minimized elastic energy spend in each of mechanical deformations:

1. Bending
2. Twisting
3. Shear
4. Compression
5. Tension

The planar section hypothesis renders the elastic potential energy, equation (3-14.1), more practical to deal with as follows

$$dW = \tfrac{1}{2}\left((\sigma_x\varepsilon_{xx} + \sigma_y\varepsilon_{yy} + \sigma_z\varepsilon_{zz}) \quad + \quad \tau_{xy}\varepsilon_{xy} + \tau_{yz}\varepsilon_{yx} + \tau_{zx}\varepsilon_{zx} \right)dxdydz$$
$$= \tfrac{1}{2}\left(\frac{M^2}{EI} + \frac{F_s^2}{EF} + \frac{Q^2}{GF\alpha} + \frac{M_t^2}{GC} \right)dxdy \qquad\qquad (12\text{-}28)$$

Example of approximate variation solution of general elasticity problems

Lamé's problem of rectangular prism

(i) Postulated series expansion for general stresses

446

We will use an approach similar to Lagrange's solution adopted in solving the three-dimensional arbitrary body, equation (12-13.4), where two sets of arbitrary functions, comprise the general forms of **stresses** as follows

$$\sigma_x = \sigma_{x,0} + \sum_{m=1}^{\infty} a_m \sigma_{x,m}$$

$$\sigma_y = \sigma_{y,0} + \sum_{m=1}^{\infty} a_m \sigma_{y,m}$$

$$\sigma_z = \sigma_{z,0} + \sum_{m=1}^{\infty} a_m \sigma_{z,m}$$

$$\tau_{xy} = \tau_{xy,0} + \sum_{m=1}^{\infty} a_m \tau_{xy,m}$$

$$\tau_{yz} = \tau_{yz,0} + \sum_{m=1}^{\infty} a_m \tau_{yz,m}$$

$$\tau_{zx} = \tau_{zx,0} + \sum_{m=1}^{\infty} a_m \tau_{zx,m}$$

(12-29.1)

The **first set** of stresses denotes **state subscript "0"**, in the above equation, namely

$$\sigma_{x,0}, \sigma_{y,0}, \sigma_{z,0}, \tau_{xy,0}, \tau_{zy,0}, \tau_{xz,0}$$

(12-29.2)

This set will satisfy the two conditions:

1. Navier's equation of equilibrium (1-4)
2. Surface equations (1-8)

The **second set** of stresses denotes **state subscript "m"**, in equation (12-29.1), namely

$$\sigma_{x,m}, \sigma_{y,m}, \sigma_{z,m}, \tau_{xy,m}, \tau_{zy,m}, \tau_{xz,m}$$

(12-29.3)

This set **vanishes on the surface**, such that equations (1-8), become

$$\sigma_{x,m}\, l + \tau_{xy,m}\, m + \tau_{xz,m}\, n = 0 \qquad \text{(12-29.4a)}$$
$$\tau_{yx,m}\, l + \sigma_{y,m}\, m + \tau_{yz,m}\, n = 0 \qquad \text{(12-29.4b)}$$
$$\tau_{zx,m}\, l + \tau_{zy,m}\, m + \sigma_{z,m}\, n = 0 \qquad \text{(12-29.4c)}$$

Hence, we constructed three arbitrary sets of variables that comprise the elastic stresses such that:

(1) One arbitrary set of variables satisfies the boundary conditions and equilibrium.
(2) Second arbitrary set of variables vanishes on the surface.

447

(3) Third arbitrary set of variables "a_m" is varied to satisfy the conditions of equilibrium and surface conditions, through the **minimization of variation** of elastic energy, equation (12-27.3).

(ii) Tensor presentation of series expansion of arbitrary stresses functions

Equations (12-29.1) can be represented more meaningfully as follows

$$
\begin{pmatrix} \sigma_x & \tau_{xy} & \tau_{xz} \\ \tau_{yx} & \sigma_y & \tau_{yz} \\ \tau_{zx} & \tau_{zy} & \sigma_z \end{pmatrix} = \begin{pmatrix} \sigma_{x,0} & \tau_{xy,0} & \tau_{xz,0} \\ \tau_{yx,0} & \sigma_{y,0} & \tau_{yz,0} \\ \tau_{zx,0} & \tau_{zy,0} & \sigma_{z,0} \end{pmatrix} + \sum_{m=1}^{\infty} a_m \begin{pmatrix} \sigma_{x,m} & \tau_{xy,m} & \tau_{xz,m} \\ \tau_{yx,m} & \sigma_{y,m} & \tau_{yz,m} \\ \tau_{zx,m} & \tau_{zy,m} & \sigma_{z,m} \end{pmatrix}
$$

$$\quad\text{general tesnor}\qquad\qquad\text{basic tensor}\qquad\qquad\qquad\text{correcting\ \ tensor}$$

(12-30.1)

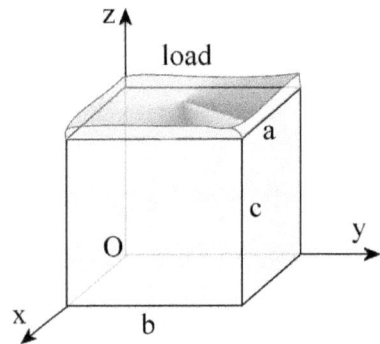

Figure 12-3. Rectangular prism loaded on two opposite faces.

(iii) Boundary conditions

Figure 12-3 shows the geometry of the rectangular prism under consideration.

1. Upper surface $z = c$

shear $\tau_{xz} = \tau_{yz} = 0$

load $\sigma_z = F(x, y)$

(12-30.2)

2. Lower surface $z = 0$

shear $\quad \tau_{.xz} = \tau_{.yz} = 0$

load $\quad \sigma_{.z} = -F(x, y)$

(12-30.3)

3. All lateral surfaces **free from loads**

(iv) Fourier's cosines of basic and correcting tensors

For the 3D rectangular prism, the following structure of Fourier's cosines serves the purpose delineated in equation (12-29.1) of **basic stress functions** that **vanish on the six surfaces** of the prism and **correcting stress functions** that satisfy the remaining conditions of equilibrium and continuity.

$$
\begin{aligned}
X_m(x) &= \cos\frac{m\pi x}{a} - \cos\frac{(m+2)\pi x}{a}, & m &= 0,1,2,.. \\
Y_n(y) &= \cos\frac{n\pi y}{b} - \cos\frac{(n+2)\pi y}{b}, & n &= 0,1,2,.. \\
Z_k(z) &= \cos\frac{k\pi z}{c} - \cos\frac{(k+2)\pi z}{c}, & k &= 0,1,2,..
\end{aligned}
$$

(12-31.1)

The system of three functions (12-31.1) vanish on the edges of the prism, Figures 12-4 and 12-5.

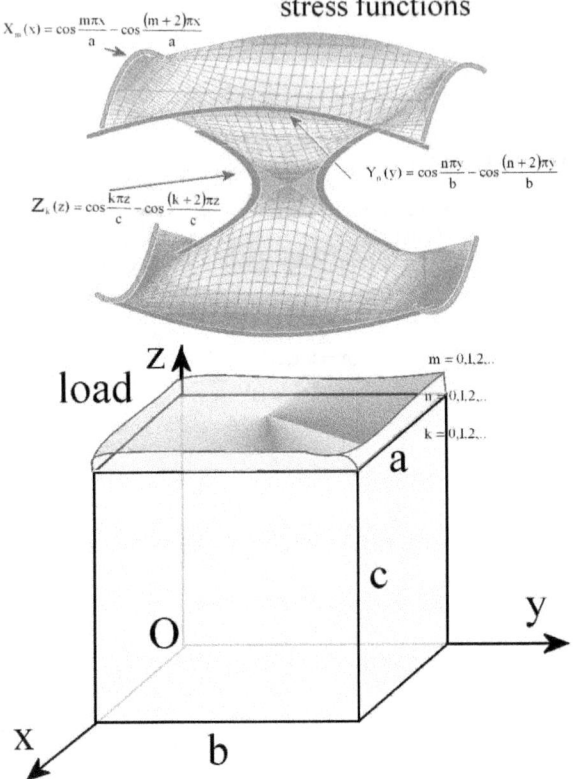

$$X_m(x) = \cos\frac{m\pi x}{a} - \cos\frac{(m+2)\pi x}{a}$$

stress functions

$$Y_n(y) = \cos\frac{n\pi y}{b} - \cos\frac{(n+2)\pi y}{b}$$

$$Z_k(z) = \cos\frac{k\pi z}{c} - \cos\frac{(k+2)\pi z}{c}$$

load

$m = 0,1,2,...$

$n = 0,1,2,...$

$k = 0,1,2,...$

z

a

c

y

O

x

b

Figure 12-4. Fourier's series expansions for rectangular prism showing the vanishing of the functions on the six faces of the rectangular prism.

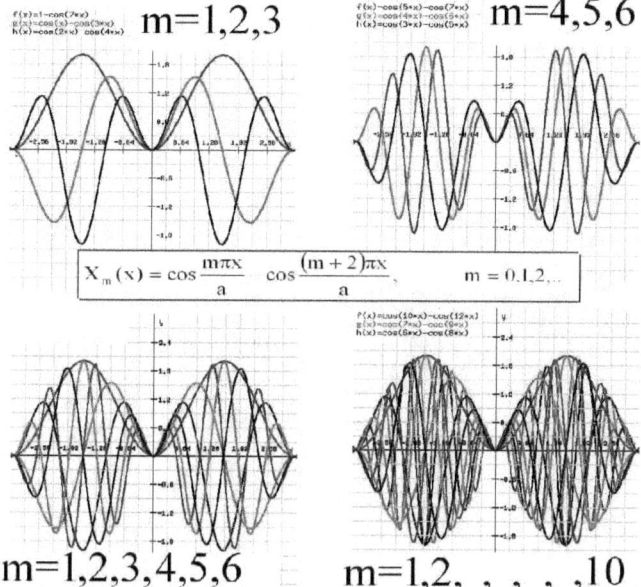

Figure 12-5. Graphing the system of equations (12-31.1) shows the closeness of solution as the index "m" increases.

(v) Maxwell's stress functions

The system of function in (12-31.1) are used to express **Maxwell's stress functions**, equations (10-5.1), in terms of expansion series of Fourier's cosines which are both **close to perfect numerical solution** and in complete compliance with Navier's equations of **equilibrium** and **surface conditions**, as follows.

$$\phi_1 = \sum_{m=1}^{\infty}\sum_{n=1}^{\infty}\sum_{k=1}^{\infty} A_{mnk} X_m(x)Y_n(y)Z_k(z)$$

$$\phi_2 = \sum_{m=1}^{\infty}\sum_{n=1}^{\infty}\sum_{k=1}^{\infty} B_{mnk} X_m(x)Y_n(y)Z_k(z) \qquad (12\text{-}31.2)$$

$$\phi_3 = \sum_{m=1}^{\infty}\sum_{n=1}^{\infty}\sum_{k=1}^{\infty} C_{mnk} X_m(x)Y_n(y)Z_k(z)$$

(vi) Correcting tensor
The components of the **correcting tensor** in equation (12-30.1) vanishes on all surfaces.

The three stress functions, φ_1, φ_2, and φ_3, were used previously to define elastic stresses, equations (10-5.2 through 5.6), as follows

$$\tau_{yx} = -\frac{\partial^2 \varphi_3}{\partial x \partial y} = -\sum_{m=1}^{\infty}\sum_{n=1}^{\infty}\sum_{k=1}^{\infty} C_{mnk}\left(\frac{dX_m(x)}{dx}\right)\left(\frac{dY_n(y)}{dy}\right)Z_k(z) \qquad (12\text{-}32.1)$$

$$\tau_{zx} = -\frac{\partial^2 \varphi_2}{\partial x \partial z} = -\sum_{m=1}^{\infty}\sum_{n=1}^{\infty}\sum_{k=1}^{\infty} B_{mnk}\left(\frac{dX_m(x)}{dx}\right)Y_n(y)\left(\frac{dZ_k(z)}{dz}\right) \qquad (12\text{-}32.2)$$

$$\tau_{yz} = -\frac{\partial^2 \varphi_1}{\partial y \partial z} = -\sum_{m=1}^{\infty}\sum_{n=1}^{\infty}\sum_{k=1}^{\infty} C_{mnk}X_m(x)\left(\frac{dY_n(y)}{dy}\right)\left(\frac{dZ_k(z)}{dz}\right) \qquad (12\text{-}32.3)$$

$$\sigma_x = \sum_{m=1}^{\infty}\sum_{n=1}^{\infty}\sum_{k=1}^{\infty}\left(B_{mnk}Y_n(y)\left(\frac{d^2Z_k(z)}{dz^2}\right)+C_{mnk}Z_k(z)\left(\frac{d^2Y_n(y)}{dy^2}\right)\right)X_m(x) \qquad (12\text{-}32.4)$$

$$\sigma_y = \sum_{m=1}^{\infty}\sum_{n=1}^{\infty}\sum_{k=1}^{\infty}\left(A_{mnk}X_m(x)\left(\frac{d^2Z_k(z)}{dz^2}\right)+C_{mnk}Z_k(z)\left(\frac{d^2X_m(x)}{dx^2}\right)\right)Y_n(y) \qquad (12\text{-}32.5)$$

$$\sigma_z = \sum_{m=1}^{\infty}\sum_{n=1}^{\infty}\sum_{k=1}^{\infty}\left(A_{mnk}X_m(x)\left(\frac{d^2Y_n(y)}{dy^2}\right)+B_{mnk}Y_n(y)\left(\frac{d^2X_m(x)}{dx^2}\right)\right)Z_k(z) \qquad (12\text{-}32.6)$$

Since all six stresses in equations (12-32) contain one undifferentiated term among the three terms, X_m, Y_n, Z_k, comprising each of the three Maxwell's functions, φ_1, φ_2, φ_3, equations (12-31.2), therefore, equations (12-32) comprise the elements of the **correcting tensor**.

(vii) Basic tensor

The **basic tensor**, equation (12-30.1), accounts for the **surface loading** and permits us to compose the **general tensor** with infinite numbers of the constants A_{mnk}, B_{mnk}, and C_{mnk} varied to accommodate varied load distributions on any surface of the prism.

(viii) Minimization of vibration of elastic energy

Substituting by the six stresses from equations of stresses (12-32) into the equation of elastic energy (12-16.4), we obtain a set of simultaneous linear equations, with the same numbers of unknowns of A_{mnk}, B_{mnk}, and C_{mnk}.

Determination of arbitrary constants in one-dimensional problem

Before we get into the three-dimensional solution of Lamé's prism, we will tackle the simpler one-dimensional equations (12-29.1).

Substituting by the six stresses from equations of stresses (12-29.1) into the equation of elastic energy (12-16.4), we get

$$W = \tfrac{1}{2}\left(\left(\sigma_x^{\,2} + \sigma_y^{\,2} + \sigma_z^{\,2}\right) - 2\nu\left(\sigma_x\sigma_y + \sigma_y\sigma_z + \sigma_z\sigma_x\right) + 2(1+\nu)\left(\tau_{xy}^{\,2} + \tau_{yz}^{\,2} + \tau_{xz}^{\,2}\right)\right)$$

$$= \sum_{m=1}^{\infty}\sum_{n=1}^{\infty} A_{mn}\, a_m a_n + \sum_{m=1}^{\infty} B_m\, a_m \qquad (12\text{-}33.1)$$

Where the constants A_{mn} and B_m sum the products of stresses according to the structure delineated in the squaring and mixed products of stresses.

The partial derivatives of W with respect to the optimization constants, a_m and a_n are **equated to zero**, giving rise to $(m + n)$-equations as follows:

$$\frac{\partial W}{\partial a_m} = \sum_{n=1}^{\infty} A_{mn}\, a_n + B_m = 0$$

$$\frac{\partial W}{\partial a_n} = \sum_{m=1}^{\infty} A_{mn}\, a_m + B_n = 0 \qquad (12\text{-}33.2)$$

We note by differentiating the above two sets equations that $A_{mn} = A_{nm}$, which implies that the two sets of equations are **canonical** (each variable designates particular events taking place at particular coordinates and in particular time frame).

We should also note that a more precise definition of **canonical coordinates** is given as

$$\left(q_m, p_n\right) = \delta_{mn} \qquad (12\text{-}34)$$

Where q_m denotes the coordinates of the event, the momentum p_m of amount of motion of the event. Thus, events that are canonical must associate their momenta with their coordinates, so as to render those events unique in time, place, and state of motion (stress, deformation, shear, bending, or twisting).

Determination of arbitrary constants in three-dimensional prism of Lamé

The three indices m, n, k, and the three arbitrary constants A, B, C, render the three-dimensional problem more complicated. In the **first approximation**, the three indices are assigned zeros, i.e., $m = n = k = 0$. Thus, we have three constants A_{000}, B_{000}, and C_{000}. The **Nth-approximation** involves $3N^3$ unknown arbitrary constants. For example, the second approximation involves $(3 \times 2^3 = 24)$ unknowns.

A more realistic approximation involves dropping **repeated indices** such that we could reduce the number of unknowns by half.

(ix) Surface loading
Equations (12-30.2) and (12-30.3) define a simple state of loading, Figure 12-3. We assign the load F(x,y), in equation (12-30.2), the form of $(X_0 Y_0)$, of equation (12-31.1) as follows

$$F(x, y) = \left(1 - \cos\frac{m\pi x}{a}\right)\left(1 - \cos\frac{m\pi y}{b}\right)$$

(12-35.1)

Thus, the general tensor, equation (12-30.1), changes only in the normal stress, equation (12-32.6), along the z-axis as follows

$$\sigma_z = F(x, y) + \sum_{m=1}^{\infty}\sum_{n=1}^{\infty}\sum_{k=1}^{\infty}\left(A_{mnk} X_m(x)\left(\frac{d^2 Y_n(y)}{dy^2}\right) + B_{mnk} Y_n(y)\left(\frac{d^2 X_m(x)}{dx^2}\right)\right)Z_k(z)$$

$$= \sum_{m=1}^{\infty}\sum_{n=1}^{\infty}\sum_{k=1}^{\infty}\left(A_{mnk} X_m(x)\left(\frac{d^2 Y_n(y)}{dy^2}\right) + B_{mnk} Y_n(y)\left(\frac{d^2 X_m(x)}{dx^2}\right)\right)Z_k(z)$$

(12-35.2)

$$+ \left(1 - \cos\frac{2\pi x}{a}\right)\left(1 - \cos\frac{2\pi y}{b}\right)$$

=== End ====

Index

www.ingramcontent.com/pod-product-compliance
Lightning Source LLC
Chambersburg PA
CBHW051437170526
45166CB00001B/19